Principles and Practices of Plant Genomics

Volume 3: Advanced Genomics

Editors

Chittaranjan Kole

Albert G. Abbott

Department of Genetics and Biochemistry
Clemson University
South Carolina
USA

CRC Press is an imprint of the
Taylor & Francis Group, an **informa** business

CRC Press
Taylor & Francis Group
6000 Broken Sound Parkway NW, Suite 300
Boca Raton, FL 33487-2742

First issued in paperback 2017

ISBN-13: 978-1-57808-683-2 (hbk)
ISBN-13: 978-1-138-11649-8 (pbk)

Library of Congress Cataloging-in-Publication Data

Principles and practices of plant genomics/editors, Chittaranjan Kole, Albert G. Abbott.

p. cm.

Includes bibliographical references and index.

ISBN 978-1-57808-683-2 (hardcover)

1. Plant genomes. 2. Plant genetics. I. Kole, Chittaranjan. II. Abbott, Albert G.

QK981.P75 2010

572.8'62--dc22

2007047428

Visit the Taylor & Francis Web site at
http://www.taylorandfrancis.com

and the CRC Press Web site at
http://www.crcpress.com

Dedicated to

Prof. Blannie Bowen

Vice Provost for Academic Affairs
The Pennsylvania State University

The Prologue

Dear Prof. Kole

Having read the preface and contents of your series of volumes on plant genetics, I wish to thank you to have undertaken together with Professor Abbott the heavy work to produce an updated encyclopedia on molecular plant genetics. I am sure that your series on "Principles and Practices of Plant Genomics" will have a tremendous, worldwide impact on plant sciences and their applications.

Our human society has benefited for thousands of years from plants found in their natural environments. Soon people learned to bring about improvements by breeding, as well as by choosing most appropriate soils to grow the plants serving for their nutrition, as well as for medical and ornamental uses.

A true paradigmatic change has started a few decades ago with the progress of scientific investigations at the molecular level. Novel opportunities have become available by a combination of genetic engineering and classical breeding to obtain even more appropriate plant variants for the service to the human population. With novel research strategies and with steadily improving knowledge on gene structure and functions, we can expect that already in the near future many cultivated plants will benefit themselves of an improved health, be more resistant to pests and, what will be of great importance, will provide improved nutritional values to our daily diets. Agriculture will enter a new, more sustainable era.

Your publication series can greatly help the scientists working on this field to reach this goal and to carry out their work, with the required care and responsibility, to the benefit both of the human society and of its environment.

Please take these words as an encouragement to complete and also to steadily update your important project.

With my best regards.

Yours Sincerely,

January 23, 2008

Prof. Werner Arber
Professor Emeritus for Molecular Microbiology
University of Basel
Nobel Laureate in Medicine 1978

Foreword

The elucidation of the double-helix structure of the Deoxyribose Nucleic Acid (DNA) molecule in 1953 by Drs. James Watson, Francis Crick, Maurice Wilkins and Rosalind Franklin marked the beginning of what is now known as *the new genetics*. Research during the last 54 years in the fields of molecular genetics and recombinant-DNA technology has opened up new opportunities in agriculture, medicine, industry, and environment protection.

Availability of DNA-based molecular markers since 1980s has facilitated construction of complete genetic maps in several plants of academic and economic importance. Several computer softwares also have paved the way for handling huge numbers of markers and segregating individuals to frame these maps, also for detection of positions of genes and gene clusters, known as quantitative trait loci, on them.

Molecular marker-based genetic mapping has provided not only comprehensive depictions of the genomes but also facilitated elucidation of phylogenetic relationships and evolutionary pathways, map-based cloning of genes and above all use of molecular markers in various schemes of crop improvement, including germplasm characterization, marker-aided selection and introgression of genes through marker-assisted backcross breeding.

The strategies of construction of BAC and YAC libraries and advent of high-throughput sequencing of DNA have built the platform for physical mapping of chromosomal regions and even whole genomes. The new millennium has already witnessed complete sequencing of the genomes of the model plant *Arabidopsis* and the leading crop plant rice followed by the genomes of a forest tree poplar and a fruit tree peach. Genome initiatives in many more crop plants individually and families comprising important crop plants are indeed progressing fast.

One of the spectacular happening during this period is the merging of information technology and biological science leading to a new discipline called as bioinformatics. This subject is now playing a pivotal role in warehousing biological data on sequences and functions of genes, transcripts, expressed sequence tags and proteins, and their searching to utilize for various purposes. We have now several new branches with the 'omics' sciences such as genomics, transcriptomics, proteomics and metabolomics. However, all of them relate to the genomes and reside under the common roof of genomics.

There is little doubt that the genomics has opened up uncommon opportunities for enhancing the productivity, profitability, sustainability and stability of major cropping systems. It has also created scope for developing crop varieties resistant/tolerant to biotic and abiotic stresses through an appropriate blend of Mendelian and molecular breeding techniques. It has led to the possibility of undertaking anticipatory breeding to meet potential changes in temperature, precipitation and sea level as a result of global warming. There are new opportunities for fostering pre-breeding and farmer-participatory breeding methods in order to continue the merits of genetic efficiency with genetic diversity.

In the coming decades, farm women and men in population-rich but land-hungry countries like India and China will have to produce more food and other agricultural commodities to meet home needs and to take advantage of export opportunities, under conditions of diminishing per capita availability of arable land and irrigation water and expanding abiotic and biotic stresses. The enlargement of the gene pool with which breeders work will be necessary to meet these challenges. Genomics supplemented by transgenics provide breeders with a powerful tool for enlarging the genetic base of crop varieties and to pyramid genes for a wide range of economically important traits.

Genetics can boast of being the subject that progressed with the most rapid speed within the shortest period of time and contributed stupendously in unraveling the nature and function of genes and also towards the genetic improvement of useful microbes, animals and plants in the last century. In fact, the *green revolution* of the sixties was the contribution of classical genetics and conventional breeding. Another revolution in agriculture, medicine and environment is in the wing and we are ready to welcome it as *gene revolution* to be drived by genomics-aided breeding. However, for a meaningful and sustainable application of plant genomics, importance of *orphan* crops which have immense potential for improving food and nutrition security should be taken into account.

The book series "Principles and Practices of Plant Genomics" is thus a timely contribution. It provides an authoritative account of recent progress in plant genomics including structural and functional genomics and their use in molecular- and genomics-assisted breeding. It also provides a road map for genomic research in the present century. I express our gratitude to Profs. Chittaranjan Kole and Albert Abbott for this labor of love in spreading knowledge among both students and scientists on the opportunities as well as challenges opened up by the genomics era.

M.S. Swaminathan
Chairman
M.S. Swaminathan Research Foundation

Preface to the Series

It is an unequivocal fact that genomics has emerged as the leading discipline in plant sciences. The enormous global efforts in the last two and a half decades have led to a very rapid development of this subject. Currently genetic linkage mapping employing molecular markers has become routine for development of chromosome maps in genetically refractory species, mapping of economic genes and gene clusters, studies on evolution and phylogenetic relationships, map-based cloning of useful genes, and marker-aided breeding. Physical mapping of plant genomes using artificial chromosomes and integration of physical and genetic maps have provided more power of resolution to depict the structural basis of genes and also to elucidate genome colinearity and evolution facilitating exploration of homologous desired genes employing comparative genomics. Whole-genome sequencing of the model plant *Arabidopsis* followed by rice and poplar has already demonstrated the power of structural genomics in elucidating focused understanding of genome organization using sequence data that represents the ultimate level of genetic information. The success stories of genome initiatives in microbes, animals, and plants have inspired 'initiatives' at national and international levels to sequence genomes of an array of other plant genomes. Information on genetic and physical maps, colinearity and whole-genome sequencing serve as the starting point for assignment of biological meaning to putative genes with no known phenotype. For this precise purpose, functional genomics emerged to address the function of genes discovered through sequencing efforts. Employment of concepts and strategies including expressed sequence tags, reverse genetics and transcriptional profiling have facilitated identification and discovery of genes and their expressions in specific tissues and stages.

Plant genomics has already found its deserved place in various courses under agriculture, medicine, and environment sciences. Application of molecular markers for several purposes in crop improvement has fostered the development of another discipline, molecular breeding. Recent developments in functional genomics, transgenics and bioinformatics have vastly enriched the resources of today's plant breeding. Information on a large number of genes handled simultaneously by genomics and thereby genetic integration of diverse processes, tissues, and organisms will highly benefit future plant improvement endeavors. Therefore, future molecular breeding will encompass marker-assisted breeding, transgenic breeding, and genomics-aided breeding. Plant genomics and molecular breeding are going to play the pivotal role in the fields of agriculture, medicine, environment, and ecology in this and hopefully in the coming centuries.

However, through academic and research related interactions with fellow research workers on plant genomics, we recognized the need of a handbook for ready reference. Textbooks explaining the fundamentals and applications of plant genomics are of significant value to students, teachers and scientists of the public and private sectors. Plant genomics is basically the frontier branch of genetics but is complemented strongly by biochemistry, microbiology, computational biology, and bioinformatics; with supplementation from various other disciplines including crop production, statistics, physiology, pathology, entomology, horticulture, just to name a few. Most interestingly in today's world, people from all these fields learn, teach and/or practice plant genomics. Thus, it is essential that academics in these broad disciplines have available the lucid deliberations of the basic concepts and strategies of plant genomics to enhance their own studies, training, and expertise. This was the main driving force for taking up this endeavor to popularize the science and art of plant genomics.

The subject of plant genomics has grown really rapidly within a short period of two and a half decades but with distinct phases. Since the mid-eighties, basic structural genomics emerged comprising construction of genetic linkage maps using molecular markers and mapping of genes and quantitative trait loci. The following phase included application of molecular markers for various purposes of crop improvement, popularly known as molecular breeding. And the third phase involved physical genome mapping with artificial chromosomes culminating in advanced structural genomics including whole-genome mapping, and functional genomics. We have tried to follow almost the same trend while organizing the chapters on different topics under the three volumes of this textbook series.

In the first volume we briefly introduce the historical background and overview on genome mapping in the first chapter. Subsequently, we present 10 chapters deliberating on different types of molecular markers, their detection, relative merits, shortcomings and applications; types of mapping populations, methods of their generation, applications; basic concepts and schematic depiction of construction of genetic linkage maps; concepts and strategies of mapping genes controlling qualitative and quantitative traits on framework genetic linkage maps; rationale, methodologies and implications of comparative mapping; principles, strategies, and outcome of map-based cloning; overviews on the recent advances on plant genomics and genome initiatives; and finally computer strategies and software employed in plant molecular mapping and breeding.

In the second volume, we emphasize areas of application of molecular mapping. We present a general deliberation on the funda-mentals of molecular breeding and applications of molecular markers for germplasm characterization; concepts and application of molecular mapping and breeding for yield, quality, and their related component traits; biotic and abiotic stresses; and physiological traits of economic importance. Some plant groups, for example fruit crops, forest trees, fodder, forage, and turf grasses, and polyploid crop plants exhibit unique breeding problems and necessarily require special breeding strategies. We have dedicated independent chapters for them. We include a chapter on transgenic breeding considering its growing popularity and contribution in crop improvement. We also include a chapter on intellectual property rights (IPR) considering the legal implications in using, developing and commercializing plant varieties and genotypes.

In the third volume on advanced genomics, we present an overview on the advances of plant genomics made in the last century; deliberations on the genomics resources; concepts, tools, strategies, and achievements of comparative, evolutionary, and functional genomics and whole-genome sequencing. We also present critical reviews on the already completed genome initiatives and glimpses on the currently progressing genome initiatives. We also have a deliberation on application of the genomics information in genetic improvement of crop plants. We include a chapter on facilitation of the progress of plant genomics researches considering the requirements of generous funding, state-of-the-art facilities and well-trained human resources. Finally, we present depiction of a map of the road for plant genomics to be traveled by us and scientists of our next generations in the twenty-first century.

Articles and reviews on all these concepts, strategies, tools, and achievements in genome analysis and improvement abound but are scattered mainly on the pages of periodicals. Most of them deal with a focus on a particular crop plant or a group of related crop plants. Recently, a seven-volume series on 'Genome Mapping and Molecular Breeding in Plants' was published by Springer that encompasses 82 chapters on more than 100 plant species of economic importance. Interested readers may enrich their knowledge from this and other resources as well. However, in these three books, we have paid particular attention to the basic principles, strategies, and requirements and applications of various facets of plant genomics useful for students to develop a comprehensive understanding of the subjects and to assist academicians & researchers in disseminating and using the knowledge.

We express our deep gratitude to Prof. Werner Arber, Nobel Laureate in Medicine 1978 for kindly contributing the Prologue for this series and his generous comments about this endeavor. We are grateful to Prof. M.S. Swaminathan for kindly scripting the Foreword for this treatise. Views of these science missionaries and social visionaries will be immensely useful for the present and future travelers in the world of plant genomics and will remain as words of inspiration for the present and future generations of plant scientists.

We must mention here that working with the authors of the chapters of these three volumes has been highly enriching, enlightening, and entertaining. We wish to express our thanks and gratitude to these devoted and reputed scientists for their excellent chapters and above all, their affection and cooperation during the pleasant and painful periods of gestation and delivery of these three book volumes. They appreciated the importance of these textbooks and donated their time to contribute these clear and concisely written chapters.

The authors of some chapters of our series have included a few previously published illustrations in original or modified versions. We believe they have obtained permission for such reproduction from the competent authors or publishers of the articles containing these illustrations. However, as editors we believe that it is our moral duty to keep in record our acknowledgement for all these authors, publishers, and sources for the helps that facilitated the improvement of the contents and formats of the concerned deliberations.

We look forward to constructive criticisms and suggestions from all corners for future improvement of the contents and format of these three book volumes.

We must express our thanks and gratitude to the publishers for their constant cooperation extended to us and the authors. We remember that on many occasions we failed to stick to the dead lines of submission of manuscripts, proof corrections and other required inputs due mainly to the conflicts of schedules and other unforeseen assignments and commitments. However, the concerned staff did bear with us with laudable patience.

Chittaranjan Kole
Albert G. Abbott

Preface to the Volume

In recent years, genomics arguably ranks first in the arena of technological advances in life sciences that have spurred unprecedented rapid acquisition of new knowledge. Initially, the term 'genomics' was popularly used to define DNA sequencing and fine-scale mapping in any organism, however all the concepts, strategies and tools related to the molecular elucidation of the structure and function of all the genes in a genome currently are encompassed under the 'genomics' umbrella. Within this broad definition of genomics, we have explored: the current fundamental concepts, strategies and requirements of molecular genetic linkage mapping in Volume 1 'Genome Mapping'; the employment of molecular markers for the purposes of crop improvement in Volume 2 'Molecular Breeding'; and in Volume 3 'Advanced Genomics' we present the basic and applied aspects of sequencing of genes and genomes and their implication in the fine-scale elucidation of the plant genomes.

The difficulties of DNA sequencing and sequence assembly of the large and complex genomes of higher plants and animals eclipses those encountered in sequencing prokaryotic counterparts. Whole-genome sequencing has been completed or is nearly complete for about a dozen plant genomes in contrast to the myriad complete genome sequences available from plant- and animal-associated microbes and higher animal systems. The advent of cheaper and faster techniques of sequencing reinforced by advances in bioinformatics software development will facilitate successful whole-genome sequencing in many other plants in near future.

The basic and applied aspects of advanced plant genomics, presented in numerous publications in peer reviewed journals, proceedings, reports and reviews, have not yet been showcased in a comprehensive book volume that includes the basic concepts, strategies, tools and methods of advanced genomics presented with a class-room approach. The 17 chapters of this volume are designed to address this need and to serve as a reference book for students, scholars and scientists in academia, industry and government.

Chapter 1 of this volume presents a lucid overview on the concepts and strategies of advanced plants genomics and the potential benefits of its various applications. Physical mapping is the structural delineation of a genome either in local regions or in its entirety. The requirements, principles and methods of physical mapping have been presented in Chapter 2. ESTs are tools for development of genetic maps using gene-based markers and also to serve as substrates to study expression of known and unknown genes; Chapter 3 deals with techniques of developing and using ESTs in functional genomics studies.

Whole-genome sequencing in any higher plant involves sequencing, assembly and annotation of the DNA sequences for all members of its haploid set of chromosomes. In the past this had required a huge investment of time and money. The strategies for sequencing, interpretation and storage of the sequence data should be judiciously employed based upon need, cost, timing and precision. Chapter 4 provides a deliberation on these aspects and also exemplifies their employment in whole-genome initiatives in some model and crop plants.

Studies on genome homology using chromosome maps and gene sequences elucidate the origin, evolutionary pathways and phylogenetic relationships among close taxa. The already completed and currently progressing different genome initiatives have provided huge genomic resources and additional tools for highlighting the genomic changes experienced by the plant species over millions of years as presented in Chapter 5.

During the period that we were formulating the contents of this volume, whole-genome physical maps had been accomplished in a small number of plant genomes. Of these we chose to dedicate Chapters 6 to 9 to *Arabidopsis*, rice, poplar and peach and a general chapter (10) to the-then on-going other genome initiatives. Fortunately for plant science and unfortunately for this volume, whole-genome sequencing of grape, papaya, sorghum and soybean has been completed during the period of preparation of this book volume and we have only one option left that is to include chapters on these genome initiatives in the next edition of this volume. Genome initiatives on many other model and crop plants, for example *Medicago*, *Lotus*, maize, wheat, *Brassica rapa*, *Brassica oleracea*, tomato, potato, melon and eucalyptus, are progressing readily and thus future chapters in volumes of a separate series will need to be written for these species!

Various strategies of reverse genetics, particularly transposon tagging, activation tagging, expression traps and TILLING facilitate gene discovery and characterization. Chapter 11 discusses their application in a few selected species representing the major food, forestry and bioenergy crops of the world and compares advantages and disadvantages of these technologies especially for developing model bioenergy crops. Patterns of SNPs and indels reflect the nature of genetic variation in a genome. SNPs can be detected by many technologies and they have several applications ranging from association mapping to genome-wide searches for natural selection. Chapter 12 provides a lucid description of the patterns of SNPs in the genomes of the model plant *Arabidopsis*, fruitfly and human and also their applications on genome-wide SNP studies in crops. Knowledge gained from fruitfly and human SNP analysis will hopefully provide a broad but clear historical perspective for similar work in other species. Understanding the patterns of genome-wide gene expression in time, space and under different environmental conditions, is at the center of transcriptomics research. Results from this work promise to elucidate the critical evolutionary trends in growth and development of organisms. Chapter 13 deals with the strategies and applications of transcriptional profiling.

The potential of genomics in the depiction of genomes and genes in high resolution is well proven, however, applications of genomics tools for crop improvement remain the "Holy Grail" of modern plant breeding. Chapter 14 discusses the currently available genomic tools and approaches applicable in plant breeding with pragmatic examples of how to improve input and output traits, in light of the limitations of genomic versus conventional tools for crop improvement.

The enormous amount of data from genome sequences, protein structures, whole-genome expression studies, and population dynamics requires extensive computational biology and bioinformatics. We included two chapters on computing strategies and software (Chapter 11) and bioinformatics (Chapter 9) in our first volume dedicated to 'Genetic Mapping' that included commentaries of the marriage and honey-moon of plant biology and information technology. Chapter 15 of this volume describes the first few anniversaries of the two contrasting but complimentary disciplines and the potential impacts of this partnership in the years to come. The benefits, limitations and improvisation requirements of bioinformatics tools for genomics studies have also been addressed.

All human efforts and practices including science are targeted towards the benefit of humankind and their societies. Introduction of new technologies has always led to controversies whose resolutions lie in alternative scientific strategies, policies and regulations to compromise on the conflicts of public concerns and necessity of employment. Genomics is a recent candidate on the list of these technologies seconding only to transgenics. Chapter 16 presents a deliberation on the response towards plant genomics from scientific and social communities and the steps being taken to comply with societal concerns enabling the utilization of current and emerging genomic tools for the benefit of mankind.

The concluding chapter (17) of this volume and this book project assesses the achievements of plant genomics research and presents the road map of the future.

We can appreciate that this volume will appear outdated in the near future particularly for cases of the fast emerging sequencing and bioinformatics techniques to be developed and the whole-genome sequences to be completed. However, we do feel satisfied that this volume will remain as a source of the basic concepts, strategies and tools that have led to the foundation of understanding and practicing plant genomics today and will serve as the platform for tomorrow.

We express our thanks and regards to the 39 scientists from 6 countries for their chapters contributed to this volume and their constant cooperation from submission of the first drafts to revisions and updating and final fine-tuning of their chapters commensurate with the reviews and fast development of databases.

Finally, we wish to extend our thanks to the Science Publishers, Inc. and all its staff involved in publication and promotion of the three volumes of this book project that will hopefully be useful to the students, scientists and industries who are practising and contributing to the growth of plant genomics.

Chittaranjan Kole
Albert G. Abbott

Contents

The Prologue v

Foreword vii

Preface to the Series xi

Preface to the Volume xvii

List of Contributors xxiii

List of Abbreviations xxix

1. **An Overview of Advances in Plant Genomics in the New Millennium** 1
   Robert J. Henry

2. **Fundamentals of Physical Mapping** 24
   Perumal Azhaguvel, Yiqun Weng, Raman Babu, Alagu Manickavelu, Dhanasekaran Vidya Saraswathi and *Harindra Singh Balyan*

3. **Comparative and Evolutionary Genomics** 63
   Amy L. Lawton-Rauh, Cynthia R. Climer and *Brad L. Rauh*

4. **ESTs and their Role in Functional Genomics** 104
   Kalpalatha Melmaiee and *Venu Kalavacharla*

5. **Whole Genome Sequencing** 120
   Swarup K. Parida and *T. Mohapatra*

6. ***Arabidopsis* Genome Initiative** 175
   Ramesh Katam, Dilip Panthee, Evelina Basenko, Rajib Bandopadhyay, Sheikh M. Basha, Kokiladevi Eswaran and *Chittaranjan Kole*

7. **Rice Genome Initiative** 205
   Kevin L. Childs and *Shu Ouyang*

8. **The *Populus* Genome Initiative** 243
Stephen DiFazio

9. **Peach Genome Initiative** 275
Albert G. Abbott

10. **Current Status of On-going Genome Initiatives** 305
Jaya R. Soneji, Madhugiri Nageswara Rao, Padmini Sudarshana, Jogeswar Panigrahi and *Chittaranjan Kole*

11. **Concepts and Strategies for Reverse Genetics in Field, Forest and Bioenergy Crop Species** 354
Kazuhiro Kikuchi, Claire Chesnais, Sharon Regan and *Thomas P. Brutnell*

12. **Genetic Variation Identified through Gene and Genome Sequencing** 399
Erica G. Bakker

13. **Transcriptional Profiling** 424
Sam R. Zwenger and *Chhandak Basu*

14. **Advanced Bioinformatics Tools and Strategies** 447
Matthew A. Hibbs

15. **Application of Genomics to Plant Breeding** 494
Thomas Lübberstedt and *Madan K. Bhattacharyya*

16. **People, Policy and Plant Genomics** 528
Emma Frow and *Steven Yearley*

17. **Roadmap of Genomics Research in the 21st Century** 571
Andrew H. Paterson

Subject Index 583

Color Plate Section 591

List of Contributors

Alagu Manickavelu

Kihara Institute for Biological Research, Yokohama City University, Yokohama 244-0813, Japan; Phone: +81-45-820-2401; Fax: +81-45-820-1901; e-mail: agromanicks@rediffmail.com

Albert G. Abbott

Clemson University, Department of Genetics and Biochemistry, 116 Jordan Hall, Clemson, SC 29634, USA; Phone: +1-864-656-3060; Fax: +1-864-656-6879; e-mail: aalbert@clemson.edu

Amy Lawton-Rauh

Clemson University, Department of Genetics and Biochemistry, 101 Jordan Hall, Clemson, SC 29534, USA; Phone: +1-864-656-4660; Fax: +1-864-656-6879; e-mail: AmyLR@clemson.edu

Andrew H. Paterson

Plant Genome Mapping Laboratory, University of Georgia, 111 Riverbend Road, Room 228, Athens, GA 30602, USA; Phone +1-706-583-0162; Fax: +1-706-583-0160; e-mail: paterson@uga.edu

Brad Rauh

Clemson University, Department of Genetics and Biochemistry, 101 Jordan Hall, Clemson, SC 29534, USA; Phone: +1-864-656-1507; Fax: +1-864-656-6879; e-mail: brauh@clemson.edu

Chhandak Basu

School of Biological Sciences, University of Northern Colorado, Greeley, Colorado 80639, USA; Phone: +1-970-351-2716; Fax: +1-970-351-2335; e-mail: chhandak.basu@unco.edu

Chittaranjan Kole

Department of Genetics and Biochemistry and Institute of Nutraceutical Research, Clemson University, 109 Jordan Hall, Clemson, SC 29634, USA; Phone: +1-864-656-3060; Fax: +1-864-656-6879; e-mail: ckole@clemson.edu

Claire Chesnais

Department of Biology, Queen's University, Kingston, Ontario, K7L 3N6, Canada; Phone: +1-613-533-6000 ex. 77330; Fax: +1-613-533-6617; e-mail: 3cc18@queensu.ca

Cynthia Climer

Clemson University, Department of Genetics and Biochemistry, 101 Jordan Hall, Clemson, SC 29534, USA; Phone: +1-864-656-1507; Fax: +1-864-656-6879; e-mail: cclimer@clemson.edu

Dhanasekaran Vidya Saraswathi

Texas Veterinary Medical Diagnostic Laboratory, Amarillo Blvd West, Amarillo, TX 79106, USA; Phone: +1-806-353-7478; Fax: +1-806-359-0636; e-mail: vdhanasekaran@tvmdl.tamu.edu

Dilip R. Panthee

Department of Horticultural Science, North Carolina State University, Mountain Horticultural Crops Research and Extension Center, Mills River NC 28759, USA; Phone: +1-828-684-8590; Fax: +1-828-684-8715; e-mail: dilip_panthee@ncsu.edu

Emma Frow

ESRC Genomics Policy and Research Forum, University of Edinburgh, St John's Land, Holyrood Road, Edinburgh EH8 8AQ, UK; Phone: +44-131-651-4745; Fax: +44-131-651-4748; e-mail: emma.frow@ed.ac.uk

Erica G. Bakker

Center for Genome Research and Biocomputing, Department of Horticulture, Oregon State University. Present Address: Dow AgroSciences, 16160 SW Upper Boones Ferry Road, Portland, OR 97224, USA; Phone: +1-503-213-2077; Fax: +1-503-670-7703; e-mail: Bakker2@dow.com

Evelina Y. Basenko

Department of Genetics, University of Georgia, Athens, GA 30602, USA; Phone +1-706-542-1427; Fax +1-706-542-3910; e-mail: ebasenko@uga.edu

Harindra Singh Balyan

Department of Genetics and Plant Breeding, Ch. Charan Singh University, Meerut 250 004, India; Phone: +91-121-2763564; Fax: +91-121-2768195; e-mail: hsbalyan@gmail.com

Jaya R. Soneji

University of Florida, IFAS, Citrus Research & Education Center, 700 Experiment Station Road, Lake Alfred, FL 33850, USA; Phone: +1-863-956-1151; Fax: +1-863-956-4631; e-mail: jrs@crec.ifas.ufl.edu

Jogeswar Panigrahi

Biotechnology Unit, Sambalpur University, Jyoti Vihar, Burla 768019, Orissa, India; e-mail: drjogeswar_panigrahi@ymail.com

Kalpalatha Melmaiee

Department of Agriculture & Natural Resources, Delaware State University, 1200 N. DuPont Highway, Dover, DE 19901, USA; Phone: +1-302-857-6461; Fax: +1-302-857-6402; e-mail: kmelmaiee@desu.edu

Kazuhiro Kikuchi

Boyce Thompson Institute for Plant Research, Cornell University, Ithaca NY, USA; Ph: +1-607-254-6747; Fax: +1-607-254-1242; e-mail: kk376@cornell.edu

Kevin L. Childs

166 Plant Biology Building, Department of Plant Biology, Michigan State University, East Lansing 48824, USA; Phone: +1-517-353-5969; Fax: +1-517-353-1926; e-mail: kchilds@plantbiology.msu.edu

Kokiladevi Eswaran

Tamil Nadu Agricultural University, Coimbatore, 641003, Tamil Nadu, India; Phone: +91-422-6611353; Fax: +91-422-6611462; e-mail: cmkokila @yahoo.com

Madan K. Bhattacharyya

Iowa State University, Department of Agronomy, G303 Agronomy Hall, Ames, IA 50011, USA; Phone: +1-515-294-2505; Fax: +1-515-294-2299; e-mail: mbhattac@iastate.edu

Madhugiri Nageswara Rao

University of Florida, IFAS, Citrus Research & Education Center, 700 Experiment Station Road, Lake Alfred, FL 33850, USA; Phone: +1-863-956-1151; Fax: +1-863-956-4631; Email: mnrao@crec.ifas.ufl.edu

Matthew A. Hibbs

The Jackson Laboratory, 600 Main Street, Bar Harbor, ME 04609, USA; Phone: +1-207-288-6944; Fax: +1-207-288-6847; e-mail: matt.hibbs@jax.org

Padmini Sudarshana

Monsanto Research Center, #44/2A, Vasant's Business Park, Bellary Road, NH-7, Hebbal, Bangalore 560092, India; Phone: +91-80-2362-2525; Fax: +91-80-2362-4343; e-mail: padmini.sudarshana@monsanto.com

Perumal Azhaguvel

Texas AgriLife Research, Texas A&M University, 6500 Amarillo Blvd. W., Amarillo, TX 79106, USA; Phone: +1-806-677-5653; Fax: +1-806-677-5644; e-mail: *azhagu@ag.tamu.edu*

Rajib Bandopadhyay

Department of Biotechnology, Birla Institute of Technology, Mesra, Ranchi 835215, India; Phone: +91-9430378406; Fax: +91-651-227-5401; e-mail: rajib_bandopadhyay@bitmesra.ac.in

Raman Babu

International Maize and Wheat Improvement Center (CIMMYT), Apdo. Postal 6-641, 06600 Mexico, DF, Mexico; Phone: +52 (55) 5804 2004 x 1121; Fax: +52(55) 5804-7558; e-mail: rbabu_2002@gmail.com

Ramesh Katam

Plant Biotechnology Laboratory, 6505 Mahan Drive, Florida A&M University, Tallahassee FL 32317, USA; Phone: +1-850-412-5190; Fax: +1-850-561-2617; e-mail: katamr@yahoo.com

Robert J. Henry

Centre for Plant Conservation Genetics, Southern Cross University, PO Box 157, Lismore, NSW 2480, Australia; Phone: +61-2-6620-3010; Fax: +61-2-222080; e-mail: robert.henry@scu.edu.au

Sam R. Zwenger

School of Biological Sciences, University of Northern Colorado, Greeley, Colorado 80639, USA; Phone: +1-970-351-2555; Fax: +1-970-351-2335; e-mail: sam.zwenger@gmail.com

Sharon Regan

Department of Biology, Queen's University, Kingston, Ontario, K7L 3N6, Canada; Phone: +1-613-533-3153; Fax: +1-613-533-6617; e-mail: sharon.regan@queensu.ca

Sheikh M. Basha

Plant Biotechnology Laboratory, 6505 Mahan Drive, Florida A&M University, Tallahassee FL 32317, USA; Phone: +1-850-412-5189; Fax: +1-850-561-2617; e-mail: mehboob.sheikh@famu.edu

Shu Ouyang

CambridgeSoft Corporation, 1003 W. 7th Street, Suite 205, Frederick, MD 21701, USA; Phone: +1-301-228-4094; Fax: +1-301-846-6183; e-mail: ouyangsn@mail.nih.gov

Stephen DiFazio

Department of Biology, West Virginia University, Morgantown, West Virginia 26506-6057, USA; Phone: +1-304-293-5201 x 31512; Fax: +1-304-293-6363; e-mail: spdifazio@mail.wvu.edu

Steven Yearley

ESRC Genomics Policy and Research Forum, University of Edinburgh, St John's Land, Holyrood Road, Edinburgh EH8 8AQ, UK; Phone: +44-131-651-4747; Fax: +44-131-651-4748; e-mail: steve.yearley@ed.ac.uk

Swarup Kumar Parida

National Research Centre on Plant Biotechnology, Indian Agricultural Research Institute, New Delhi 110012, India; Phone: +91-11-25841787, Ext. 237; Fax: +91-11-25843984; e-mail: swarupdbt@gmail.com

Thomas Lübberstedt

Iowa State University, Department of Agronomy, 1204 Agronomy Hall, Ames, IA 50011, USA; Phone: +1-515-294-5356; Fax: +1-515-294-5506; e-mail: thomasl@iastate.edu

Thomas P. Brutnell

Boyce Thompson Institute for Plant Research, Cornell University, Ithaca NY 14853, USA; Phone: +1-607-254-8656; Fax: +1-607-254-1242; e-mail: tpb8@cornell.edu

Trilochan Mohapatra

National Research Centre on Plant Biotechnology, Indian Agricultural Research Institute, New Delhi 110012, India; Phone: +91-11-25841787 x 237; Fax: +91-11-25843984; e-mail: tm@nrcpb.org

Venu Kalavacharla

Department of Agriculture & Natural Resources, Delaware State University, 1200 N. DuPont Highway, Dover, DE 19901, USA; Phone: +1-302-857-6492/ 302-857-6461; Fax: +1-302-857-6402; e-mail: vkalavacharla@desu.edu

Yiqun Weng

Vegetable Crop Research Unit, USDA-ARS, University of Wisconsin, 1575 Linden Drive, Madison, WI 53706, USA; Phone: +1-608-262-0028; Fax: +1-608- 262-4743; e-mail: weng4@wisc.edu

List of Abbreviations

4CL	Hydroxycinnamate CoA ligase
4DTV	Four-fold degenerate transversion
ABRC	*Arabidopsis* Biological Resource Center
ACC	1-Aminocyclopropane-1-carboxylic acid
Ac/Ds	*Activator/Dissociation* transposable elements
ACPFG	Australian Centre for Plant Functional Genomics
ACS	ACC synthase
AFLP	Amplified fragment length polymorphism
AGI	*Arabidopsis* Genome Initiative
APG	Angiosperm phylogeny group
APHIS	Animal and Plant Health Inspection Service
AtDB	*Arabidopsis thaliana* Data Base
AZ	Azacytidine
BAC	Bacterial artificial chromosome
BBS	Biological, Behavioral, and Social Sciences Directorate
BBSRC	Biotechnology and Biological Sciences Research Council
BC	Backcross
BES	BAC-end sequence
BGI	Beijing Genomics Institute
BIBAC	Binary vector BAC
BLAST	Basic local alignment search tool
BLASTN	Basic local alignment search tool: nucleotide
BLASTP	Basic local alignment search tool: protein
Bt	*Bacillus thuringiensis*
C3H	Coumarate hydroxylase
C4H	Cinnamate hydroxylase

CAD	Cinnamyl alcohol dehydrogenase
CAP	Coordinated Agricultural Project
CBCS	Cot-based cloning and sequencing
CBD	Convention on Biological Diversity
CBOL	Consortium for the Barcoding of Life
CCR	Cinnamoyl CoA reductase
cDNA	Complementory DNA
CDS	Coding sequence
CGH	Comparative genome hybridization
CGIAR	Consultative Group on International Agricultural Research
COMT	Caffeic O-methyltransferase
COS	Conserved orthologous set
cpDNA	Chloroplast DNA
CPT	Conditional probability table
CR	Chilling requirement
DAG	Directed acyclic graph
DArT	Diversity array technology
DB	Database
DBA	Dot blot assay
dCAPS	Derived cleaved amplified polymorphic sequence
DCL-1	Dicer-like 1RNase
DDBJ	DNA Data Bank of Japan
DH	Doubled haploid
DHPLC	Denaturing high-performance liquid chromatography
dNTP	Deoxyribonucleotide triphosphate
DOE	Department of Energy
dsRNA	Double stranded RNA
EMAIL	Endonucleolytic mutation analysis by internal labeling
EMS	Ethylmethane sulfonate
EPA	Environmental Protection Agency
eQTL	Expression QTL
EST	Expressed sequence tag
EU	European Union
EVG	Evergrowing (gene)
FA5H	Ferulate hydroxylase
FAO	Food and Agriculture Organization of the United Nations
FISH	Fluorescence in situ hybridization

FL-cDNA	Full length cDNA
FM	Functional marker
FN	False negative
FP	False positive
FR	Functional relationship
FST	Flanking sequence tag
GDP	Gross domestic product
GDR	Genome Database for Rosaceae
GFP	Green fluorescence protein
GIGO	Garbage in, garbage out
GLP	Germin like protein
GM	Genetically modified
GMO	Genetically modified organism
GO	Gene Ontology
GUS	β-Glucuronidase
HapSTR	Haplotype simple-tandem repeat
HICF	High information content fingerprinting
HKA test	Hudson-Kreitmam-Agaudé test
HLA	Human leucocyte antigen
HMPR	Hypomethylated partial restriction library
HMW	High molecular weight
HSP	Heat-shock protein
IHP	International HapMap Project
Indels	Insertions and deletions
IP	Intellectual Property
IRGSP	International Rice Genome Sequencing Project
ISSR	Inter simple sequence repeat
ITPGRFA	International Treaty for Plant Genetic Resources for Food and Agriculture
JA	Jasmonate
JGI	Joint Genome Institute
KEGG	Kyoto Encyclopedia of Genes and Genomes
KOME	Knowledge-based *Oryza* Molecular Biological Encyclopedia
LCR	Ligation chain reaction
LD	Linkage disequilibrium
LG	Linkage group

LINE	Long interspersed transposable element
LRR	Leucine-rich repeat
LUC	Luciferase
MAB	Marker-assisted backcrossing breeding
MARS	Marker-assisted recurrent selection
MAS	Marker-assisted selection
MAYG	Mapping as you go
MBC	Map-based cloning
MF	Methylation filtration
MIAME	Minimum information about a microarray experiment
MIPS	Munich Information Center for Protein Sequences
miRNA	Micro-RNA
MITE	Miniature inverted-repeat transposable element
MNU	N-Methyl-N-nitrosourea
MPSS	Massively parallel sequencing signatures
mRNA	Messenger-RNA
MSLL	Methylation spanning linker library
MSU	Michigan State University
MTA	Material Transfer Agreement
mtDNA	Mitochondrial-DNA
MTP	Minimum tiling path
MYA	Million years ago
NAD	Nicotinamide adenine dinucleotide
NB-LRR	Nucleotide binding leucine-rich repeat
NBS	Nucleotide binding site
NCBI	National Center for Biotechnology Information
NIAB	National Institute of Agricultural Biotechnology
NIH	National Institute of Health
NP	Non-polynomial
NPGI	National Plant Genome Initiative
NRI	National Research Initiative
NSF	National Science Foundation
ORF	Open reading frame
OTU	Operational taxonomic unit
P1	P1 phase genome vector
PAC	P1-derived artificial chromosome
PAL	Phenylalanine ammonia-lyase

PASA	Program for the alignment of sequence assemblies
PBC	Plasmid based cloning vector
PCA	Principal component analysis
PCR	Polymerase chain reaction
PFGE	Pulse-field gel electrophoresis
PHB	Polyhydroxy butyrate
PPV	Plum pox virus
PRCD	Plate, row, column and diagonal
PR-curve	Precision-recall curve
PTGS	Post-translational gene silencing
PVR	Plant Variety Rights
qRT-PCR	Quantitative real time PCR
QTIndel	Quantitative trait insertion-deletion
QTL	Quantitative trait loci
QTN	Quantitative trait nucelotide
RACE	Rapidly amplified cDNA ends
RAP	Rice Annotation Project
RAPD	Random(ly) amplified polymorphic DNA
rDNA	Ribosomal-DNA
RFLP	Restriction fragment length polymorphism
RGA	Resistance gene analog
RH	Radiation hybrid
RI	Recombinant inbred
RiceGAAS	Rice Genome Automated Annotation System
RIL	Recombinant inbred line
RISC	RNA-induced silencing complex
RITS	RNA-induced transcriptional silencing complex
RLK	Receptor-like kinase
RNAi	RNA interference
RNase A	Ribonuclease A
ROC	Receiver operating characteristic
RPA	Ribonuclease protection assay
rRNA	Ribosomal-RNA
RT PCR	Real time PCR
RT-PCR	Reverse transcription-polymerase chain reaction
SAGE	Serial analysis of gene expression
SCA	Specific combining ability

SCAR	Sequence characterized amplified region
SCNL	Single copy nuclear loci
SFP	Single-feature polymorphism
siRNA	Small interfering-RNA
smRNA	Small-RNA
SMS	Single-molecule sequencing
SNP	Single nucleotide polymorphism
snRNA	Small nuclear-RNA
SOM	Self-organizing map
SRAP	Sequence-related amplified polymorphism
SSH	Suppression subtractive hybridization
SSIIa	Starch synthase IIa
SSLP	Simple sequence length polymorphism
SSR	Simple sequence repeat
STM	Sequence tagged microsatellite
STR	Simple tandem repeat
STS	Sequence tagged site
SVD	Singular value decomposition
SVM	Support vector machine
SVP	Short vegetative phase
TAC	T1 phage artificial chromosome
TAC	Transformation-competent artificial chromosome
TAIR	The *Arabidopsis* Information Resource
T-DNA	Transfer-DNA
TE	Transposable element
Ti plasmid	Tumor inducing plasmid
TIGR	The Institute for Genomic Research
TILLING	Targeting induced local lesions in genomes
TN	True negative
TP	True positive
TRAP	Target region amplified polymorphism
tRNA	Transfer-RNA
TSD	Target site duplication
UAS	Upstream activator sequence
UGMS	Unigene derived microsatellite sequence
UPOV	International Union for the Protection of New Varieties of Plants

URL	Uniform resource locators
USDA	United States Department of Agriculture
UTR	Untranslated region
VIGS	Virus-induced gene silencing
WGA	Whole-genome arrays
WGS	Whole-genome shotgun (sequencing)
WHO	World Health Organization of the United Nations
WTO	World Trade Organization
YAC	Yeast artificial chromosome

1 An Overview of Advances in Plant Genomics in the New Millennium

Robert J. Henry
Centre for Plant Conservation Genetics, Southern Cross University, PO Box 157, Lismore, NSW 2480, Australia
Email: robert.henry@scu.edu.au

1 INTRODUCTION

The technologies available for genomics have continued to develop rapidly since 2000, accelerating their application in plant biology and plant breeding. These technologies are providing the tools to allow cost-effective collection of much larger volumes of data on plant genomes. This, in turn, requires the development of bioinformatics capabilities to analyze these data and apply them successfully in plant improvement. Genomics, in a general sense, is advancing by improvements in technology at each of the levels from studies of DNA (genomics) and RNA (transcriptomics), which are often considered a core part of genomics, to proteins (proteomics), metabolites (metabolomics) and the phenotypes (phenomics) (Table 1).

1.1 Genomics

Genomics is the study of all or most of the genes in an organism. Higher plants are challenging targets for genomics because of their large genomes. However, genomics remains a dominant area of science and a powerful tool for plant scientists with complementation from the emer-

Table 1 Genomics and related 'omics'.

Omic	*Level*
Genomics	DNA
Transcriptomics	RNA
Proteomics	Protein
Metabolomics	Metabolite
Phenomics	Phenotype

ging associated rechk from cd 'omics' still lagging significantly behind genomics in applications to plants but likely to increase in the near future.

1.1.1 Genome Polymorphism Discovery and Analysis

Differences between plant genomes are analyzed in phylogenetic and evolutionary research, population genetics, gene discovery and functional analysis and plant breeding and in plant identification. The discovery of genetic polymorphism between plants has been based upon a range of mutation detection techniques and DNA sequencing (Table 2). The Angiosperm Phylogeny Group developed phylogenies of flowering plants based upon DNA sequence rather than morphology. The Consortium for the Barcode of Life (CBOL) has defined target sequences (Kress et al. 2005) for use in species distinctions. Genomics will ultimately allow the APG and CBOL analysis to be extended from analysis of a few genes to include the whole genome in phylogenetic analysis. DNA sequencing has recently become a more cost-effective method for discovery of sequence differences (Wicker et al. 2006; Mardis 2008).

Transcriptome sequencing (Emrich et al. 2007) has been developed as a tool for discovery of polymorphism in the coding parts of genes

Table 2 Mutation discovery approaches.

Re-sequencing (di-deoxy)	Bradbury et al. 2005
TILLING	Henikoff et al. 2004
EcoTILLING	Comai et al. 2004
454 pyrosequencing	Margulies et al. 2005
Solexa/Illumina sequencing by synthesis	Bennett et al. 2005
ABI SOliD ligation based sequencing	Hudson 2008
EMAIL	Cross et al. 2008

(exons). Expressed sequence tag (EST) sequencing has been widely used for this purpose but is being replaced by more cost-effective methods. Approaches employing sequencing include the use of cDNA, purification of the exons by hybridization to long oligonucleotide microarrays or by enrichment using methylation sensitive restriction enzymes.

Rarer mutations can be discovered by techniques using nucleases to detect mismatches in DNA pools (Till et al. 2006; Cross et al. 2008).

1.1.2 Genotyping

The genotyping of plants has progressed through a range of technologies in the past decade (Table 3). The genotyping of plants in the 1990s was based largely upon the analysis of microsatellites or simple sequence repeats (SSRs) (Henry 2001). Since 2000, single nucleotide polymorphism (SNP) markers have increasingly replaced the SSR markers as more sequence data and techniques for SNP discovery and analysis have advanced (Henry 2008). SNP analysis methods with greater repeatability compared to SSR marker methods have been developed (Jones et al. 2007) facilitating much more widespread adoption of SNP methods. These technical developments have contributed to an expansion of the range of applications in plant genotyping. In addition to research applications in plant improvement, plant identification based upon DNA analysis is widely applied; in protection of intellectual property (disputes over ownership of protected plant varieties), commercial transactions (to confirm seed is true-to-the-type and pure, and to verify

Table 3 Advances in molecular marker applications in plants.

Pre-PCR	RFLP	
Post-PCR	Arbitrary or random markers	RAPD
		AFLP
		DART
	Sequence-based markers	SSR
		EST-derived SSR
		SNP

Marker technology has advanced in stages. The development of PCR made earlier methods redundant. The increasing availability of DNA sequence data is now resulting in sequence based markers replacing earlier markers based upon arbitrary or random sequences. SSR markers have been replaced by SNP markers because of their relative ease of assay.

the identity of products purchased for processing) and in the analysis of the composition of competitors food products.

1.1.3 SNP Analysis

SNP discovery is now mostly *in silico* or by new sequencing approaches (Henry and Edwards 2009) Large-scale SNP analysis is now possible in plants using a range of technology platforms (Table 4). Simple methods have been developed for use in laboratories with limited facilities (Bundock et al. 2006) together with very highly automated options for large-scale low-cost analysis (Henry 2008). Many plant genomes contain duplications due to polyploidy during the evolution of the genomes. New methods for large-scale quantitative analysis of SNP in polyploidy or complex genomes provide tools for exploring the relationships between allele copy number and gene expression.

Table 4 Technologies for SNP analysis (as defined by Henry 2008).

Platform	*Reference*
Gel electrophoresis	Bundock et al. 2006
Real-time PCR	Kennedy et al. 2006a, b
Capillary electrophoresis	Batley et al. 2003
Mass-spectrometry	Ragoussis 2006
Microarray	Baner et al. 2003
Flow cytometry	Lee et al. 2004

1.1.4 Genome Mapping

Conventional plant genome mapping has involved linkage analysis in a segregating population often generated by crossing two divergent individuals or two genotypes that differed in the trait that was to be mapped. The increasing ease of sequencing and automated genotyping has made the use of association mapping a more attractive option in plants. The association mapping approach analyzes loci in diverse populations and associates them with one another and with phenotypes. This trend is likely to continue as sequencing of genomes increases.

1.1.5 Genome Sequencing

The rate of sequencing of plant genomes is continuing to accelerate (Parida and Mohapatra 2009). The sequencing of the genome of the model plant *Arabidopsis* (Katam et al. 2009) at the end of the last century was followed after 2000 by the completion of a rice genome sequence (Goff et al. 2002; Childs and Quyang 2009)). Poplar and grape followed soon after. Other genomes being actively sequenced include peach (Abbott 2009), sorghum (Paterson et al. 2009) and eucalyptus. The first approach was to sequence overlapping clones covering the entire genome that had been isolated and located on a physical map. Shot-gun sequencing and end-to-end sequencing of bacterial artificial chromosome (BAC) tiles provide complementary routes to a complete genome sequence. Shot-gun sequencing has emerged as an alternative in which large numbers of random clones are sequenced. This helps to avoid the large effort required to develop a tilled array of clones to sequence but creates considerable difficulty in genome assembly, especially in repetitive parts of the genome. In practice, some combination of these two approaches has been used. The more recent development of next generation sequencing technologies is likely to dramatically increase the cost effectiveness of genome sequencing. The advances in technology are allowing experiments to be considered that are based upon comparisons of the whole-genome sequences of individual species within a group of related species, such as the genus *Oryza* or even individuals within a population of a species. We can expect new approaches to many biological questions to be investigated using these new approaches.

1.1.6 Integration of Physical and Genetic Maps

The integration of physical and genetic maps has been achieved in many species by analysis of genetic markers in BAC libraries that have been assembled to produce a physical map. Genotyping of whole-genome sequencing provides the ultimate integration of physical and genetic maps. Synteny in the grasses has been very valuable because so many species are of economic importance. The availability of the rice genome sequence has been complemented by extensive data in other species (especially sorghum) allowing detailed comparative analysis of physical and genetic maps. This has been extended to compare across the monocotyledons (Lothithaswa et al. 2007).

1.1.7 Conservation Genomics

The study of wild and domesticated plants to support biodiversity conservation has been supported by the discipline of conservation genetics (Henry 2006). Genomics approaches are now beginning to be applied in conservation genetics to produce the new field of conservation genomics. Our understanding of plant diversity in wild populations at the genome and phenome level (Henry 2005) is accelerating with the wider availability of genomics techniques for cost-effective application to non-model organisms. The association of genetic variation with climate has been reported for specific genes (Cronin et al. 2007). Genome-scale analysis of genome and environment variation will reveal the parts of the genome under adaptive selection and those parts under diversifying or purifying selection.

1.1.8 Functional Genomics—Gene Discovery Strategies

The determination of gene function has been a key area of advancement in the recent years. The availability of whole-genome sequences facilitates new approaches to gene discovery and functional genomics. The discovery of genes for specific traits can be approached in several ways at the genome level using current tools. The starting point is often either:

i) a biochemical candidate (a gene with an annotation suggesting sequence homology with a gene that has an appropriate biochemical function)
ii) a positional candidate (a gene located on the chromosome in a region showing genetic linkage to the trait)
iii) or an expression candidate (a gene showing a pattern of expression consistent with the a expected for the gene of interest).

Recent examples of the successful application of these strategies are given below.

1.1.8.1 Gelatinization Temperature Gene in Rice: A Biochemical and Positional Candidate

The temperature at which rice cooks is associated with the gelatinization temperature of the starch in the seed. The trait is important in rice and has been widely measured using an alkaline spreading test resulting in the gene being named *alk*. This trait had been mapped using conventional approaches. Fine-mapping indicated that the gene co-segregated with starch synthase IIa (*SSIIa*). This gene was a strong biochemical candidate because of biochemical evidence that SSII influenced the amylopectin structure. Waters et al. (2006) identified the entire sequence of the gene by analysis of the rice genome sequence and explored variation in sequence over the entire gene. This resulted in the identification of an SNP and a two base pair change both of which caused amino acid substitutions in the encoded protein. These conservative changes near the active site of the enzyme were both associated with a reduction in gelatinization temperature of 8°C. The combination of mapping data, biochemical evidence and the rice genome sequence allowed identification of the molecular basis of this commercially important quantitative trait loci (QTL).

1.1.8.2 Fragrance Gene in Rice: A Positional Candidate

The fragrance of rice has been associated with the accumulation 2-acetyl-1-pyroline and is a highly desirable trait in some regions. The *fgr* gene had been mapped to chromosome 8 using SSR and SNP markers. When the rice genome sequence became available, Bradbury et al. (2005a) identified that the gene was between two flanking markers only 387,000 bp apart. Re-sequencing of the part of the genome from fragrant rice genotypes quickly identified a gene annotated as a betaine aldehyde dehydrogenase as having deletions likely to result in loss-of-function in all fragrant but not non-fragrant genotypes. In this example, the combination of mapping data and genome sequence data identified a gene with previously unknown biochemical function.

1.2 Transcriptomics

Sequencing of expressed genes has allowed discovery of the expressed parts of plant genomes. EST analysis allows comparison of different tissues or developmental stages. Serial analysis of gene expression (SAGE) analysis has extended the depth of this analysis. Differential analysis of expressed genes has become a major route to gene discovery. Microarray analysis (Li et al. 2006) allows the large-scale comparison of genes expressed in different tissues or genotypes and their association with possible function (Gregory et al. 2008). Techniques such as RT-PCR for quantitative expression analysis have been widely applied. Expression has been usually related to that of house-keeping genes that are expressed at a constant level. The difficultly of finding true house-keeping genes may be overcome by using techniques that measure absolute levels of expression (Turkulov et al. 2007).

1.3 Small Regulatory RNA

The role of small RNA in the regulation of expression of the genome has become more widely understood in the past decade (Groβhans and Filipowicz 2008). The analysis of the small RNA transcriptome of most plant species is at an early stage. Non-coding regulatory RNAs include siRNAs (small interfering RNAs) formed by cleavage of long double-stranded RNA and miRNAs (microRNAs) encoded by specific genes and functioning by interaction with mRNA. The use of these mechanisms to engineer plants is likely to increase with our understanding of the mechanisms of action and role of these small regulatory RNAs. They are likely to be important in the immediate future contributing to our understanding of the relationship between plant genomes and plant phenotype.

1.4 Proteomics

The application of proteomics has lagged genomics in plants. Newer technologies are improving the sampling and fractionation of proteins with higher resolution mass-spectrometry. Increasing population of databases is a key limiting factor in the adoption of proteomic approaches in plants. Recently advances in DNA-sequencing will provide much improved data sets for interpretation of proteomic data. The analysis of peptides in plants is also being accelerated by the availability of more genomic data (Farrokhi et al. 2008). Applications of proteomics include analysis of the composition of nutritionally important plant materials (Mak et al. 2005) and identification of biochemical events associated with stress or disease (Mak et al. 2006).

1.5 Metabolomics

Metabolomics is allowing the analysis of complex pathways of plant metabolism to be dissected (Guy et al. 2008). Improving techniques (Fiehn et al. 2008) could see much wider application of this approach. Metabolomics has been successfully applied to the study of the impact of stresses on plant metabolism (Shulaev et al. 2008). However, the potential to reliably engineer plant metabolism remains some way off (Sweetlove et al. 2008). The integration of metabolomic data, proteomic data and genomic data will support significant advances. These approaches have found special application in attempts to genetically improve the nutritional value of plants for human and animal diets (Davies 2007).

1.6 Phenomics

Methods for automated analysis of plant phenotype have become important tools for functional genomics. Growth can be monitored automatically. Despite these advances, analysis of phenotype often remains a limiting step in attempts to associate genes and traits. High-throughput phenotyping technologies also allow rapid detection of desired mutants (Grainer et al. 2006).

1.7 Transgenics

Techniques for the efficient transformation of plants have been continuously improved in efficiency. Direct DNA transfer has been widely applied. The range of species for which *Agrobacterium*-mediated transformation is possible has been progressively expanded to include the important cereal crop species.

Methods for homologous recombination have been perfected for use in altering the genomes of economic species (D'Halluin et al. 2008). These advances offer the potential to make targeted alterations to genomes to achieve plant improvement objectives.

The development of crops with multiple transgenes is a key to the widespread adoption of this technology. This requires the availability of a range of promoters for use in driving transgene expression in the target tissues. The use of the same promoter on more than one gene is considered undesirable because of the potential for this to lead to gene silencing. This can be avoided by sourcing promoters from other related species. However the exact specificity of the promoter may not be maintained when introduced into even very closely related species (Furtado et al. 2008). Much more research on promoter discovery and function will be required to support the deployment of crops with stacked transgenes for a wide range of useful traits.

Important applications of transgenic technology (Table 5) being delivered with increased efficiency and effectiveness, include improving the nutritional value of crops for animals and humans (Frizzi et al. 2008), manipulating plants for the production of pharmaceuticals (Allen et al. 2008) in molecular pharming (Table 6) and improving their resistance to pests (Shao et al. 2008).

1.8 Bioinformatics

Data analysis in support of genomics has been made more difficult by advances in the capacity of DNA-sequencing instruments. Currently DNA-sequence file sizes exceed the capacity on individual personal computers for the first time. Rapid advances in computer hardware and software are expected to correct this situation. Edwards and Batley (2008) reviewed the applications of bioinformatics in plant genetics and breeding.

Table 5 Examples of applications of transgenic approaches in plant improvement.

Trait	*Genes/product*	*Reference*[1]
Photosynthesis	*ictB* trehalose	Lieman-Hurwitz et al. 2003 Pellny et al. 2004
Insect resistance	*Bt* *gp* TSP14 Snowdrop lectin Protease inhibitor *mpi* ASAL	Naimov et al. 2003 Kater et al. 2003 Maiti et al. 2003 Nagadhara et al. 2003 Outchkourov et al. 2004 Vila et al. 2005 Dutta et al. 2005
Fungal resistance	endochitanase antimicrobial peptide	Emani et al. 2003 Rajasekaran et al. 2005
Abiotic stress tolerance	transcription factor glycinebetaine	Kim et al. 2004 Quan et al. 2004
Nematode resistance	cystatin	Lilley et al. 2004
Phytosterols	HMGR	Harker et al. 2003
Chilling tolerance	OC-1 PPO WCOR410	Van der Vyver et al. 2003 Zhou et al. 2003 Houde et al. 2004
Amino acid content	$tRNS^{lys}$ cystathionine synthase	Wu et al. 2003 Avraham et al. 2005
Fatty acids	*ADS1* desaturase	Yao et al. 2003 de Gyves et al. 2004
Lignin	CAD	Gill et al. 2003
Heavy metal tolerance	*merA9*	Che et al. 2003
Phosphorus nutrition	phytase	George et al. 2005
Flowering time	*OsSOC1*	Tadege et al. 2003
Digestibility	CAD	Chen et al. 2003
Herbicide tolerance	scFv glutathione transferase	Almquist et al. 2004 Skipsey et al. 2005
Protein content	PEP carboxylase	Rolletschek et al. 2004
Starch granule size	SBD2	Ji et al. 2004
High amylase	branching enzymes	Hofvander et al. 2004
High sucrose	sucrose isomerase	Wu and Birch 2007

Contd.

Table 5 continued

Trait	*Genes/product*	*Reference*[1]
High oil	glycerol phosphate dehydrogenase	Vigeolas et al. 2007
Fruit size	*codA*	Park et al. 2007
Male sterility	glutamine synthetase	Ribarits et al. 2007
Low allergy peanuts	*Ara h2*	Dodo et al. 2008

[1]The examples given in this table came from a survey of reports in the Plant Biotechnology Journal.

Table 6 Examples of applications of molecular pharming.

Product	*Reference*
Human serum albumin	Fernandez-San Millan et al. 2003
Fungal laccase	Hood et al. 2003
Monoclonal antibody	Bardor et al. 2003 Triguero et al. 2005 Schahs et al. 2007
Viral animal vaccine	Molina et al. 2004
Fructans	Weyens et al. 2004
Spider silk	Menassa et al. 2004
Protein sweetener	Lamphear et al. 2005
p-hydroxybenzoic acid	McQualter et al. 2005
Cyanophycin (polyaspartate)	Neumann et al. 2005
Human interleukin-4	Ma et al. 2005
Polyhydroxybutyrate	Kourtz et al. 2005
Oral immunotherapy	Takagi et al. 2005 Ruhlman et al. 2007 Wu et al. 2007
Human growth hormone	Gils et al. 2005
Human insulin	Nykiforuk et al. 2006
Provitamin A	Baisakh et al. 2006
Blue dye indigo	Warzecha et al. 2007
Human iduronidase	Downing et al. 2006
Glucocerebrosidase	Shaaltiel et al. 2007
Cellulase	Hood et al. 2007
Alkaloids	Apuya et al. 2008

2 PLANT IMPROVEMENT APPLICATIONS

Advances in plant genomics are providing new tools and approaches for use in plant improvement programs. Characterization of available genetic resources is facilitated by molecular tools and the availability of DNA in banks. Genetic improvement can be accelerated by applying molecular selection tools and by improved understanding of the key traits to be selected in domesticating wild material.

2.1 DNA-banking

The assembly of large collections of plant DNA (Rice et al. 2006) provides a new resource for use in plant gene discovery complementing available genetic resources in seed banks and living collections. These collections (biobank.com) may include wild crop relatives, landraces and cultivated varieties. DNA-banks provide a back up for seed banks. Ultimately DNA-banks may contain species of germplasm that is extinct, providing a source of novel alleles for biotechnology. In practice, the main application is to allow screening of accessions in seed banks by many different researchers of plant breeders without the need for repeated DNA extraction and the associated burden on seed stocks.

2.2 Capture of Genes from Wild Genetic Resources

Analysis of the genomes of wild relatives of major crops (Dillon et al. 2007) allows the capture of novel alleles for use in crop improvement. Recent developments providing improved understanding of the genes selected during domestication (Li et al. 2006) allow greater access to wild germplasm. The insights provided by genomic data are also providing explanations for the current genetic structure and diversity in cultivated plants (Olsen et al. 2006).

2.3 Accelerated Domestication

An improved understanding of the molecular basis of plant domestication provides the tools for accelerated domestication of new species. Plant domestication predominated in certain areas with a critical mass of suitable species. Plants from other regions may have been overlooked but may be amenable to domestication using genomics tools. An example of the key domestication genes that have been identified can be given by the maize genes *tga1* and *barren stalk1*. The naked grains of maize are explained by the *tga1* locus, which encodes a transcription factor (Wang et al. 2005). The architecture of the maize plant is determined by the *barren stalk1* locus, which encodes a protein that is associated with lateral meristem formation and explains dramatically different plant forms when maize is compared with its ancestor, teosinte (Gallavotti et al. 2004).

2.4 Marker-assisted Selection

Plant improvement programs can now access an increasing range of very cost-effective marker screening technologies. The application of markers will be greatly accelerated as larger numbers of desirable functional alleles are identified through functional genomics. These markers can be applied with more confidence than the more traditional markers that have been linked to the trait by mapping. The stage of application of markers is still evolving. The first applications of markers in breeding selection involved adding markers on to an existing breeding process. More recently, the benefits of markers are being enhanced by a complete restructuring of breeding approaches to take account of the availability of markers. Modern plant breeding requires an analysis of the costs and benefits of markers relative to alternative selection tools for each trait and at each stage in the breeding cycle.

Examples of Difficult-to-Measure Traits in Rice

Molecular markers offer special advantages in analysis of traits that are difficult to measure. Rice quality traits provide good examples of the opportunity that a complete genome sequence can offer in breeding for these difficult-to-measure characteristics. The fragrance trait detailed earlier requires assessment of a flavor that is difficult for humans to repeatedly assess on individual seeds (required for this recessive trait) and present in very low concentrations that are challenging for reliable chemical detection. However, a simple PCR-based marker that can detect the functional and non-functional alleles can be used for routine screening including the detection of heterozygotes (Bradbury et al. 2005b). Grain quality traits in rice, as in other cereals, often requires large sample sizes and complex instrumentation for conventional phenotyping. Starch traits in rice are a good example. Pasting properties and cooking properties such as paste viscosity, gelatinization temperature and retrogradation require a range of instruments and protocols to be applied for large samples. When reduced to SNP assays for the specific alleles associated with the traits, these tests can now be multiplexed and assayed in a single reaction using a small amount of tissue from a seed or leaf.

2.5 Targeted Mutagenesis

The difficulty of gaining public support for the widespread application of transgenic plants has continued to place emphasis on the use of genomics for accelerated selection in plant improvement. Techniques for targeted mutagenesis have been developed allowing the manipulation of any gene for which sequence information is available. This technique is best suited to the development of traits that can be achieved by eliminating or reducing expression of the target gene. Target-induced local lesions in genomes (TILLING; Till et al. 2004) and EMAIL (Cross et al. 2008) allow rapid screening of mutant populations for the discovery of rare mutations.

2.6 Targets for Plant Improvement

Plant improvement has targeted the improvement of plant production (quantity) and the quality of the plant product (e.g., food nutritional or functional value or processing performance). Recently, the focus has moved to include more emphasis on non-traditional uses of plants as in molecular pharming (producing high value molecules for medical applications in plants) and in the development of energy crops (e.g., biofuel crops). Genomic research is providing the key tools for these new targets.

References

Abbott AG (2009) Peach genome initiative. In: Kole C, Abbott AG (eds) Principles and Practices of Plant Genomics. Vol 3: Advanced Genomics. Science Publ, New Hampshire, Jersey, USA

Allen RS, Miller JAC, Chitty JA, Fist AJ, Gerlach WL, Larkin PJ (2008) Metabolic engineering of morphinan alkaloids by over-expression and RNAi suppression of salutaridinol 7-O-acetyltransferase in opium poppy. Plant Biotechnol J 6: 22-30

Almquist KC, Niu Y, McLean MD, Mena FL, Yau KYF, Brown K, Brandle JE, Hall JC (2004) Immunomodulation confers herbicide resistance in plants. Plant Biotechnol J 2: 189-197

Angiosperm Phylogeny Group II (2003) Bot J Linn Soc 141: 399-436

Apuya NR, Park J-H, Zhang L, Ahyow M, Davidow P, Van Fleet J, Rarang JC, Hippley M, Johnson TW, Yoo H-D, Trieu A, Krueger S, Wu C-y, Lu Y-p, Flavell RB, Bobzin SC (2008) Enhancement of alkaloid production in opium and California poppy by transactivation using heterologous regulatory factors. Plant Biotechnol J 6: 160-175

Avraham T, Badani H, Galili S, Amir R (2005) Enhanced levels of methionine and cysteine in transgenic alfalfa (*Medicago sativa* L.) plants over-expressing the Arabidopsis cystathionine y-synthase gene. Plant Biotechnol J 3: 71-79

Baisakh N, Rehana S, Rai M, Oliva N, Tan J, Mackill DJ, Khush GS, Datta K, Datta SK (2006) Marker-free transgenic (MFT) near-isogenic introgression lines (NIILs) of 'golden' indica rice (cv. IR64) with accumulation of provitamin A in the endosperm tissue. Plant Biotechnol J 4: 467-475

Banner J, Isaksson A, Waldenstrom E, Jarvius J, Landegren U, Nilsson M (2003) Parallel gene analysis with allele-specific padlock probes and tag micro-arrays. Nucl Acids Res 31: e103

Bardor M, Loutelier-Bourhis C, Paccalet T, Cosette P, Fitchette A-C, Vezina L-P, Trepanier S, Dargis M, Lemieux R, Lange C, Faye L, Lerouge P (2003) Monoclonal C5-1 antibody produced in transgenic alfalfa plants exhibits a N-glycosylation that is homogenous and suitable for glyco-engineering into human-compatible structures. Plant Biotechnol J 1: 451-462

Batley J, Mogg R, Edwards D, O'Sullivan H, Edwards KJ (2003) A high-throughput SnuPE assay for genotyping SNPs in the flanking regions of *Zea mays* sequence tagged simple sequence repeats. Mol Breed 11: 111-120

Bennett S, Barnes C, Cox A, Davies L, Brown C (2005) Toward the $1,000 human genome. Pharmacogenomics 6: 373-382

Bradbury L, Fitzgerald T, Henry R, Jin Q, Waters DLE (2005a) The gene for fragrance in rice. Plant Biotechnol J 3: 363-370

Bradbury L, Henry RJ, Jin Q, Reinke RF, Waters DLE (2005b) A perfect market for fragrance genotyping in rice. Mol Breed 16: 279-283

Bundock P, Cross MJ, Shapter F, Henry RJ (2006) Robust allele-specific PCR markers developed for SNP's in expressed barley sequences. Theor Appl Genet 112: 358-365

Che D, Meagher RB, Heaton ACP, Lima A, Rugh CL, Merkle SA (2003) Expression of mercuric ion reductase in Eastern cottonwood (*Populus deltoides*) confers mercuric ion reduction and resistance. Plant Biotechnol J 1: 311-319

Chen L, Auh C-K, Dowling P, Bell J, Chen F, Hopkins A, Dixon RA, Wang Z-Y (2003) Improved forage digestibility of tall fescue (*Festuca arundinacea*) by transgenic down-regulation of cinnamyl alcohol dehydrogenase. Plant Biotechnol J 1: 437-449

Childs K, Quyang S (2009) Rice genome initiative In: Kole C, Abbott AG (eds) Principles and Practices of Plant Genomics. Vol 3: Advanced Genomics. Science Publ, New Hampshire, Jersey, USA

Comai L, Young K, Till BJ, Reynolds SH, Greene EA, Codomo CA, Enns LC, Johnson JE, Burtner C, Odden AR, Henikoff S (2004) Efficient discovery of DNA polymorphisms in natural populations by ecotilling. Plant J 37: 778-786

Cronin J, Bundock P, Henry RJ, Nevo E (2007) Adaptive Climatic Molecular Evolution in Wild Barley at the *Isa* defense locus. Proc Natl Acad Sci USA 104: 2773-2778

Cross M, Waters D, Lee LS, Henry RJ (2008) Endonucleolytic mutation analysis by internal labeling. Electrophoresis 29: 1291-1301

Davies KM (2007) Genetic modification of plant metabolism for human health benefits. Mutat Res 622: 122-137

de Gyves EM, Sparks CA, Sayanova O, Lazzeri P, Napier JA, Jones HD (2004) Genetic manipulation of y-linolenic acid (GLA) synthesis in a commercial variety of evening primrose (*Oenothera* sp.). Plant Biotechnol J 2: 351-357

D'Halliun KD, Vanderstraeten C, Stals E, Cornelissen M, Ruiter R (2008) Homologous recombination: A basis for targeted genome optimization in crop species such as maize. Plant Biotechnol J 6: 93-102

Dillon SL, Shapter FM, Henry RJ, Cordeiro G, Izquierdo L, Lee LS (2007) Domestication to crop improvement: Genetic resources for *Sorghum* and *Saccharum* (Andropogoneae). Ann Bot: Open Access (10.1093/aob/mcm192)

Dodo HW, Konan KN, Chen FC, Egnin M, Viquez OM (2008) Alleviating peanut allergy using genetic engineering: the silencing of the immunodominant allergen *Ara h 2* leads to its significant reduction and a decrease in peanut allergenicity. Plant Biotechnol J 6: 135-145

Downing WL, Galpin JD, Clemens S, Lauzon SM, Samuels AL, Pidkowich MS, Clarke LA, Kermode AR (2006) Synthesis of enzymatically active human alpha-l-iduronidase in Arabidopsis *cgl* (complex glycan-deficient) seeds. Plant Biotechnol J 4: 169-181

Dutta I, Saha P, Majumder P, Sarkar A, Chakraborti D, Banerjee S, Das S (2005) The efficacy of a novel insecticidal protein, *Allium sativum* leaf lectin (ASAL), against homopteran insects monitored in transgenic tobacco. Plant Biotechnol J 3: 601-611

Edwards D, Batley J (2008) Bioinformatics: Fundamentals and applications in plant genetics, mapping and breeding. In: Kole C, Abbott AG (eds) Principals and Practices of Plant Genomics. Vol 1: Genome Mapping. Science Publ, New Hampshire, Jersey, USA, pp 269-301

Emani C, Garcia JM, Lopata-Finch E, Pozo MJ, Uribe P, Kim D-J, Sunilkumar G, Cook DR, Kenerley CM, Rathore KS (2003) Enhanced fungal resistance in transgenic cotton expressing an endochitinase gene from *Trichoderma virens*. Plant Biotechnol J 1: 321-336

Emrich SJ, Barbazuk WB, Li L, Schnable PS (2007) Gene discovery and annotation using LCM_454 transcriptome sequencing. Genome Res 17: 69-73

Farroki N, Whitelegge JP, Brusslan JA (2008) Plant peptides and peptidomics. Plant Biotechnol J 6: 105-134

Fernandez-San Millan A, Mingo-Castel A, Miller M, Daniell H (2003) A chloroplast transgenic approach to hyper-express and purify Human Serum Albumin, a protein highly susceptible to proteolytic degradation. Plant Biotechnol J 1: 71-79

Fiehn O, Wohlgemuth G, Scholz M, Kind T, Lee DY, Lu Y, Moon S, Nikolau B (2008) Quality control for plant metabolomics: Reporting MSI-compliant studies. Plant J 53: 691-704

Frizzi A, Huang S, Gilbertson LA, Armstrong TA, Luethy MH, Malvar TM (2008) Modifying lysine biosynthesis and catabolism in corn with a single bifunctional expression/silencing transgene cassette. Plant Biotechnol J 6: 13-21

Furtado A, Henry RJ, Takaiwa F (2008) Comparison of promoters in transgenic rice. Plant Biotechnol J 6(7): 679-693

Gallavotti A, Zhao Q, Kyozuka J, Meely RB, Ritter MK, Doebley JF, Enrico Pe M, Schmidt RJ (2004) The role of *barren stalk1* in the architecture of maize. Nature 432: 630-635

George TS, Simpson RJ, Hadobas PA, Richardson AE (2005) Expression of a fungal phytase gene in *Nicotiana tabacum* improves phosphorus nutrition of plants grown in amended soils. Plant Biotechnol J 3: 129-140

Gill GP, Brown GR, Neale DB (2003) A sequence mutation in the cinnamyl alcohol dehydrogenase gene associated with altered lignification in loblolly pine. Plant Biotechnol J 1: 253-258

Gils M, Kandzia R, Marillonnet S, Klimyuk V, Gleba Y (2005) High-yield production of authentic human growth hormone using a plant virus-based expression system. Plant Biotechnol J 3: 613-620

Grainer C, Aguirrezabal L, Chenu K, Cookson SJ, Dauzat M, Hamard P, Thioux JJ, Rolland G, Bouchier-Combaud S, Lebaudy A, Muller B, Simonneau T, Tardieu F (2006) PHENOPSIS, an automated platform for reproducible phenotyping of plant responses to soil water deficit in Arabidopsis thaliana permitted the identification of an accession with low sensitivity to water deficit. New Phytol 169: 623-635

Goff SA, Ricke D, Lan TH, Presting G, Wang R, Dunn M, Glazebrook J, Sessions A et al. (2002) A draft sequence of the rice genome (*Oryza sativa* L. ssp. *japonica*). Science 296: 92-100

Gregory BD, Yazaki J, Ecker JR (2008) Utilizing tiling microarrays for whole-genome analysis in plants. Plant J 53: 636-644

Groβhans H, Filipowicz W (2008) The expanding world of small RNAs. Nature 451: 414-416

Guy C, Kopka J, Moritz T (2008) Plant metabolomics coming of age. Physiol Planta 132: 113-116

Harker M, Holmberg N, Clayton JC, Gibbard CL, Wallace AD, Rawlins S, Hellyer SA, Lanot A, Safford R (2003) Enhancement of seed phytosterol levels by expression of an N-terminal truncated *Hevea brasiliensis* (rubber tree) 3-hydroxy-3-methylglutaryl-CoA reductase. Plant Biotechnol J 1: 113-121

Henikoff S, Till BJ, Comai L (2004) TILLING Traditional mutagenesis meets functional genomics. Plant Physiol 135: 630-636

Henry RJ (2001) Plant Genotyping: the DNA Fingerprinting of Plants. CABI Publ, Oxon, UK.

Henry RJ (2005) Plant Diversity and Evolution: Genotypic and Phenotypic

Variation in Higher Plants. CABI Publ, Oxon, UK

Henry RJ (2006) Plant Conservation Genetics. Haworth Press, Binghamton NY, USA

Henry RJ (2008) Plant Genotyping II: SNP Technology. CABI Publ, Oxon, UK

Henry RJ, Edwards K (2009) New tools for single nucleotide polymorphism (SNP) discovery and analysis accelerating plant biotechnology. Plant Biotechnol J 7: 311

Hofvander P, Andersson M, Larsson C-T, Larsson H (2004) Field performance and starch characteristics of high-amylose potatoes obtained by antisense gene targeting of two branching enzymes. Plant Biotechnol J 2: 311-320

Hood EE, Bailey MR, Beifuss K, Magallanes-Lundback M, Horn ME, Callaway E, Drees C, Delaney DE, Clough R, Howard JA (2003) Criteria for high-level expression of a fungal laccase gene in transgenic maize. Plant Biotechnol J 1: 129-140

Hood EE, Love R, Lane J, Bray J, Clough R, Pappu K, Drees C, Hood KR, Yoon S, Ahmad A, Howard JA (2007) Subcellular targeting is a key condition for high-level accumulation of cellulase protein in transgenic maize seed. Plant Biotechnol J 5: 709-719

Houde M, Dallaire S, N'Dong D, Sarhan F (2004) Over expression of the acidic dehydrin WCOR410 improves freezing tolerance in transgenic strawberry leaves. Plant Biotechnol J 2: 381-387

Hudson M (2008) Sequencing breakthroughs for genomic ecology and evolutionary biology. Mol Ecol Resour 8: 3-17

Ji Q, Oomen RJFJ, Vincken J-P, Bolam DN, Gilbert HJ, Suurs LCJM, Visser RGF (2004) Reduction of starch granule size by expression of an engineered tandem starch-binding domain in potato plants. Plant Biotechnol J 2: 251-260

Jones ES, Sullivan H, Bhattramakki D, Smith JSC (2007) A comparison of simple sequence repeat and single nucleotide polymorphism marker technologies for the genotypic analysis of maize (*Zea mays* L.) Theor Appl Genet 115: 361-371

Katam R et al. (2009) Arabidopsis genome initiative In: Kole C, Abbott AG (eds) Principles and Practices of Plant Genomics. Vol 3: Advanced Genomics. Science Publ, New Hampshire, Jersey, USA

Kater MM, Franken J, Inggamer H, Gretenkort M, van Tunen AJ, Mollema C, Angenent GC (2003) The use of floral homeotic mutants as a novel way to obtain durable resistance to insect pests. Plant Biotechnol J 1: 123-127

Kennedy B, Arar K, Reja V, Henry RJ (2006a) LNA for optimising strand displacement probes for quantitative real-time PCR. Anal Biochem 348: 294-299

Kennedy B, Waters DLE, Henry RJ (2006b) Screening for the rice blast resistance gene *Pi-ta* using LNA displacement probes and real-time PCR. Mol Breed 18: 185-193

Kim J-B, Kang J-Y, Kim SY (2004) Over-expression of a transcription factor regulating ABA-responsive gene expression confers multiple stress tolerance. Plant Biotechnol J 2: 459-466

Kourtz L, Dillon K, Daughtry S, Madison LL, Peoples O, Snell KD (2005) A novel thiolase-reductase gene fusion promotes the production of polyhydroxybutyrate in Arabidopsis. Plant Biotechnol J 3: 435-447

Kress WJ, Wudack KJ, Zimmer EA, Weigh LA, Janzen DH (2005) Use of DNA barcodes to identify flowering plants. Proc Natl Acad Sci USA 102: 8369-8374

Lamphear BJ, Barker DK, Brooks CA, Delaney DE, Lane JR, Beifuss K, Love R, Thompson K, Mayor J, Clough R, Harkey R, Poage M, Drees C, Horn ME, Streatfield SJ, Nikolov Z, Woodard SL, Hood EE, Jilka JM, Howard JA (2005) Expression of the sweet protein brazzein in maize for production of a new commercial sweetener. Plant Biotechnol J 3: 103-114

Lee S-H, Walker DR, Cregan PB, Boerma HR (2004) Comparison of four flow cytometric SNP detection assays and their use in plant improvement. Theor Appl Genet 110: 167-174

Li CB, Zhou AL, Sang T (2006) Rice domestication by reduced shattering. Science 311: 1936-1939

Li L, Wang XF, Stolc V, Li XY, Su N, Tongprasit W, Li SG, Cheng ZX, Wang J, Deng XW (2006) Genome-wide transcription analysis in rice using tiling microarrays. Nat Genet 38: 124-129

Lieman-Hurwitz J, Rachmilevitch S, Mittler R, Marcus Y, Kaplan A (2003) Enhanced photosynthesis and growth of transgenic plants that express *ictB*, a gene involved in HCO3 accumulation in cyanobacteria. Plant Biotechnol J 1: 43-50

Lilley CJ, Urwin PE, Johnston KA, Atkinson HJ (2004) Preferential expression of a plant cystatin at nematode feeding sites confers resistance to *Meloidogyne incognita* and *Globodera pallida*. Plant Biotechnol J 2: 3-12

Lohithaswa HC, Feltus FA, Singh HP, Bacon CD, Bailey CD, Paterson AH (2007) Leveraging the rice genome sequence for monocot comparative and translational genomics. Theor Appl Genet 115: 237-243

Ma S, Huang Y, Davis A, Yin Z, Mi Q, Menassa R, Brandle JE, Jevnikar AM (2005) Production of biologically active human interleukin-4 in transgenic tobacco and potato. Plant Biotechnol J 3: 309-318

Maiti IB, Dey N, Pattanaik S, Dahlman DL, Rana RL, Webb BA (2003) Antibiosis-type insect resistance in transgenic plants expressing a teratocyte secretory protein (TSP14) gene from a hymenopteran endoparasite (*Microplitis croceipes*). Plant Biotechnol J 1: 209-219

Mak L, Skylas DJ, Willows R, Connolly A, Cordell SJ, Wrigley CW, Sharp PJ, Copeland L (2005) A proteomic approach to the identification and characterization of protein composition in wheat germ. Funct Integr Genom: DOI 10.1007/s10142-005-0018-9

Mak Y, Willows RD, Roberts TH, Wrigley CW, Sharp PJ, Copeland L (2006) Black point is associated with reduced levels of stress and disease- and defence-related proteins in wheat germ. Mol Plant Pathol 3: 177-189

Mardis, E.R. (2008) The impact of next-generation sequencing technology on genetics. Trends Genet 24: 133-141

Margulies M, Egholm M, Altman WE, Attiya S, Bader JS, Bemben LA, Berka J, Braverman MS, Chen YJ et al. (2005) Genome sequencing in micro-fabricated high-density piolitre reactors Nature 437: 376-380

McQualter RB, Chong BF, Meyer K, Van Dyk DE, O'Shea MG, Walton NJ, Viitanen PV, Brumbley SM (2005) Initial evaluation of sugarcane as a production platform for p-hydroxybenzoic acid. Plant Biotechnol J 3: 29-41

Menassa R, Zhu H, Karatzas CN, Lazaris A, Richman A, Brandle J (2004) Spider dragline silk proteins in transgenic tobacco leaves: accumulation and field production. Plant Biotechnol J 2: 431-438

Molina A, Hervas-Stubbs S, Daniell H, Mingo-Castel AM, Veramendi J (2004) High-yield expression of a viral peptide animal vaccine in transgenic tobacco chloroplasts. Plant Biotechnol J 2: 141-153

Nagadhara D, Ramesh S, Pasalu IC, Rao YK, Krishnaiah NV, Sarma NP, Bown DP, Gatehouse JA, Reddy VD, Rao KV (2003) Transgenic indica rice resistant to sap-sucking insects. Plant Biotechnol J 1: 231-240

Naimov S, Dukiandjiev S, de Maagd RA (2003) A hybrid *Bacillus thuringiensis* delta-endotoxin gives resistance against a coleopteran and a lepidopteran pest in transgenic potato. Plant Biotechnol J 1: 51-57

Neumann K, Stephan DP, Ziegler K, Huhns M, Broer I, Lockau W, Pistorius EK (2005) Production of cyanophycin, a suitable source for the biodegradable polymer polyaspartate, in transgenic plants. Plant Biotechnol J 3: 249-258

Nykiforuk CL, Boothe JG, Murray EW, Keon RG, Goren HJ, Markley NA, Moloney MM (2006) Transgenic expression and recovery of biologically active recombinant human insulin from *Arabidopsis thaliana* seeds. Plant Biotechnol J 4: 77-85

Olsen KM, Caicedo AL, Polato N, McClung A, McCouch S, Purugganan MD (2006) Selection under domestication: Evidence for a sweep in the rice *Waxy* genomic region. Genetics 173: 975-983

Outchkourov NS, de Kogel WJ, Wiegers GL, Abrahamson M, Jongsma MA (2004) Engineered multidomain cysteine protease inhibitors yield resistance against western flower thrips (*Frankliniella occidentalis*) in greenhouse trials. Plant Biotechnol J 2: 449-458

Parida SK, Mohapatra T. (2009) Whole genome sequencing. In: Kole C, Abbott AG (eds) Principles and Practices of Plant Genomics. Vol 3: Advanced Genomics. Science Publ, New Hampshire, Jersey, USA

Park E-J, Jeknic Z, Chen THH, Murata N (2007) The *codA* transgene for glycinebetaine synthesis increases the size of flowers and fruits in tomato. Plant Biotechnol J 5: 422-430

Paterson AH, Bowers JE, Bruggmann R, Dubchak I, Grimwood J, Gundlach H, Haberer G, Hellsten U, Mitros T, Poliakov A, Schmutz J, Spannagl M, Tang H, Wang X, Wicker T, Bharti AK, Chapman J, Feltus FA, Gowik U, Grigoriev IV, Lyons E, Maher CA, Martis M, Narechania A, Otillar RP, Penning BW, Salamov AA, Wang Y, Zhang L, Carpita NC, Freeling M, Gingle AR, Hash CT, Keller B, Klein P, Kresovich S, McCann MC, Ming R, Peterson DG, Mehboob ur R, Ware D, Westhoff P, Mayer KFX, Messing J, Rokhsar DS (2009) The *Sorghum bicolor* genome and the diversification of grasses. Nature 457: 551-556

Pellny TK, Ghannoum O, Conroy JP, Schluepmann H, Smeekens S, Andralojc J, Krause KP, Goddijn O, Paul MJ (2004) Genetic modification of photosynthesis with *E. coli* genes for trehalose synthesis. Plant Biotechnol J 2: 71-82

Quan R, Shang M, Zhang H, Zhao Y, Zhang J (2004) Engineering of enhanced glycine betaine synthesis improves drought tolerance in maize. Plant Biotechnol J 2: 477-486

Ragoussis J, Elvidge GP, Kuar K, Colella S (2006) Matrix-assisted laser desorption/ionisation time of flight mass spectrometry in genomics research. PLOS Genet 2: 920-929

Rajasekaran K, Cary JW, Jaynes JM, Cleveland TE (2005) Disease resistance conferred by the expression of a gene encoding a synthetic peptide in transgenic cotton (*Gossypium hirsutum* L.) plants. Plant Biotechnol J 3: 545-554

Ribarits A, Mamun ANK, Li S, Resch T, Fiers M, Heberle-Bors E, Liu C-M, Touraev A (2007) Combination of reversible male sterility and doubled haploid production by targeted inactivation of cytoplasmic glutamine synthetase in developing anthers and pollen. Plant Biotechnol J 5: 483-494

Rice N, Cordeiro GM, Shepherd M, Bundock P, Bradbury LME, Watson L, Crawford AC, Pacey-Miller T, Furtado A, Henry RJ (2006) DNA Banks and their role in facilitating the application of genomics to plant germplasm. Plant Genet Resour 4: 64-70

Rolletschek H, Borisjuk L, Radchuk R, Miranda M, Heim U, Wobus U, Weber H (2004) Seed-specific expression of a bacterial phosphoenolpyruvate carboxylase in *Vicia narbonensis* increases protein content and improves carbon economy. Plant Biotechnol J 2: 211-219

Ruhlman T, Ahangari R, Devine A, Samsam M, Daniell H (2007) Expression of cholera toxin B-proinsulin fusion protein in lettuce and tobacco chloroplasts—oral administration protects against development of insulitis in non-obese diabetic mice. Plant Biotechnol J 5: 495-510

Schahs M, Strasser R, Stadlmann J, Kunert R, Rademacher T, Steinkellner H (2007) Production of a monoclonal antibody in plants with a humanized N-glycosylation pattern. Plant Biotechnol J 5: 657-663

Shaaltiel Y, Bartfeld D, Hashmueli S, Baum G, Brill-Almon E, Galili G, Dym O, Boldin-Adamsky SA, Silman I, Sussman JL, Futerman AH, Aviezer D (2007)

Production of glucocerebrosidase with terminal mannose glycans for enzyme replacement therapy of Gaucher's disease using a plant cell system. Plant Biotechnol J 5: 579-590

Shao M, Wang J, Dean RA, Lin Y, Gao X, Hu S (2008) Expression of a hairpin-encoding gene in rice confers durable nonspecific resistance to *Magnaporthe grisea*. Plant Biotechnol J 6: 73-81

Shulaev V, Cortes D, Miller G, Mittler R (2008) Metabolomics for plant stress response. Physiol Planta 132: 199-208

Skipsey M, Cummins I, Andrews CJ, Jepson I, Edwards R (2005) Manipulation of plant tolerance to herbicides through co-ordinated metabolic engineering of a detoxifying glutathione transferase and thiol cosubstrate. Plant Biotechnol J 3: 409-420

Sweetlove LJ, Fell D, Fernie AR (2008) Getting to grips with the plant metabolic network. Biochem J 409: 27-41

Tadege M, Sheldon CC, Helliwell CA, Upadhyaya NM, Dennis ES, Peacock WJ (2003) Reciprocal control of flowering time by *OsSOC1* in transgenic *Arabidopsis* and by *FLC* in transgenic rice. Plant Biotechnol J 1: 361-369

Takagi H, Saito S, Yang L, Nagasaka S, Nishizawa N, Takaiwa F (2005) Oral immunotherapy against a pollen allergy using a seed-based peptide vaccine. Plant Biotechnol J 3: 521-533

Till BJ, Burtner C, Comai L, Henikoff S (2004) Mismatch cleavage by single-strand specific nucleases. Nucl Acids Res 32: 2632-2641

Triguero A, Cabrera G, Cremata JA, Yuen C-T, Wheeler J, Ramirez NI (2005) Plant-derived mouse IgG monoclonal antibody fused to KDEL endoplasmic reticulum-retention signal is N-glycosylated homogeneously throughout the plant with mostly high-mannose-type N-glycans. Plant Biotechnol J 3: 449-457

Turakulov R, Nontachaiyapoom S, Mitchelson KR, Gresshoff PM (2007) Ultra-sensitive determination of absolute mRNA amounts at attomole levels of nearly identical plant genes with high-throughput mass spectroscopy (MassARRAY). Plant Cell Rep 48: 1379-1384

Van der Vyver C, Schneidereit J, Driscoll S, Turner J, Kunert K, Foyer CH (2003) Oryzacystatin I expression in transformed tobacco produces a conditional growth phenotype and enhances chilling tolerance. Plant Biotechnol J 1: 101-112

Vigeolas H, Waldeck P, Zank T, Geigenberger P (2007) Increasing seed oil content in oil-seed rape (*Brassica napus* L.) by over-expression of a yeast glycerol-3-phosphate dehydrogenase under the control of a seed-specific promoter. Plant Biotechnol J 5: 431-441

Vila L, Quilis J, Meynard D, Breitler JC, Marfa V, Murillo I, Vassal JM, Messeguer J, Guiderdoni E, San Segundo B (2005) Expression of the maize proteinase inhibitor (*mpi*) gene in rice plants enhances resistance against the striped stem borer (*Chilo suppressalis*): effects on larval growth and insect gut proteinases. Plant Biotechnol J 3: 187-202

Wang H, Nussbaum-Wagler T, Li B, Zhao Q, Vigouroux Y, Faller M, Bomblies K, Lukens L, Doebley JF (2005) The origin of the naked grains of maize. Nature 436: 714-719

Warzecha H, Frank A, Peer M, Gillam EMJ, Guengerich FP, Unger M (2007) Formation of the indigo precursor indican in genetically engineered tobacco plants and cell cultures. Plant Biotechnol J 5: 185-191

Waters DLE, Henry RJ, Reinke RF, Fitzgerald MA (2006) Gelatinisation temperature of rice explained by polymorphisms in starch synthase. Plant Biotechnol J 4: 115-122

Weyens G, Ritsema T, Van Dun K, Meyer D, Lommel M, Lathouwers J, Rosquin I, Denys P, Tossens A, Nijs M, Turk S, Gerrits N, Bink S, Walraven B, Lefebvre M, Smeekens S (2004) Production of tailor-made fructans in sugar beet by expression of onion fructosyltransferase genes. Plant Biotechnol J 2: 321-327

Wicker T, Schlagenhauf E, Graner A, Close TJ, Keller B, Stein N (2006) 454 sequencing put to the test using the complex genome of barley. BMC Genom 7: 275

Wu J, Yu L, Li L, Hu J, Zhou J, Zhou X (2007) Oral immunization with transgenic rice seeds expressing VP2 protein of infectious bursal disease virus induces protective immune responses in chickens. Plant Biotechnol J 5: 570-578

Wu L, Birch RG (2007) Doubled sugar content in sugarcane plants modified to produce a sucrose isomer. Plant Biotechnol J 5: 109-117

Wu XR, Chen ZH, Folk WR (2003) Enrichment of cereal protein lysine content by altered tRNAlys coding during protein synthesis. Plant Biotechnol J 1: 187-194

Yao K, Bacchetto RG, Lockhart KM, Friesen LJ, Potts DA, Covello PS, Taylor DC (2003) Expression of the *Arabidopsis ADS1* gene in *Brassica juncea* results in a decreased level of total saturated fatty acids. Plant Biotechnol J 1: 221-229

Zhou Y, O'Hare TJ, Jobin-Decor M, Underhill SJR, Wills RBH, Graham MW (2003) Transcriptional regulation of a pineapple polyphenol oxidase gene and its relationship to blackheart. Plant Biotechnol J 1: 463-478

2 Fundamentals of Physical Mapping

Perumal Azhaguvel[1*], Yiqun Weng[2], Raman Babu[3], Alagu Manickavelu[4], Dhanasekaran Vidya Saraswathi[5] and Harindra Singh Balyan[6]

[1]Texas AgriLife Research, Texas A&M University, 6500 Amarillo Blvd. W., Amarillo, TX 79106, USA

[2]Vegetable Crops Research Unit, USDA-ARS, University of Wisconsin, Madison, WI 53706, USA

[3]International Maize and Wheat Improvement Center (CIMMYT), 06600 Mexico, DF, Mexico

[4]Kihara Institute for Biological Research, Yokohama City University, Yokohama 244-0813, Japan

[5]Texas Veterinary Medical Diagnostic Laboratory, Amarillo Blvd West, Amarillo, TX 79106, USA

[6]Department of Genetics and Plant Breeding, Ch. Charan Singh University, Meerut 250 004, India

*Corresponding author: aperumal@noble.org

*Present address: Forage Improvement Division, 2510 Sam Noble Parkway, The Samuel Roberts Noble Foundation, Ardmore, OK 73401, USA

1 INTRODUCTION

Since the rediscovery of Mendelian principles, the field of genetics has witnessed many landmarks in understanding genetic principles. Better understanding of genetic principles using molecular approaches led to a new subdivision of genetics called 'Genomics'. The development and abundant availability of molecular markers such as restriction fragment length polymorphism (RFLP), amplified fragment length polymorphism

(AFLP), sequence tagged site (STS), microsatellites or simple sequence repeats (SSRs) and sequence based markers such as single nucleotide polymorphism (SNP) during the last two decades has opened up new avenues to locate and measure the effects of the gene(s) in an unprecedented manner. Molecular markers are now widely used to track loci and genomic regions in plant breeding programs. Genetic maps are built based on the meiotic recombination between homologous chromosomes. Though the recombination has been used to order the genes, the application of molecular markers made significant impact to developing saturated linkage map, which is one prerequisite for map-based gene cloning. On the other hand, physical mapping reflects the actual physical distance in base pairs between molecular markers, which is becoming increasingly important to understand the molecular insights of the genes. In principle, genetic and physical maps should have the same order of markers and genes and equivalent distances among the loci. However, the genetic distances in recombination units (cM) and the physical distances (bp) may not be the same due to uneven distribution of crossing-over events along chromosome arms. Genetic and physical mapping are the components of forward genetic approach, in which the ultimate goal is to physically locate the gene(s) of interest through high-density markers using large segregating mapping population(s).

This chapter will provide an overview of physical mapping in plants and its eventual application in map-based gene cloning. We will also furnish a brief overview of cytogenetics-based physical mapping strategies, approaches currently used and the lessons learnt from the success stories. Finally, we will also discuss comparative physical mapping approaches and the future direction of physical mapping as we anticipate the fully sequenced genomes of many major crop plants in the near future.

2 GENETIC VERSUS PHYSICAL MAPPING

Genetic mapping is also called linkage mapping, which is a linear designation of genes or loci within a chromosome based on recombination events. The genetic mapping concept was first reported as early as in 1913 (Sturtevant 1913) demonstrated by placing five sex-linked genes in a linear fashion on Y chromosome of fruit fly (*Drosophila melanogaster*), but a whole-genome genetic map with molecular markers was not available until 1987 (Doniskeller et al. 1987). Technology advancements, especially in DNA sequencing now make development of molecular markers relatively easy (for a review on molecular markers, please see

Chapter 2 of Volume 1). Genetic map is currently available for most of the important crops and model plant species.

The recombination-based linkage mapping can only use those loci that demonstrate polymorphims between the parental lines, which makes it less efficient in many crops with low genetic diversity. Genetic mapping allows establishing linkage relationships among genes or molecular markers and is the first step for physical mapping, which ultimately opens up the door for map-based gene isolation. Although linkage mapping can clarify the mutual relationship between a set of polymorphic genetic marker loci or genes, it does not allow their assignment to a particular chromosome. Also, the genetic distance between two markers computed based on the number of recombination events often does not reflect their actual physical distance. To address these issues, physical mapping has to be performed.

A physical map is defined as a map that consists of linearly ordered set of DNA fragments encompassing the whole genome or a particular genomic region of interest. Physical map is basically of two types, macro-restriction maps and ordered clone maps. The macro-restriction maps provide information regarding the DNA fragments at chromosome level, whereas the ordered clone map consists of overlapping collection of cloned DNA fragments such as yeast artificial chromosome (YAC) and bacterial artificial chromosome (BAC). The physical map is considered as a real map expressed in million base-pairs or Mbps, whereas the genetic map is expressed in cM (centi-Morgan) units. The genetic-physical map ratio is dependent on the nature of chromosome regions and the frequency of recombination in that region, which may vary significantly. For instance the estimated genetic to physical distance ratio of wheat homoeologous group-7 chromosomes was between 442 Kb/cM to 21,687 Kb/cM (Erayman et al. 2004).

Genetic and physical mapping are often iterative processes in a map-based gene cloning project. A genetic map is constructed for the target gene. Closely linked molecular markers are employed to screen a large-insert library to identify positive clones, which are used to construct a physical contig. New molecular markers are identified from these clones for fine genetic mapping. This chromosome walking process will be iterated until a candidate gene is identified. Genetic and physical map integration provides all about the genome sequences, new opportunity to develop DNA markers, identify genes, quantitative trait loci (QTLs), expressed sequence tags (ESTs), regulatory sequences, and repeat elements. The physical map allows us to conduct structural, functional, and evolutionary genomics research.

3 CHROMOSOME-BASED PHYSICAL MAPPING

Today, when we are talking about physical mapping, we refer to the ordering of large-insert DNA fragments or clones. However, physical mapping started even from the early times when the field genetics was still in its infancy. The earliest work on physical mapping was mostly carried out in fruit fly and the corn crop. Morphological characteristics in the chromosomes that are viewable under light microscope were used as landmarks. Physical mapping was performed at the chromosomal level until approximately two decades ago when molecular tools especially molecular marker technologies were introduced. In this section, we will present a brief review of the classical cytogenetics-based physical mapping strategies, some of which are still relevant and routinely used.

3.1 Physical Mapping through Aneuploid and Deletion Stock Analysis

The chromosomal or cytogenetic maps are the classical physical mapping methods, where the position of the markers on the chromosome is determined using cytogenetic stocks. Cytogenetic maps were developed by using aneuploid and substitution lines (Gill et al. 1993; Joppa 1993; Endo and Gill 1996; Young 2000). Aneuploids are organisms that have an increase or decrease in the normal number of chromosomes or their arms. The change in chromosome (arm) number can involve as little as a portion of a chromosome (e.g., telosomics) or may involve more than one chromosome (i.e., nullisomic-tetrasomics). The first set of wheat lines missing entire chromosomes or chromosome arms was described as early as in 1954 (Sears 1954).

These aneuploid stocks can effectively be used to place genes or molecular markers to a given chromosome or chromosomal arm region. The hybridization or PCR amplification of a genetically mapped marker using DNA from aneuploid lines indicates the chromosomal location of the clone (e.g., Fig. 1). Aneuploid analysis can only assign a gene or marker to a particular chromosome or its arm. To localize to a sub-chromosome arm region, a deletion mapping strategy could be used. In wheat, with the help of the gametocidal system (Endo 1998, 1990), many deletion lines involving all the 21 wheat chromosomes were developed (Endo and Gill 1996) in which the segments of a chromosome were sequentially deleted. By comparing the profile of a series of deletion lines from the same chromosome arm, target genes or markers could be

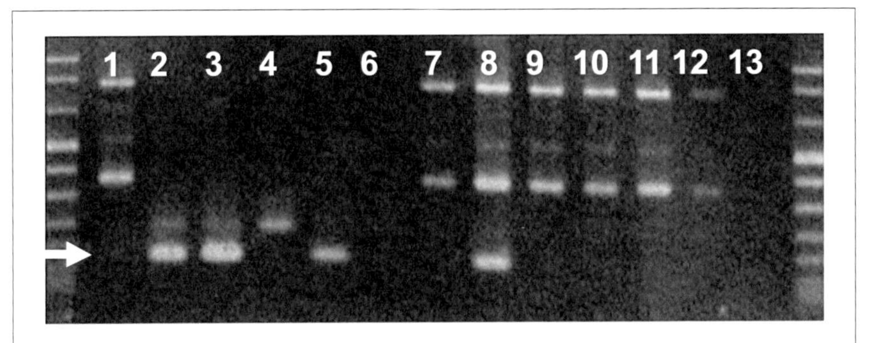

Fig. 1 Wheat-Barley chromosome additional lines—A useful resource for the chromosome identification. Agarose gel image showing the barley chromosome 3H specific PCR marker. 1 = Wheat (Chinese Spring), 2 = Barley (Betzes), 3 = Barley (Kanto Nakate Gold), 4 = Barley (Azumamugi), 5 = Wild barley (OUH602), 6 = Barley (Morex), 7 = 2H (Wheat + Barley 2H), 8 = 3H (Wheat + Barley 3H), 9 = 4H (Wheat + Barley 4H), 10 = 5H (Wheat + Barley 5H), 11 = 6H (Wheat + Barley 6H), 12 = 7H (Wheat + Barley 7H) and 13 = Control (-ve control).

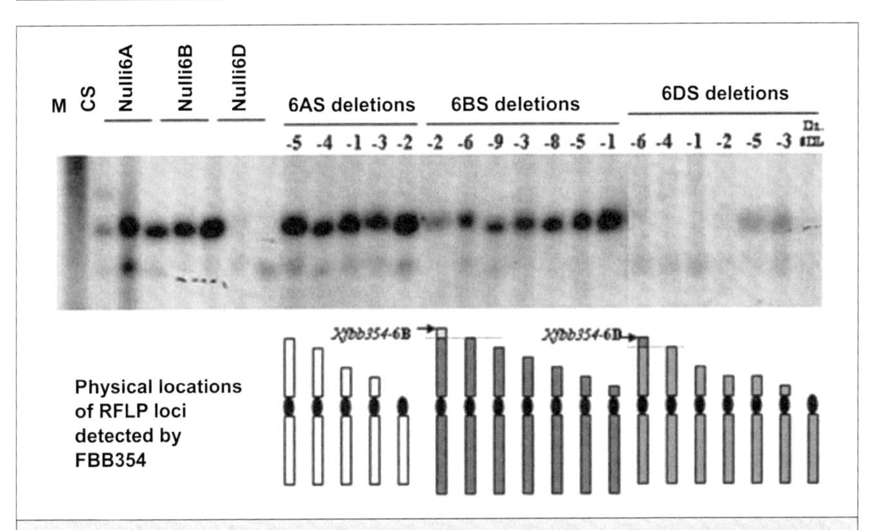

Fig. 2 Physical mapping of RFLP loci with deletion lines in wheat (deletion mapping). RFLP marker FBB354 was used as the probe in Southern hybridization with a blot containing restriction enzyme digested genomic DNAs from wheat group-6 chromosome aneuploid stocks and deletion lines. Two loci of FBB354 can be physically located in the telomeric regions of 6BS and 6DS respectively based on the hybridization patterns. CS = Wheat cultivar "Chinese Spring", Nulli = Nullisomic, T = tetrasomic, d.t = ditelosomic.

(Color image of this figure appears in the color plate section at the end of the book).

assigned into a specific region in the chromosome arm. Hundreds of molecular markers were deletion mapped using this strategy, which provided significant insights into the structure and organization of the wheat genome (e.g., Fig. 2). Wheat is the only crop plant species in which deletion mapping was successfully used. However, a major disadvantage of this technique is the unavailability of viable aneuploids in other plant species especially in diploid plants.

3.2 Radiation Hybrid (RH) Mapping

Physical mapping methods that do not rely on meiotic recombination are necessary for complex polyploid genomes with large genome size and uneven distribution of recombination and significant variation in genetic to physical distance ratios (Kalavacharla et al. 2006). For instance nearly 30% of wheat genes reside in the recombination-poor regions, which pose enormous challenges in map-based cloning projects (Erayman et al. 2004). RH mapping is a recombination-independent mapping strategy. RH map was first developed by Goss and Harris (1975, 1977) for mapping human chromosomes, and was reported to be an alternate mapping approach, if the gene resides in a recombination-poor region. RH maps are extremely useful for ordering markers in regions where highly polymorphic genetic markers are limited. RH maps are often used as a bridge between linkage maps (e.g., Stewart et al. 1997).

RH mapping rely on irradiating donor cells. A random subset of the chromosomal fragments is then rescued by fusing the irradiated cells with normal cells. The resulting clone may contain none, one or more chromosome fragments. Thus, the irradiated chromosomes could be studied in the background of the recipient's cells. A panel of radiation hybrids with random breaks on the chromosomes of the donor can be used to precisely place and order molecular markers or genes along a chromosome or entire genome (Cox et al. 1990). Different doses of irradiation can be used to construct maps with varying levels of resolution (Gyapay et al. 1996; Lunetta et al. 1996; Schuler et al. 1996; Stewart et al. 1997). The unit of measuring map distance in RH mapping is Ray, where the distance of one ray indicates an average of one break per chromosome between two loci. A Centi-Ray (cR) is analogous to a centi-Morgan, the unit measured in linkage mapping. Since cR does not represent the actual physical distance between loci, an approximation is made to represent in terms of kilobases (Warrington and Wasmuth 1996). RH mapping has been widely exploited in mammalian genomes, but successful reports in plants are few. The first plant RH panel was constructed by γ-irradiation on maize chromosome 9 in an oat background (oat-maize addition lines) and subsequent characterization of these lines with maize-specific molecular markers (Riera-Lizarazu et al. 1996, 2000; Ananiev et al. 1997). In recent years, RH mapping has been extended to wheat, barley and cotton (Wardrop et al. 2002, 2004; Gao et al. 2004; Kalavacharla et al. 2006).

3.3 HAPPY Mapping

RH mapping requires tremendous amount of efforts to produce a RH panel, which is considerably affected because of the interaction of the donor fragments in the host cell lines leading to biased segregation. The presence of host genome in the hybrids excludes the use of molecular markers such as AFLPs. HAPPY mapping (mapping based on the analysis of approximately HAPloid DNA samples using the PolYmerase chain reaction) is another variant approach for physical mapping of genomes by overcoming the difficulties of RH mapping. HAPPY mapping involves breaking genomic DNA randomly by irradiation or shearing followed by an optional size fractionation step. Markers which lie next to each other generally tend to remain together on the same random fragments. Markers then are segregated by diluting the resulting fragments and dispense a 'panel' of very small samples into the wells of a microtitre plate to give aliquots containing one haploid genome equivalent. Markers are detected using PCR and the linked markers tend to be found together in an aliquot. The map order of markers, and the distance between them are deduced from the frequency with which they cosegregate. Since HAPPY mapping is not based on meiotic recombination or hybrid cell formation, it is not subjected to the distortion and the results are accurate and obtained fast. A high-resolution HAPPY map of human chromosome 14 was constructed by Dear et al. (1998). Biunno et al. (2000) cloned the human *SEL1L* homolog of *C. elegans sel-1* using HAPPY mapping approach.

3.4 Mapping by FISH

A major breakthrough in mapping efforts was provided by the development of in situ hybridization of isotopically labeled probes, which helped in visualizing the hybridization of a DNA probe to chromosome. FISH (fluorescence in situ hybridization) is a technique used to detect and localize the presence or absence of specific DNA sequences in chromosomes. It uses fluorescent probes that bind to only those parts of the chromosome with which they show a high degree of sequence similarity. The method comprises of three basic steps: fixation of a specimen on a microscope slide, hybridization of fluorescent labeled probe to homologous fragments of genomic DNA, and enzymatic detection of the tagged target hybrids. The observation of the hybridized sequences is made using epifluorescence microscopy. FISH has a large number of applications in molecular biology and medical science, including gene mapping, diagnosis of chromosomal abnormalities, and studies of cellular structure and function. High resolution FISH (fiber FISH) has become a popular method for ordering genes or DNA markers within chromosomal regions of interest. Given the availability of fluorescent probe-labeling systems and detection reagents, FISH provides an efficient and powerful technique for ordering loci both on metaphase chromosomes and in less condensed interphase chromatin. Two-color metaphase FISH can be used to order pairs of loci relative to the centromere; two- and three-color interphase FISH can be used to accurately order trios of loci spaced within 1 Mb relative to one another.

4 LARGE INSERT GENOMIC LIBRARIES

A perfect blend of molecular biology with cloning technology has revolutionized the cloning of large-sized DNA fragments. During the late 1980s, the development of YACs for cloning of megabase-size DNA fragments became possible and library-based exploration of even larger genomes appeared practical (Burke et al. 1987). Since then, large-insert genomic DNA libraries have become the choice of genomic resources for creation of physical maps. Nowadays, most of the major crop plants and model species have genomic libraries (BACs or YACs) (Table 1). These large-insert genomic DNA BAC libraries form the foundation for constructing physical maps and provide the framework for genome sequencing, gene cloning, functional and evolutionary analysis of genomes. In general, there are three cloning systems (i) BAC-based plasmid cloning system, (ii) YAC-based cloning system, and (iii) bacteriophage P1-based cloning system (Sternberg 1990), which include cosmids (Collins and Hohn 1978) and fosmids (Kim et al. 1992), bacteriophage P1-derived artificial chromosomes (PACs, Ioannou et al. 1994), binary BACs (Hamilton 1997), conventional plasmid based cloning vectors (PBCs; Tao and Zhang 1998) and transformation-competent artificial chromosomes (TACs; Liu et al. 1999). So far, several TAC libraries have been constructed from rice genomic DNA, and it has been shown that large fragments of genomic DNA from TAC clones could be

Table 1 The available genomic information of model species and other major crop plants.

Species	*Genome size*[1]	*Ploidy*	*BAC/YAC library*			*EST sequences*[2]
			Genome coverage	*Number of clones*	*Reference*	
Arabidopsis thaliana	0.15×10^9	2×	4.5×	9,000	Wang et al. 1996	1,527,298
Oryza sativa	0.4×10^9	2×	6.0×	6,932	Umehara et al. 1995	1,248,660
Brachypodium sylvaticum	0.3×10^9	2×	6.6×	30,228	Foote et al. 2004	20,449
Medicago trancatula	5.0×10^8	2×	5.0×	30,720	Nam et al. 1999	260,238
Sorghum bicolor	0.8×10^9	2×	2.6×	13,440	Woo et al. 1994	209,814
Pennisetum glaucum	2.5×10^9	2×	4.7×	159,100	Allouis et al. 2001	2,914
Zea mays	3.0×10^9	2×	3.0×	79,000	Edwards et al. 1992	2,018,337
Hordeum vulgare	5.5×10^9	2×	4.0×	100,000	Kleine et al. 1997	501,366
Triticum boeticum	5.7×10^9	2×	3.0×	170,000	Chen et al. 2002	11,190
Triticum dicoccoides	10.2×10^9	4×	5.0×	516,000	Cenci et al. 2003	9,343
Triticum aestivum	17.9×10^9	6×	3.1×	656,640	Nilmalgoda et al. 2003	1,064,005
Ae. tauschii	11.7×10^9	2×	3.7×	144,000	Moullet et al. 1999	4,315
Glycine max	1.2×10^9	4×	9.0×	73,728	Tomkins et al. 1999	1,386,618
Brassica rapa	1.1×10^9	2×	12×	56,592	Park et al. 2005	149,395
Helianthus annuus	2.8×10^9	2×	8.9×	192,000	Feng et al. 2006	133,682
Solanum tuberosum	1.5×10^9	4×	3.7×	23,808	Song et al. 2000	236,549
Saccharum officinarum	2.5×10^9	2×	4.5×	103,296	Tomkins et al. 1999	246,379
Gossypium hirsutum	2.1×10^9	4×	8.3×	129,024	Tomkins et al. 2001	268,779
Arachis hypogaea	2.8×10^9	4×	6.5×	182,784	Yüksel and Paterson 2005	52,279
Avena sativa	11.7×10^9	6×	-	-	-	7,633
Secale cereale	9.4×10^9	2×	-	-	-	9,298

[1]Arumuganathan and Earle (1991)
[2]dbEST release 030609, *http://www.ncbi.nlm.nih.gov/dbEST/dbEST_summary.html*

efficiently integrated into the rice genome (Liu et al. 2002). In wheat, TAC library was constructed (Liu et al. 2000) and used to find the gene for pre-harvest sprouting tolerance, the clones for a TAC contig covering a target gene can be used to transform and complement mutants and for other genetic manipulations (Torada et al. Unpublished). Of these many variants only a few cloning systems are popular due to their applicability in current genomics research. The major cloning systems (YACs, BACs and Binary vectors-BIBACs) are described below.

4.1 YACs

Yeast artificial chromosome cloning system was first reported in 1987 (Burke et al. 1987). This system has a capacity of cloning dot blot assay (DBA) fragments up to 1,000 kb. YAC libraries were made and used in human physical mapping (Chumakov et al. 1995) and in rice (Wu et al. 2002) and *Arabidopsis* (Schmidt et al. 1995) physical mapping projects. Though this system has an excellent capacity to clone larger DNA fragments when compared to other cloning sytems, YACs are no longer used as a major cloning system of choice. YAC libraries contain high percentage of chimeric clones and the inserts are composed of two or more unrelated genomic fragments that become co-ligated in an undefined manner. It is mandatory to identify the chimeric clones before generating a physical map. The instability of the YACs in the host cells and isolation of the yeast insert DNA are some additional issues related with these YAC-based genomic libraries.

4.2 BACs

Bacterial artificial chromosome (BAC) was first reported in 1992 (Shizuya et al. 1992). A capacity of cloning up to 300 kb size DNA fragments, high stability of the clones for many generations and fewer chimeras made BACs a system of choice over other cloning systems. BAC vectors are based on the *E. coli* fertility (F-factor) plasmid. The F-factor plasmid naturally occurs as a 100-kb circular DNA. Since the replication of the plasmid in *E. coli* is strictly controlled, it is usually maintained in a low copy number. The pBAC108L was the first BAC vector reported with a capacity of maintaining 300 kb in *E. coli*. Later there were many improved BAC vectors with a selectable marker to identify the recombinants and additional cloning sites (e.g., pBeloBAC11, Kim et al. 1996; pBeloBAC1, Frijters et al. 1997). The pBeloBAC1 vector has three restriction sites (*Hin*dIII, *Bam*HI and *Eco*RI) within the *lacZ* gene. The vectors having the DNA insert disrupt the *lacZ* and are identified as white colonies while growing on a medium containing X-gal and IPTG. The non-transformants clones are blue colored.

4.3 Binary Vectors BAC libraries (BIBAC)

The binary vector is a combination of a BAC plasmid (pBAC108L) and the *Agrobacterium tumefaciens* Ri plasmid, which can replicate a single copy plasmid in both *E. coli* and *A. tumefaciens*. Similar to BAC vectors, it contains minimal sequences needed for the autonomous replication and copy-number control of the F-factor plasmid. The lcosN and P1 loxP sites from pBAC108L and the T7 and Sp6 promoters from bacteriophage are also incorporated into the BIBAC. The recombinant selection marker of the BIBAC vector is the *Sac*B gene from the bacteriophage P1 vector pAd10sacBII, in which the insert cloning site *Bam*HI is situated. The antibiotic selectable marker in BIBAC is the *Km*R gene, which confers resistance to the antibiotic kanamycin. The large-size insert clones (YACs or BACs) are difficult to subclone into transformation-competent binary vectors. Once the gene of interest is physically mapped, the next step will be the complementation test through transformation or mutant analysis. It is often desirable to construct large-insert DNA libraries in binary vectors, which are able to directly transform the insert DNA fragment into a desirable system. Along with the BAC-based genomic libraries, binary vector based genomic libraries are also available for many crop plants (e.g., Rice: Li et al. 2007; Soybean: Wu et al. 2004) and have been used for functional analysis and genetic engineering through transformation of the DNA inserts.

4.4 Chromosome Specific BAC Libraries

Molecular marker development and genetic mapping in polyploidy species (e.g., wheat and onion) are formidable tasks when compared to diploid species. Most of the physically mapped genes from wheat invariably used the BAC or cosmid libraries of diploid species (*Ae. tauschii* or *T. monococcum*). Strategically, diploid species are used for chromosome walking and BAC contig assembly. Molecular markers developed are then mapped in the polypoid species (Stein et al. 2000; Yahiaoui et al. 2003). A major break-through in flowcytometry and chromosome sorting facilitated molecular dissection of chromosomes. The genome analysis at molecular level could be simplified by dissecting the genome to small parts and creating sub-genomic molecular resources such as chromosome arm-specific BAC libraries (Vrána et al. 2000; Kubaláková et al. 2002; Doležl et al. 2007; Kovářová et al. 2007; Šimková et al. 2008). This resource is considered as a valuable tool for physical mapping of polyploidy species with large and complex genomes such as wheat and cotton where more than 70% sequences are of repeats and retrotransposable elements (Flavell et al. 1974). The chromosome and chromosome arm specific libraries facilitate screening and mapping of the targeted region in complex genomes. The field bean (*Vicia faba* L.) was the first plant species, in which the whole genome has been fractionated into chromosome-specific DNA libraries (Macas et al. 1996). The first successful wheat chromosome 3B sorting (Šafář et al. 2004) has revolutionized the wheat physical mapping, since then many wheat chromosome arm specific libraries are available or to be available in the near future (J. Doležel, Pers. comm.).

4.5 Random Shear BAC Libraries

Standard BAC library construction methods are based on partial restriction digestion of high molecular weight genomic DNA. Because restriction enzyme sites are not evenly distributed across a genome, the resulting libraries are substantially biased, with numerous gaps in the resulting physical contigs developed. A recent development in BAC library construction is to use random shear large insert DNA fragments, which are believed to be an unbiased representative of the genome. Therefore, even with lower genome coverage, random shear BAC libraries will significantly reduce the finishing cost of a physical mapping project (*http://www.lucigen.com*).

5 PHYSICAL MAPPING WITH LARGE INSERT DNA BAC LIBRARIES

5.1 BAC Filters and BAC DNA Pooling

Once the genomic library has been constructed using any of the cloning systems, it is ready for characterization to check the copy number and the genome coverage. The average optimal genome coverage ranges from 8x to 20x. The viable bacterial cells are robotically printed, from 96 or 384 well plates onto high-density nylon filters. After overnight growth of the clones, the filters are processed and ultimately, the DNA from each clone is permanently fixed to a known position on the filter. These clones can be screened using standard DNA hybridization procedures. Positive

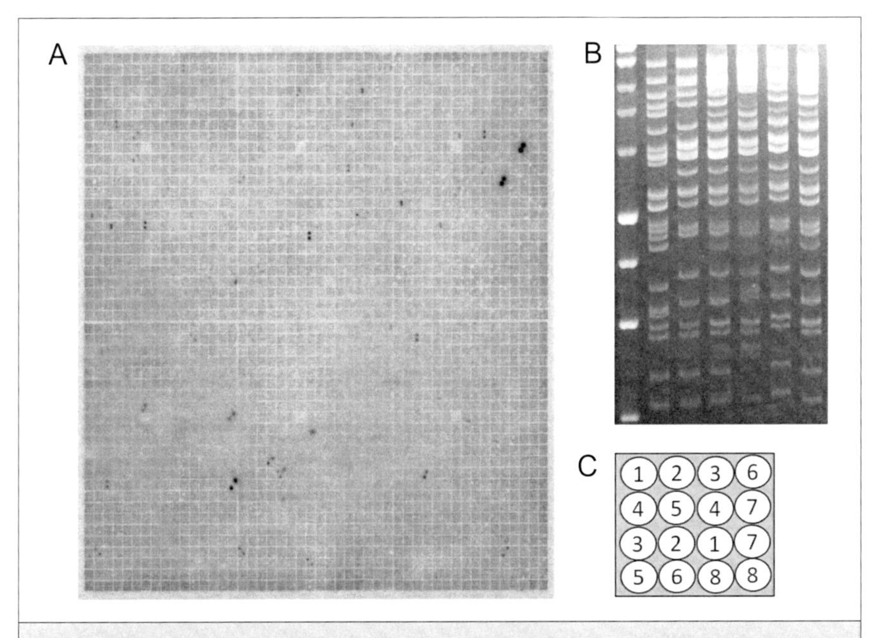

Fig. 3 (A) Southern hybridization image of a high-density colony filter of *Aegilops tauschii* accession AL8/78 BAC library using four pooled overgo probes. Each filter contains 9,216 double spotted BAC clones (Weng et al. unpublished data). (B) *Hind*III fingerprinting of clones to identify the protruding clone in the contig. (C) The arrangements of BAC DNA on the filter.

clones are identified and can be retrieved from the original plates. Typically, 22 cm × 22 cm filters at two densities: 9,216 clones/filter or 18,432 clones/filter are the commonly used systems. All clones are printed in duplicate for the unambiguous selection (Fig. 3).

Besides the large-insert genomic library on the nylon filter, the BAC DNA can be extracted and used for PCR screening. The pooling and super-pooling system enables researchers to use PCR for screening a BAC library and identify the specific plate and the well containing a BAC clone with a sequence of interest. There are many pooling strategies available (Ma et al. 2000). The general strategy is as follows. The screening is done in two separate rounds of PCR on pooled BAC clones (Round I and Round II). The Round I PCR is performed on all the super-pools (containing all BAC clones in the library). Each Super-pool contains 'N' individual BAC clones. The results from Round I PCR will identify which Super-pool(s) contains BAC clone(s) with the sequence of interest (there may be more than one Super-pool identified). The Round

II PCR is performed on the Plate, Row, Column and Diagonal (PRCD) pools for the specific Super-pool being investigated. The results from Round II of PCR should allow identifying the exact plate (384-well plate in general) and well position for one or several positive hits in the particular Super-pool under investigation. This allows identifying the specific BAC clone or clones harboring the gene of interest, out of the complete genomic BAC library.

5.2 BAC Library Screening

In any physical mapping project, identification of low-copy sequences is the key factor to extend the chromosome walking. Low-pass sequencing is the most important method in identifying low copy probes (Stein et al. 2000; Wicker et al. 2008). There are many strategies such as plasmid-rescue and inverse PCR, shotgun subcloning of BAC (Yahiaoui et al. 2004) to identify the low-copy sequences. The following are the commonly used strategies in BAC screening and chromosome walking towards physical mapping of the gene of interest.

5.2.1 Chromosome Walking

Chromosome walking is the first step used for construction of the clone contigs (contiguous array of overlapping clones) of a physical map. A genomic library (YAC or BAC) is screened with a DNA marker (Probe or PCR based marker) and set of positive clones is isolated. All positive clones from BAC/YAC, can be size fractioned after digestion using restriction enzymes (e.g. *Hind*III) by pulse-field gel electrophoresis (PFGE) or run a 30 cm long, 1% agarose gel at 35-40 V overnight. The resulted fragments are scored and fragments at both ends of each insert are isolated rapidly by several standard protocols (Riley et al. 1990; Zoghbi and Chinault 1994). It is often advisable to transfer the gel to nlylon membrane for future use such as cross-hybridization experiments. End fragments from each clone should be used as probes to perform an initial test of the possibility of chimerism. If the two end fragments show complete concordance in transmission, this can be taken as strong evidence for non-chimerism; in contrast, two or more recombination events would be highly suggestive of a chimeric clone. If multiple clones have been obtained from a screen with a DNA marker, end fragments from each should be used in cross-hybridization experiments to identify the particular clones that extend furthest (protruding ends) in each direction along the chromosome. Often this approach will reduce number of clones worth pursuing to just two. Next round of chromosome walking using the farthest end fragments for rescreening, and then analyzing the resulting clones in the same manner described above. In this manner, a 'contig' will be built over the genomic region that contains the locus of interest (Zoghbi and Chinault 1994). The major problem in chromosome walking is the presence of repeat sequences, which will hybridize to the clones located in different regions of the genome. The following subtitles describe the strategies to identify low-copy probes or increase the throughput or DNA markers specifically amplify the genomic region.

5.2.2 Overgo Probing

Overgo probing approach is used in many mapping projects to map ESTs on a physical map. This strategy will certainly help to reduce the contig numbers in a physical map. In general, overgo probe consists of two 24-mer oligonucleotides, derived from genomic sequence, which share eight base pairs of complimentary sequence at their 3' ends. After synthesis and annealing, they create a 16 base overhang. This overhang is 'filled in' using Klenow fragment thereby incorporating appropriate radionucleotides. Typically, ^{32}P-dATP and ^{32}P-dCTP are used and the GC-content of the oligonucleotides is kept between 40-60%. The resulting product is a double-stranded 40-mer with, on an average eight radiolabeled nucleotides incorporated per single-stranded product. This probe is sequence specific with high specific activity (40-70% ^{32}P dNTP incorporation rate). OvergoMaker software is used to design specific overlapping oligonucleotide probes (*http://www.genome.wustl.edu/tools/software/overgo.cgi*). Pooling two or more mapped markers in and around the gene of interest and use as an overgo probe in the hybridization process facilitate the high-throughput identification of positive BAC clones or contigs (Fig. 3).

When comparative mapping approach is used to screen across the species (for instance using wheat markers to screen *Ae. tauschii* BACs), overgo probes are most commonly used. This can be a radio-labeled (Ross et al. 1999) or non-radiolabeled method (Hilario et al. 2007). Comparisons are based on computational detection of orthologous sequences between species. 'Universal' overgo hybridization probes can also be used for the efficient construction of BAC-based physical maps of orthologous chromosome segments from multiple species in parallel. 'Universal' overgo hybridization probes can therefore facilitate the assembly of deep and diverse collections of experimental and computational comparative genomic resources corresponding to specific segments of the genome. The design of 'universal' overgo probes is dependent on the presence of sequences that are highly conserved within a group of species. Such conserved sequences can be readily identified using local- or whole-genome interspecies sequence alignments. Once 'universal' overgo hybridization probes are designed, simple and uniform labeling and hybridization conditions can be carried out to exploit the utility of these probes for targeted comparative physical mapping.

5.2.3 Subcloning

Screening of BAC clones using the tightly linked markers identifies BAC clone(s) carrying the linked markers. Development of DNA markers from the BAC end is not always successful due the fact that the BAC end sequences are not necessarily unique particularly when the BAC library is from a complex large genome. If the BAC end sequences (less than 1.0 kb in each end) failed to generate new molecular marker(s), subcloning or trimming of BAC insert must be performed. Generally, subcloning is performed by digesting (e.g., *Hind*III) the BAC clone, run by electrophoresis, elutes, clone the fragment and sequenced. There are many modified methods (Ohmi et al. 2004) and subcloning kits are available from various providers (Cambuio Ltd; Epicenter, Inc.; Invitrogen). For example, in the process of physical mapping of row type gene (*vrs1*) in barley, wherever the BAC end sequences did not provide any useful makers, BAC subcloning was done to get the internal sequences to complete the contig construction from a low-copy sequence based PCR marker (Komatsuda et al. 2007).

5.2.4 Genetic and Physical Map Integration

The integration of cytogenetic, genetic and physical maps is essential to identify and characterize the genes. In all these three fields of mapping, significant progress has been made over the past two decades and several novel techniques have been developed (Bray-Ward et al. 1996; Marra et al. 1997; Chen et al. 2002; Luo et al. 2003). A large number of assembled BAC contigs are now available for some important crops and model species (e.g., rice, *Arabidopsis*, barley, wheat D genome). With the exception of the model species (rice and *Arabidopsis*), majority of the BAC contigs are not linked to a genetic map position and also large number of BAC clones do not assemble into any contigs. The high-density genetic maps with an average of one marker per 50 kb are necessary to anchor the assembled BAC contigs. The BAC anchoring can be done by using the genetically mapped markers (RFLPs, STS, ESTs, etc). The EST or EST-derived SSR markers are used to screen the BACs will identify the BACs positively contain the gene rich regions (Varshney et al. 2006). The six-dimensional pooling strategy of BACs in combination with the an AFLP screening helps up to 40 genetic map positions to BAC contigs for each AFLP primer combinations (Klein et al. 2000). The integrated genetic and physical mapping by using AFLP fingerprinting has been developed by KeyGene (*http://www.keygene.com/keygene/techs-apps/technologies_keymaps.php*) and utilized in many mapping programs (Srinivasan et al. 2003).

5.2.5 Exon Trapping and Annotation

Once BAC clones harboring the candidate gene are obtained, exon trapping can be performed as a method of transcript selection even before annotation and characterization. Trapped exons are then used for expressional and functional studies (Buckler et al. 1991). This exon trapping relies on the conserved sequences at intron-exon boundaries in all eukaryotic species. It is based on the empirical finding that the vast majority of splice recognition sites are not cell type specific. Instead, general splicing machinery present in all cells can act with precision upon endogenous as well as foreign transcripts. Exon trapping is one of the methods for transcript identification. In the process of chromosome walking, exon trapping is an excellent tool to identify the low-copy sequences, which can generate molecular markers, if the BAC ends failed to generate.

As sequencing becomes highly automated and more accurate, the feasibility of stepping nucleotide by nucleotide across an entire BAC or YAC clone becomes realistic. The basic approach would be to begin sequencing across the insert from both sides with initial primers facing in from the two BAC or YAC arms. The shotgun library of BAC or YAC yields shorter fragments of 2-3 kb. In general 8x to 10x coverage will be sequenced and contigs are made. With a long-range sequence, it becomes possible to use computational methods alone to ferret out coding regions. Since the physical mapping provides the genomic sequences which should be used for identification of genes by gene prediction programs such as GENSCAN 10.0 (*http://genes.mit.edu/GENSCAN.html*) or FGENSH (*http://www/softberry.com*) or Rice Genome Automated Annotation System (RiceGAAS; *http://ricegaas.dna.affrc.go.jp*). Sophisticated computer programs based on neural nets have been developed that can identify 90% of all exons with a 20% false positive rate (Martin et al. 1993). Once a putative exon has been identified, it can be used as a probe to search for the tissue in which its expression takes place, and with further studies, it becomes possible to identify the remaining portions of the transcription unit with which it is associated.

6 PHYSICAL MAPPING OF TRAITS: LESSONS, STRATEGIES AND SUCCESS STORIES

Map based isolation of genes basically involves (i) genetic and fine mapping and (ii) physical mapping, identification and isolation of candidate gene. Then the identified annotated gene(s) is further characterized by mutant or transformation analysis. Physical mapping of gene(s) for a map-based cloning (also called as positional cloning) is a forward genetic approach to understand the molecular basis of traits. There are many reviews on this approach (Leyser and Chang 1996; Jander et al. 2002; Peters et al. 2003; Stein and Graner 2004) which basically describe the following approaches and strategies for physical mapping and isolating the gene of interest. The general scheme followed in physical mapping and map-based gene isolation is given in Fig. 4.

6.1 Mapping Population and Recombination

The genetic mapping of any gene whose phenotypic effects can be followed in a segregating population is the first step for physical mapping. The success of gene isolation is directly related to the mapping population size. In general, a large segregating mapping population has to be screened for meiotic recombination events in the region of interest. The genetic distance is roughly estimated by dividing one by number of gametes (size of the population × 2) multiplied by 100 to get a recombination fraction. For instance, if there is one recombination between two markers in 1,000 F_2 plants (2,000 gametes), the genetic distance between markers is 0.05 cM. The rough estimation of genetic-physical distance ratio is an indicator to know the amount of recombination in the regions of interest. For instance, wheat leaf rust resistance gene was physically mapped by using 520 F_2s (Huang et al. 2003), whereas 3,095 F_2s were used to physically delimit the *Vrn1* gene (Yan et al. 2003). Uneven recombination frequencies are reported in many species (e.g., Barley-Kunzel et al. 2000; Wheat-Akhunov et al. 2003) and the recombination rates vary up to ten-fold within a species (Stein et al. 2000). For instance the barley yellow mosaic and barley mild mosaic resistance gene (*rym4/5*) physical mapping showed significant variation of genetic-physical

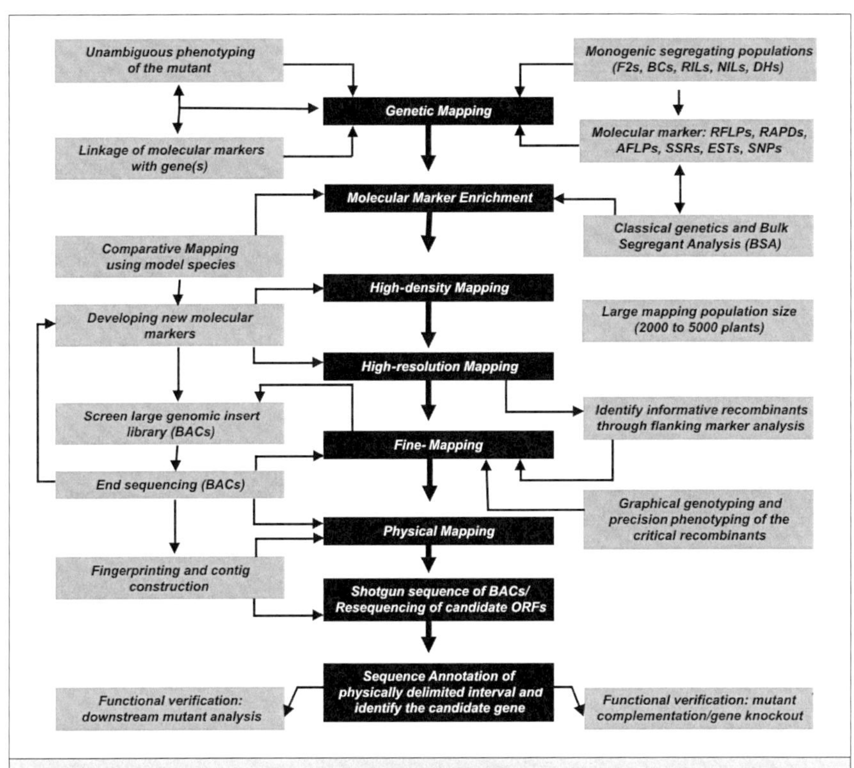

Fig. 4 General schemes of physical mapping and map-based gene isolation. The steps and strategies involved are connected with arrows.

distance ratio from 0.8 Mb/cM to 30 Mb/cM (Stein et al. 2005), with uneven distribution of recombination in several chromosomes as well as within the narrow region of the chromosome. It is often necessary to have more than one mapping population using different parental combinations for genetic mapping. For instance, three different mapping populations (6,269 gametes) were used for genetic mapping of barley row type gene (*vrs1*) and eventually the candidate gene was isolated (Komatsuda et al. 2007).

6.2 Marker Saturation and High-Resolution Mapping

In general AFLP markers are being used for marker saturation and fine mapping. AFLP markers (Vos et al. 1995) are abundant in nature and have been used for construction of primary genetic linkage maps, high-density linkage map of a targeted region, identification of QTLs controlling complex traits and studies on genetic diversity. The bulked sergeant analysis using AFLP markers (Quarrie et al. 1999) were popularly used for marker saturation, since AFLP requires no specific prior knowledge about the genome. AFLP still remains inaccessible for locus-specific applications in large segregating populations. Yet another limitation of AFLP is its dominant nature. For a high-resolution or fine mapping, the AFLP markers closely linked to the targeted traits were then converted into sequence-tagged sites (STSs) and utilized in many mapping programs (He et al. 2004; Azhaguvel et al. 2006). Since the release of model genome sequences, many genetic mapping programs have effectively utilized this information for the targeted region.

6.3 Comparative Mapping

Comparative mapping has been the primary approach for analyzing genomes of divergent species because gene-rich regions are often conserved over evolutionary period of time. This phenomenon is known as synteny or colinearity. Comparative mapping using molecular marker systems among major grass species in Poaceae (Barley, rice, maize, sorghum, wheat) revealed conserved marker order over large chromosomal regions (Devos et al. 1992; Ahn and Tanksley 1993; Gale and Devos 1998a,b; Devos and Gale 2000), also to some extent within Solanaceae species (Tanksley et al. 1992). Rice is the model genome for cereals, the rice genome browser has rice physical map aligned with ESTs of cereals and related plant species (*http://www.gramene.org*; *http://rpg.dna.affrc.go.jp/giot/INE.html*). The model species *Arabidopsis* showed high genomic colinearity with *Brassica* species (Lagercrantz 1998). The major chromosomal rearrangements between *Arabidopsis* and Asteraceae (Timms et al. 2006) and the structural divergence of the genic regions between *Arabidopsis* and *Brassica* (Suwabe et al. 2006; Town et al. 2006) also have been demonstrated by using the complete *Arabidopsis* reference sequence. The sorghum genome was compared to the rice genome sequence using two sorghum physical maps integrated with genetic markers and BAC hybridization data (Bowers et al. 2005). Various local rearrangements between other cereals and rice have been reported in sequence-level comparisons using the rice genome sequence as the reference (Chen et al. 1998; Goff et al. 2002; Bennetzen and Ma 2003; Sorrells et al. 2003; Salse et al. 2004; Rice Chromosome 3 Sequencing Consortium 2005). Using the abundant availability of EST markers, in silco mapping of the targeted species with the model species has opened up the resource for marker enrichment. Physical mapping projects for large genome size species (wheat and barley) are fully utilizing the rice genomic resources (e.g., Stein et al. 2005; Komatsuda et al. 2007). For example, rice chromosome 1 and barley chromosome 3H are highly colinear (Smilde et al. 2001). Though high colinearity was observed between the rice chromosome 4L and the barley chromosome 2L and yielded many useful markers for the *vrs1* mapping in barley, there is significant breakdown of micro-colinearity (Pourkheirandish et al. 2007). There are many strategies and established pipelines to use the molecular markers anchored on the model species. For instance an automated bioinformatics pipeline that combines multi-species ESTs and genome sequence data (Fredslund et al. 2006) has been reported. However, comparisons between genetic maps of distantly related species are usually difficult, since genomes often undergo chromosomal re-arrangements, such as inversions, translocations, duplications, deletions and cycles of polyploidization followed by diploidization (Coghlan et al. 2005). Over millions of years of genome evolution have led to disruptive colinearity by having colinear and non-colinear segments. Despite the frequent observation of disturbed microsynteny between the genomes of Triticeae species and rice, wheat and barley has greatly facilitated the isolation of genes via map-based cloning approaches (see reviews Stein and Graner 2004; Varshney et al. 2007a, b). In recent years *Brachypodium distachyon* has been found to be an ideal model system, which is comparable to rice and it is considered as a species for colinearly based gene isolation in Triteceae plants. This weedy species has a number of advantages such as annual growth habit, self fertility, short generation time, small genome size (~355 Mb) (Bennett and Leitch 2005) and five pairs of readily identifiable chromosomes (*www.brachypodium.org*).

It is evident that comparative genome analysis is fast becoming an important tool, not only to discover genes and understand their functions, but also to unravel the evolutionary relationships between species. As most of the related species are derived from recent common ancestors, it is not surprising that in both dicotyledons and monocotyledons extensive genetic colinearity was found when related species were mapped using common RFLP probe sets (Bonierbale et al. 1988; Hulbert et al. 1990; Ahn and Tanksley 1993; Jena et al. 1994). While extensive colinearity seems to be limited to the genus level among dicots, sufficient colinearity exists to allow rough alignment of genetic maps across entire genomes throughout the entire cereal clade (Moore et al. 1995).

6.4 Physical Mapping in Positional Cloning of genes

Map-based cloning (MBC, also called positional cloning) refers to the isolation of DNA sequences responsible for the trait of interest. The logical naming is map-based isolation, since it is not always necessary to clone the gene for functional verification. Compared with any other gene isolation approaches, map-based isolation gives the direct relationship of DNA sequences and the phenotype (trait) of interest. There are two basic steps in the process of map-based isolation. The first step is to use formal linkage analysis and other genetic approaches, as tools, to find flanking DNA markers that lie very close to the locus of interest for genetic mapping. The second step is to physically map the tightly linked marker by physical mapping through screening of the large-insert libraries (Fig. 3). Success of map-based cloning in crop plants depends on the target gene being localized to a short genetic interval (co-segregates with the gene of interest or 0.1-0.01 cM) that the markers and target gene are separated from regions rich in repeated sequences. Fine mapping can be accomplished 10-50 fold more efficiently with a physical map that encompasses the whole genome than by marker saturation and chromosome walking or landing (Zhang and Wing 1997). Some successful map-based gene isolation examples in different plants are listed in Table 2.

6.5 Physical Mapping Projects

Genome-wide physical maps have provided powerful tools and infrastructure for advanced genomics research of plant species. They are not only crucial for large-scale genome sequencing but also provide powerful platforms required for many other aspects of genome research, including targeted marker development, efficient positional cloning, and high-throughput EST mapping. Three approaches are used to physical map construction. In the first approach, low resolution YAC-based physical maps are constructed by DNA marker-based chromosome landing followed by a cosmid sub-cloning from the YAC contiguous clone set. In the second approach, physical maps are developed from random BAC, cosmid or bacteriophage lambda clones by fingerprinting to identify contiguous overlapping clones. In the third approach, a physical map may be developed by random sequencing about 10% of the genome, or about 500 bp from the two ends of enough BACs to encompass 15-fold the haploid genome. Among the three approaches, only BAC-based physical mapping is successful in plant species, whereas the other two approaches are efficiently utilized in animal systems. Among plants, only the *Arabidopsis* (The *Arabidopsis* Genome

Table 2 Examples of successful physical mapping and map-based cloning.

Species	*Gene*	*Mapping strategies*	*Reference*
Arabidopsis–*ABI3*	Seed development and germination	Cosmid library construction	Giraudat et al. 1992
Barley–*Mlac*	Powdery mildew	Single transient assay	Wei et al. 1999
Barley–*Mlo*	Powdery mildew	RT-PCR; Allele analysis	Buschges et al. 1997
Barley–*Rpg1*		RT-PCR; Transgenic	Brueggeman et al. 2002
Barley–*rym4/rym5*	Barley yellow mosaic and barley mild mosaic	Rice-barley comparative mapping; Allelic variation	Stein et al. 2005
Barley–*vrs1*	Six-row type	Rice-barley comparative mapping; Sub-cloning; Mutant analysis	Komatsuda et al. 2007
Maize–*a1-sh2*	Meiotic recombination related	YAC screening	Civardi et al. 1994
Maize–*RTCS*	Rootless concerning crown and seminal roots	BAC screening; RT-PCR; Rice-maize comparative mapping	Taramino et al. 2007
Rice–*d1*	Dwarfing gene	YAC and PAC screening; Mutant analysis	Ashikari et al. 1999
Rice–*d2*	Dwarfing gene	RT-PCR; Transgenic	Hong et al. 2003
Rice–*D11*	Dwarfing gene	RT-PCR; Transgenic	Tanabe et al. 2005
Rice–*Adh1-Adh2*	Alcohol dehydrogenase	BAC screening; Comparative map	Tarchini et al. 2000
Rice–*Hd6*	Photoperiod sensitivity	YAC and PAC screening	Takahashi et al. 2001
Rice–*Pi36(t)*	Rice blast resistance	BSA-RAPDs; BAC; PAC screening; *In silico* mapping	Liu et al. 2005
Rice–*Rf-1*	Fertility restorer gene	cDNA library construction and screening	Komori et al. 2003
Rice–*Pib*	Blast resistance	RNA gel blot analysis	Wang et al. 1999

Contd.

Table 2 continued

Species	*Gene*	*Mapping strategies*	*Reference*
Rice–*Pi-ta*	Blast resistance	BAC screening; RFLP and RAPD marker	Bryan et al. 2000
Rice–*Xa1*	Bacterial blight resistance	cDNA; Cosmid libraries; RT-PCR	Yoshimura et al. 1998
Rice–*Xa21*	Bacterial leaf blight resistance	BAC screening	Song et al. 1995
Wheat–*Gpc-B1*	Grain protein content	Southern blotting; Functional marker development	Distelfeld et al. 2006
Wheat–*Lr10*	Leaf rust	Sub-cloning; Mutagenesis	Feuillet et al. 2003
Wheat–*Lr21*	Leaf rust	RT-PCR; Transgenic	Huang et al. 2003
Wheat–*Ph1*	Chromosome pairing locus	RT-PCR, multiplex PCR	Griffiths et al. 2006
Wheat–*Ph2*	Chromosome pairing locus	*In silico*	Sutton et al. 2003
Wheat–*Pm3b*	Powdery mildew	RT-PCR; Mutagenesis; Single transient assay	Yahiaoui et al. 2004
Wheat–*Q*	Shattering	Mutagenesis; Transgenic; allelic variation; RT-PCR	Faris et al. 2003 Simons et al. 2005
Wheat–*Vrn1*	Vernalization	RT-PCR; Allelic variation;	Yan et al. 2003
Wheat–*Vrn2*	Vernalization	RT-PCR; Allelic variation;	Yan et al. 2004
Wheat–*Lr34*	Lear rust	RT-PCR; Allelic variation	Krattinger et al. 2009
Wheat–*Yr36*	Stripe rust	TILLING; Transgenic	Fu et al. 2009

Table 3 List of major physical mapping database and the genome browsers.

Name	*URL*
Rice	*http://www.tigr.org/tigr-scripts/osa1_web/gbrowse/rice/* *http://www.gramene.org/Oryza_sativa_japonica/Info/Index*
Arabidopsis	*http://atidb.org/cgi-perl/gbrowse/atibrowse/* *http://atensembl.arabidopsis.info/Arabidopsis_thaliana_TAIR/index.html* *http://arabidopsis.org/*
Barley	*http://phymap.ucdavis.edu:8080/barley/index.jsp*
wheat D genome	*http://phymap.plantsciences.ucdavis.edu:8080/wheatdb/*
Brachypodium	*http://www.brachybase.org/cgi-bin/gbrowse/brachy_core/?name=super_1%3A34970724..34988224;width=800*
Soybean	*http://soybeanphysicalmap.org/* *http://www.phytozome.net/cgi-bin/gbrowse/soybean/?name=Gm09* *http://soybase.org/SequenceIntro.php*
Maize	*http://www.maizegdb.org/*
Lotus japonicus	*http://www.kazusa.or.jp/lotus/*
Medicago	*http://www.medicago.org/genome/*
Tomato	*http://www.sgn.cornell.edu/* *http://solanaceae.plantbiology.msu.edu/projects_potato_bacs_summary.php*
Brassica rapa	*http://www.brassica-rapa.org/BGP/NC_brgp.jsp*
Potato	*https://gabi.rzpd.de/PoMaMo.html* *http://www.potatogenome.net/*
Cotton	*http://cottondb.org/*
Mouse	*http://www.informatics.jax.org/*
Human	*http://www.ornl.gov/sci/techresources/Human_Genome/posters/chromosome/*
BLAST	*http://blast.ncbi.nlm.nih.gov/Blast.cgi* *http://blast.jcvi.org/euk-blast/index.cgi?project=tae1* *http://wheat.pw.usda.gov/GG2/index.shtml* *http://www.shigen.nig.ac.jp/wheat/komugi/top/top.jsp* *http://blast.brachybase.org/* *http://www.arabidopsis.org/Blast/*

Initiative 2000), rice (International Rice Genome Sequencing Project 2005) and Sorghum (Sasaki and Antonio 2009) genome sequences have been completed so far. The first report of genome-wide, BAC and BIBAC-based physical map of the soybean genome was published in 2004 (Wu et al. 2004). The list of some available genomics information of major crop plants and major genomic database information are given in Tables 2 and 3. For more details please refer Chapters 5 to 10 in this volume.

7 CONCLUSION AND PERSPECTIVES

The high-throughput technologies in the genome analysis have made significant impact in recent years. The 'next generation' sequencing systems such as Genome Sequencer 20/FLX (454/Roche), Solexa 1G (Illiumina/Solexa), SOLiD™ system (Applied Biosystems) and Polonator G.007 (Dover Systems) are commercially made available. Besides, there are many more in the technology development pipeline for commercialization in near future. The examples are SMRT—Single-molecule real-time sequencing (Pacific Bioscience), single-molecule sequencing (SMS) and True single-molecule sequencing (tSMS, Helicos Biosciences); for a complete overview about these technologies, readers may consult the review by Gupta (2008). These newly developed DNA sequencing systems not only handle large volumes of samples and data but also reduce the cost involved per base pair. With this present pace of genomic revolution, soon all the major crops, animal and microbial species will be fully sequenced. These sequences will certainly provide in-depth knowledge in the understanding of the metabolic pathways at the nucleotide level and also will complement genetic and physical mapping projects.

Excellent bioinformatics tools are currently available for the integration of genetic and physical mapping. These tools play a major role in maintaining and logical visualization of the physical mapping database. The major community standard tool is Generic Genome Browser (GBrowse, *http://www.gmod.org*; Stein et al. 2002). For contig construction, FPCV4.7 is the commonly used tool to align the fingerprints of large-insert genomic libraries (BACs/YACs). iMap (Fang et al. 2003) presents the genetic map, associated BAC contigs of the physical map, and access to dynamic links for information about the contigs and the markers. With the development of such tools, the whole genome physical mapping and physical mapping of the gene have become a routine reality. The ever evolving computational tools together with the high-throughput genomic technologies (molecular marker genotyping such as SNP detection system and sequencing) will certainly usher in a new generation for the genetic and physical mapping projects.

References

Ahn S, Tanksley SD (1993) Comparative linkage maps of rice and maize genomes. Proc Natl Acad Sci USA 90: 7980-7984

Akhunov ED, Goodyear AW, Geng S, Qi L-L, Echalier B, Gill BS, Miftahudin, Gustafson JP, Lazo G, Chao S, Zhang D, Nguyen HT, Kalavacharla V, Hossain K, Kianian SF, Peng J, Lapitan NLV, Gonzalez-Hernandez JL, Anderson JA, Choi D-W, Close TJ, Dilbirligi M, Gill KS, Walker-Simmons MK, Steber C, McGuire PE, Qualset CO, Dvorak J (2003) The organization and rate of evolution of wheat genomes are correlated with recombination rates along chromosome arms. Genome Res 13: 753-763

Allouis S, Qi X, Lindup S, Gale MD, Devos KM (2001) Construction of a BAC library of pearl millet, *Pennisetum glaucum*. Theor Appl Genet 102: 1200-1205

Ananiev EV, Riera-Lizarazu O, Rines HW, Phillips RL (1997) Oat-maize chromosome addition lines: A new system for mapping the maize genome. Proc Natl Acad Sci USA 94: 3524-3529

Arumuganathan K, Earle ED (1991) Nuclear DNA content of some important plant species. Plant Mol Biol Rep 9: 208-219

Ashikari M, Wu J, Yano M, Sasaki T, Yoshimura A (1999) Rice gibberellin-insensitive dwarf mutant gene *Dwarf 1* encodes the a-subunit of GTP-binding protein. Proc Natl Acad Sci USA 96: 10284-10289

Azhaguvel P, Vidya Saraswathi D, Komatsuda T (2006) High-resolution linkage mapping for the non-brittle rachis locus *btr1* in cultivated x wild barley (*Hordeum vulgare*). Plant Sci 170: 1087-1094

Bennett MD, Leitch IJ (2005) Nuclear DNA amounts in angiosperms: progress, problems and prospects. Ann Bot 95: 45-90

Bennetzen JL, Ma J (2003) The genetic colinearity of rice and other cereals on the basis of genomic sequence analysis. Curr Opin Plant Biol 6: 128-133

Biunno I, Bernard L, Dear P, Cattaneo M, Volorio S Zannini L, Bankier A, Zollo M (2000) SEL1L, the human homolog of *C. elegans* sel-1: refined physical mapping, gene structure and identification of polymorphic markers. Hum Genet 106: 227-235

Bonierbale MW, Plaisted RL Tanksley SD (1988) RFLP maps based on a common set of clones reveal modes of chromosomal evolution in potato and tomato. Genetics 120: 1095-1103

Bowers JE, Arias MA, Asher R, Avise JA, Ball RT, Brewer GA, Buss RW, Chen AH, Edwards TM, Estill JC, Exum HE, Goff VH, Herrick KL, James SCL, Karunakaran S, Lafayette GK, Lemke C, Marler BS, Masters SL, McMillan JM, Nelson LK, Newsome GA, Nwakanma CC, Odeh RN, Phelps CA, Rarick EA, Rogers CJ, Ryan SP, Slaughter KA, Soderlund CA,Tang H, Wing RA, Paterson AH (2005) Comparative physical mapping links conservation of microsynteny to chromosome structure and recombination in grasses. Proc Natl Acad Sci USA 102: 13206-13211

Bray-Ward P, Menninger J, Lieman J, Desai T, Mokady N, Banks A, Ward DC (1996) Integration of the cytogenetic, genetic, and physical maps of the human genome by FISH mapping of CEPH YAC clones. Genomics 32: 1-14

Brueggeman R, Rostoks N, Kudrna D, Kilian A, Han F, Chen J, Druka A, Steffenson B, Kleinhofs A (2002) The barley stem rust-resistance gene *Rpg1* is a novel disease resistance gene with homology to receptor kinases. Proc Natl Acad Sci USA 99: 9328-9333

Bryan GT, Wu KS, Farrall L, Jia Y, Hershey HP, McAdams SA, Faulk KN, Donaldson GK, Tarchini R, Valent B (2000) A single amino acid difference distinguishes resistant and susceptible alleles of the rice blast resistance gene *Pi-ta*. Plant Cell 12: 2033-2046

Buckler AJ, Chang DD, Graw SL, Brook JD, Haber DA, Sharp PA, Housman DE (1991) Exon amplification: A strategy to isolate mammalian genes based on RNA splicing. Proc Natl Acad Sci USA 88: 4005-4009

Burke DT, Carle GF, Olson MV (1987) Cloning of large segments of exogenous DNA into yeast by means of artificial chromosome vectors. Science 236: 806-812

Buschges R, Hollricher K, Panstruga R, Simons G, Wolter M, Frijters A, van Daelen R, vanderLee T, Diergaarde P, Groenendijk J, Topsch S, Vos P, Salamini F, Schulze-Lefert P (1997) The barley *mlo* gene: A novel control element of plant pathogen resistance. Cell 88: 695-705

Cenci A, Chantret N, Kong X, Gu Y, Anderson OD, Fahima T, Distelfeld A, Dubcovsky J (2003) Construction and characterization of a half million clone BAC library of durum wheat (*Triticum turgidum* ssp. durum). Theor Appl Genet 107: 931-939

Chen M, Presting G, Barbazuk WB, Goicoechea JL, Blackmon B, Fang G, Kim H, Frisch D, Yu Y, Sun S, Higingbottom S, Phimphilai J, Phimphilai D, Thurmond S, Gaudette B, Li P, Liu J, Hatfield J, Main D, Farrar K, Henderson C, Barnett L, Costa R, Williams B, Walser S, Atkins M, Hall C, Budiman MA, Tomkins JP, Luo M, Bancroft I, Salse J, Regad F, Mohapatra T, Singh NK, Tyagi AK, Soderlund C, Dean RA, Wing RA (2002) An integrated physical and genetic map of the rice genome. Plant Cell 14: 537-545

Chen FG, Zhang XY, Xia GM, Jia JZ (2002) Construction and characterization of a bacterial artificial chromosome library for *Triticum boeoticum*. Acta Bot Sin 44: 451-456

Chen MS, SanMiguel P, Bennetzen JL (1998) Sequence organization and conservation in sh2/a1-homologous regions of sorghum and rice. Genetics 148: 435-443

Chumakov IM, Rigault P, Le Gall I, Bellanne-Chantelot C, Billault A, Guillou S, Soularue P, Guasconi G, Poullier E, Gros I et al. (1995) A YAC contig map of the human genome. Nature 377: 157-298

Civardi L, Xia Y, Edwards KJ, Schnable PS Nikolau BJ (1994) The relationship between genetic and physical distances in the cloned *a1-sh2* interval of the *Zea mays* L. genome. Proc Natl Acad Sci USA 91(17): 8268-8272

Coghlan A, Eichler EE, Oliver SG, Paterson AH, Stein L (2005) Chromosome evolution in eukaryotes: a multi-kingdom perspective. Trends Genet 21: 673-682

Collins J, Hohn B (1978) Cosmids: a type of plasmid gene-cloning vector that is packageable in vitro in bacteriophage l heads. Proc Natl Acad Sci 75: 4242-4246

Cox DR, Burmeister M, Roydon Price E, Kim S, Myers RM (1990) Radiation hybrid mapping: A somatic cell genetic method for constructing high-resolution maps of mammalian chromosomes. Science 250: 245-250

Dear PH, Bankier AT, Piper MB (1998) A high-resolution metric HAPPY map of human chromosome 14. Genomics 48: 232-241

Devos KM, Gale MD (2000) Genome relationships: The grass model in current research. Plant Cell 12: 637-646

Devos KM, Atkinson MD, Chinoy CN, Liu CJ, Gale MD (1992) RFLP based genetic map of the homoeologous group 3 chromosomes of wheat and rye. Theor Appl Genet 83: 931-939

Distelfeld A, Uauy C, Fahima T, Dubcovsky J (2006) Physical map of the wheat high-grain protein content gene *Gpc-B1*and development of a high-throughput molecular marker. New Phytol 169: 753-763

Doležl J, Kubaláková M, Paux E, Bartoš J, Feuillet C (2007) Chromosome-based genomics in the cereals. Chrom Res 15: 51-66

Doniskeller H, Green P, Helms C, Cartinhour S, Weiffenbach B, Stephens K, Keith TP, Bowden DW, Smith DR, Lander ES, Botstein D, Akots G, Rediker KS, Gravius T, Brown VA, Rising MB, Parker C, Powers JA, Watt DE, Kauffman ER, Bricker A, Phipps P, Mullerkahle H, Fulton TR, Ng S, Schumm JW, Braman JC, Knowlton RG, Barker DF, Crooks SM, Lincoln SE, Daly MJ, Abrahamson J (1987) A genetic-linkage map of the human genome. Cell 51(2): 319-337

Edwards KJ, Thompson H, Edwards D, de Saizien A, Sparks C, Thompson JA, Greenland AJ, Eyers M, Schuch W (1992) Construction and characterisation of a yeast artificial chromosome library containing three haploid maize genome equivalents. Plant Mol Biol 19: 299-308

Endo TR (1988) Induction of chromosomal structural changes by a chromosome *Aegilops cylindrica* L. in common wheat. J Hered 79: 366-370.

Endo TR (1990) Gc chromosomes and their induction of chromosome mutations in wheat. Jpn J Genet 65: 135-152

Endo TR, Gill BS (1996) The deletion stocks of common wheat. J Hered 87: 295-307

Erayman M, Sandhu D, Sidhu D, Dilbirligi M, Baenziger PS, Gill KS (2004) Demarcating the gene-rich regions of the wheat genome. Nucl Acids Res 32: 3546-3565

Fang Z, Cone K, Sanchez H, Polacco ML, Mcmullen MD, Schroeder S, Gardiner J, Davis G, Seth A, Yim Y (2003) Imap: a tool for the integration of genetic and physical maps. Bioinformatics 19(16): 2105-2111

Faris JD, Fellers JP, Brooks SA, Gill BS (2003) A bacterial artificial chromosome contig spanning the major domestication locus Q in wheat and identification of a candidate gene. Genetics 64: 311-321

Feng J, Brady AV, Kyung LM, Zhang H-B, Jan CC (2006) Construction of BAC and BIBAC libraries from sunflower and identification of linkage group-specific clones by overgo hybridization. Theor Appl Genet 113: 23-32

Feuillet C, Travella S, Stein N, Albar L, Nublat A, Keller B (2003) Map-based isolation of the leaf rust disease resistance gene *Lr10* from the hexaploid wheat (*Triticum aestivum* L.) genome. Proc Natl Acad Sci USA 100: 15253-15258

Flavell RB, Bennett MD, Smith JB, Smith DB (1974) Genome size and the proportion of repeated nucleotide sequence DNA in plants. Biochem Genet 12: 257-269

Foote T, Griffiths S, Allouis S, Moore G (2004) Construction and analysis of a BAC library in the grass *Brachypodium sylvaticum*: its use as a tool to bridge the gap between rice and wheat in elucidating gene content. Funct Integr Genom 4: 26-33

Fredslund J, Madsen LH, Hougaard BK, Nielsen AM, Bertioli D, Sandal N, Stougaard J, Schauser L (2006) A general pipeline for the development of anchor markers for comparative genomics in plants. BMC Genom 7: 207

Frijters ACJ, Zhang Z, van Damme M, Wang GL, Ronald PC, Michelmore RW (1997) Construction of a bacterial artificial chromosome library containing large EcoRI and Hind III genomic fragments of lettuce. Theor Appl Genet 94: 390-399

Fu D, Uauy C, Distelfeld A, Blechl A, Epstein L, Chen X, Sela H, Fahima T, Dubcovsky J (2009) A kinase-START gene confers temperature-dependent resistance to wheat stripe rust. Science 323: 1357-1359

Gale MD, Devos KM (1998a) Plant comparative genetics after 10 years. Science 282: 656-658

Gale MD, Devos KM (1998b) Comparative genetics in the grasses. Proc Natl Acad Sci USA 95: 1971-1974

Gao W, Chen ZJ, Yu JZ, Raska D, Kohel RJ, Womack JE, Stelly DM (2004) Wide-cross whole genome radiation hybrid mapping of cotton (*Gossypium hirsutum* L.). Genetics 167: 1317-1329

Gill KS, Gill BS, Endo TR (1993) A chromosome region-specific mapping strategy reveals gene-rich telomeric ends in wheat. Chromosoma 102: 374-381

Giraudat J, Hauge BM, Valon C, Smalle J, Parcy F, Goodman HM (1992) Isolation of the Arabidopsis ABI3 gene by positional cloning. Plant Cell 4(10): 1251-1261

Goff SA, Ricke D, Lan TH, Presting G, Wang R, Dunn M, Glazebrook J, Sessions A, Oeller P, Varma H, Hadley D, Hutchison D, Martin C, Katagiri F, Lange

BM, Moughamer T, Xia Y, Budworth P, Zhong J, Miguel T, Paszkowski U, Zhang S, Colbert M, Sun W, Chen L, Cooper B, Park S, Wood TC, Mao L, Quail P, Wing R, Dean R, Yu Y, Zharkikh A, Shen R, Sahasrabudhe S, Thomas A, Cannings R, Gutin A, Pruss D, Reid J, Tavtigian S, Mitchell J, Eldredge G, Scholl T, Miller RM, Bhatnagar S, Adey N, Rubano T, Tusneem N, Robinson R, Feldhaus J, Macalma T, Oliphant A, Briggs S (2002) A draft sequence of the rice genome (*Oryza sativa* L. ssp. *japonica*). Science 296: 92-100

Goss SJ, Harris H (1975) New method for mapping genes in human chromosomes. Nature 255: 680-684

Goss SJ, Harris H (1977) Gene transfer by means of cell fusion. II. The mapping of 8 loci on human chromosome 1 by statistical analysis of gene assortment in somatic cell hybrids. J Cell Sci 25: 17-20

Griffiths S, Sharp R, Foote TN, Bertin I, Wanous M, Reader S, Colas I, Moore G (2006) Molecular characterization of *Ph1* as a major chromosome pairing locus in polyploid wheat. Nature 439: 749-752

Gupta PK (2008) Single-molecule DNA sequencing technologies for future genomics research. Trends Biotechnol 26(11): 602-611

Gyapay G, Schmitt K, Fizames C, Jones H, Vega-Czarny N Spillett D, Muselet D, Prud'homme JF, Dib C, Auffray C, Morissette J, Weissenbach J, Goodfellow PN (1996) A radiation hybrid map of the human genome. Hum Mol Genet 5: 339-346

He CF, Sayed-Tabatabaei BE, Komatsuda T (2004) AFLP targeting of the 1-cM region conferring the *vrs1* gene for six-rowed spike in barley, *Hordeum vulgare* L. Genome 47: 1122-1129

Hilario E, Bennell TF, Rikkerink E (2007) Screening a BAC library with non-radioactive overlapping oligonucleotide (Overgo) probes. In: Hilario E, Mackay J (eds) Methods in Molecular Biology: Protocols for Nucleic Acid Analysis by Nonradioactive Probes.Vol 353. 2nd edn, pp 79-91

Hong Z, Ueguchi-Tanaka M, Umemura K, Uozu S, Fujioka S, Takatsuto S, Yoshida S, Ashikari M, Kitano H, Matsuoka M (2003) A rice brassinosteroid-deficient mutant, ebisu dwarf (*d2*), is caused by a loss of function of a new member of cytochrome P450. Plant Cell 15: 2900-2910

Huang L, Brooks SA, Li W, Fellers JP, Trick HN, Gill BS (2003) Map-based cloning of leaf rust resistance gene Lr21 from the large and polyploid genome of bread wheat. Genetics 164: 655-664

Hulbert SH, Richter TE, Axtell JD, Bennetzen JL (1990) Genetic mapping and characterization of sorghum and related crops by means of maize DNA probes. Proc Natl Acad Sci USA 87: 4251-4255

International Rice Genome Sequencing Project (2005) The mapbased sequence of the rice genome. Nature 436: 793-800

Ioannou PA, Amemiya CT, Garnes J, Kroisel PM, Shizuya H, Chen C, Batzer MA, de Jong PJ (1994) A new bacteriophage P1-derived vector for the propagation of large human DNA fragments. Nat Genet 6:84-89

Jander G, Norris SR, Rounsley SD, Bush DF, Levin IM, Last RL (2002) Arabidopsis map based cloning in the post-genome era. Plant Physiol 129: 440-450

Jena KK, Khush GS, Kochert G (1994) Comparative RFLP mapping of a wild rice, *Oryza officinalis*, and cultivated rice, *O. sativa*. Genome 37: 382-389

Joppa LR (1993) Chromosome engineering in tetraploid wheat. Crop Sci 33: 908-913

Kalavacharla V, Hossain K, Gu Y, Riera-Lizarazu O, Vales MI, Bhamidimarri S, Gonzalez-Hernandez JL, Maan SS, Kianian SF (2006) High-resolution radiation hybrid map of wheat chromosome 1D. Genetics 173(2): 1089-1099

Kim UJ, Shizuya H, de Jong PJ, Birren B, Simon MI (1992) Stable propagation of cosmid sized human DNA inserts in an F factor based vector. Nucl Acids Res 20: 1083-1085

Kim UJ, Birren B, Slepak T, Mancino V, Boysen C, Kang HL, Simon MI H. Shizuya (1996) Construction and characterization of a human bacterial artificial chromosome library. Genomics 34: 213-218

Klein PE, Klein RR, Cartinhour SW, Ulanch PE, Dong J, Obert JA, Morishige DT, Schlueter SD, Childs KL, Ale M, Mullet JE (2000) A high-throughput AFLP-based method for constructing integrated genetic and physical maps: progress toward a sorghum genome map. Genome Res 10: 789-807

Kleine M, Michalek W, Diefenthal T, Dargatz H, Jung C (1997) Construction of a MluI-YAC library from barley (*Hordeum vulgare* L.) and analysis of YAC insert terminal regions. Genome 40: 896-902

Komatsuda T, Pourkheirandish M, He C, Azhaguvel P, Kanamori H, Perovic D, Stein N, Graner A, Wicker T, Lundqvist U, Fujimura T, Matsuoka M, Matsumoto T, Yano M (2007) Six-rowed barley originated from a mutation in a homeodomain-leucine zipper I-class homeobox gene. Proc Natl Acad Sci USA 104: 1424-1429

Komori T, Ohta S, Murai N, Takakura Y, Kuraya Y, Suzuki S, Hiei Y, Imaseki H, Nitta N (2003) Map-based cloning of a fertility restorer gene, *Rf-1*, in rice (*Oryza sativa* L.). Plant J 37: 315-325

Kovářová P, Navrátilová A, Macas J, Doležel J (2007) Chromosome analysis and sorting in *Vicia sativa* using flow cytometry. Biol Plant 51: 43-48

Krattinger SG, Lagudah ES, Spielmeyer W, Singh RP, Espino JH, McFadden H, Bossolini E, Selter LL, Keller B (2009) A putative ABC transporter confers durable resistance to multiple fungal pathogens in wheat. Science 323: 1360-1363

Kubaláková M, Vrána J, Číhalíková J, Šimková H, Doležel J (2002) Flow karyotyping and chromosome sorting in bread wheat (*Triticum aestivum* L.). Theor Appl Genet 104: 1362-1372

Kunzel G, Korzun L, Meister A (2000) Cytologically integrated physical restriction fragment length polymorphism maps for the barley genome based on translocation breakpoints. Genetics 154: 397-412

Lagercrantz U (1998) Comparative mapping between *Arabidopsis thaliana* and

Brassica nigra indicates that Brassica genomes have evolved through extensive genome replication accompanied by chromosome fusions and frequent rearrangements. Genetics 150: 1217-1228

Leyser O, Chang C (1996) Chromosome walking. In: Foster GD, Twell D (eds) Plant Gene Isolation. John Wiley, Chichester, USA, pp 248-271

Li Y, Uhm T, Ren C, Wu C, Santos TS, Lee M-K, Yan B, Santos F, Zhang A, Scheuring C, Sanchez A, Millena AC, Nguyen HT, Kou H, Liu D, Zhang H-B (2007) A plant-transformation-competent BIBAC/BAC-based map of rice for functional analysis and genetic engineering of its genomic sequence. Genome 50: 278-288

Liu XQ, Wang L, Chen S, Lin F, Pan QH (2005) Genetic and physical mapping of *Pi36(t)*, a novel rice blast resistance gene located on rice chromosome 8. Mol Genet Genom 274: 394-401

Liu YG, Shirano Y, Fukaki H, Yanai Y, Tasaka M, Tabata S, Shibata D (1999) Complementation of plant mutants with large genomic DNA fragments by a transformation-competent artificial chromosome vector accelerates positional cloning. Proc Natl Acad Sci USA 96: 6535-6540

Liu YG, Nagaki K, Fujita M, Kawaura K, Uozumi M, Ogihara Y (2000) Development of an efficient maintenance and screening system for large-insert genomic DNA libraries of hexaploid wheat in a transformation-competent artificial chromosome (TAC) vector. Plant J 23(5): 687-695

Liu YG, Liu H, Chen L, Qiu W, Zhang Q, Wu H, Yang C, Su J, Wang Z, Tian D, Mei M (2002) Development of new transformation-competent artificial chromosome vectors and rice genomic libraries for efficient gene cloning. Gene 282: 247-255

Lunetta KL, Boehnke M, Lange K, Cox DR (1996) Selected locus and multiple panel models from radiation hybrid mapping. Am J Hum Genet 59: 717-725

Luo MC, Thomas C, You FM, Hsiao J, Ouyang S, Buell CR, Malandro M, McGuire PE, Anderson OD, Dvorak J (2003) High-throughput fingerprinting of bacterial artificial chromosomes using the snapshot labeling kit and sizing of restriction fragments by capillary electrophoresis. Genomics 82(3): 378-389

Ma Z, Weining S, Sharp PJ, Liu C (2000) Non-gridded library: a new approach for BAC (Bacterial Artificial Chromosome) exploitation in hexaploid wheat (*Triticum aestivum*). Nucl Acid Res 28:e106

Macas J, Gualberti G, Nouzová M, Samec P, Lucretti S, Doležel J (1996) Construction of chromosome-specific DNA libraries covering the whole genome of field bean (*Vicia faba* L.). Chrom Res 4: 531-539

Marra MA, Kucaba TA, Dietrich NL, Green ED, Brownstein B, Wilson RK, McDonald KM, Hillier LW, McPherson JD, Waterston RH (1997) High throughput fingerprint analysis of large-insert clones. Genome Res 7(11): 1072-1084

Martin GB, Brommonschenkel SH, Chunwongse J, Frary A, Ganal MW, Spivey R, Wu T, Earle ED, Tanksley SD (1993) Map-based cloning of a protein kinase gene conferring disease resistance in tomato. Science 262: 1432-1436

Moore G, Devos KM, Wang Z, Gale MD (1995) Cereal genome evolution: grasses, line up and form a circle. Curr Biol 5: 737-739

Moullet O, Zhang HB, Lagudah ES (1999) Construction and characterization of a large DNA insert library from the D genome of wheat. Theor Appl Genet 99: 305-313

Nam YW, Penmetsa RV, Endre G, Uribe P, Kim D, Cool DR (1999) Construction of a bacterial artificial chromosome library of *Medicago truncatula* and identification of clones containing ethylene-response gene. Theor Appl Genet 98: 638-646

Nilmalgoda SD, Cloutier S, Walichnowski AZ (2003) Construction and characterization of a bacterial artificial chromosome (BAC) library of hexaploid wheat (*Triticum aestivum* L.) and validation of genome coverage using locus-specific primers. Genome 46: 870-878

Ohmi Y, Sato M, Ohtsuka M (2004) Direct subcloning of target region from BAC insert using restriction enzymes that produce non-identical cohesive ends. Nucl Acids Symp Sr 48(1):141-142

Park JY, Koo DH, Hong CP, Lee SJ, Jeon JW, Lee SH, Yun PY, Park BS, Kim HR, Bang JW, Plaha P, Bancroft I, Lim YP (2005) Physical mapping and microsynteny of *Brassica rapa* ssp. *pekinensis* genome corresponding to a 222 kb gene-rich region of Arabidopsis chromosome 4 and partially duplicated on chromosome 5. Mol Genet Genom 274: 579-588

Peters JL, Cnudde F, Gerats T (2003) Forward genetics and map-based cloning approaches. Trends Plant Sci 8: 484-491

Pourkheirandish M, Wicker T, Stein N, Fujimura T, Komatsuda T (2007) Analysis of the barley chromosome 2 region containing the six-rowed spike gene vrs1 reveals a breakdown of rice-barley micro collinearity by a transposition. Theor Appl Genet 114: 1357-1365

Quarrie S, Lazic-Jancic V, Kovacevic D, Steed A, Pekic S (1999) Bulk segregant analysis with molecular markers and its use for improving drought resistance in maize. J Exp Bot 50: 1299-1306

Rice Chromosome 3 Sequencing Consortium (2005) Sequence, annotation, and analysis of synteny between rice chromosome 3 and diverged grass species. Genome Res 15: 1284-1291

Riera-Lizarazu O, Rines HW, Phillips RL (1996) Cytological and molecular characterization of oat x maize partial hybrids. Theor Appl Genet 93: 123-135

Riera-Lizarazu O, Vales MI, Ananiev EV, Rines HW, Philips RL (2000) Production and characterization of maize chromosome 9 radiation hybrids derived from an oat-maize addition line. Genetics 156: 327-339

Riley J, Butler R, Ogilvie D, Finniear R, Jenner D, Powell S, Anand R, Smith JC, Markham AF (1990) A novel rapid method for the isolation of terminal sequences from yeast artificial chromosome (YAC) clones. Nucl Acid Res 18: 2887-2890

Ross MT, LaBrie S, McPherson J, Stanton VP Jr (1999) Screening large insert libraries by hydridization. In: Boyl A (ed.) Current Protocols in Human Genetics. Wiley, New York, USA, pp 5.6.1-5.6.52

Šafář J, Bartoš J, Janda J, Bellec A, Kubaláková M, Valárik M, Pateyron S, Weiserová J, Tušková R, Číhalíková J, Vrána J, Šimková H, Faivre-Rampant P, Sourdille P, Caboche M, Bernard M, Doležel J, Chalhoub B (2004) Dissecting large and complex genomes: flow sorting and BAC cloning of individual chromosomes from bread wheat. Plant J 39: 960-968

Salse J, Piegu B, Cooke R, Delseny M (2004) New *in silico* insight into the synteny between rice (*Oryza sativa* L.) and maize (*Zea mays* L.) highlights reshuffling and identifies new duplications in the rice genome. Plant J 38: 396-409

Sasaki T, Antonio BA (2009) Plant genomics: Sorghum in sequence. Nature 457: 547-548

Schmidt R, West J, Love K, Lenehan Z, Lister C, Thompson HY, Bouchez D, Dean C (1995) Physical map and organization of *Arabidopsis thaliana* chromosome 4. Science 270: 480-483

Schuler GD, Boguski MS, Stewart EA, Stein LD, Gyapay G Rice K, White RE, Rodriguez-Tomé P, Aggarwal A, Bajorek E, Bentolila S, Birren BB, Butler A, Castle AB, Chiannilkulchai N, Chu A, Clee C, Cowles S, Day PJR, Dibling T, Drouot N, Dunham I, Duprat S, East C, Edwards C, Fan J-B, Fang N, Fizames C, Garrett C, Green L, Hadley D, Harris M, Harrison P, Brady S, Hicks A, Holloway E, Hui L, Hussain S, Louis-Dit-Sully C, Ma J, MacGilvery A, Mader C, Maratukulam A, Matise TC, McKusick KB, Morissette J, Mungall A, Muselet D, Nusbaum HC, Page DC, Peck A, Perkins S, Piercy M, Qin F, Quackenbush J, Ranby S, Reif T, Rozen S, Sanders C, She X, Silva J, Slonim DK, Soderlund C, Sun W-L, Tabar P, Thangarajah T, Vega-Czarny N, Vollrath D, Voyticky S, Wilmer T, Wu X, Adams MD, Auffray C, Walter NAR, Brandon R, Dehejia A, Goodfellow PN, Houlgatte R, Hudson Jr JR, Ide SE, Iorio KR, Lee WY, Seki N, Nagase T, Ishikawa K, Nomura N, Phillips C, Polymeropoulos MH, Sandusky M, Schmitt K, Berry R, Swanson K, Torres R, Venter JC, Sikela JM, Beckmann JS, Weissenbach J, Myers RM, Cox DR, James MR, Bentley D, Deloukas P, Lander ES, Hudson TJ (1996) A gene map of the human genome. Science 274: 540-546

Sears ER (1954) The aneuploids of common wheat. Res Bull Mo Agri Exp Stn 572: 1-58

Shizuya H, Birren B, Kim U-J, Mancino V, Slepak T, Tachiiri Y, Simon M (1992) Cloning and stable maintenance of 300-kilobase fragments of human DNA in *Escherichia coli* using and F-factor-based vector. Proc Natl Acad Sci USA 89: 8794-8797

Šimková H, Šafář J, Suchánková P, Kovářová P, Bartoš J, Kubaláková M, Janda J, Číhalíková J, Mago R, Lelley T, Doležel J (2008) A novel resource for genomics of Triticeae: BAC library specific for the short arm of rye (*Secale cereale* L.) chromosome 1R (1RS). BMC Genom 9: 237

Simons K, Fellers J, Trick H, Zhang Z, Tai Y-S, Gill B, Faris J (2006) Molecular characterization of the major wheat domestication gene *Q*. Genetics 172: 547-555

Smilde WD, Haluskova J, Sasaki T, Graner A (2001) New evidence for the synteny of rice chromosome 1 and barley chromosome 3H from rice expressed sequence tags. Genome 44: 361-367

Song J, Dong F, Jiang J (2000) Construction of a bacterial artificial chromosome (BAC) library for potato molecular cytogenetics research. Genome 43: 199-204

Song W-Y, Wang G-L, Chin L-L, Kim H-S, Pi L-Y, Holsten T, Gardner J, Wang B, Zhai W-X, Zhu L-H, Fauquet C, Ronald P (1995) A receptor kinase-like protein encoded by the rice disease resistance gene, *Xa21*. Science 270: 1804-1806

Sorrells ME, La Rota M, Bermudez-Kandianis CE, Greene RA, Kantety R, Miftahudin, Mahmoud A, Gustafson JP, Qi LL, Echalier B (2003) Comparative DNA sequence analysis of wheat and rice genomes. Genome Res 13: 1818-1827

Srinivasan J, Sinz W, Jesse T, Wiggers-Perebolte L, Jansen K, Buntjer J, van der Meulen M, Sommer RJ (2003) An integrated physical and genetic map of the nematode *Pristionchus pacificus*. Mol Genet Genom 269: 715-722

Stein N, Graner A (2004) Map-based gene isolation in cereal genomes. In: Gupta PK, Varshney RK (eds) Cereal Genomics. Kluwer Academic Publ, Dordrecht, Netherlands, pp 331-360

Stein N, Feuillet C, Wicker T, Schlagenhauf E, Keller B (2000) Subgenome chromosome walking in wheat: A 450-kb physical contig in *Triticum monococcum* L. spans the Lr10 resistance locus in hexaploid wheat (*Triticum aestivum* L.). Proc Natl Acad Sci USA 97: 13436-13441

Stein LD, Mungall C, Shu S, Caudy M, Mangone M, Day A, Nickerson E, Stajich JE, Harris TW, Arva A, Lewis S (2002) The generic genome browser: a building block for a model organism system database. Genome Res 12: 1599-1610

Stein N, Perovic D, Kumlehn J, Pellio B, Stracke S, Streng S, Ordon F, Graner A (2005) The eukaryotic translation initiation factor 4E confers multiallelic recessive Bymovirus resistance in *Hordeum vulgare* (L.). Plant J 42: 912-922

Sternberg N (1990) Bacteriophage P1 cloning system for the isolation, amplification, and recovery of DNA fragments as large as 100 kilobase pairs. Proc Natl Acad Sci USA 87: 103-107

Stewart EA, McKusick KB, Aggarwal A, Bajorek E, Brady S, Chu A, Fang N, Hadley D, Harris M, Hussain S, Lee R, Maratukulam A, O'Connor K, Perkins S, Piercy M, Qin F, Reif T, Sanders C, She X, Sun WL, Tabar P, Voyticky S, Cowles S, Fan JB, Mader C, Quackenbush J, Myers RM, Cox DR (1997) An STS-based radiation hybrid map of the human genome. Genome Res 7: 422-433

Sturtevant AH (1913) The linear arrangement of six sex-linked factors in

Drosophila, as shown by their mode of association. J Exp Zool 14: 43-59

Sutton T, Whitford R, Baumann U, Dong C, Able JA, Langridge P (2003) The *Ph2* pairing homoeologous locus of wheat (*Triticum aestivum*): identification of candidate meiotic genes using a comparative genetics approach. Plant J 36: 443-456

Suwabe K, Tsukazaki H, Iketani H, Hatakeyama K, Kondo M, Fujimura M, Nunome T, Fukuoka H, Hirai M, Matsumoto S (2006) Simple sequence repeat-based comparative genomics between *Brassica rapa* and *Arabidopsis thaliana*: the genetic origin of clubroot resistance. Genetics 173: 309-319

Takahashi Y, Shomura A, Sasaki T, Yano M (2001) Hd6, a rice quantitative trait locus involved in photoperiod sensitivity, encodes the α subunit of protein kinase CK2. Proc Natl Acad Sci USA 98(14): 7922-7927

Tanabe S, Ashikari M, Fujioka S, Taketsuto S, Yoshida S, Yano M, Yoshimura A, Kitano H, Matsuoka M, Fujisawa Y, Kato H, Iwasaki Y (2005) A novel cytochrome P450 is implicated in brassinosteroid biosynthesis via the characterization of a rice dwarf mutant, *dwarf11*, with reduced seed length. Plant Cell 17: 776-790

Tanksley SD, Ganal MW, Prince JP, de Vicente MC, Bonierbale MW, Broun P, Fulton TM, Giovannoni JJ, Grandillo S, Martin GB, Messeguer R, Miller JC, Miller L, Paterson AH, Pineda O, Roder MS, Wing RA, Wu W, Young ND (1992) High density molecular linkage maps of the tomato and potato genomes. Genetics 132(4): 1141-60

Taramino G, Sauer M, Stauffer Jr JL, Multani D, Niu X, Sakai H, Hochholdinger F (2007) The maize (*Zea mays* L.) RTCS gene encodes a LOB domain protein that is a key regulator of embryonic seminal and post-embryonic shoot-borne root initiation. Plant J 50: 649-659

Tarchini R, Biddle P, Wineland R, Tingey S, Rafalski A (2000) The complete sequence of 340 kb of DNA around the rice *Adh1–Adh2* region reveals interrupted colinearity with maize chromosome 4. Plant Cell 12: 381-392

The Arabidopsis Genome Initiative (2000) Analysis of the genome sequence of the flowering plant *Arabidopsis thaliana*. Nature 408: 796-815

Timms L, Jimenez R, Chase M, Lavelle D, McHale L, Kozik A, Lai Z, Heesacker A, Knapp S, Rieseberg L, Michelmore R, Kesseli R (2006) Analyses of synteny between *Arabidopsis thaliana* and species in the Asteraceae reveal a complex network of small syntenic segments and major chromosomal rearrangements. Genetics 173: 2227-2235

Tomkins JP, Yu Y, Miller-Smith H, Frisch DA, Woo SS, Wing RA (1999) A bacterial artificial chromosome library for sugarcane. Theor Appl Genet 99: 419-424

Tomkins JP, Peterson DG, Yang TJ, Main D, Wilkins TA, Paterson AH and Wing RA (2001) Development of genomic resources for cotton (*Gossypium hirsutum* L.): BAC library construction, preliminary STC analysis, and identification of clones associated with fiber development. Mol Breed 8: 255-261

Town CD, Cheung F, Maiti R, Crabtree J, Haas BJ, Wortman JR, Hine EE, Althoff R, Arbogast TS, Tallon LJ, Vigouroux M, Trick M, Bancroft I (2006) Comparative genomics of *Brassica oleracea* and *Arabidopsis thaliana* reveal gene loss, fragmentation, and dispersal after polyploidy. Plant Cell 18: 1348-1359

Umehara Y, Inagaki A, Tanoue H, Yasukochi Y, Nagamura Y, Saji S, Otsuki Y, Fujimura T, Kurata N, Minobe Y (1995). Construction and characterization of a rice YAC library for physical mapping. Mol Breed 1: 79-89

Varshney RK, Grosse I, Hahnel U, Siefken R, Prasad M, Stein N, Langridge P, Altschmied L, Graner A (2006) Genetic mapping and BAC assignment of EST-derived SSR markers shows non-uniform distribution of genes in the barley genome. Theor Appl Genet 113: 239-250

Varshney RK, Langridge P, Graner A (2007a) Application of genomics to molecular breeding of wheat and barley. Adv Genet 58: 121-155

Varshney RK, Mahendar T, Aggarwal RK, Börner A (2007b) Genic molecular markers in plants: Development and applications. In: Varshney RK, Tuberosa R (eds) Genomics-Assisted Crop Improvement. Vol 1: Genomics Approaches and Platforms. Springer, Dordrecht, Netherlands, pp 13-29

Vos P, Hogers R, Bleeker M, Reijans M, van de Lee T, Hornes M, Frijters A, Pot J, Peleman J, Kuiper M, Zabeau M (1995) AFLP: a new technique for DNA fingerprinting. Nucl Acids Res 23: 4407-4414

Vrána J, Kubaláková M, Šimková H, Cíhalíková J, Lysák MA, Doležel J (2000) Flow-sorting of mitotic chromosomes in common wheat (*Triticum aestivum* L.). Genetics 156: 2033-2041

Wang GL, Warren R, Innes G, Osborne B, Baker B, Ronald PC (1996) Construction of an Arabidopsis BAC library and isolation of clones hybridizing with disease-resistance, gene-like sequences. Plant Mol Biol Rep 14: 107-114

Wang Z-X, Yamanouchi U, Katayose Y, Sasaki T, Yano M (1999) Expression of the *Pib* rice-blast-resistance gene family is up-regulated by environmental conditions favouring infection and by chemical signals that trigger secondary plant defenses. Plant Mol Biol 47(5): 653-661

Wardrop J, Snape J, Powell W, Machray GC (2002) Constructing plant radiation hybrid panels. Plant J 31: 223-228

Wardrop J, Fuller J, Powell W Machray GC (2004) Exploiting plant somatic radiation hybrids for physical mapping of expressed sequence tags. Theor Appl Genet 108: 343-348

Warrington JA, Wasmuth JJ (1996) A contiguous high-resolution radiation hybrid map of 44 loci from the distal portion of the long arm of human chromosome 5. Genome Res 6: 628-632

Wei F, Gobelman-Werner K, Morroll SM, Kurth J, Mao L, Wing R, Leister D, Schulze-Lefert P, Wise RP (1999) The *Mla* (powdery mildew) resistance cluster is associated with three NBS-LRR gene families and suppressed recombination within a 240-kb DNA interval on chromosome 5S (1HS) of barley. Genetics 153: 1929-1948

Wicker T, Narechania A, Sabot F, Stein J, Vu GTH, Graner A, Ware D, Stein N (2008) Low-pass shotgun sequencing of the barley genome facilitates rapid identification of genes, conserved non-coding sequences and novel repeats. BMC Genom 9: 518

Woo SS, Jiang JM, Gill BS, Paterson AH, Wing RA (1994) Construction and characterization of a bacterial artificial chromosome library of *Sorghum bicolor*. Nucl Acid Res 22: 4922-4931

Wu C, Sun S, Nimmakayala P, Santos FA, Meksem K, Springman R, Ding K, Lightfoot DA, Zhang HB (2004) A BAC- and BIBAC-based physical map of the soybean genome. Genome Res 14: 319-326

Wu J, Maehara T, Shimokawa T, Yamamoto S, Harada C, Takazaki Y, Ono N, Mukai Y, Koike K, Yazaki J, Fujii F, Shomura A, Ando T, Kono I, Waki K, Yamamoto K, Yano M, Matsumoto T, Sasaki T (2002) A comprehensive rice transcript map containing 6591 expressed sequence tag sites. Plant Cell 14: 525-535

Yahiaoui N, Srichumpa P, Dudler R, Keller B (2003) Genome analysis at different ploidy levels allows cloning of the powdery mildew resistance gene *Pm3b* from hexaploid wheat. Plant J 37: 528-538

Yahiaoui N, Srichumpa P, Dudler R, Keller B (2004). Genome analysis at different ploidy levels allows cloning of the powdery mildew resistance gene *Pm3b* from hexaploid wheat. Plant J 37: 528-538

Yan L, Loukoianov A, Tranquilli G, Helguera M, Fahima T, Dubcovsky J (2003) Positional cloning of the wheat vernalization gene *VRN1*. Proc Natl Acad Sci USA 100: 6263-6268

Yan L, Loukoianov A, Blechl A, Tranquilli G, Ramakrishna W, SanMiguel P, Bennetzen JL, Echenique V, Dubcovsky J (2004) The wheat *VRN2* gene is a flowering repressor down-regulated by vernalization. Science 303(5664): 1640-1644

Yoshimura S, Yamanouchi U, Katayose Y, Toki S, Wang ZX, Kono I, Kuruta N, Yano M, Iwata N, Sasaki T (1998) Expression of Xa1, a bacterial blight resistance gene in rice, is induced by bacterial inoculation. Proc Natl Acad Sci USA 95: 1663-1668

Young ND (2000) Constructing a plant genetic linkage map with DNA markers. In: Phillips RL, Vasil JK (eds) DNA-Based Markers in Plants. Kluwer Academic Publ, Dordrecht, Netherlands, pp 31-47

Yüksel B, Paterson AH (2005) Construction and characterization of a peanut *Hin*dIII BAC library. Theor Appl Genet 111: 630-639

Zhang H-B, Wing RA (1997) Physical mapping of the rice genome with BACs. Plant Mol Biol 35: 115-127

Zoghbi HY, Chinault AC (1994) Generation of YAC contigs by walking. In: Nelson DL, Brownstein BH (eds) YAC Libraries, A User's Guide. W.H. Freeman and Co, New York, USA

3 Comparative and Evolutionary Genomics

Amy L. Lawton-Rauh*, Cynthia R. Climer and Brad L. Rauh

Department of Genetics and Biochemistry, Clemson University, 100 Jordan Hall, Clemson, SC 29634-0318, USA

*Corresponding author: amylr@clemson.edu

1 INTRODUCTION

Science, by its very nature, relies upon the comparison and contrast of observations and measurements and thus is always 'comparative' in approach. The *hierarchical level* of possible comparisons that can be examined using a hypothesis-testing framework is restricted to some extent by imagination but more profoundly by the tools that are available to record and test the very attributes under investigation. The field of genomics is becoming increasingly more amenable to larger scale comparisons that were previously restricted to within-individual or at specific timepoints to examine environmental impact on a trait. For population geneticists studying the processes which mediate long term molecular evolutionary patterns, the development of new comparative approaches facilitiates an unprecedented opportunity to examine finer-scale interactions between genotype frequency and phenotypic variation. For functional genomicists searching for mutations in genes that have the greatest impact on target phenotypes, comparative genomics unveils an amazing tool to test hypotheses generated by single-taxon observations in other species that examine consistency in most probable associations between phenotype and genotype. This list could continue, however the bottom line is that comparative genomics approaches utilizing the evolutionary genetics framework are emerging as powerful tools directly impacting both theoretical knowledge and practical applications. In this chapter, we discuss how to establish the evolutionary context. We then discuss inter-genic and regional genome dynamics as well as intra-genic dynamics of different temporal sensitivities and then describe key points in parameter estimation. In the final section, we discuss how the combination of these approaches is being used to investigate co-expression networks and building towards a larger eco-genomics effort.

2 EVOLUTIONARY CONTEXT

2.1 Establishing the Evolutionary Context

In addition to the implicit value of describing the processes underlying the relationships amongst species and populations, understanding the evolutionary dynamics of genes and genomes provides the framework for examining the consistency of form and function. In the age of 'omics' the definitions of 'form' and 'function' take on additional meanings by including the component of change ascribed to how the genetic code is transcribed and translated into the machinery of life at the molecular level. If we were to only rely upon single examples of the genetic basis for processes responsible for form and function, we lose not only the statistical power, but also the framework of how well patterns are consistent across species. Without an evolutionary context, historical events and processes cannot be inferred and there is no basis for predicting the positive and negative consequences of actions taken today. From a more direct practical perspective, without a comparative evolutionary context, it is impossible to know if tools developed in one species can be successfully utilized in other species. Thus, establishing the evolutionary context of genome dynamics is very important and necessary because it provides the foundation for insightful comparisons that make it possible to transfer knowledge and requisite resources across species as well as cultivars and their relatives. In this section, we discuss the basics of phylogenetic trees, gene trees versus species trees, and haplotype trees because these models provide a significant tool for subsequent comparative evolutionary analyses.

To develop an appropriate framework and interpret results in a biologically relevant and insightful context, it is important to know the difference between two different types of phylogenetic trees: gene trees and species trees. Here, we include a brief overview of the important

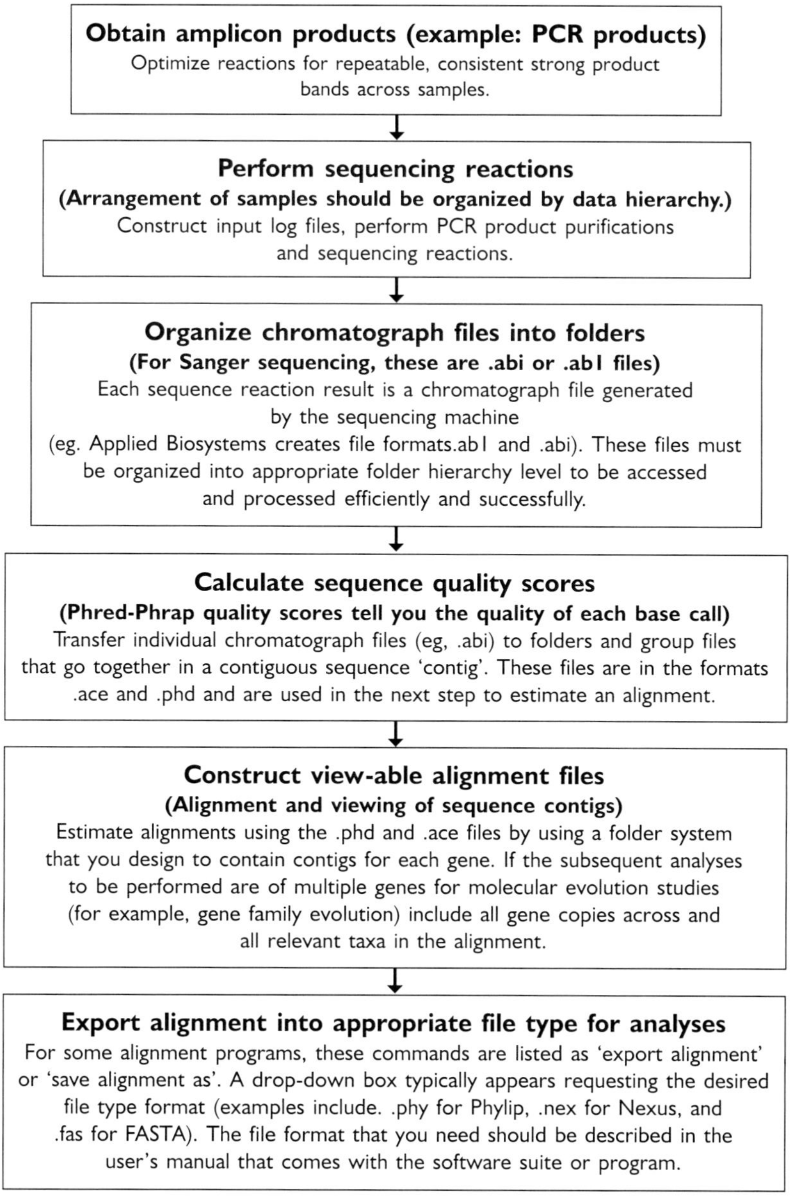

Obtain amplicon products (example: PCR products)
Optimize reactions for repeatable, consistent strong product bands across samples.

↓

Perform sequencing reactions
(Arrangement of samples should be organized by data hierarchy.)
Construct input log files, perform PCR product purifications and sequencing reactions.

↓

Organize chromatograph files into folders
(For Sanger sequencing, these are .abi or .ab1 files)
Each sequence reaction result is a chromatograph file generated by the sequencing machine (eg. Applied Biosystems creates file formats.ab1 and .abi). These files must be organized into appropriate folder hierarchy level to be accessed and processed efficiently and successfully.

↓

Calculate sequence quality scores
(Phred-Phrap quality scores tell you the quality of each base call)
Transfer individual chromatograph files (eg, .abi) to folders and group files that go together in a contiguous sequence 'contig'. These files are in the formats .ace and .phd and are used in the next step to estimate an alignment.

↓

Construct view-able alignment files
(Alignment and viewing of sequence contigs)
Estimate alignments using the .phd and .ace files by using a folder system that you design to contain contigs for each gene. If the subsequent analyses to be performed are of multiple genes for molecular evolution studies (for example, gene family evolution) include all gene copies across and all relevant taxa in the alignment.

↓

Export alignment into appropriate file type for analyses
For some alignment programs, these commands are listed as 'export alignment' or 'save alignment as'. A drop-down box typically appears requesting the desired file type format (examples include. .phy for Phylip, .nex for Nexus, and .fas for FASTA). The file format that you need should be described in the user's manual that comes with the software suite or program.

Fig. 1 Example data flow for Sanger-sequenced PCR products used in a comparative evolutionary genetics experiment. Note that the specific file extensions listed may change due to updates in programs involved.

Data Acquisition and Management Questions

Quality of results from procedure used:
- What are the usable components?
- What results must be omitted?
- What is the usable/non-usable ratio?
- How might this procedure be improved to increase the usable/non-usable data?
- Should I wait until the procedure is better optimized to use these data?
- How do I integrate optimized & sub-optimized data into the same dataset?

Identification of results:
- Where did these raw data originate?
- How do results relate to one another?
- How do I partition results into appropriate group memberships for further analyses?

Assessment of dataset status: group membership results status check
- How do I identify 'holes' in the dataset?
- How do I label dataset groups as 'incomplete' or 'complete'?
- How do I compare incomplete datasets to assess strategy (related to 'Quality' above)
- How do I fill in gaps with subsequent data inputs?

Fig. 2 Questions to address when managing large numbers of raw data output files typically used in comparative and evolutionary genomics projects.

differences between these types of phylogenetic trees and also provide an overview of an alternative approach primarily used at the population level: haplotype trees. For a more thorough discussion of the differences between gene trees and species trees, refer to (Rosenberg 2002; Oliver 2008; Rosenberg and Tao 2008). See Fig. 1 for an outline of data acquisition and management flow and Fig. 2 for a series of questions that researchers must consider when setting up the logistics for a comparative evolutionary genomics experiment. For more technical details of how to take raw data to estimate best fit phylogenetic trees, refer to general phylogeny textbooks and especially up-to-date web tutorials such as the Molecular Evolution Workshop (*www.workshop.molecularevolution.org*). To learn more about generating alignments and finding homologous gene sequences amongst species, you will find

Cannon and Young (2003), Hampson et al. (2003), Hampson et al. (2005), Leebens-Mack et al. (2006), and Rannala and Yang (2008) very helpful.

2.2 Gene Trees and Species Trees

The most common reference to phylogenetic trees is the species tree. A species tree represents the best fit model of divergence amongst taxonomic divisions defined as operational taxonomic units (OTUs). Most often, OTUs are indicated as species and a phylogenetic tree is used to describe the evolutionary relationships and overall distances. Species trees are very helpful because they provide a framework for discerning trends in traits, genetic distances, and when overlaid with other characters such as phenotype values species trees can be particularly useful for examining most likely longer term evolutionary patterns and processes involved in speciation or for specific trait adaptations.

A gene tree is the best fit model of the relationships among homologous copies of a gene encoded in the genome. To estimate the best fit model of relationships, gene trees utilize a nucleotide substitution model to estimate genetic distances by sorting sequence polymorphisms and parsing out ancestral versus derived mutations (see Section 5.5). If a gene has not had a strong signature of selection and is sampled from species that have been separated for a long period of time, the gene tree will closely resemble the species tree. However, gene trees for different genes sampled from a set of species will be incongruent due to different rates, patterns, and modes of natural selection as well as the relative impact of genetic drift on mutation and allele frequencies for each gene.

Examination of multiple gene trees representing loci around the genome can be used to help detect demographic effects on a genome. For example, gene trees constructed from samples of a recently expanded population will have a 'younger' most recent common ancestor than the gene tree from a population whose size has remained steady. Variance in single gene tree estimates makes demographic events difficult to detect without a large number of unlinked loci because of low statistical power in finding best-fit models that differ from patterns expected under models of natural selection. Because of this low statistical power, it is necessary to examine multiple, unlinked (and therefore presumably independent) loci representing multiple points around the genome. One study has found that an accurate estimate of the population mutation parameter (θ) requires at least 25 independent loci (Carling and Brumfield 2007).

2.3 Haplotype Trees

Phylogenetic trees are excellent models demonstrating the relationships among taxa that best fit the data for deeper divergence times. For more recently-diverged taxa, it is generally more informative to estimate best fit networks amongst haplotypes. These networks are known as 'haplotype networks' or 'haplotype trees'. Haplotype trees are statistical parsimony-based models representing the fewest number of changes

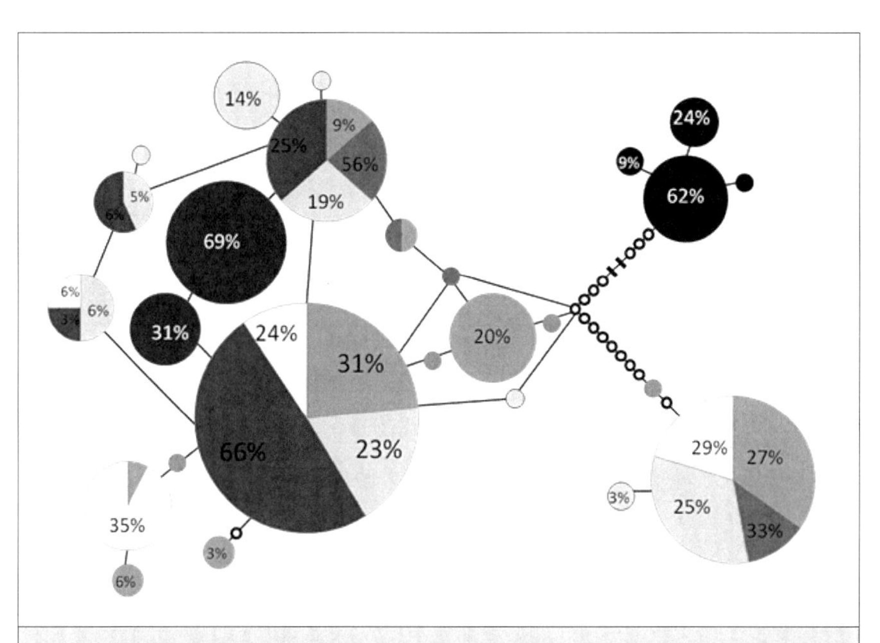

Fig. 3 Maximum parsimony haplotype network constructed in TCS version 1.21 with a fixed connection limit of 50 steps. Each circle represents a distinct haplotype with size correlated to the number of copies represented in the total dataset. Each haplotype circle is divided into segments based on what proportion of that haplotype was found in each of 6 populations (denoted in shades of gray). The number included in each circle is the percentage of that particular haplotype in each population's total number of haplotypes. The smaller circles represent haplotypes that are present in one copy in the dataset and the open circles represent missing haplotypes (or changes between haplotypes that are not represented in the dataset). Each connection is a single base pair substitution between the two haplotypes. The double lines in the connection of the top right haplotypes to the network represent over 20 substitutions between these two groups of haplotypes. The haplotypes on the top right (in black) are from a closely related species. The network is rooted because of the inclusion of this closely related species, allowing for the distinction between ancestral (old) and derived (new) haplotypes.

amongst the haplotypes included in a dataset. An example haplotype network is provided in Fig. 3. Generally, these best fit models use a percentage cutoff criterion for the number of inferred missing mutational steps that would be required to connect haplotypes to one another. The number of missing steps and the number of steps between extant haplotypes indicates the progression of mutations that accumulated since the divergence of extant haplotypes from their common ancestor, presumably identical by descent. The most widely-utilized software program for finding the best fit haplotype tree for a dataset is TCS v.1.21 (Clement et al. 2000). Recently, additional applications have incorporated haplotype trees for genotype-phenotype associations including the program TreeScan ver. 1.0 (Posada et al. 2005).

3 INTER-GENIC AND REGIONAL DYNAMICS

3.1 Whole-Genome Duplication and Comparative Colinearity

Models like phylogenetic trees have been used to characterize the relationships amongst species and often reveal genome dynamics at both the small and large scales. For plants in particular, evolutionary relationships can be examined using the patterns of large-scale and small-scale duplication events. In this section, we discuss several duplication and divergence processes that occur at these two scales and lead to within-gene, within-region, amongst small regions, and across-genome duplications. Following a discussion of duplication processes, we discuss gene family evolution by gene duplication and the alternative fates of gene duplication.

On a larger spatial scale, entire genome duplication has been a recurring event throughout the diversification of plants where some estimates suggest that up to 70% of all land plants have genomes shaped by polyploidization (Lawton-Rauh 2003; Doyle et al. 2004; Mable 2004; Cui et al. 2006; Leebens-Mack et al. 2006). Although recent polyploidization has been detected in specific recently-derived lineages such as within the Brassicaceae, polyploidization apparently also had a strong impact in shaping several ancestral genomes (Byrne and Blanc 2006). Gene duplication due to polyploidization often results in divergence in evolutionary trajectories between gene copies (Adams et al. 2000; Lawton-Rauh 2003; Lawton-Rauh et al. 2003; Blanc and Wolfe 2004; Adams and Wendel 2005; Moore et al. 2005; Moore and Purugganan 2005). Some research indicates that polyploid-derived gene duplicates (i.e., homeologs) are impacted differently with respect to progenitor genome origin (Cronn et al. 1999; Cronn et al. 2002; Wang et al. 2005), while other studies provide examples suggesting that progenitor genome origin does not impact a consistent evolutionary trajectory (Lawton-Rauh 2003; Lawton-Rauh et al. 2003). Differences between gene copies arising from polyploidization are not only due to alternative mutation accumulation. Several studies suggest rapid activation of transposable element activity (Li et al. 2005; Hazzouri et al. 2008) and epigenetic changes impacting gene expression (Chen 2007).

Smaller scale rearrangements leading to inversion polymorphisms can occur following hybridization between species as documented in *Helianthus* (Baack and Rieseberg 2007; Hoffmann and Rieseberg 2008). Several tools have been developed specifically to scan selected genome sequences to uncover segmental duplications and synteny amongst genomes such as DAGchainer (Haas et al. 2004). Other tools have been published as resources to search for paralogous chromosomal regions resulting from recent and historical genome duplication events such as the *Arabidopsis* paralogon search website *http://wolfe.gen.tcd.ie/athal/dup* (Blanc et al. 2003; Blanc and Wolfe 2004). Tools like these have been used to detect and examine paralogous protein family members in rice (Lin et al. 2008) and undoubtedly inspired the extensive plant comparative genomics webtool PlantGDB (*http://www.plantgdb.org/*).

3.2 Duplication: Gene Families and Alternative Fates

As discussed above, gene duplications can arise from local and large-scale events. Following duplication, copies of the same gene accumulate independent mutations. The rate and pattern of mutation accumulation can differ between gene copies and this topic has long been the subject of discussion and research (Ohta 1992; Force et al. 1999; Charlesworth et al. 2001; Hughes 2002; Zhang 2003; Doyle et al. 2008; Demuth and Hahn 2009). In this section, we highlight the impact of gene duplication on the origin and proliferation of gene family members and then outline the alternative fates of duplicated genes.

The origin, proliferation, and occasional loss of duplicated genes are responsible for the presence of gene families. A gene family is a group of genes that have a common ancestor and are likewise derived from an ancestral function. The persistence of this original function is not a necessary hallmark of gene families, but rather the fact that gene family members have recognizable sequence similarities derived from identity by descent. Gene duplication of gene family members sometimes results in a relaxation of natural selection constraint, as detected in floral regulatory genes in the Lamiales plant order (Aagaard et al. 2006). Several examples of gene families include the pathogen resistance *R* gene family, the MYB-factor gene family, and the extensive MADS-box gene family of transcription factors that often regulate pivotal developmental switches (Alvarez-Buylla et al. 2000; Lawton-Rauh 2003; Demuth and Hahn 2009). Many gene family members are the result of polyploidization, while others appear to arise through more local events such as tandem duplication, the most famous tandem duplication-derived family is the animal *HOX* gene cluster. In plants, evidence in insect tolerance gene clusters suggests that functional divergence can occur quickly following tandem duplication in *Arabidopsis* (Clauss and Mitchell-Olds 2004) and other species (Leister 2004).

The gene birth and death process involved in the origin, proliferation and occasional loss of gene family members center around the probabilities of alternative fates for duplicate gene copies (Force et al. 1999; Lawton-Rauh 2003; Li et al. 2005; Moore and Purugganan 2005; Cui et al. 2006; Leebens-Mack et al. 2006; Semon and Wolfe 2007; Kalisz and Kramer 2008). The probabilities of alternative fates for specific gene duplicates depend upon the rate and pattern of mutations as they accumulate in different functional gene regions (Force et al. 1999). The possible alternatives include subfunctionalization, neofunctionalization, and nonfunctionalization. Subfunctionalization occurs when duplicate gene copies accumulate compensatory mutations in different gene regions, resulting in the requirement of gene products from both gene copies to perform the original function. Neofunctionalization is the result of mutations that lead to the recruitment of one of the gene copies into a separate functional pathway. When mutations accumulate in gene regions that make a gene copy completely lose functional gene products, nonfunctionalization occurs and the impacted gene copy becomes a pseudogene.

4 INTRA-GENIC DYNAMICS

4.1 Temporal Sensitivities of Different Genetic Markers

The genome is comprised of regions that range in size and have different temporal sensitivities due to alternative mutation rates and structural constraints impacting the origin and persistence of each marker type. Examples of such regions include fast-evolving microsatellites, functional genes, organelle genomes, and nuclear genomes. As described in more details below, the alternative origins and sensitivities to mutations among these types of genetic markers can be very useful for comparative evolutionary analyses. Because these regions have different origins and retain mutations at different rates over time, genetic markers can be selected to find the best fit evolutionary model at alternative timescales. Each genetic marker can be assessed for the most applicable nucleotide substitution model unique to the origin and structural constraints in these markers, thereby affording a snapshot of the dynamic evolutionary history of the genome at different temporal depths. There are many available molecular markers in plant systems. In this section, we discuss a few extensively used markers that are most applicable to comparative genomics: microsatellites or simple sequence repeats (SSRs), candidate genes, single copy nuclear loci (SCNL), conserved orthologous sets (COS), and HapSTRs/SNPSTRs. We then discuss how genomes from different sources provide alternative utilities. Information about other genetic markers can be found in reviews (Schlotterer 2004; Agarwal et al. 2008; Doveri et al. 2008) as well as a very helpful 'survival guide' to population genetics software that explores related issues (Excoffier and Heckel 2006) and several textbooks (Avise 2004; Lowe et al. 2004).

4.2 Microsatellites

Microsatellites or simple sequence repeats (SSRs) are codominant markers that consist of repeats of 1-6 base pair nucleotide motifs. They are abundant throughout eukaryotic genomes (Schlotterer 2000) and are characterized by a high level of polymorphism. These markers are easily amplified using PCR and scored using gel electrophoresis (band size change due to change in the number of repeats) or by direct DNA sequencing. Finding SSRs in the genome can be costly and difficult without sequence information. When sequence information is available, databases such as SciRoKo (Kofler et al. 2007) or Sputnik (Morgante et al. 2002) can be utilized to search for informative SSRs in the sequence. Some researchers have used expressed sequence tag (EST) sequences to search for microsatellites that can often be used across species within the same genus due to high sequence conservation in EST regions (Ellis and Burke 2007). Recently, microsatellite markers have been independently developed for many plant species (Agarwal et al. 2008).

SSRs often arise as a result of DNA replication error (Schlotterer and Tautz 1992) but may also be caused by retrotransposable element activities (Nadir et al. 1996). Once the repeat has been established, mutation in the size of the repeat is due to DNA slippage during replication. These mutations can either result in larger SSRs or smaller SSRs than the parental microsatellite. This mutation rate is very high (10^{-2}-10^{-6} mutations per locus per meiosis; Zhang and Hewitt 2003), making SSRs extremely useful in studying differences in recently-diverged species or in intraspecies comparisons when there is not enough mutational data at other loci. This high mutation rate, along with the simplified step-wise mutation model used to analyze microsatellites, more appropriate for recently-derived taxa and makes these markers uninformative for comparing species that have been diverged for a long period of time due to mutation saturation and short persistence times of microsatellites.

Care needs to be taken when analyzing SSRs for a variety of reasons. The first is that the mutation rate of these markers can be highly variable within and between SSR loci. There is also evidence of inconsistencies in SSR mutation rates over time due to impact of microsatellite length on both mutation rate and direction (Xu et al. 2000). Another issue that must be considered is that microsatellites that are identical by size (state) are not always identical by descent, a situation referred to as homoplasy. This causes an underestimation of genetic distance between taxa because mutation events are not 'recorded' in the nucleotide sequence. Neutrality of SSR loci has also been called into question with evidence of functional importance for or conservation of some repeats (Rothenburg et al. 2001). Like other markers, microsatellites should be tested for Hardy-Weinberg Equilibrium, independence amongst markers (i.e., linkage disequilibrium) to verify the effective number of genomic reference points represented, and selective neutrality before analyses are performed (Selkoe and Toonen 2006).

4.3 Candidate Genes

Many comparative studies start with a gene (or targeted gene) that is a candidate or functional gene of interest (Lawton-Rauh et al. 1999; Hedrick and Verrelli 2006; Parmakelis et al. 2008). The goal of these studies is to see if these potential target genes fit models of non-neutral evolution (indicating evidence of a type of natural selection) either on the population level or interspecies level. These candidate or functionally relevant genes are typically previously characterized as having some known or putative function such as pesticide resistance or flower development. Patterns consistent with natural selection increasing or decreasing frequencies of particular alleles of the gene may help unravel the role of that gene in the population, species, or higher taxonomic levels. The level examined is commiserate with the information content available and this is highly dependent upon the mutation model and divergence times. Calibration of the hypothesis being tested necessarily, therefore, requires the selection of taxa and genes that have a very low likelihood of having mutational saturation, or multiple mutational hits at the same site over time, because such saturation decreases the estimated number of events separating the lineages examined in a study.

In early studies, allozymes (alternate alleles of a protein) were distinguished by charge through differential migration rates on an electrophoresis gel. Nucleotide substitutions that resulted in an amino acid change (non-synonymous mutations) were the only changes detected by this process and this method therefore can seriously underestimate the number of actual mutation events thus genetic distance estimates can be inaccurate. Today, target genes usually are first PCR amplified and then nucleotide sequenced, either directly from PCR products or from subclones. Nucleotide sequencing allows all extant nucleotide substitutions (synonymous and non-synonymous) to be detected, generating a larger dataset with higher resolution than previous methods particularly when a best fit nucleotide substitution model can be utilized. These sequences are processed using the quality score indices Phred and Phrap (Ewing et al. 1998; Ewing and Green 1998), aligned using programs such as BioLign (Tom Hall, North Carolina State University, Raleigh, North Carolina, United States) and then analyzed for level of selection using multiple programs (see the section Detecting Diversity). For excellent sources of the most recent DNA analysis software pertaining to everything from examination of raw sequence data to performing statistical tests to finding the best fit models at the molecular evolution and population genetics levels, refer to the following internet resources: Biology Software Net (*http://en.bio-soft.net/dna.html*) and the exhaustive 'Felsenstein Phylogeny Programs' website (*http://evolution.genetics.washington.edu/phylip/software.html*).

One important consequence to consider in these studies is that the magnitude and direction of natural selection varies between species and even between different populations of various sizes because of differences in allele frequency variance amongst generations. This requires testing multiple loci to estimate so-called empirical null distributions and then testing to see if the target gene is in a statistically unique coordinate in the distribution. In addition to gene-specific processes, other population dynamics may lead to patterns consistent with natural selection model expectations. Genetic hitchhiking is a phenomenon in which regions surrounding a locus under strong selection have diversity and divergence patterns consistent with natural selection due solely to their close proximity and insufficient time to break up non-random associations of alleles amongst loci via effective recombination. One recent example of this was found in regions around a gene, *Rht-B1*, involved in height of wheat (Raquin et al. 2008).

4.4 Single Copy Nuclear Loci

Single copy nuclear loci (SCNL) are regions of DNA that exist in low copy number in a genome. In plants, it is critical to establish a useful set of SCNLs for a species group due to heterozygosity in outcrossing species and the high frequency of genome duplications rampant in plant evolutionary history (from recent and ancient events). The duplication of alleles and gene copies via heterozygosity and gene/genome duplication make the detection of allele copies versus gene copies nearly impossible and also present the possibility of PCR recombination errors. Each SCN-locus should contain long exons in order to be useful in analysis because of the decreased likelihood of complicated indel polymorphisms. The use of multiple unlinked loci is recommended in order to capture the variation across the genome. The use of these markers has proven useful in phylogenetic analyses as well as population genetics studies (Li et al. 2007; Song and Mitchell-Olds 2007). Other studies have combined SCNLs with chloroplast markers to further resolve phylogenies (Cronn et al. 2002) as well as to test putative progenitors of allopolyploidization events (Slotte et al. 2006; Straub et al. 2006). SCNL can also set a standard for the level of variation in genes across a genome in empirical null distributions, providing a framework to test for unique patterns of polymorphism and divergence in target genes relative to the rest of the genome.

The development of an SCNL set can be rather costly and time consuming. This process becomes more cost effective with the help of a sequenced genome for a close relative of the system of interest. Once the markers are designed, data collection can be relatively simple. PCR is utilized to amplify the region of interest. The products can either be subcloned to separate haplotypes or directly sequenced and then haplotype-phased using haplotype partitioning programs such as PHASE (Stephens et al. 2001; Stephens and Donnelly 2003). Direct sequencing greatly reduces cost and provides a direct means of assessing PCR-induced recombination errors but makes haplotype determination a bottleneck in the data preparation pipeline. When working with polyploids, cloning becomes paramount due to more than two base calls appearing in some regions of the sequence. To reduce the number of necessary subcloning reactions, some researchers utilize a PCR based labeling system (Bierne et al. 2007). This group found that they could mix PCR products from multiple individuals into one cloning reaction and separate based on their 5′ labels after sequencing. To bypass the costly development phase, one group has found that the *WRKY* gene family, which has a highly conserved DNA binding domain and intron size polymorphisms, can be used to create SCNL sets for phylogenetic analyses (Borrone et al. 2007). This approach becomes inapplicable if the goal is to test variation across the genome. Because these markers are all in highly constrained regions (DNA binding domains), they may not accumulate mutations in patterns similar to the 'average' genome region. If mutations do not accumulate in patterns similar to 'most' genes, then these highly constrained regions do not represent the baseline evolutionary dynamics impacting the entire genome. As genomes continue to be sequenced, the development of SCNL will become more accessible for many plant systems.

4.5 Conserved Orthologous Sets

Conserved orthologous set (COS) markers are a group of gene markers that are designed to be utilized across species made possible because of the presence of regions conserved enough to design successful PCR experiments amongst more distantly related species. Once an initial set of COS markers are developed, they can be screened for informativeness by assessing diversity and divergence estimates afforded by the inclusion of intron regions. Development of these markers for a system requires a large amount of sequence information from several species within a family to design the conserved primer sets, but once created COS markers are very useful for comparative evolutionary analyses between most species within a family. Like SCNL, many of these markers are designed from EST information and are predicted to contain an intron. They also should occur in low copy number. One important difference is that these markers are meant to amplify across a taxonomic family. To design primer that amplify markers consistently across taxa, sequence information must be available from species that are diverged enough to uncover conserved regions yet not so diverged that multiple mutations could have accumulated at the same site (mutation saturation). Mutation saturation creates significant problems in divergence estimates because it leads to an underestimation of the number of actual mutation events. An example COS marker set used ESTs from lettuce and sunflower combined with genomic sequences from *Arabidopsis* to create COS markers for the Asteraceae plant family (Chapman et al. 2007).

4.6 Compound Markers: HapSTRs and SNPSTRs

HapSTRs and SNPSTRs are compound markers that include a microsatellite (simple tandem repeat, 'STR', also referred to as SSRs) linked to a gene (haplotype) or one or more single nucleotide polymorphism (SNP) marker(s). The strength in these markers lies in the use of combined mutation models that have different temporal sensitivities. STRs mutate quickly and are therefore highly polymorphic, but homoplasy is common and hard to detect. Mutations in gene regions are easier to model, but also occur less frequently. By combining the two marker types in one linked marker, the evolutionary history of mutational accumulation and retention in both types of markers results in the detection of older processes shaping variation in the gene region followed by more recent processes that result in patterns in the fast evolving STRs. These compound markers are most useful in adaptive radiations or in species that have quickly and recently diverged because of the ephemeral persistence of origin and loss of signal in STRs/SSRs. The design of these markers can either start from known microsatellites that have shown enough diversity to be informative and then PCR amplification of the flanking region or by starting with gene sequences and searching for microsatellites in the introns. The end product is a primer set that flanks the microsatellite and the surrounding region that contains one or more polymorphism(s) in the population. The substitution models unique to both types of markers can then be incorporated into subsequent molecular evolution and population genetics analyses. One group has designed these markers for the adaptively radiated African Cichlids (Hey et al. 2004; Won et al. 2005).

4.7 Different Genomes Have Different Utilities

The nuclear genome differs from both mitochondrial and chloroplast genomes in several ways that impact how informative gene sequences are for comparative evolutionary analyses. In this section, we highlight the basis for these differences and outline how to employ these differences in evolutionary genomics analyses.

The mitochondrial genome (mtDNA) is circular, single stranded, and resides in the eukaryotic mitochondrial organelles. The inheritance pattern of mtDNA is usually maternal but can be paternally inherited in some cases (paternal leakage). The higher rate of nucleotide substitutions in the mitochondrial genome in animals makes it very useful in evolutionary and population genetic studies. In plants, however, the rate of nucleotide substitution is much slower and outpaced by plant nuclear and chloroplast genomes. Plant mtDNA harbors increased recombination rates, leading to large variation in genome sizes and genetic organization even within a single individual (Kubo and Newton 2008). This variation in size combined with low nucleotide sequence variation in mitochondrial genomes makes mtDNA markers less informative for evolutionary and population genetic studies in plants. Evidence also suggests recent and abundant transfers of mitochondrial genes to the nucleus in plants (Adams et al. 2000). Additionally, maternal inheritance and haploid state results in a four-fold reduction in effective population size when compared to nuclear markers. Overall, recent studies of plant mtDNA have found it difficult to establish a null model (Barr et al. 2007) and have at best been only able to use sequence comparison studies to test hypotheses regarding the expected patterns of polymorphism and the mutation rate of mitochondrially-encoded genes.

Chloroplasts are unique to plants and contain circular DNA. Like plant mtDNA, the chloroplast genome (cpDNA) is characterized by a low nucleotide substitution rate. In contrast to the plant mitochondrial genome, the chloroplast genome appears to have a very low rate of recombination and gene rearrangement. The chloroplast genome also harbors a four-fold reduction in effective population size compared to the nuclear genome. While mitochondrial genes generally have low resolution, chloroplast markers can be used in addition to other nuclear markers to further resolve events and relationships at higher taxonomic levels (Cronn et al. 2002) because of their reduced rate of nucleotide substitution (thus affording the detection of patterns resulting from processes that occurred further back in time than potentially mutationally-saturated nuclear genes). For example, the use of these markers recently grouped Hydatellacea in the angiosperm clade rather than with the monocots as previously thought (Saarela et al. 2007).

Another compound marker type similar in design to the HapSTRs mentioned above is organellar microsatellite markers. Microsatellite markers have more recently been developed using plant organelle DNA as a means to capture intermediate timepoints. Chloroplast SSRs have been particularly useful (Provan et al. 2001). The increased substitution rates associated with microsatellites make the use of chloroplast genomes more useful in recent divergence events than regions lacking microsatellites. These markers have been used to uncover population structure for many tree species (Nasri et al. 2008; Pakkad et al. 2008) and one green algae species (Provan et al. 2005). They have also been used in estimating phylogenies (Bucci et al. 2007; Magri et al. 2007). Transferability of a particular marker into distant taxa is limited for these markers, but it may be possible to develop more conserved PCR primer pairs (Weising and Gardner 1999).

Nuclear genomes reside in the nucleus of all eukaryotes and makes up most of the heritable genetic information in both plants and animals. Nuclear DNA is inherited biparently with one copy of each chromosome coming from each parent. For plants, nuclear DNA accumulates substitutions faster than both mitochondrial and chloroplast genomes. However, due to differences in inheritance patterns and the varying rates of substitutions, many studies have found it useful to include both nuclear and organelle markers.

An important factor to keep in mind when analyzing genes (whether they are nuclear or organellar in origin) is to understand that the level of polymorphism not only varies across the genome, but also within a single gene. Inter-genic, coding, introns and flanking regions all have different levels of selection and therefore different levels of variation. The substitution rate in synonymous sites is higher in exons than in introns which have a higher rate than flanking regions and this can be used as a powerful tool to detect signatures of natural selection that are unique to specific gene regions, particularly important in studies of target genes mentioned above (see the section Detecting Diversity for more information).

5 PARAMETER ESTIMATION: MOLECULAR EVOLUTION AND DEMOGRAPHY

5.1 Estimating Diversity and Divergence Indices for Comparative Analyses

Comparative genomics is fundamentally driven by the estimation of diversity and divergence parameters amongst genes and gene categories. There are many ways to describe patterns of polymorphisms comprising alleles and haplotypes. Thus most studies examine sequence datasets for patterns across the expected associations amongst polymorphisms in a sequence that best fit particular models of neutral-equilibrium or natural selection model expectations. Whether it is on the population, species, taxa, or even kingdom level, nucleotide polymorphism can be defined and then used to estimate whether the data fits an expected pattern better than other models and can be used to estimate the divergence patterns between taxa and/or examine target genes for unique evolutionary histories. The number and location of substitutions can provide information about the level of selection on a particular gene and the expected 'null' pattern serves as the basis for determining if data fit natural selection models better than what is expected under neutral-equilibrium expectations. In this section, we discuss the neutral theory as the operational statistical null hypothesis, the estimation of evolutionary genomics parameters, and the necessity to examine demography as an impact on sequence variation.

5.2 Neutral and Nearly Neutral Theory

According to the neutral theory of population genetics (Kimura 1983), the majority of extant polymorphism observed within and among species is selectively neutral. This theory is based on the idea that any mutations that result in a negative or deleterious impact on fitness will be removed from the population. On the other hand, mutations that result in a positive impact on fitness are exceedingly rare. While neutral mutations (mutations that have no impact on fitness) would be as common as deleterious mutations, they would persist in the population and therefore become detectable. According to this theory, the fate of a neutral mutation is controlled mostly by genetic drift (the random change in allele frequency over time).

The nearly neutral theory (Ohta 1973) includes mutations that have only a slight affect on fitness. This effect can be negative or positive, but in both cases the fate of these mutations is determined by effective population size. In small populations, drift leads to a higher variance in allele frequency changes amongst generations thus will overpower selection in nearly neutral mutations making them essentially neutral. In large populations, however, the variance of allele frequency change amongst generations is significantly reduced, thus selection can overcome drift and remove slightly deleterious mutations and retain slightly positive mutations more effectively.

The concept that detectable polymorphisms are primarily neutral (or nearly neutral) continues to be a heated debate. Another theory suggests that most mutations have a direct impact on population fitness and emphasizes that natural selection drives allele frequency changes more efficiently than what the neutral and nearly-neutral theories indicate. This selectionist theory suggests that mutations are shaped more directly by natural selection. Genetic drift is a secondary component, and has little impact on allele frequency maintenance in populations. While this argument persists to this day (Chamary et al. 2006; Nei 2007), many on both sides operationally use the neutral theory as a null model to test for evidence of natural selection. Although the nearly neutral theory can account for some of the observed stochasticity that can't be accounted for when using neutral theory expectations as a null model, the additional parameters involved to fit data to nearly neutral theory expectations makes the implementation of this theory for hypothesis testing significantly less practical at this point. If the pattern of polymorphism is significantly different than what would be expected from the neutral-equilibrium model, then selection or change in demography can be inferred (see below).

5.3 Polymorphism

Polymorphism is an estimate of variation and has been described and studied on many levels. The first descriptions were phenotypic polymorphisms such as color, height, or seed production. These visual polymorphisms are often caused by the interaction of many genes and not just a change in one protein. They are also more strongly affected by natural selection. For example, the plant that produces more seeds will be more likely to pass its alleles on to the next generation and therefore alleles that increase seed production will be maintained or accumulate more quickly in a population over generations. With the discovery that allozymes (different variations of a protein) could be separated based on charge and scored, polymorphism was then measured without this strong dependence upon natural selection. However, this quantification of polymorphism was an underestimate of mutation events that occurred in the genome because only mutations that changed a protein's charge could be scored. With advances in molecular biology (PCR, restriction endonucleases, and DNA sequencing) the amount of detectable polymorphism available in a single population is seemingly unlimited. The focus of this section is polymorphism on the gene level, but genomic rearrangements, duplications, and transposable elements can also generate genome-wide polymorphisms (see the following references and Section 3 above on Global Dynamics (Haas et al. 2004; Kellogg and Bennetzen 2004; Koszul et al. 2004; Leister 2004; Stracke et al. 2004; Adams and Wendel 2005; Wang et al. 2005; Cui et al. 2006; Wang et al.

2006; Ziolkowski et al. 2006; Hahn 2007; Liu and Xue 2007; Roulin et al. 2008).

The most common genetic polymorphism is single nucleotide polymorphism (SNP). SNPs are scattered across all genomes and are defined as a single base pair position that has more than one allele or nucleotide represented in a population with the least abundant allele being present in 1% or more of the individuals in the population. SNPs can be detected using restriction enzymes if they occur in a recognition site of a particular enzyme. However, they are most commonly scored from sequence data, which can detect a much higher number of these polymorphisms. Genes from many individuals in a population are sequenced, then aligned and base calls that are polymorphic can be easily visualized from that data.

SNPs are exceedingly abundant in plant genomes with estimates of 15 SNPs per 1,000 base pairs of DNA in rice (Yu et al. 2005). The level of polymorphism, however, is varied across the genome and even within a single gene. These differences in levels of polymorphism are due to differences in constraints across gene regions. Introns and exons have markedly different levels of constraint with exons being more highly conserved than introns simply because a mutation in the former can often result in a change in the gene product (which is often deleterious and quickly removed from a population). There are also some exons that are more highly conserved than others within the same gene; these are often part of functionally or structurally canalized domains of the protein (such as DNA binding motifs of transcription factors, particularly those involved in development). Changes in these regions often have deleterious impacts on fitness versus changes in other regions of the protein and therefore have a fewer number of extant polymorphisms present. As a tool, the level of polymorphism present in a gene region can reflect the level of selection on that particular region or protein.

5.4 Natural Selection

Natural selection is the non-neutral removal or retention of specific alleles in the genome resulting in changes in allele frequencies over time. Such allele frequency changes can eventually lead to complete loss or fixation of mutations and alleles, leading to longer term patterns of molecular evolution. The concept of selection dates back to Darwin's theory of evolution, which states that the most fit individuals will remain in the population while the least fit will be removed. We will focus on selection at the gene level rather than the organismal level, however the same concept applies; where alleles that are more fit for the environment will remain in the population and those that are less fit are lost. Natural selection is a dynamic process in which slight changes in the environment, changes at other genes, or the frequency of the allele of interest in the population can all impact which allele will be the most fit at any given time. Natural selection can act in several directions - removing a particular allele (purifying selection), increasing the frequency of a particular allele (positive selection) or retaining a restricted set of alleles (balancing selection). In small populations, selection must be strong enough to overcome genetic drift, while in large populations even weak selective pressures can overpower random drift.

Positive selection is often referred to as Darwinian selection or directional selection and acts by increasing the frequency of an advantageous allele in the population. There are many examples of positive selection in both plants and animals with the most well studied cases being genes involved in host defense or immune response and reproduction. In fact, positive selection may be over-represented in the literature because target candidate genes are often the genes involved in processes that are strongly influenced by positive selection. Determining the exact locus shaped by positive selection can sometimes be difficult because the same molecular signature can be detected in regions around the gene (genetic hitchhiking). Furthermore, the ability to detect positive selection also depends greatly on time since the selective sweep might have taken place (Wright and Gaut 2005) due to the accumulation of other mutations and the breaking up of associations with markers by effective recombination. These caveats, along with the difficulty of distinguishing between selection and demographic effects on the genome, can hamper detection of positive selection. This is especially relevant when without some a priori knowledge of what might be under selection.

Balancing selection acts to maintain diversity at an individual locus or by keeping alleles from dropping to low frequency in a population. In plants, the self-incompatibility locus is an example of a locus with extant polymorphisms strongly impacted by balancing selection caused by a frequency-dependent mechanism. This locus maintains outcrossing in plant species by impeding fertilization if the parental genotypes are too similar. Consequently, rare alleles are at a higher reproductive advantage than common alleles at the locus, resulting in fluctuations of allele frequencies but never fixation of one allele over another. Other mechanisms of balancing selection are heterozygote advantage (overdominance) and variable selective pressures. Unlike positive selection, detection of balancing selection can be maintained over long periods of time. However, a high rate of effective recombination can result in the breaking up of marker associations, making detection of balancing selection difficult.

5.5 Models of Nucleotide Substitution

Models such as phylogenetic trees, neutral-equilibrium, and the natural selection models mentioned earlier utilize summary statistics. Thus, to understand the processes described in these models, it is important to examine the basics regarding how these statistics are used to indicate sequence polymorphism. Here we present a very brief overview of some of the most often utilized summary statistics. For a more exhaustive discussion, consider the following references (Doyle and Gaut 2000; Emerson et al. 2001; Nordborg and Innan 2002; Nordborg and Tavare 2002; Rosenberg and Nordborg 2002; Wolfe and Li 2003; Nielsen 2005; Excoffier and Heckel 2006; Weir et al. 2006).

Nucleotide sequence diversity (π) is the average number of differences per site between any two alleles in a population. To get another estimate of the total number of differences between any two alleles (Π or θ_{Π}), multiply π by the length of the sequence. Theta (θ) can be estimated via the average number of segregating sites per nucleotide site and is designated θ_S. If a gene is not impacted by selection (is selectively neutral), we expect both of these estimates of θ to be equal to the true value of θ. This true value of θ is proportional to the effective population size times the mutation rate ($\theta = 4N\mu$, otherwise known as the population-mutation parameter). If a gene is under positive selection, θ_S will be greater than θ_{Π} because the low frequency alleles will be removed from the population thereby decreasing the total number of segregating sites while having little effect on the average number of differences between any two alleles. If the gene is under balancing selection, θ_S will be greater than θ_{Π} because the total number of segregating sites will change little but the frequency of each allele will be intermediate, increasing the average differences between any two alleles. Tajima's D is a test statistic that compares these two different estimates of sequence diversity (Tajima 1989). Tajima's D is zero when $\theta_S = \theta_{\Pi}$, negative when $\theta_S > \theta_{\Pi}$, and positive when $\theta_S < \theta_{\Pi}$. Because θ is dependent on effective population size and mutation rate, a change in either of these parameters will alter the sensitivity of the test.

The Hudson-Kreitmam-Agaudé test (HKA test; Hudson et al. 1987) evaluates for patterns consistent with natural selection on a locus by testing for a significant difference between loci with respect to the level of sequence diversity within a species compared to the level of divergence between species. This test is computationally complex, but program suites such as DnaSP (Rozas and Rozas 1999; Rozas et al. 2003) can be used to quickly perform the analysis. Like Tajima's D, the HKA test assumes a constant effective population size and mutation rate. It also requires an absence of effective recombination within loci, because that would directly impact the level of polymorphism seen at that locus.

Nucleotide substitutions are expected to occur at different rates according to structural constraints. Some mutations change the resulting amino acid and thus change the protein for which that gene codes. These kinds of substitutions are called non-synonymous or replacement substitutions because they result in a change in the end product. Other substitutions result in no change on the amino acid level. These substitutions usually occur in introns or in the 3rd codon position and are called synonymous or silent substitutions. Introns do not code for amino acids and are processed out during translation, so substitutions in these regions are almost always synonymous. There are rare instances when substitutions in these regions result in a change at the protein level such as a change in the starting position for the next exon. Substitution rates in the 3rd codon position depend upon the redundancy of the genetic code. Some amino acids are coded for by a few different combinations of nucleic acids. This redundancy usually occurs at the 3rd or "wobble" position but is sometimes also present at the second coding position. There are different constraints on synonymous and non-synonymous sites in a gene. Because most mutations that result in an amino acid change are deleterious to the protein, non-synonymous sites are more constrained than synonymous sites. Changes in non-synonymous sites are the mutations that selection acts upon. Therefore, the proportion of mutations in these two different kinds of sites is useful for inferring if natural selection has shaped sequence variation in a gene.

Because the number of synonymous and non-synonymous sites is different from gene to gene, the number of substitutions for each is compared based on frequency. Synonymous substitutions per synonymous site is designated Ks while non-synonymous substitutions per non-synonymous site is designated Ka. The ratio of these two measures can be used to test for selection on gene. The Ka/Ks ratio should be less than 1 for most genes because most non-synonymous mutations are often deleterious and are therefore not retained (Nei and Gojobori 1986). Therefore, most polymorphism should be seen at the synonymous sites.

If a gene is completely neutral, or there is no constraint on it, the Ka/Ks ratio should be equal to 1. This pattern can help determining a gene from a pseudogene, which is no longer functional and therefore has no constraint. A Ka/Ks ratio of greater than 1 is a rare phenomenon which indicates adaptive selection for changes in the amino acid code. This indicates that the protein is under diversifying selection.

Another signature of nucleotide substitution that is often incorporated into models is transitions vs. transversions. Nucleotide substitutions are called transitions when the substitution is from a purine to another purine or a pyrimidine to another pyrimidine. A transversion is when the substitution is from pyrimidine to purine or purine to pyrimidine. Transitions are more common than transversions, but this is due the replication machinery catching this sort of mutation less often. The purines are adenine (A) and guanine (G) and are larger than the pyrimidines cytosine (C) and thymine (T). Purines base pair with pyrimidines (A with T and C with G) resulting in the space between the DNA backbone being even throughout the strand. When a transversion occurs, the resulting base pair is either larger (if the change is from pyrimidine to purine) or smaller (if the change is from purine to pyrimidine) and can be more easily detected and molecularly repaired than a transition which results in the space remaining even.

Nucleotide substitution models range greatly in the number of parameters included. One of the most basic models is the Jukes Cantor model, which assumes equal transition rates and equal equilibrium frequencies for all nucleotides. The Kimura 2 parameter model gives different weights to transitions and transversions, while the Felsenstein model includes the equilibrium frequency of each nucleotide. Many other models have integrated these two nucleotide substitution models and increase model complexity by adding additional parameters such as weighting specific substitutions according to different expected distributions. The multitude of models available makes it difficult to determine which model is most appropriate for analysis. One approach routinely implemented is to run a gene alignment through algorithms such as Modeltest (Posada and Crandall 1998) to find which model or set of models best fit each sequenced gene or gene fragment via the use of information criteria to estimate likelihoods for tested models. The substitution model with the best likelihood score can then be used for subsequent analyses to calibrate expected rates and patterns of mutations and polymorphisms. If several models have similar likelihood scores, the general practice is to find the model that is most closely related to the top models and has fewer parameters and use the simplified common model. To determine the best-fit model amongst several possibilities, a likelihood ratio test should be employed.

5.6 Demography and Natural Selection

Demographic factors can shape gene trees and polymorphism levels in a population and impact the rate of species differentiation and can leave footprints across the genome that resemble patterns expected for natural selection acting on specific genes. These factors include size, structure and distribution of populations as well as the changes in any of these components over space and/or time. Demography acts in a genome-wide manner and has more recently been incorporated into models in an effort to decrease the false inference of natural selection on a target gene. To distinguish between the impact of natural selection and various demographic factors, sequence data from multiple, presumably neutral loci or at least in a distribution representing gene categories across the genome are employed. The demographic processes impacting genetic variation that are most directly examined using multi-locus and genomic data are discussed here and include changes in effective population size, population structure, and the rates and patterns of gene flow and introgression between species and populations.

Effective population size refers to the number of breeding individuals contributing alleles to the next generation in a population. Only alleles that are passed on to the next generation impact the fundamental unit of natural selection: changes in allele frequency over time. Effective population size is reduced with increased inbreeding because the alleles passed on are from alleles that are identical by state and descent. Changes in population size can result in significant changes in the genome. When a population experiences a bottleneck (a drastic reduction in population size) there is a loss of rare alleles from the population and therefore a decrease in overall polymorphism. The gene trees that result from this are sparse with little to no branching. A drastically expanded population will have many singleton mutations and an overall increase in polymorphism. These gene trees have many short branches.

5.7 Genetic Connectivity between Taxa: Gene Flow and Population Structure

Gene flow between species and populations can result in the introduction of new alleles into a population at a rate superseding the expected slow rate of new alleles arising from mutations. Hybridization between two species does not necessarily lead to introgression of all genes, but seems to result in the exchange of those genes that do not negatively affect the fitness of backcross hybrids (Morjan and Rieseberg 2004; Lawton-Rauh et al. 2007a, b; Lawton-Rauh 2008). This is because the backcrossing of hybrids to one or both of the parental genotype(s) is the only mechanism by which introgression of new genes can take place. This process is often uneven with respect to which parental genotype receives more new alleles due to differences in the ability of the hybrid to backcross with one genotype as well as differences in the fitness of the resulting cross. The exchange of alleles for only some genes is important in maintaining species differentiation. This also makes the process detectable. Genetic data from many random and independent loci from two hybridizing species would reveal some alleles that have introgressed uniquely from one species into another. In order to detect this, however, data are needed from the hybridizing populations as well as populations of those two species that do not hybridize. These other two populations should not be exchanging alleles with the populations that are hybridizing. The introgressed alleles are those that are more similar between the hybridizing populations than they are to the same species that are not interbreeding. Further study of the alleles that are shared between separate species is needed to understand exactly what drives and maintains speciation in the face of hybridization. There are several research programs that have examined introgression during hybridization between species at the chromosomal level, which suggest that particular genomic regions may be more labile for introgression (Baack and Rieseberg 2007; Hoffmann and Rieseberg 2008).

Similar to the phylogenetic tree context, it is necessary to establish the population structure context when examining rates and patterns of shared and private genomic processes amongst closely related species and populations. One of the major reasons is because inferences of diversity and divergence, both at the gene sequence and functional product level, rely upon comparisons of shared versus private and common versus unique patterns. If sampling does not accurately represent assumed relationships, then it is impossible to know if an inference is correct. In population genetics, it is widely known that inaccurate sampling can significantly impact parameter estimates. For example, a sample that includes multiple populations rather than a single population which is used to estimate linkage disequilibrium would most likely result in a higher estimated linkage amongst polymorphisms and Tajima's D estimates could indicate balancing selection. This is because alleles that were mutationally diverged between populations

```
ATCGTGTGCCCATAAACCCTTTACTCCTTACGGGC
..................................
...C......T....................T..
...C......T....................T..
.....A..........G...G.............
.....A..........G...G.............
..................................
..................................
..................................
..................................
..................................
.....A..........G...G.............
```

Single presumed population, no substructure

Fig. 4a An alignment of the same DNA sequence fragment obtained from 13 individuals presumed to be from the same population or taxon. The estimated nucleotide diversity parameter for this alignment: $\theta \sim 0.057$.

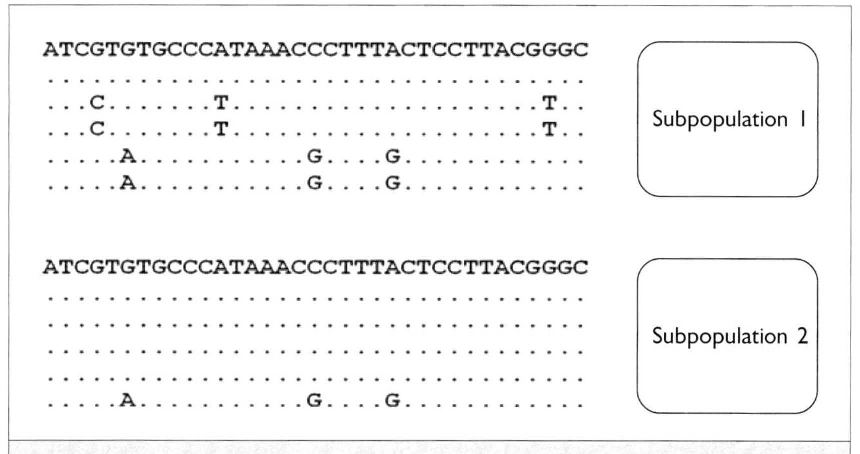

Fig. 4b The same alignment in Fig. 4a partitioned by accurate assignment to the two subpopulations. The estimated nucleotide diversity for each subpopulation: Subpop1 $\theta \sim 0.075$, Subpop2 $\theta \sim 0.038$.

over time would be detected as separate high-frequency haplotypes. This type of sampling impact on parameter estimation is known as the Wahlund Effect (See Figs. 4a and 4b; Lowe et al. 2004).

The importance of incorporating population structure in comparative evolutionary genomics analyses cannot be overstated as highlighted in the trend to increase sampling within species to include multiple ecotypes and individuals representing multiple populations and then estimating the best fit number of populations (Schmid et al. 2005, 2006). The best fit number of populations given a dataset can be estimated, statistically compared using likelihood ratio tests, and then visually represented using haplotype analysis programs such as STRUCTURE (Pritchard et al. 1999, 2000; Rosenberg et al. 2005). For research incorporating datasets including phenotype and expression data, it is subsequently possible to use these haplotype assignment models with mixed association model approaches to test for statistical significance between genotypes and traits (Helgason et al. 2005; Stich et al. 2008; Weber et al. 2008).

6 REGULONS, NEXT GENERATION SEQUENCING AND ECO-GENOMICS

6.1 Using the Comparative Approach to Study Regulation of Expression

Until recently, most statistically comprehensive comparative approaches have focused primarily on the examination of large-scale genome fragments or smaller-scale sequenced gene fragments. Additionally, most analyses of gene expression have centered on representative individuals sampled at specific developmental time points or in targeted environmental conditions. Combining these two approaches has become increasingly more feasible, both experimentally and statistically. While not without issues primarily due to estimates with compounded variance, the hypotheses that can be tested by merging these two perspectives are transforming our understanding of the relationship between inheritance and phenotypic variance. These connections are being examined more explicitly using comparative evolutionary genomics approaches for a variety of taxonomic levels, from within-individual, population and species levels to across all eukaryotes. In this section, we discuss expression analyses afforded by studying regulons, gene co-expression networks, and the merging of expression and ecology in so-called 'eco-genomics'.

6.2 Expression Analyses: Regulons and Structural Genomics

While it is known that gene regulation is a pivotal component for organismal development and adaptation to the environment, regulation mediated by genomic factors is largely unclear even for the relatively exhaustively well characterized species *Arabidopsis thaliana* (Tirosh et al. 2007). Examination of *A. thaliana* transcriptome using probes derived from microarray chips has recently provided important insights into what lies ahead as frameworks for future comparative approaches and subsequently more hierarchically extensive analyses across additional species and taxonomic levels. Using this approach, the number of so-called 'regulons' have been identified and suggest that the *A. thaliana* genome is comprised of 998 regulons that range in size from 1 to 1,623 genes (Mentzen and Wurtele 2008). These regulons include mitochondrial and chloroplast genes that were shuttled to the nuclear genome as well as mitochondrially-encoded genes that were once encoded by the nuclear genome. Taken together, the pattern and organization of regulons in the *A. thaliana* genome suggest that transcriptional regulation occurs through concerted function controlled by higher-level organization in the genome. As soon as the genome of *A. thaliana*'s closest relatives are assembled, edited, and publicly available, it will be very exciting to test for higher order organizations that are shared versus derived in the *A. thaliana* genome, particularly because it is known that the *A. thaliana* genome is significantly reduced in content versus all other *Arabidopsis* species and the Brassicaceae plant family in general (Oyama et al. 2008).

6.3 Gene Co-expression Networks

Gene co-expression networks are powerful new tools which use microarray expression profiles and Pearson correlations to 'link' genes that are co-expressed in different cell types or life stages (Zhang et al. 2007; Mentzen and Wurtele 2008). The nodes of the networks represent each gene and the lines indicated a correlation to another gene. These networks can be used for genome annotation because highly connected genes are often involved in the same cellular pathway. This 'guilt by association' principle is based on the idea that the most genes that are involved in the same pathway also share the same regulatory system. They can also be used to find candidate genes for a particular function. The network can be overlaid with genes that are up- or down-regulated in a particular cell type, then scanned (visually or computationally) for clusters (sub-networks) that contain a large number of those genes, which are likely involved in that function. The resulting candidate genes need to be tested further for involvement in the function. Co-expression networks are also used to identify the central players of a given pathway. The most highly connected genes (known as hubs) in a cluster are thought to be the most important. For example, transcription factors will be highly correlated with all the genes that they regulate and therefore will show more connections than the other genes involved in a particular function like flower development. These connections can also include genes that are not directly linked, but are associated through a third gene. Clustering coefficients are used to determine the overall associations and structure of a cluster to determine if genes in that cluster are all truly connected.

A comparative analysis of co-expression networks between different species is an important tool for understanding how gene interactions are involved in the differences between species. This approach was used to determine that gene connectivity was less conserved between humans and chimpanzees than gene expression, indicating that connectivity differences are responsible for the phenotypic differences seen (Oldham et al. 2006). Comparative analysis of plant co-expression networks may offer insight into how genome duplication events can facilitate neofunctionalization of whole gene networks by differential expression (Veron et al. 2007). One problem with comparative analysis of co-expression networks lies in the data collection. Microarray chips designed for one species or population may not capture the mRNA from another species as efficiently due to SNPs present in the sequence. This would result in a reduction of detected expression for that gene causing false positives and negatives for interactions. Another problem is microarrays from different platforms that have different outputs further complicating comparisons. Microarrays from the same platform but are run at different times are generally considered to be comparable. Despite problems with data collection, co-expression networks produce hefty amounts of information and will only become more useful as more genomic and functional data are incorporated.

6.4 Re-sequencing

Amazing accomplishments in gene and genome sequencing technology beyond Sanger sequencing have significantly improved the capacity to rapidly sequence large numbers of genome fragments in what is known as 'next generation sequencing'. Current next generation sequence platforms include 454 pyrosequencing technology, Solexa's reversible terminator technology, and SOLiD ligation-based technology (Shendure and Ji 2008; Imelfort et al. 2009; Pettersson et al. 2009). Discussion of this technology is beyond the scope of this chapter and there are several helpful reviews that are good references for a more extensive background (Rokas and Abbot 2009). It is clear that the genomics community has just seen the tip of the iceberg regarding the unveiling of large scale and small scale structural-based trends in genome dynamics. We are moving beyond the scale of single genomes representing a species towards an increased representation of within-species diversity (Luikart et al. 2003). This within-species diversity examined at the genome-wide level, either based on the resequencing and reference-aligned entire genomes from several individuals or the sequencing of entire transcriptomes for comparative analyses of expressed genes (Morozova and Marra 2008) present incredible opportunities for both so-called 'model' organisms and for less well-characterized systems such as recent work in *Eucalyptus grandis* (Novaes et al. 2008).

6.5 Ecological Genomics (Eco-Genomics)

Ecological genomics is a branch of interdisciplinary science, which combines genomics and ecology to discover and understand the genetic mechanisms regulating adaptation to the environment. To investigate the interaction between genotype and the environment, functional genomics approaches are used to identify candidate genes and further characterize the relative roles of alternative alleles and specific genes identified in the context of adaptation and environmental interaction. The obvious model system for this approach would be one that already has an extensive genomic database as a reference as well as an interesting ecological context. The genus *Arabidopsis* is well suited as a model ecological genomics system for these reasons, because at this point of time this genus fits these criteria better than any other plant system. *Arabidopsis thaliana* boasts a sequenced genome, large amount of functional proteomic work, and recent evolutionary and ecological studies on several ecotypes. Additionally, the genome of *A. lyrata*, the closest relative to *A. thaliana,* is currently being sequenced as part of a comparative genomics project that also includes another member of the same plant family, *Capsella rubella* (for more information about this project, see the DOE Joint Genome Institute website: *http://www.jgi.doe.gov/sequencing/why/3066.html*). Several close wild relatives to *A. thaliana* including *A. lyrata*, *A. halleri*, and *A. arenosa* have important ecological differences from *A. thaliana* in breeding system, habitat, and life history thus serve as an excellent group for comparative genomics analyses in the ecological context. Genomic tools developed in *A. thaliana* are easily transferable to these wild relatives (Mitchell-Olds 2001). As more genomic information becomes available for other plant systems, we will likely see how universal trends inferred from this system are amongst other plant systems.

7 CONCLUSIONS

With the increased feasibility of sampling multiple genomes from both model and non-model species, the field of comparative and evolutionary genomics will only get closer to merging longer term molecular evolution investigations using the population genomics context. While studies in model systems are clearly at an advantage with respect to streamlining the process of multiple marker development (especially through resequencing and next generation sequencing tools), systems presently considered 'non-model' systems will soon be significantly more amenable to comparative genomics analyses. As more systems are sampled at ecotype and population levels, the genomics field will yield results that will examine just how transferable generalized trends of trait divergence are across organisms. Some of the most exciting emerging comparative approaches that will greatly benefit from such increasingly integrative projects include comparative analyses of gene co-expression networks, gene expression, and the integration of ecology and the genetic versus environmental basis of phenotype plasticity.

Dedication

This chapter is dedicated to William Frederick Lawton (1928-2009), whose curiosity, stubborn optimism, and sense of adventure was matched only by his ability to enjoy the moment.

References

Aagaard JE, Willis JH, Phillips PC (2006) Relaxed selection among duplicate floral regulatory genes in Lamiales. J Mol Evol 63: 493-503

Adams KL, Wendel JF (2005) Polyploidy and genome evolution in plants. Curr Opin Plant Biol 8: 135-141

Adams KL, Daley DO, Qiu YL, Whelan J, Palmer JD (2000) Repeated, recent and diverse transfers of a mitochondrial gene to the nucleus in flowering plants. Nature 408: 354-357

Agarwal M, Shrivastava N, Padh H (2008) Advances in molecular marker techniques and their applications in plant sciences. Plant Cell Rep 27: 617-631

Alvarez-Buylla ER, Pelaz S, Liljegren SJ, Gold SE, Burgeff C, Ditta GS, de

Pouplana LR, Martinez-Castilla L, Yanofsky MF (2000) An ancestral MADS-box gene duplication occurred before the divergence of plants and animals. Proc Natl Acad Sci USA 97: 5328-5333

Avise JC (2004) Molecular Markers, Natural History, and Evolution. Sinauer Assoc, Sunderland, MA, USA

Baack EJ, Rieseberg LH (2007) A genomic view of introgression and hybrid speciation. Curr Opin Genet Dev 17: 513-518

Barr CM, Keller SR, Ingvarsson PK, Sloan DB, Taylor DR (2007) Variation in mutation rate and polymorphism among mitochondrial genes of *Silene vulgaris*. Mol Biol Evol 24: 1783-1791

Bierne N, Tanguy A, Faure M, Faure B, David E, Boutet I, Boon E, Quere N, Plouviez S, Kemppainen P, Jollivet D, Moraga D, Boudry P, David P (2007) Mark-recapture cloning: a straightforward and cost-effective cloning method for population genetics of single-copy nuclear DNA sequences in diploids. Mol Ecol Notes 7: 562-566

Blanc G, Wolfe KH (2004) Functional divergence of duplicated genes formed by polyploidy during Arabidopsis evolution. Plant Cell 16: 1679-1691

Blanc G, Hokamp K, Wolfe KH (2003) A recent polyploidy superimposed on older large-scale duplications in the Arabidopsis genome. Genome Res 13: 137-144

Borrone JW, Meerow AW, Kuhn DN, Whitlock BA, Schnell RJ (2007) The potential of the WRKY gene family for phylogenetic reconstruction: An example from the Malvaceae. Mol Phylogenet Evol 44: 1141-1154

Bucci G, Gonzalez-Martinez SC, Le Provost G, Plomion C, Ribeiro MM, Sebastiani F, Alia R, Vendramin GG (2007) Range-wide phylogeography and gene zones in *Pinus pinaster* Ait. revealed by chloroplast microsatellite markers. Mol Ecol 16: 2137-2153

Byrne KP, Blanc G (2006) Computational analyses of ancient polyploidy. Curr Bioinformat 1: 131-146

Cannon SB, Young ND (2003) OrthoParaMap: Distinguishing orthologs from paralogs by integrating comparative genome data and gene phylogenies. BMC Bioinformat 4: 35

Carling MD, Brumfield RT (2007) Gene sampling strategies for multi-locus population estimates of genetic diversity (θ). PLoS ONE 2: e160

Chamary JV, Parmley JL, Hurst LD (2006) Hearing silence: non-neutral evolution at synonymous sites in mammals. Nat Rev Genet 7: 98-108

Chapman MA, Chang J, Weisman D, Kesseli RV, Burke JM (2007) Universal markers for comparative mapping and phylogenetic analysis in the Asteraceae (Compositae). Theor Appl Genet 115: 747-755

Charlesworth D, Charlesworth B, McVean GAT (2001) Genome sequences and evolutionary biology, a two-way interaction. Trends Ecol Evol 16: 235-242

Chen ZJ (2007) Genetic and epigenetic mechanisms for gene expression and phenotypic variation in plant polyploids. Annu Rev Plant Biol 58: 377-406

Clauss MJ, Mitchell-Olds T (2004) Functional divergence in tandemly duplicated *Arabidopsis thaliana* trypsin inhibitor genes. Genetics 166: 1419-1436

Clement M, Posada D, Crandall KA (2000) TCS: A computer program to estimate gene genealogies. Mol Ecol 9: 1657-1659

Cronn RC, Small RL, Wendel JF (1999) Duplicated genes evolve independently after polyploid formation in cotton. Proc Natl Acad Sci USA 96: 14406-14411

Cronn RC, Small RL, Haselkorn T, Wendel JF (2002) Rapid diversification of the cotton genus (*Gossypium*: Malvaceae) revealed by analysis of sixteen nuclear and chloroplast genes. Am J Bot 89: 707-725

Cui LY, Wall PK, Leebens-Mack JH, Lindsay BG, Soltis DE, Doyle JJ, Soltis PS, Carlson JE, Arumuganathan K, Barakat A, Albert VA, Ma H, dePamphilis CW (2006) Widespread genome duplications throughout the history of flowering plants. Genome Res 16: 738-749

Demuth JP, Hahn MW (2009) The life and death of gene families. Bioessays 31: 29-39

Doveri S, Lee D, Maheswaran M, Powell W (2008) Molecular markers: History, features and applications. In Kole C, Abbott AG (eds) Principles and Practices of Plant Genomics. Vol 1: Genome Mapping. Science Publ, Enfield (NH), Jersey, USA, pp 23-67

Doyle JJ, Gaut BS (2000) Evolution of genes and taxa: A primer. Plant Mol Biol 42: 1-23

Doyle JJ, Doyle JL, Rauscher JT, Brown AHD (2004) Diploid and polyploid reticulate evolution throughout the history of the perennial soybeans (*Glycine* subgenus *Glycine*). New Phytol 161: 121-132

Doyle JJ, Flagel LE, Paterson AH, Rapp RA, Soltis DE, Soltis PS, Wendel JF (2008) Evolutionary genetics of genome merger and doubling in plants. Annu Rev Genet 42: 443-461

Ellis JR, Burke JM (2007) EST-SSRs as a resource for population genetic analyses. Heredity 99: 125-132

Emerson BC, Paradis E, Thebaud C (2001) Revealing the demographic histories of species using DNA sequences. Trends Ecol Evol 16: 707-716

Ewing B, Green P (1998) Base-calling of automated sequencer traces using phred. II. Error probabilities. Genome Res 8: 186-194

Ewing B, Hillier L, Wendl MC, Green P (1998) Base-calling of automated sequencer traces using phred. I. Accuracy assessment. Genome Res 8: 175-185

Excoffier L, Heckel G (2006) Computer programs for population genetics data analysis: a survival guide. Nat Rev Genet 7: 745-758

Force A, Lynch M, Pickett FB, Amores A, Yan YL, Postlethwait J (1999) Preservation of duplicate genes by complementary, degenerative mutations. Genetics 151: 1531-1545

Haas BJ, Delcher AL, Wortman JR, Salzberg SL (2004) DAGchainer: A tool for mining segmental genome duplications and synteny. Bioinformatics 20: 3643-3646

Hahn MW (2007) Detecting natural selection on cis-regulatory DNA. Genetica 129: 7-18

Hall BG (2001) Phylogenetic Trees Made Easy: A How-to Manual for Molecular Biologists. Sinauer Assoc, Sunderland, MA, USA

Hampson S, McLysaght A, Gaut B, Baldi P (2003) LineUp: Statistical detection of chromosomal homology with application to plant comparative genomics. Genome Res 13: 999-1010

Hampson SE, Gaut BS, Baldi P (2005) Statistical detection of chromosomal homology using shared-gene density alone. Bioinformatics 21: 1339-1348

Hazzouri RM, Mohajer A, Dejak SI, Otto SP, Wright SI (2008) Contrasting patterns of transposable-element insertion polymorphism and nucleotide diversity in autotetraploid and allotetraploid Arabidopsis species. Genetics 179: 581-592

Hedrick PW, Verrelli BC (2006) 'Ground truth' for selection on CCR5-Delta 32. Trends Genet 22: 293-296

Helgason A, Yngvadottir B, Hrafnkelsson B, Gulcher J, Stefansson K (2005) An Icelandic example of the impact of population structure on association studies. Nat Genet 37: 90-95

Hey J, Won YJ, Sivasundar A, Nielsen R, Markert JA (2004) Using nuclear haplotypes with microsatellites to study gene flow between recently separated *Cichlid* species. Mol Ecol 13: 909-919

Hoffmann AA, Rieseberg LH (2008) Revisiting the impact of inversions in evolution: From population genetic markers to drivers of adaptive shifts and speciation? Annu Rev Ecol Evol Syst 39: 21-42

Hudson RR, Kreitman M, Aguade M (1987) A test of neutral molecular evolution based on nucleotide data. Genetics 116: 153-159

Hughes AL (2002) Adaptive evolution after gene duplication. Trends Genet 18: 433-434

Imelfort M, Duran C, Batley J, Edwards D (2009) Discovering genetic polymorphisms in next-generation sequencing data. Plant Biotechnol J 7: 312-317

Kalisz S, Kramer EM (2008) Variation and constraint in plant evolution and development. Heredity 100: 171-177

Kellogg EA, Bennetzen JL (2004) The evolution of nuclear genome structure in seed plants. Am J Bot 91: 1709-1725

Kimura M (1983) The Neutral Theory of Molecular Evolution. Cambridge Univ Press, Cambridge, UK (reprinted 1986)

Kofler R, Schlotterer C, Lelley T (2007) SciRoKo: A new tool for whole genome microsatellite search and investigation. Bioinformatics 23: 1683-1685

Koszul R, Caburet S, Dujon B, Fischer G (2004) Eucaryotic genome evolution through the spontaneous duplication of large chromosomal segments. EMBO J 23: 234-243

Kubo T, Newton KJ (2008) Angiosperm mitochondrial genomes and mutations. Mitochondrion 8: 5-14

Lawton-Rauh A (2003) Evolutionary dynamics of duplicated genes in plants. Mol Phylogenet Evol 29: 396-409

Lawton-Rauh A (2008) Demographic processes shaping genetic variation. Curr Opin Plant Biol 11: 103-109

Lawton-Rauh AL, Buckler ESIV, Purugganan MD (1999) Patterns of molecular evolution among paralogous floral homeotic genes. Mol Biol Evol 16: 1037-1045

Lawton-Rauh A, Robichaux RH, Purugganan MD (2003) Patterns of nucleotide variation in homoeologous regulatory genes in the allotetraploid Hawaiian silversword alliance (Asteraceae). Mol Ecol 12: 1301-1313

Lawton-Rauh A, Friar EA, Remington DL (2007a) Collective evolution processes and the tempo of lineage divergence in the Hawaiian silversword alliance adaptive radiation (Heliantheae, Asteraceae). Mol Ecol 16: 3993-3994

Lawton-Rauh A, Robichaux RH, Purugganan MD (2007b) Diversity and divergence patterns in regulatory genes suggest differential gene flow in recently derived species of the Hawaiian silversword alliance adaptive radiation (Asteraceae). Mol Ecol 16: 3995-4013

Leebens-Mack JH, Wall K, Duarte J, Zheng ZG, Oppenheimer D, Depamphilis C (2006) A genomics approach to the study of ancient polyploidy and floral developmental genetics. In: Advances in Botanical Research: Incorporating Advances in Plant Pathology. Vol 44. Academic Press, London, UK, pp 527-549

Leister D (2004) Tandem and segmental gene duplication and recombination in the evolution of plant disease resistance genes. Trends Genet 20: 116-122

Li CH, Orti G, Zhang G, Lu GQ (2007) A practical approach to phylogenomics: The phylogeny of ray-finned fish (Actinopterygii) as a case study. BMC Evol Biol 7: 44

Li WH, Yang J, Gu X (2005) Expression divergence between duplicate genes. Trends Genet 21: 602-607

Lin HN, Ouyang S, Egan A, Nobuta K, Haas BJ, Zhu W, Gu X, Silva JC, Meyers BC, Buell CR (2008) Characterization of paralogous protein families in rice. BMC Plant Biol 8: 18

Liu QP, Xue QZ (2007) Computational identification and phylogenetic analysis of the MAPK gene family in *Oryza sativa*. Plant Physiol Biochem 45: 6-14

Lowe AR, Harris S, Ashton P, Pub B (2004) Ecological Genetics: Design, Analysis, and Application. Blackwell Publ, Malden, MA, USA

Luikart G, England PR, Tallmon D, Jordan S, Taberlet P (2003) The power and promise of population genomics: From genotyping to genome typing. Nat Rev Genet 4: 981-994

Mable BK (2004) 'Why polyploidy is rarer in animals than in plants': Myths and mechanisms. Biol J Linn Soc 82: 453-466

Magri D, Fineschi S, Bellarosa R, Buonamici A, Sebastiani F, Schirone B, Simeone MC, Vendramin GG (2007) The distribution of *Quercus suber* chloroplast haplotypes matches the palaeogeographical history of the western Mediterranean. Mol Ecol 16: 5259-5266

Mentzen WI, Wurtele ES (2008) Regulon organization of Arabidopsis. BMC Plant Biol 8: 99

Mitchell-Olds T (2001) *Arabidopsis thaliana* and its wild relatives: A model system for ecology and evolution. Trends Ecol Evol 16: 693-700

Moore RC, Purugganan MD (2005) The evolutionary dynamics of plant duplicate genes. Curr Opin Plant Biol 8: 122-128

Moore RC, Grant SR, Purugganan MD (2005) Molecular population genetics of redundant floral-regulatory genes in *Arabidopsis thaliana*. Mol Biol Evol 22: 91-103

Morgante M, Hanafey M, Powell W (2002) Microsatellites are preferentially associated with nonrepetitive DNA in plant genomes. Nat Genet 30: 194-200

Morjan CL, Rieseberg LH (2004) How species evolve collectively: Implications of gene flow and selection for the spread of advantageous alleles. Mol Ecol 13: 1341-1356

Morozova O, Marra MA (2008) Applications of next-generation sequencing technologies in functional genomics. Genomics 92: 255-264

Nadir E, Margalit H, Gallily T, BenSasson SA (1996) Microsatellite spreading in the human genome: Evolutionary mechanisms and structural implications. Proc Natl Acad Sci USA 93: 6470-6475

Nasri N, Bojovic S, Vendramin GG, Fady B (2008) Population genetic structure of the relict Serbian spruce, *Picea omorika*, inferred from plastid DNA. Plant Syst Evol 271: 1-7

Nei M (2007) The new mutation theory of phenotypic evolution. Proc Natl Acad Sci USA 104: 12235-12242

Nei M, Gojobori T (1986) Simple methods for estimating the numbers of synonymous and nonsynonymous nucleotide substitutions. Mol Biol Evol 3: 418-426

Nielsen R (2005) Molecular signatures of natural selection. Annu Rev Genet 39: 197-218

Nordborg M, Innan H (2002) Molecular population genetics. Curr Opin Plant Biol 5: 69-73

Nordborg M, Tavare S (2002) Linkage disequilibrium: what history has to tell us. Trends Genet 18: 83-90

Novaes E, Drost DR, Farmerie WG, Pappas GJ, Jr., Grattapaglia D, Sederoff RR, Kirst M (2008) High-throughput gene and SNP discovery in *Eucalyptus grandis*, an uncharacterized genome. BMC Genom 9: 312

Ohta T (1973) Slightly deleterious mutant substitutions in evolution. Nature 246: 96-98

Ohta T (1992) The nearly neutral theory of molecular evolution. Annu Rev Ecol Syst 23: 263-286

Oldham MC, Horvath S, Geschwind DH (2006) Conservation and evolution of gene colexpression networks in human and chimpanzee brains. Proc Natl Acad Sci USA 103: 17973-17978

Oliver JC (2008) AUGIST: Inferring species trees while accommodating gene tree uncertainty. Bioinformatics 24: 2932-2933

Oyama RK, Clauss MJ, Formanova N, Kroymann J, Schmid KJ, Vogel H, Weniger K, Windsor AJ, Mitchell-Olds T (2008) The shrunken genome of *Arabidopsis thaliana*. Plant Syst Evol 273: 257-271

Pakkad G, Ueno S, Yoshimaru H (2008) Genetic diversity and differentiation of *Quercus semiserrata* Roxb. in northern Thailand revealed by nuclear and chloroplast microsatellite markers. For Ecol Manag 255: 1067-1077

Parmakelis A, Slotman MA, Marshall JC, Awono-Ambene PH, Antonio-Nkondjio C, Simard F, Caccone A, Powell JR (2008) The molecular evolution of four anti-malarial immune genes in the *Anopheles gambiae* species complex - art no 79. BMC Evol Biol 8: 79-79

Pettersson E, Lundeberg J, Ahmadian A (2009) Generations of sequencing technologies. Genomics 93: 105-111

Posada D, Crandall KA (1998) MODELTEST: Testing the model of DNA substitution. Bioinformatics 14: 817-818

Posada D, Maxwell TJ, Templeton AR (2005) TreeScan: A bioinformatic application to search for genotype/phenotype associations using haplotype trees. Bioinformatics 21: 2130-2132

Pritchard JK, Stephens M, Donnelly PJ (1999) Correcting for population stratification in linkage disequilibrium mapping studies. Am J Hum Genet 65: 528

Pritchard JK, Stephens M, Donnelly P (2000) Inference of population structure using multilocus genotype data. Genetics 155: 945-959

Provan J, Powell W, Hollingsworth PM (2001) Chloroplast microsatellites: New tools for studies in plant ecology and evolution. Trends Ecol Evol 16: 142-147

Provan J, Murphy S, Maggs CA (2005) Tracking the invasive history of the green alga *Codium fragile* ssp. *tomentosoides*. Mol Ecol 14: 189-194

Rannala B, Yang Z (2008) Phylogenetic inference using whole genomes. Annu Rev Genom Hum Genet 9: 217-231

Raquin AL, Brabant P, Rhone B, Balfourier F, Leroy P, Goldringer I (2008) Soft selective sweep near a gene that increases plant height in wheat. Mol Ecol 17: 741-756

Rokas A, Abbot P (2009) Harnessing genomics for evolutionary insights. Trends Ecol Evol 24: 192-200

Rosenberg NA (2002) The probability of topological concordance of gene trees and species trees. Theor Popul Biol 61: 225-247

Rosenberg NA, Nordborg M (2002) Genealogical trees, coalescent theory and the analysis of genetic polymorphisms. Nat Rev Genet 3: 380-390

Rosenberg NA, Tao R (2008) Discordance of species trees with their most likely gene trees: The case of five taxa. Syst Biol 57: 131-140

Rosenberg NA, Mahajan S, Ramachandran S, Zhao CF, Pritchard JK, Feldman MW (2005) Clines, clusters, and the effect of study design on the inference of human population structure. PLoS Genet 1: 660-671

Rothenburg S, Koch-Nolte F, Rich A, Haag F (2001) A polymorphic dinucleotide repeat in the rat nucleolin gene forms Z-DNA and inhibits promoter activity. Proc Natl Acad Sci USA 98: 8985-8990

Roulin A, Piegu B, Wing RA, Panaud O (2008) Evidence of multiple horizontal transfers of the long terminal repeat retrotransposon RIRE1 within the genus *Oryza*. Plant J 53: 950-959

Rozas J, Rozas R (1999) DnaSP version 3: An integrated program for molecular population genetics and molecular evolution analysis. Bioinformatics 15: 174-175

Rozas J, Sanchez-DelBarrio JC, Messeguer X, Rozas R (2003) DnaSP, DNA polymorphism analyses by the coalescent and other methods. Bioinformatics 19: 2496-2497

Saarela JM, Rai HS, Doyle JA, Endress PK, Mathews S, Marchant AD, Briggs BG, Graham SW (2007) Hydatellaceae identified as a new branch near the base of the angiosperm phylogenetic tree. Nature 446: 312-315

Schlotterer C (2000) Evolutionary dynamics of microsatellite DNA. Chromosoma 109: 365-371

Schlotterer C (2004) The evolution of molecular markers-just a matter of fashion? Nat Rev Genet 5: 63-69

Schlotterer C, Tautz D (1992) Slippage synthesis of simple sequence DNA. Nucl Acids Res 20: 211

Schmid KJ, Ramos-Onsins S, Ringys-Beckstein H, Weisshaar B, Mitchell-Olds T (2005) A multilocus sequence survey in *Arabidopsis thaliana* reveals a genome-wide departure from a neutral model of DNA sequence polymorphism. Genetics 169: 1601-1615

Schmid K, Torjek O, Meyer R, Schmuths H, Hoffmann MH, Altmann T (2006) Evidence for a large-scale population structure of *Arabidopsis thaliana* from genome-wide single nucleotide polymorphism markers. Theor Appl Genet 112: 1104-1114

Selkoe KA, Toonen RJ (2006) Microsatellites for ecologists: A practical guide to using and evaluating microsatellite markers. Ecol Lett 9: 615-629

Semon M, Wolfe KH (2007) Consequences of genome duplication. Curr Opin Genet Dev 17: 505-512

Shendure J, Ji H (2008) Next-generation DNA sequencing. Nat Biotechnol 26: 1135-1145

Slotte T, Ceplitis A, Neuffer B, Hurka H, Lascoux M (2006) Intrageneric phylogeny of *Capsella* (Brassicaceae) and the origin of the tetraploid *C. bursa-pastoris* based on chloroplast and nuclear DNA sequences. Am J Bot 93: 1714-1724

Song BH, Mitchell-Olds T (2007) High genetic diversity and population differentiation in *Boechera fecunda*, a rare relative of Arabidopsis. Mol Ecol 16: 4079-4088

Stephens M, Donnelly P (2003) A comparison of bayesian methods for haplotype reconstruction from population genotype data. Am J Hum Genet 73: 1162-1169

Stephens M, Smith NJ, Donnelly P (2001) A new statistical method for haplotype reconstruction from population data. Am J Hum Genet 68: 978-989

Stich B, Mohring J, Piepho H, Heckenberger M, Buckler ES, Melchinger AE (2008) Comparison of mixed-model approaches for association mapping. Genetics 178: 1745-1754

Stracke S, Sato S, Sandal N, Koyama M, Kaneko T, Tabata S, Parniske M (2004) Exploitation of colinear relationships between the genomes of *Lotus japonicus*, *Pisum sativum* and *Arabidopsis thaliana*, for positional cloning of a legume symbiosis gene. Theor Appl Genet 108: 442-449

Straub SCK, Pfeil BE, Doyle JJ (2006) Testing the polyploid past of soybean using a low-copy nuclear gene - Is Glycine (Fabaceae: Papilionoideae) an auto- or allopolyploid? Mol Phylogenet Evol 39: 580-584

Tajima F (1989) Statistical method for testing the neutral mutation hypothesis by DNA polymorphism. Genetics 123: 585-595

Tirosh I, Bilu Y, Barkai N (2007) Comparative biology: Beyond sequence analysis. Curr Opin Biotechnol 18: 371-377

Veron AS, Kaufmann K, Bornberg-Bauer E (2007) Evidence of interaction network evolution by whole-genome duplications: A case study in MADS-box proteins. Mol Biol Evol 24: 670-678

Wang HB, Yu LJ, Lai F, Liu LS, Wang JF (2005) Molecular evidence for asymmetric evolution of sister duplicated blocks after cereal polyploidy. Plant Mol Biol 59: 63-74

Wang XY, Shi XL, Li Z, Zhu QH, Kong L, Tang W, Ge S, Luo JC (2006) Statistical inference of chromosomal homology based on gene colinearity and applications to Arabidopsis and rice. BMC Bioinformat 7: 447

Weber AL, Briggs WH, Rucker J, Baltazar BM, de Jesus Sanchez-Gonzalez J, Feng P, Buckler ES, Doebley J (2008) The genetic architecture of complex traits in Teosinte (*Zea mays* ssp. *parviglumis*): New evidence from association mapping. Genetics 180: 1221-1232

Weir BS, Anderson AD, Hepler AB (2006) Genetic relatedness analysis: Modern data and new challenges. Nat Rev Genet 7: 771-780

Weising K, Gardner RC (1999) A set of conserved PCR primers for the analysis of simple sequence repeat polymorphisms in chloroplast genomes of dicotyledonous angiosperms. Genome 42: 9-19

Wolfe KH, Li WH (2003) Molecular evolution meets the genomics revolution. Nat Genet 33: 255-265

Won YJ, Sivasundar A, Wang Y, Hey J (2005) On the origin of Lake Malawi cichlid species: A population genetic analysis of divergence. Proc Natl Acad Sci USA 102: 6581-6586

Wright SI, Gaut BS (2005) Molecular population genetics and the search for adaptive evolution in plants. Mol Biol Evol 22: 506-519

Xu X, Peng M, Fang Z, Xu XP (2000) The direction of microsatellite mutations is dependent upon allele length. Nat Genet 24: 396-399

Yu J, Wang J, Lin W, Li SG, Li H, Zhou J, Ni PX, Dong W, Hu SN, Zeng CQ, Zhang JG, Zhang Y, Li RQ, Xu ZY, Li ST, Li XR, Zheng HK, Cong LJ, Lin L, Yin JN, Geng JN, Li GY, Shi JP, Liu J, Lv H, Li J, Wang J, Deng YJ, Ran LH, Shi XL, Wang XY, Wu QF, Li CF, Ren XY, Wang JQ, Wang XL, Li DW, Liu DY, Zhang XW, Ji ZD, Zhao WM, Sun YQ, Zhang ZP, Bao JY, Han YJ, Dong LL, Ji J, Chen P, Wu SM, Liu JS, Xiao Y, Bu DB, Tan JL, Yang L, Ye C, Zhang JF, Xu JY, Zhou Y, Yu YP, Zhang B, Zhuang SL, Wei HB, Liu B, Lei M, Yu H, Li YZ, Xu H, Wei SL, He XM, Fang LJ, Zhang ZJ, Zhang YZ, Huang XG, Su ZX, Tong W, Li JH, Tong ZZ, Li SL, Ye J, Wang LS, Fang L, Lei TT, Chen C, Chen H, Xu Z, Li HH, Huang HY, Zhang F, Xu HY, Li N, Zhao CF, Li ST, Dong LJ, Huang YQ, Li L, Xi Y, Qi QH, Li WJ, Zhang B, Hu W, Zhang YL, Tian XJ, Jiao YZ, Liang XH, Jin JA, Gao L, Zheng WM, Hao BL, Liu SQ, Wang W, Yuan LP, Cao ML, McDermott J, Samudrala R, Wang J, Wong GKS, Yang HM (2005) The Genomes of *Oryza sativa:* A history of duplications. PLoS Biol 3: 266-281

Zhang DX, Hewitt GM (2003) Nuclear DNA analyses in genetic studies of populations: Practice, problems and prospects. Mol Ecol 12: 563-584

Zhang JZ (2003) Evolution by gene duplication: an update. Trends Ecol Evol 18: 292-298

Zhang S, Jin G, Zhang X, Chen L (2007) Discovering functions and revealing mechanisms at molecular level from biological networks. Proteomics 7: 2856-2869

Ziolkowski PA, Kaczmarek M, Babula D, Sadowski J (2006) Genome evolution in *Arabidopsis/Brassica*: Conservation and divergence of ancient rearranged segments and their breakpoints. Plant J 47: 63-74

4 ESTs and their Role in Functional Genomics

Kalpalatha Melmaiee and Venu Kalavacharla*

Department of Agriculture & Natural Resources,
Delaware State University, 1200 N. DuPont Highway,
Dover, DE 19901, USA

Corresponding author: vkalavacharla@desu.edu

1 INTRODUCTION

An expressed sequence tag (EST) is a short complementary DNA (cDNA) sequence derived from messenger RNA (mRNA) (Adams et al. 1991). ESTs are produced by single read sequencing of cDNA clones from either the 3′- or 5′-end untranslated regions (UTR). ESTs provide direct evidence for all sampled transcripts and are very important resources for transcriptome studies (Nagaraj et al. 2007).

Genome sequencing has been completed for *Arabidopsis*, soybean, *Medicago*, rice, poplar, sorghum and maize. Meanwhile, whole-genome sequencing is in progress for several other plant species such as *Brachypodium*, lotus, cassava, tomato and potato. Undertaking genome sequencing projects for all crop species of interest is probably not feasible given the costs and other limitations with currently available technology. Even for the crops that have the complete genome sequence available, prediction of gene expression patterns in different types of tissues, cells, and at various developmental stages is not possible based on genome sequence information. Independent of genome size, studies utilizing ESTs open exciting prospects towards gene discovery in an organism (Shoemaker et al. 2002; Van der Hoeven et al. 2002). There are several uses of ESTs which include:

a) Serving as a major source for developing DNA microarrays.
b) Providing a snap shot of expressed genes in the system under any given conditions (Rudd et al. 2004).
c) Redundant collection of ESTs allows for the selection of molecular markers (gene and SNP) with certain degree of confidence.
d) Since ESTs come from the coding region of the genome, these can easily be associated with phenotypes.

2 DEVELOPMENT OF EST RESOURCES-CDNA LIBRARY

Generation of ESTs starts with the creation of a cDNA library. Since the population of mRNAs represents the expressed genes in the cell at a given point of time, they are extracted from different tissues, developmental stages and/or different biotic/abiotic treatments to identify genes expressed under various spatial and temporal conditions. RNA molecules cannot be cloned directly, and therefore have to be converted to cDNA by reverse transcription with the enzyme called reverse transcriptase. Usually, the 3′-end poly (A) tail serves as a priming site during first strand synthesis. After first strand synthesis, the remaining RNA in the sample is removed by adding RNase followed by second strand synthesis to produce double stranded cDNA. Double stranded cDNA molecules are then cloned into a vector and transformed into non-pathogenic *Escherichia coli*. Transformed bacteria are isolated with the help of selection tools like antibiotic resistance or blue and white colony screening. EST sequences are obtained by plasmid isolation and single-pass sequencing reads. In general, cDNAs are sequenced either from the 3′- or 5′-end (Fig. 1). Quality checks of sequences are performed by removing vector and bacterial genomic sequences by using base calling

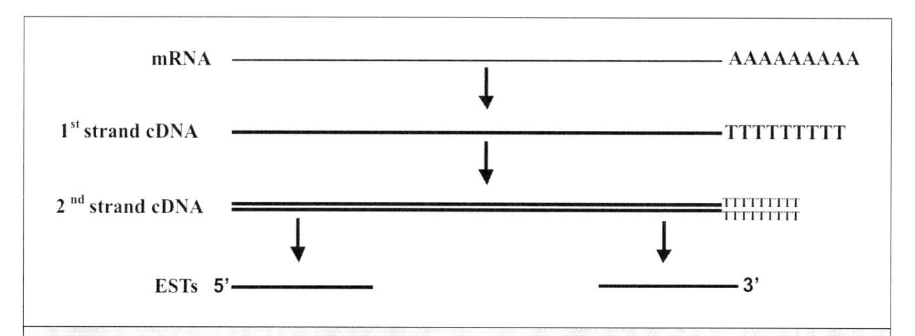

Fig. 1 Schematic diagram of cDNA synthesis from messenger RNA and EST generation from 3' and 5'ends.

Table 1 EST sequence assembly and consensus generation programs.

Name	*Web address*	*Authors*
CAP3	*http://genome.cs.mtu.edu/cap/cap3.html*	(Huang and Madan 1999)
Phrap	*http://www.phrap.org*	(Ewing and Greene 1998)
CLOBB	*http://zeldia.cap.ed.ac.uk/bioinformatics/docs/clobb.htm*	(Parkinson et al. 2002)
megaBLAST	*http://www.ncbi.nlm.nih.gov/blast/megablast.shtml*	(Zhang et al. 2000)
miraEST	*http://www.chevreux.org/projects_mira.html*	(Chevreux et al. 2004)
TIGR_ASSEMBLER	*http://www.tigr.org/software/assembler/*	(Sutton et al. 1995)

software such as "Phred" (Ewing and Greene 1998). High quality ESTs are assembled into contigs using assembly programs (Table 1).

Since the majority of high-throughput EST libraries are sequenced only once, the sequence quality may be poor due to sequencing artifacts such as base calling and base shuttering (Aaronson et al. 1996). Contamination of genomic, vector DNA or adapters are often found in EST sequences. Because of the nature of EST sequences, it is difficult to distinguish between natural sequence variation (SNPs, RNA variations in RNA editing, etc.) and sequence artifacts. Though a gene may consist of thousands of base pairs, with currently available technology only a few hundred bases can be sequenced per EST at one time from one end. Base quality of an EST is poor at the beginning (up to ~100 bp) and increases gradually and then diminishes towards the end of the sequence (~last 100 bp).

Under- and over-representation of transcripts of a cell are observed in EST libraries. Sampling of rare transcripts is difficult for a given time point or tissue, and such transcripts are usually masked by the more abundant ones. Several normalization methods are available to minimize the representation of redundant sequences and by enriching for rarely expressed transcripts. Suppression subtractive hybridization (SSH) is one such normalization method that utilizes two different steps in the isolation of uniquely expressed genes. Initially, a subtraction step is employed, which equalizes or normalizes the sequence concentrations (Diatchenko et al. 1996). In this step, ends of *Rsa*I digested cDNA populations are ligated to special adapters for each sample followed by hybridization of samples to be compared. During this hybridization, common cDNA molecules present within and between samples hybridize together to form double strands. Uniquely expressed molecules remain as single strands. The next step is based on the polymerase chain reaction (PCR), which exponentially amplifies the single strands and suppresses the amplification of double strands by forming hairpin-like loop structures at the ends (Diatchenko et al. 1996). The main drawback of using this technique is that the mRNAs are fragmented, and hence, multiple ESTs may represent the same gene.

3 APPLICATION OF ESTS FOR FUNCTIONAL GENOMICS

3.1 Expression Profiling

Gene expression studies with northern blots, quantitative real time-PCR and dot blots allow investigation of a single or few genes at a time. Biological responses to stimuli are due to the result of coordinated expression of several genes and gene networks. In order to better understand molecular changes at the global level, investigation at the systems level is necessary. Understanding of gene expression at the whole cell and systems level requires the use of new techniques, which combine both efficiency as well as accuracy. In a pioneering effort, one such method called serial analysis of gene expression (SAGE) was developed in the mid-1990s (Velculescu et al. 1995). In this method, a portion of the sequence is isolated from each cDNA strand by tagging with restriction enzymes and adapter linkers. These tagged pieces of sequences are concatamerized to create long sequences and then cloned for sequencing. The resulting sequences are then analyzed to obtain an understanding of the representation of the sequence tags and the number of copies of a particular mRNA by counting those tags. Gene expression profiles in response to a particular stimulus can be revealed by comparing sequence tag counts of a gene under experimental conditions. Results obtained from SAGE have two major problems: In the first problem, the researcher has to ensure that tags and counts are valid representation of transcripts and their level of expression, and the second problem relates to gene assessment based on sequence of tags as these tags are only 9 to 11 base pairs long.

Utilization of ESTs in functional discovery has gradually increased since the early 1990s. With the advent of microarrays, studies on transcriptome profiles in plants and other organisms is now possible.

Structural annotation of gene sequences and gene models relies on experimental evidence provided by ESTs and full-length cDNA sequences. However, functional annotation depends on development and availability of bioinformatic tools to assign gene functions for new genes based on experimental evidence. At the whole-genome level, microarrays provide a high-throughput platform to measure gene expression patterns. DNA microarray technology allows researchers to investigate thousands of genes simultaneously (Brown and Botstein 1999; Park et al. 2004). For those organisms that have complete genome sequence information available, cDNA microarrays permit the study of whole transcriptome expression patterns in a single experiment.

Microarrays are useful in answering a number of biological questions. As an example, microarray experiments can be carried out to find specific marker genes (Richmond and Somerville 2000), which are highly induced or repressed in relation to particular environmental stimuli. Microarrays can be used as exploratory tools to provide clues to the function of a particular set of genes under study (Reymond 2001). Gene regulatory networks can also be inferred for a particular organism in a given environment (Alba et al. 2004; Hashimoto et al. 2004).

Studies utilizing microarrays have been used for research in plant biology to better understand processes ranging from seed filling to cold tolerance among others (Fowler and Thomashow 2002; Alba et al. 2004). In recent years, cDNA microarrays have been used to study global gene expression in response to biotic and abiotic stresses (Thomashow 1999; Shinozaki and Yamaguchi-Shinozaki 2000; Ueda et al. 2006; Zhu 2002; Kathiresan et al. 2006; Manavella et al. 2006). Transcription profiling experiments using microarrays in *Arabidopsis* against cold acclimation revealed the existence of multiple low temperature regulatory pathways in addition to the CBF cold response pathway (Fowler and Thomashow 2002).

In a microarray experiment, thousands of cDNAs or partial and unique gene sequences (oligonucleotides) can be fixed to a solid (glass or silicon chip) surface that allows comparison of gene expression profiles on a global scale. Plant mRNA samples are collected from different experimental treatments or tissues. Equal amounts of mRNA are labeled with fluorescent dyes (such as Cy5 and Cy3) and hybridized to the microarray slides. Fluorescent intensities are then measured through corresponding channels separately. These images are super-imposed and the resulting log ratios of expression of individual genes (transcripts) are analyzed.

Lack of a unified 'language' for exchange of microarray data made it initially difficult to accurately compare data collected from different microarray platforms and from different research groups. In order to better utilize microarray data in the scientific community and to improve research progress, the microarray gene expression data society (MGED) has developed guidelines for the publication of DNA microarray data called MIAME (Minimum Information About a Microarray Experiment). Another public repository named ArrayExpress stores transcriptome data and gene-indexed expression profiles from a curetted subset of experiments in the repository.

Due to the high-throughput nature of current gene expression technologies including microarrays, the fidelity of gene expression data depends on several factors such as experimental conditions, quality of probes, hybridization techniques and stringency of statistical tools used. Validation of microarray results with Northern blots and quantitative real-time PCR is a common practice. Large scale oligonucleotide arrays require prior sequence knowledge (closed transcript profiling) and are based on strain- or genotype-specific information.

cDNA-AFLP is a PCR-based transcript profiling method where prior knowledge of the sequence is not needed and has been used in combination with high-throughput analysis to identify rare transcripts (Reijans et al. 2003). In this method, mRNAs are reverse transcribed to cDNA and digested with two restriction enzymes. Ends of these cDNA fragments are ligated to specific adapters and primers are designed from these adapters, which allow selective amplification during the PCR step.

3.2 Comparative Genomics

Studies in the area of organismal evolution often provide threads of continuity that allow comparative biological analyses to link genes, proteins, genomes, and traits across species and genera. These relational patterns can lead to new knowledge, hypotheses, and predictions about related species (Sorrells et al. 2003). DNA sequences that encode for functional RNAs (and proteins) are conserved from common ancestors and are preserved in contemporary genome sequences (Hardison 2003). Comparative genomics helps in identification and differentiation between gene homologs (genes, which are related through a common ancestor), paralogs (genes arising through duplication events), and orthologs (genes which arise from speciation events).

Cross-genome comparisons based on sequence similarity between related organisms helps in understanding the structure and function of genomes. For example, a comparative map of *Brassica oleracea* and *Arabidopsis* was developed by utilizing *Arabidopsis* ESTs (Lan et al. 2000). Crop improvement programs can apply knowledge of comparative genomics obtained through ESTs to transfer information about genes from model species to related, and more complex nonmodel species, to help identify genes controlling traits of interest, and to assess allelic diversity within species so that the best alleles can be identified and assembled in superior genotypes (Sorrells et al. 2003).

Genome sizes in plant species vary widely ranging from 150 Mb in *Arabidopsis* to 120,000 Mb in lily. Because of this variation in genome size, a direct comparison of the genomes of plant species and subsequent application of this knowledge is impractical in most cases. Therefore, ESTs have emerged as an alternative to explore the gene content of complex genomes (Adams et al. 1991). Availability of vast numbers of ESTs across plant species (Fig. 2), and sophisticated computational tools to predict consensus and contiguous sequences helps in utilizing ESTs for identifying genes and gene families, and to compare and map such ESTs across plant species (Blanca and Wolfea 2004). The Eukaryotic Gene Orthologous (EGO) database formerly called TIGR Orthologous Gene Alignment provides a cross-reference between the eukaryotic genomes.

In a comparative genomic study, wheat ESTs were compared with those from other plant genomes. Out of 119 consensus wheat ESTs considered in this study, 62% of the ESTs significantly matched to rice BACs and 22% showed homology with *Arabidopsis* genomic sequence from five different chromosomes (Miftahudin et al. 2004). In another study, the relative age of gene duplication was investigated in wheat, maize, tetraploid cotton, diploid cotton, tomato, potato, soybean, *Medicago*, and *Arabidopsis*, utilizing ESTs corresponding to unigenes (Blanca and Wolfea 2004). An EST based comparative map was constructed for *Brassica olaracea* and *Arabidopsis* utilizing *Arabidopsis* ESTs (Lan et al. 2000).

3.3 Physical Mapping

From public repositories, such as GenBank, EST sequence data is available for approximately 212 plant species, which include the majority of crop plants, and other agronomically important plant species (Table 2). For plant species with a large number of ESTs, these can be mapped to genomic sequences to provide insights into the organization of genomes and chromosomes. In many cases, most ESTs map to the euchromatic regions of the chromosomes. A large genome size, polyploidy, and the

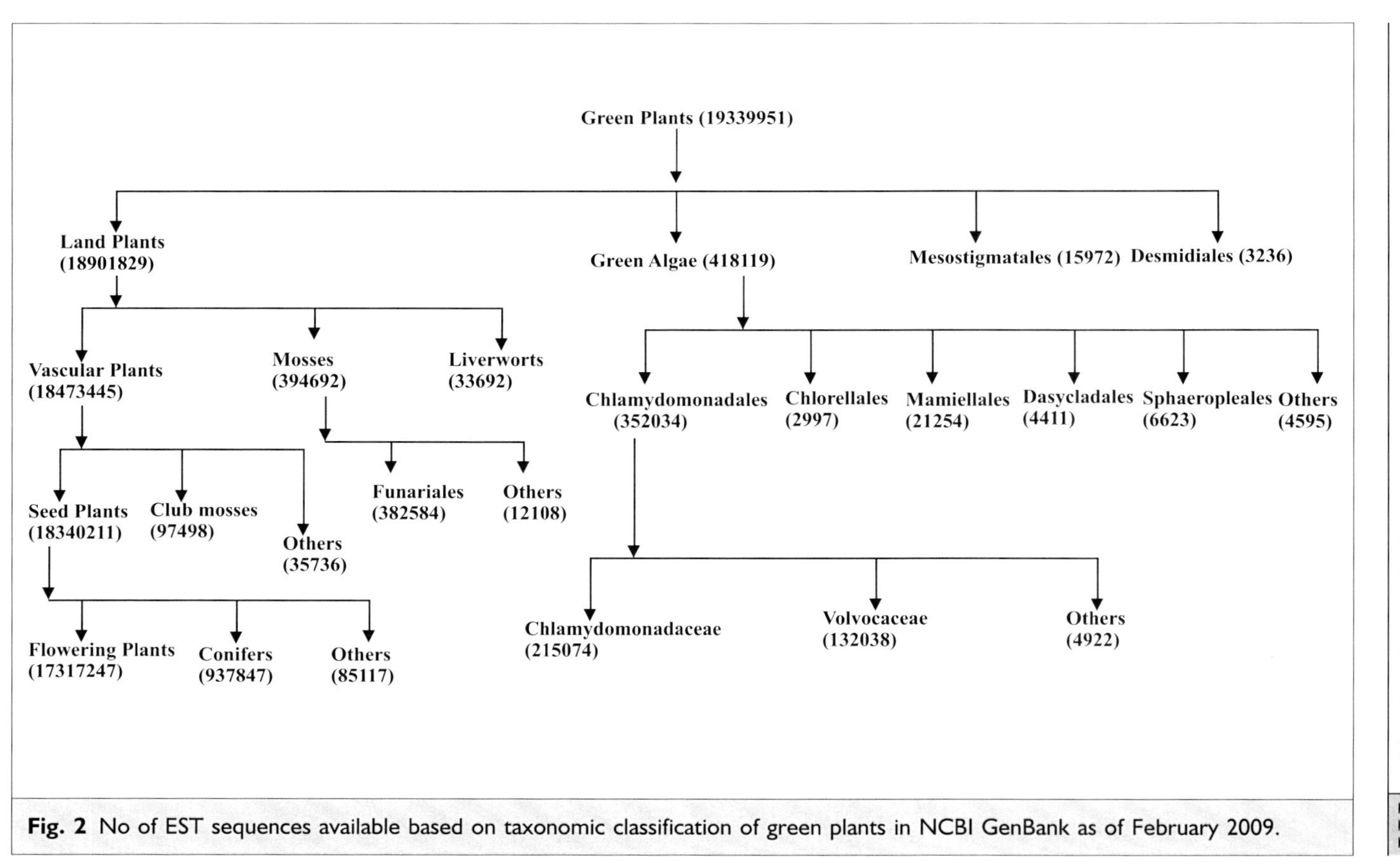

Fig. 2 No of EST sequences available based on taxonomic classification of green plants in NCBI GenBank as of February 2009.

Table 2 Number of EST sequences available for different plant species in the order of highest abundance in NCBI GenBank as of February 2009.

Organism	*Number of ESTs*
Zea mays	2018337
Arabidopsis thaliana	1527298
Glycine max	1386618
Oryza sativa	1248661
Triticum aestivum	1064009
Oryza sativa Japonica Group	985283
Brassica napus	596361
Hordeum vulgare	525527
Panicum virgatum	436535
Hordeum Vulgare subsp.*vulgare*	415133
Phaseolus coccineus	391138
Physcomitrella patens	382584
Physcomitrella patens subsp.*patens*	362131
Vitis vinfera	353705
Pinus taeda	328628
Picea glauca	284329
Gossypium hirsutum	268779
Solanum lycopersicum	260581
Medicago truncatula	260238
Malus x domestica	256249
All other taxa	7977211

presence of repetitive elements make genome analysis complicated in the majority of crop species. In wheat, despite all the above complications, EST mapping was accomplished successfully by using deletion mapping (Qi et al. 2004) to chromosomal bins. This collaborative project took advantage of the availability of aneuploid genetic stocks and several cDNA libraries (Qi et al. 2004). Nearly 6,426 ESTs representing unigenes were mapped to the wheat genome by deletion mapping. Deletion mapping was performed by hybridizing cDNA clones to Southern blots of DNA from the panel of wheat genetic stocks, each missing a terminal portion of chromosome arm (Qi et al. 2003). In a rice EST physical mapping study, nearly 6,713 EST sequences derived from 19 cDNA libraries were mapped to yeast artificial chromosome (YACs)-based map (Wu et al. 2002).

3.4 Single Nucleotide Polymorphism (SNP) Discovery

SNPs are single nucleotide or small insertion-deletion polymorphisms among haplotypes. Since DNA is made up of four nucleotides-Adenine (A), Thymine (T), Cytosine (C) and Guanine (G) at a given position in the genome, any of these four bases may be present. SNP variation occurs when a single nucleotide for example 'C' at a given position is replaced by any of the other three nucleotides. This variation is due to either transition: purine-purine (A and G) or pyrimidine-pyrimidine (C and T) exchanges, or transversion: purine-pyrimidine or pyrimidine-purine exchanges (Vignal et al. 2002).

SNPs can be used for genome-wide linkage disequilibrium and association studies (Rafalski 2002). SNPs can be identified in several ways, which include:

a) Direct sequencing of genomic PCR products from a target region in the genome; as an example, if we are trying to study candidate genes that may be responsible for a particular phenotype, sequencing and analysis for SNPs can be performed on a number of different genotypes, which differ phenotypically. However, on a large scale, this approach tends to be costly due to the need for locus-specific primers, and application of high-throughput techniques is difficult.

b) Comparing genomic sequences; as an example, comparison of sequences from overlapping bacterial artificial chromosome (BAC) clones from different individuals of the same species (Sachidanandam et al. 2001). This approach may not possible in all conditions, because in many genome sequencing projects, the genotypes utilized for sequencing are not optimally diverse (Sachidanandam et al. 2001).

c) Comparing sequences from different EST projects; this approach has been useful for the majority of crop species. EST sequences are available from a range of tissues and in some cases from different varieties. ESTs attract additional interest because of a greater chance of SNPs being in the coding region and hence having an influence on phenotype association (Picoult-Newberg et al. 1999; Remington et al. 2001). As an example, an average of one SNP for every 540 bp was found between wheat genotypes. SNPs found in wheat were then mapped to genome using segregating populations; upon validation can be applied to conventional genetic studies (Somers et al. 2003).

3.5 Gene Discovery

EST sequencing is the most common technique utilized for identifying genes from moderate to highly abundant transcripts (Gai et al. 2000). Eukaryotic mRNA sequences from 5′- and 3′-UTRs play a major role in gene regulation, expression and translation (Mignone et al. 2003). The 5′-end sequence of ESTs are useful in verifying promoter regions and finding the first exon of a gene in eukaryotes (Schimid et al. 2006). Poly (A) tails from the 3′-UTRs play a role in predicting gene boundaries and genes from genomic sequences (Kan et al. 2001), post-transcriptional regulation (Gautheret et al. 1998), and mRNA metabolism (Mignone et al. 2003).

New genes can also be identified by comparing uncharacterized mutants to the larger expression datasets of reference profiles (compendium of expression profiles), by matching to similar expression profiles observed under different conditions from the same species, or by comparing expression profiles of different species under similar experimental conditions (co-expression) (Rensink and Buell 2005).

3.6 Prediction of Alternative Splice Variants

Alternative splicing of pre-mRNAs is a powerful and versatile regulatory mechanism that can effectively control quantitative gene expression and functional diversification of proteins that play a role in major developmental decisions (Lopez 1998; Kan et al. 2001). As reported in large scale genomic studies, alternative splicing has been observed in 30 to 60% of human genes. Computational databases like the Human Alternative Splicing Database (HASDB) and 'the intronerator' predict alternative splice variants by aligning ESTs to mRNA or genomic sequences. Majority of the bioinformatic tools first identify candidate splice variants by aligning ESTs derived from the same gene to its corresponding mRNA sequence by looking for large insertions or deletions in EST sequences. Then, each candidate splice variant can be further assessed by aligning corresponding ESTs to the exact gene sequence in the reference genome (Brett et al. 2000). An alternative splice detection study that utilized TIGR Human Gene Index identified 133 alternatively spliced genes and estimated that at least 35% of human genes are alternatively spliced (Mironov et al. 1999). In another study, approximately 3,011 alternatively spliced genes were identified by aligning ESTs to mRNA and estimated that 38% of human genes are alternatively spliced (Brett et al. 2000).

4 CONCLUSIONS

A wealth of sequence information is publicly available for broad array of plant species and other organisms. As the cost of production of ESTs is inexpensive compared to whole-genome sequencing projects, there is a greater chance of obtaining expressed sequences from different plant species. Depending upon the genome size, economic value, and availability of genome sequence from closely related species, ESTs can be very useful towards the ultimate goal of a complete genome sequence, and a well-characterized transcriptome can be utilized to speed up the genome sequencing process (Nagaraj et al. 2007). With regards to EST analysis and data processing, though there are numerous tools available for clustering and assembly, majority of these tools were developed for processing genomic DNA. There is a need for developing improved tools aimed at processing EST sequences. As applications of ESTs increase, appropriate methods and tools have to be chosen to achieve specific goals.

Acknowledgements

The authors thank Dr. Sathya Elavarthi for offering helpful suggestions and reviewing this chapter. KM was supported by USDA-CSREES grant no. 2007-03-409 to VK.

References

Aaronson J, Eckman B, Blevins R, Borkowski J, Myerson J, Imran S, Elliston K (1996) Toward the development of a gene index to the human genome: an assessment of the nature of high-throughput EST sequence data. Genome Res 6: 829-845

Adams M, Kelley J, Gocayne J, Dubnick M, Polymeropoulos M, Xiao H, Merril C, Wu A, Olde B, Moreno R, Kerlavage A, McCombie W, Venter J (1991) Complementary DNA sequencing: Expressed sequence tags and Human Genome Project. Science 252: 1651-1656

Alba R, Fei Z, Payton P, Liu Y, Moore S, Debbie P, Cohn J, D'Ascenzo M, Gordon J, Rose J, Martin G, Tanksley S, Bouzayen M, Jahn M, Giovannoni J (2004) ESTs, cDNA microarrays, and gene expression profiling: tools for dissecting plant physiology and development. Plant J 39: 697-714

Blanca G, Wolfea K (2004) Widespread paleopolyploidy in model plant species inferred from age distributions of duplicate genes. Plant Cell 16: 1667-1678

Brett D, Hanke J, Lehmann G, Haase S, Delbrück S, Krueger S, Reich J, Bork P

(2000) EST comparison indicates 38% of human mRNAs contain possible alternative splice forms. FEBS Lett 474: 83-86

Brown PO, Botstein D (1999) Exploring the new world of the genome with DNA microarrays. Nat Genet 21: 33-37

Chevreux B, Pfisterer T, Drescher B, Driesel A, Müller W, Wetter T, Suhai S (2004) Using the miraEST assembler for reliable and automated mRNA transcript assembly and SNP detection in sequenced ESTs. Genome Res 14: 1147-1159

Diatchenko L, Lau YF, Campbell AP, Chenchik A, Moqadam F, Huang B, Lukyanov S, Lukyanov K, Gurskaya N, Sverdlov ED, Siebert PD (1996) Suppression subtractive hybridization: a method for generating differentially regulated or tissue-specific cDNA probes and libraries. Proc Natl Acad Sci USA 93: 6025-6030

Ewing B, Greene P (1998) Base-calling of automated sequencer traces using phred. II. error possibilities. Genome Res 8: 186-194

Fowler S, Thomashow F (2002) Arabidopsis transcriptome profiling indicates that multiple regulatory pathways are activated during cold acclimation in addition to the CBF cold response pathway. Plant Cell 14: 1675-1690

Gai X, Lal S, Xing L, Brendal V, Walbot V (2000) SNP frequency, haplotype structure and linkage disequilibrium in elite maize inbred lines. Nucl Acid Res 28: 94-96

Gautheret D, Poirot O, Lopez F, Audic S, Claverie JM (1998) Alternate polyadenylation in human mRNAs: a large-scale analysis by EST clustering. Genome Res 8: 524-530

Hardison R (2003) Comparative Genomics. PLOs Biol 1:e58

Hashimoto R, Kim S, Shmulevich I, Zhang W, Bittner M, Dougherty E (2004) Growing genetic regulatory networks from seed genes. Bioinformatics 20: 1241-1247

Huang X, Madan A (1999) CAP3: A DNA sequence assembly program. Genome Res 9: 868-877

Kan Z, Rouchka EC, Gish WR, States DJ (2001) Gene structure prediction and alternative splicing analysis using genomically aligned ESTs. Genome Res 11: 889-900

Kathiresan A, Lafitte H, Chen J, Mansueto L, Bruskiewich R, Bennett J (2006) Gene expression microarrays and their application in drought stress research. Field Crops Res 97: 101-110

Lan TH, DelMonte TA, Reischmann KP, Hyman J, Kowalski SP, McFerson J, Kresovich S, Paterson AH (2000) An EST-enriched comparative map of *Brassica oleracea* and *Arabidopsis thaliana*. Genome Res 10: 776-788

Lopez J (1998) Alternative splicing of PremRNA: Developmental consequences and mechanisms of regulation. Annu Rev Genet 32: 279-305

Manavella P, Arce A, Dezar C, Bitton F, Renou J, Crespi M, Chan R (2006) Cross-talk between ethylene and drought signalling pathways is mediated by the sunflower Hahb-4 transcription factor. Plant J 48: 125-137

Miftahudin, Ross K, Ma XF, Mahmoud AA, Layton J, Milla MAR, Chikmawati T, Ramalingam J, Feril O, Pathan MS, Momirovic GS, Kim S, Chema Y, Fang P, Haule L, Struxness H, Birkes J, Yaghoubian C, Skinner R, McAllister J, Nguyen V, Qi LL, Echalier B, Gill BS, Linkiewicz AM, Dubcovsky J, Akhunov ED, Dvorak AJ, Dilbirligi M, Gill KS, Peng JH, Lapitan NLV, Bermudez-Kandianis CE, Sorrells ME, Hossain KG, Kalavacharla V, Kianian SF, Lazo GR, Chao S, Anderson OD, Gonzalez-Hernandez J, Conley EJ, Anderson JA, Choi DW, Fenton RD, Close TJ, McGuire PE, Qualset CO, Nguyen HT, Gustafson JP (2004) Analysis of expressed sequence tag loci on wheat chromosome group 4. Genetics 168: 651-663

Mignone F, Gissi C, Liuni S, Pesole G (2003) Untranslated regions of mRNAs. Genome Biol 3: 1-10

Mironov AA, Fickett J, Gelfand M (1999) Frequent alternative splicing of human genes. Genome Res 9(12): 1288-1293

Nagaraj SH, Gasser RB, Ranganathan S (2007) A hitchhiker's guide to expressed sequence tag (EST) analysis. Brief Bioinform 8: 6-21

Park CH, Jeong HJ, Jung JJ, Lee GY, Kim SC, Kim TS, Yang SH, Chung HC, Rha SY (2004) Fabrication of high quality cDNA microarray using a small amount of cDNA. Int J Mol Med 13: 675-679

Parkinson J, Guiliano D, Blaxter M (2002) Making sense of EST sequences by CLOBBing them. BMC Bioinform 3: 31

Picoult-Newberg L, Ideker T, Pohl M, Taylor S, Donaldson M, Nickerson D, Boyce-Jacino M (1999) Mining SNPs from EST databases. Genome Res 9: 167-174

Qi L, Echalier B, Friebe B, Gill BS (2003) Molecular characterization of a set of wheat deletion stocks for use in chromosome bin mapping of ESTs. Funct Integr Genom 3: 39-55

Qi LL, Echalier B, Chao S, Lazo GR, Butler GE, Anderson OD, Akhunov ED, Dvorak J, Linkiewicz AM, Ratnasiri A, Dubcovsky J, Bermudez-Kandianis CE, Greene RA, Kantety R, La Rota CM, Munkvold JD, Sorrells SF, Sorrells ME, Dilbirligi M, Sidhu D, Erayman M, Randhawa HS, Sandhu D, Bondareva SN, Gill KS, Mahmoud AA, Ma XF, Miftahudin, Gustafson JP, Conley EJ, Nduati V, Gonzalez-Hernandez JL, Anderson JA, Peng JH, Lapitan NLV, Hossain KG, Kalavacharla V, Kianian SF, Pathan MS, Zhang DS, Nguyen HT, Choi DW, Fenton RD, Close TJ, McGuire PE, Qualset CO, Gill BS (2004) A chromosome bin map of 16,000 expressed sequence tag loci and distribution of genes among the three genomes of polyploid wheat. Genetics 168: 701-712

Rafalski A (2002) Applications of single nucleotide polymorphisms in crop genetics. Curr Opin Plant Biol 5: 94-100

Reijans M, Lascaris R, Groeneger A, Wittenberg A, Wesselink E, van Oeveren J, de Wit E, Boorsma A, Voetdijk B, van der Spek H, Grivell L, Simons G

(2003) Quantitative comparison of cDNA-AFLP, microarrays, and genechip expression data in *Saccharomyces cerevisiae*. Genomics 82: 606-618

Remington D, Thornsberry J, Matsuoka Y, Wilson L, Whitt S, Doebley J, Kresovich S, Goodman M, Buckler E (2001) Structure of linkage disequilibrium and phenotypic associations in the maize genome. Proc Natl Acad Sci USA 11: 479-484

Rensink WA, Buell CR (2005) Microarray expression profiling resources for plant genomics. Trends Plant Sci 10: 603-609

Reymond P (2001) DNA microarrays and plant defence. Plant Physiol Biochem 39: 313-321

Richmond T, Somerville S (2000) Chasing the dream: plant EST microarrays. Curr Opin Plant Biol 3: 108-116

Rudd S, Frisch M, Grote K, Meyers BC, Mayer K, Werner T (2004) Genome-wide *in silico* mapping of scaffold/matrix attachment regions in Arabidopsis suggests correlation of intragenic scaffold/matrix attachment regions with gene expression. Plant Physiol 135: 715-722

Sachidanandam R, Weissman D, Schmidt S, Kakol J, Stein L, Marth G, Sherry S, Mullikin J, Mortimore B, Willey D, Hunt S, Cole C, Coggill P, Rice C, Ning Z, Rogers J, Bentley D, Kwok P, Mardis E, Yeh R, Schultz B, Cook L, Davenport R, Dante M, Fulton L, Hillier L, Waterston R, McPherson J, Gilman B, Schaffner S, Van Etten W, Reich D, Higgins J, Daly M, Blumenstiel B, Baldwin J, Stange-Thomann N, Zody M, Linton L, Lander EDA (2001) A map of human genome sequence variation containing 1.42 million single nucleotide polymorphisms. Nature 409: 928-933

Schimid C, Perier R, Praz V, Bucher P (2006) EPD in its twentieth year: towards complete promoter coverage of selected model organisms. Nucl Acid Res 34: D82-D85

Shinozaki K, Yamaguchi-Shinozaki K (2000) Molecular responses to dehydration and low temperature: differences and cross-talk between two stress signaling pathways. Curr Opin Plant Biol 3: 217-223

Shoemaker R, Keim P, Vodkin L, Retzel E, Clifton S, Waterston R, Smoller D, Coryell V, Khanna A, Erpelding J, Gai X, Brendel V, Raph-Schmidt C, Shoop E, Vielweber C, Schmatz M, Pape D, Bowers Y, Theising B, Martin J, Dante M, Wylie T, Granger C (2002) A compilation of soybean ESTs: generation and analysis. Genome 45: 329-338

Somers DJ, Kirkpatrick R, Moniwa M, Walsh A (2003) Mining single-nucleotide polymorphisms from hexaploid wheat ESTs. Genome 46: 431-437

Sorrells M, La Rota M, Bermudez-Kandianis C, Greene R, Kantety R, Munkvold J, Miftahudin, Mahmoud A, Ma X, Gustafson P, Qi L, Echalier B, Gill B, Matthews D, Lazo G, Chao S, Anderson O, Edwards H, Linkiewicz A, Dubcovsky J, Akhunov E, Dvorak J, Zhang D, Nguyen H, Peng J, Lapitan N, Gonzalez-Hernandez J, Anderson J, Hossain K, Kalavacharla V, Kianian S, Choi D, Close T, Dilbirligi M, Gill K, Steber C, Walker-Simmons M, McGuire P, Qualset C (2003) Comparative DNA sequence analysis of wheat and rice genomes. Genome Res 13: 1818-1827

Sutton GG, White O, Adams MD, Kerlavage AR (1995) TIGR assembler: A new tool for assembling large shotgun sequencing projects. Genom Sci Technol 1: 9-19

Thomashow M (1999) Plant cold acclimation:freezing tolerance genes and regulatory mechanisms. Annu Rev Plant Physiol Mol Biol 50: 571-599

Ueda A, Kathiresan A, Bennett J, Takabe T (2006) Comparative transcriptome analyses of barley and rice under salt stress. Theor App Genet 112: 1286-1294

Van der Hoeven R, Ronning C, Giovannoni J, Martin G, Tanksley S (2002) Deductions about the number, organization, and evolution of genes in the tomato genome based on analysis of a large expressed sequence tag collection and selective genomic sequencing. Plant Cell 14: 1441-1456

Velculescu V, Zhang L, Vogelstein B, Kinzler K (1995) Serial analysis of gene expression. Science 270: 484-487

Vignal A, Milan D, SanCristobal M, Eggen A (2002) A review on SNP and other types of molecular markers and their use in animal genetics. Genet Sel Evol 34: 275-305

Wu JZ, Maehara T, Shimokawa T, Yamamoto S, Harada C, Takazaki Y, Ono N, Mukai Y, Koike K, Yazaki J, Fujii F, Shomura A, Ando T, Kono I, Waki K, Yamamoto K, Yano M, Matsumoto T, Sasaki T (2002) A comprehensive rice transcript map containing 6591 expressed sequence tag sites. Plant Cell 14: 525-535

Zhang Z, Schwartz S, Wagner L, Miller W (2000) A greedy algorithm for aligning DNA sequences. J Comput Biol 7: 203-214

Zhu J (2002) Salt and drought stress signal transduction in plants Annu Rev Plant Biol 54: 247-273

5 Whole Genome Sequencing

Swarup K. Parida and T. Mohapatra*

National Research Centre on Plant Biotechnology,
Indian Agricultural Research Institute,
New Delhi 110012, India

*Corresponding author: tm@nrcpb.org

1 INTRODUCTION

DNA sequencing is a method to determine the exact sequence of nucleotides in a sample of DNA. The most popular method for DNA sequencing is the dideoxy method or Sanger method. This method is named after its inventor, Frederick Sanger, who was awarded the 1980 Nobel Prize in chemistry for this achievement. The dideoxy method gets its name from the synthetic nucleotides that lack the -OH group at the 3′ carbon atom of the deoxyribose sugar moiety, which play a critical role. When a dideoxynucleotide (ddNTP) gets added to the growing DNA strand, the chain elongation stops because there is no 3′ -OH for the next nucleotide to be incorporated. For this reason, the dideoxy method is also known as chain termination method. In the presence of all the four normal nucleotides, chain elongation proceeds normally until the DNA polymerase incorporates a dideoxy nucleotide instead of the normal deoxynucleotide. Since the ratio of normal nucleotide to the dideoxy versions is kept high in the sequencing reaction, some DNA strands will have dideoxy version that halts further DNA polymerization only after addition of several normal nucleotides. After the sequencing reaction, the fragments are separated based on their length using a high-resolution separation method. Two fragments differing in length even by a single nucleotide get separated from each other precisely. The original Sanger's method employed radiolabeled dideoxy nucleotides that required four reactions for four nucleotides to be carried out in four different tubes and separated in four different lanes. Use of fluorescent dyes, however, allows a single tube reaction and a single lane separation enabling automation of the whole process. A standard DNA sequencing reaction gives about 500 nucleotide long sequence at one go, which is negligible when compared to large eukaryotic genomes which are million or even billion nucleotides long. Sequencing of the whole genome of an organism obviously requires different strategies while using the same basic DNA sequencing technology. The objective of this chapter is to provide an outline of whole genome sequencing, emphasizing on new generation technologies and the associated requirements of computational facilities.

2 ENABLING TECHNOLOGIES FOR WHOLE GENOME SEQUENCING

2.1 Fluorescence Based DNA Labeling and Capillary Electrophoresis

Fluorescence based DNA labeling has multiple applications in the area of genomics that includes whole genome sequencing, automated fragment analysis for high-throughput genotyping of fluorescently labeled molecular markers, and detection of mutation like single nucleotide polymorphisms (SNPs) and Insertion/Deletion. The fluorescent dyes, most commonly used in DNA sequencing, are xanthine like fluorescein and rhodamine dyes, which are mostly excited through Argon at 488 nm and most commonly used in DNA sequencers such as gel based *ABI PRISM* 373/377 (Applied Biosystems) and capillary based MegaBACE 1000/4000 (Amersham-GE Healthcare) and *ABI PRISM* 3100/3700. The other classes of fluorescent dyes like Cyanine dyes (Cy5.0 and Cy5.5) are used in ALFexpress™ and SEQ 4x4™ (Amersham-GE Healthcare), and Clipper™ and Long-Read Tower™ (Bayer Diagnostics) for sequencing purposes. Besides, Bodipy dyes (Metzker et al. 1996) and IR dyes (*IRDye*700 and *IRDye*800) are used in LiCOR DNA Analyzer for mutation detection more efficiently. For dye primer sequencing, primers are usually labeled with 5′ end of the phosphate group of one nucleotide. In contrast for dye terminator sequencing, the dyes are attached at C-5 position of the pyrimidines (ddC and ddU) and C-7 position of 7-deazapurines (7-deaza-ddA and 7-deaza-ddG) with a propargyl-amino linker. Four different fluorescent dyes namely, succinyl fluorescein labeled ddNTPs were first discovered by Prober and coworkers (Prober et al. 1987) and used in MegaBACE Automated DNA Sequencer. The first fluorescent automated DNA Sequencing using four different fluroscent dye-labeled primers were reported by Smith et al. (1986). Later on, the fluoresceins have been replaced with rhodamine dyes (Bergot et al. 1994) used in ABI Capillary DNA Sequencer. With the use of *Thermo Sequenase*™ DNA polymerase, a single color Cyanine dye-labeled terminator is highly preferred for generating high quality sequences in automated sequencer. Besides, the Cyanine dye terminators (Duthie et al. 2002), dichlororhodamine terminators and energy transfer (ET) dye-labeled terminators (Kumar et al. 2005) have also been found effective for sequencing. The rhodamine dyes including R110, REG, TAMRA and ROX are currently the most commonly used dyes for DNA sequencing. Based on sequencing band uniformity and read-length, an optimized four color ET terminator set of FAM-R110, FAM-REG, FAM-TAMRA, and FAM-ROX are now available from GE Healthcare as fluorescence resonance energy transfer (FRET) based DYEnamic ET terminators. ET terminators are usually brighter than the single dye-labeled terminators. For instance, FAM-ROX-labeled ET terminator is more than 18-fold brighter than the single ROX-labeled terminator. ET terminator has been replaced by *Big dye* terminator in ABI Sequencer, which is based on the use of 4-aminomethyl benzoic acid as linker.

The sequencing takes place in a 'cycle sequencing' reaction, in which several cycles of template denaturation, primer annealing and primer extension are performed. The primer complementary to the known sequence immediately flanking the region of interest is used to prime DNA synthesis. Each round of primer extension is stochastically terminated by the incorporation of fluorescently labeled ddNTPs. In the resulting mixture of end-labeled extension products, the label on the terminating ddNTP of any given fragment corresponds to the nucleotide identity of its terminal position. The automated sequencer/fragment analysis system uses capillary array electrophoresis to perform fragment size separation. Prior to sample run, the instrument fills the capillaries with sieving matrix. Then a voltage pulse is applied to electro-kinetically inject a portion of the fluorescent labeled samples from each well of a plate simultaneously into the capillaries. The DNA fragments present in the samples are separated on the basis of size with the shorter fragments moving faster through the matrix. The laser beam is utilized by the instrument to excite and scan fluorescent dyes in the sample, which in turn emits fluorescence. A four-color detection of emission spectra provides the readout that is represented in a Sanger sequencing 'trace'. Software translates the traces into DNA sequence with error probabilities for each base-call, which is subsequently used for quality checking, sequence assembly and detection of variation.

2.2 Automated Liquid Handling and Colony Picking

Handling and sequencing of large numbers of samples required in standard genome sequencing projects demand high level of automation in order to obtain high-throughput and consistent quality. The automation of DNA sequencing has resulted in a vast number of different commercially available robotic workstations. These range from small semi-automatic instruments for liquid handling, like a plate filler (e.g., Hydra, Robbins Scientific), to huge robots capable of performing several operations simultaneously such as the Genesis RSP (Tecan). Pipetting heads have evolved from a single tip to 384 tips, the most commonly used ones having eight tips (Biomek™ 1000, BekmanCoulter) or 96 tips (Microlab 4200, Hamilton; PlateMate™ , Matrix; Biorobot 8000, Qiagen). Most of these workstations can be used to set up cycle sequencing reactions. Many liquid-handling workstations can also be equipped with vacuum manifolds or magnetic separation stations, enabling a number of purification techniques to be automated. Several workstations are now commercially available (Magnatrix 1200, Magnetic Biosolutions AB), which utilize a movable magnet in order to perform the magnetic separation in the tips, enabling fast and efficient bead recovery. Colonies or plaques can be automatically picked and placed into microtiter plates using a robot equipped with a camera and one or several needles. This saves time and manual labor involved in picking colonies of thousands of sub-clones for sequencing. However, the currently available robots are not as good as a trained human at deciding if a colony is worth picking.

2.3 Purification of Cycle Sequencing Reaction Products

Faster handling of a large number of sub-clones also requires automated purification methods. Considerable efforts have been made by a large number of research groups to develop such methods for sequencing products, which are fast, cheap and efficient. These methods have their specific advantages and disadvantages. Suitability of a purification method depends on the sequencing chemistry and separation platform used. For instance, the use of dye terminators demands more thorough purification than the use of dye primers, since the excess labeled terminators will otherwise result in dye blobs that obscure part of the sequence. The importance of purifying the cycle sequencing products has increased with the introduction of capillary sequencers since they are more sensitive to salt and other impurities than slab gels. The different purification methods are divided into three major categories depending on whether they are based on precipitation, spin columns/ filter plates, or magnetic beads.

2.3.1 Precipitation Techniques

Traditionally, purification of cycle sequencing reaction products is carried out by ethanol or isopropanol precipitation, which is usually followed by a 70% ethanol wash, and sometimes preceded by a phenol/ chloroform extraction. Addition of salts such as sodium acetate, ammonium acetate or magnesium chloride- can improve the performance of ethanol precipitation. It is a very cheap method that does not require any expensive equipment. However, a major limitation of this method is the co-precipitation of template along with the sequencing products. Besides, excess salt and terminators may also be precipitated if the precipitation is not properly optimized. In addition, it is difficult to automate the process, due to the centrifugation steps. Another disadvantage is the variable yield of sequencing products. A modified precipitation protocol, in which n-butanol is used instead of ethanol has also been described (Tillett and Neilan 1999). It has several advantages over traditional ethanol precipitation, since it gives higher yields especially of short DNA fragments, requires shorter centrifugation steps, co-precipitates less salt, and avoids the need for a washing step.

2.3.2 Filtration Methods

Based on the principle of filtration, a large number of commercial kits have been made for the removal of dye terminators from the sequencing products. For a low throughput application, spin columns are generally used, while for medium- and high-throughput applications, filter plates with 96 or 384 wells are used. One approach employs filling of the columns or filter plate wells with a gel separation matrix consisting of spheres with uniformly sized pores. When the cycle sequencing products are passed through the matrix the small molecules, such as salt and nucleotides diffuse into the pores where they are retained. In contrast, longer DNA fragments easily pass through the matrix and are recovered in the filtrate. In another approach, the filter used acts as a molecular sieve, allowing small molecules to pass while larger DNA fragments are retained. The sequencing products can then be obtained by resuspension in the desired loading buffer.

2.3.3 Magnetic Bead Techniques

Magnetic beads are in use in a number of procedures such as solid phase sequencing (Uhlen et al. 1992; Wahlberg et al. 1992), in vitro mutagenesis (Hultman et al. 1990), gene assembly (Stahl et al. 1993), solid-phase cloning (Hultman and Uhlen 1994), immunomagnetic separation (Stark et al. 1996), diagnostic detection assays (O'Meara et al. 1998), preparation of single-stranded DNA for pyrosequencing, and purification of PCR or cycle sequencing products. Magnetic bead assays are easy to automate using liquid handling robots equipped with magnetic stations. In addition, buffer exchanges are efficient and fast, enabling thorough washes of the captured moiety. Magnetic beads have been utilized in a number of ways for the purification of sequencing products, some of which are described below.

2.3.3.1 Hybridization Based Techniques

Specific oligonucleotides coupled to magnetic beads have been used for the purification of single-stranded DNA. Due to high specificity of hybridization, this approach can be used in diagnostics for extracting viral particles that are present in very low concentrations (Millar et al. 1995; O'Meara et al. 1998). Another possible application is the purification of M13 templates for sequencing (Johnson et al. 1996).

2.3.3.2 Streptavidin-biotin Based Methods

For purification of DNA, Streptavidin-biotin system is considered highly efficient. The non-covalent interaction involved in this interaction is very stable that allows harsh washes to remove all contaminants. Several groups have described the use of biotin and streptavidin for purification of sequencing products (Fangan et al. 1999; Ju 2002). The sequencing reaction is often performed using a biotinylated primer, which is either labeled internally when dye primers are used (Tong and Smith 1992, 1993) or 5′-labeled when dye terminators are used (Fangan et al. 1999).

2.4 Capture of Dideoxynucleotides

This method works in case of dye terminator sequencing chemistry. Instead of capturing the desired sequencing products, as described above for the other methods, the unincorporated dye terminators are specifically removed (Springer et al. 2003). Since no washes of the beads are necessary, this method is faster than other magnetic bead approaches. However, it fails to remove template, salt and other impurities from the sample.

2.5 Methods for Unspecific Capture of DNA

Solid-phase reversible immobilization (SPRI) is a purification method in which DNA is precipitated onto the surface of carboxylated magnetic particles. After washing, the DNA can be eluted using water. In the original SPRI protocol, polyethylene glycol (PEG) and sodium chloride were used to precipitate the DNA onto the beads (Hawkins et al. 1994). This is not a suitable approach for the purification of sequencing products since capillary sequencers are sensitive to salt. Instead, an ethanol solution containing tetraethylene glycol (TEG) is used to precipitate the DNA onto the beads, followed by a 70% ethanol wash (Elkin et al. 2002). Large templates, e.g., BACs or rolling circle amplified DNA, will remain bound to the beads under the conditions used to elute sequencing products, while smaller templates will be co-eluted. This method can be useful for sequencing projects where, for example, primer walking is used, since there is no requirement for modified primers or universal sequence in the products.

3 WHOLE GENOME SEQUENCING APPROACHES

As exemplified by the human genome, two distinct approaches are being employed for complete sequencing of eukaryotic genomes: (i) ordered clone-by-clone approach, and (ii) whole genome shotgun approach. The public funded Human Genome Project employed the first approach, while the private company Celera Genomics followed the second strategy in case of human genome. In case of plants, the *Arabidopsis* Genome Initiative followed the same method as the Human Genome Project. The sequencing of the rice genome carried out by the International Rice Genome Sequencing Project (IRGSP) was also based on the first approach.

3.1 Clone-by-clone Approach

This approach essentially requires a complete genome map based on a large number of DNA markers and a large fragment (>100 kb) genomic library with at least ten-fold coverage of the genome of the given species. In case of rice for instance (Fig. 1), a genetic map containing more than 3,000 genetic markers had been constructed, which was used to develop a physical map of the genome using large fragment DNA libraries. Three different large fragment libraries using yeast artificial chromosomes (YACs), bacterial artificial chromosomes (BACs) and P1-derived artificial chromosomes (PACs) were constructed, which carried different rice genomic DNA fragments obtained by partial restriction digestion using enzymes such as *Mbo*I, *Sau3*A, EcoRI and *Hind*III. The BAC and PAC clones have been mostly used in sequencing, while the YACs were used for gap-filling and anchoring of expressed sequence tags (ESTs). Prior to their use in sequencing, however, the BAC/PAC clones are anchored onto the genetic map to construct physical maps of the genome. For this, the genetic markers are used as probes to screen the BAC/PAC libraries. Those clones, which hybridize are then placed against the marker(s) along the chromosome to generate a physical map of the marker region. Fingerprinting of the large fragment DNA clones helps identification of clones with partial overlap based on fragment sharing. Software, FingerPrinted Contig (FPC), helps identification such overlaps and thus putting together a contiguous set of clones in a genomic region (called a contig). This way, using markers from all the genomic regions, contigs of varying size are anchored onto the chromosomes. A saturated genetic map provides scope for construction of a whole genome physical map. However, either due to absence of DNA markers in certain genomic regions or due to poor representation of those regions in the large fragment genomic libraries, which are especially prepared by using restriction enzymes, gaps are observed in the physical maps. Such physical gaps are closed by walking from the ends of the contigs already anchored by markers flanking the gap. Alternative genomic libraries are also used to fill some physical gaps. Once all/most of the gaps are closed in the physical map, clones with minimum overlap but giving maximum physical coverage are identified. These clones constitute the minimum tiling path and thus become the substrate for sequencing. This method, although laborious, provides opportunity for chromosome sharing among the partner institutions. By mapping the genome, researchers produce, an important genetic resource at an early stage that can be used to map genes. In addition, because every DNA sequence is derived from a known region, it is relatively easy to keep track of the project and to determine where there are gaps in the sequence. Moreover, assembly of relatively short regions of DNA is an efficient step. However, mapping can be a time-consuming and costly process.

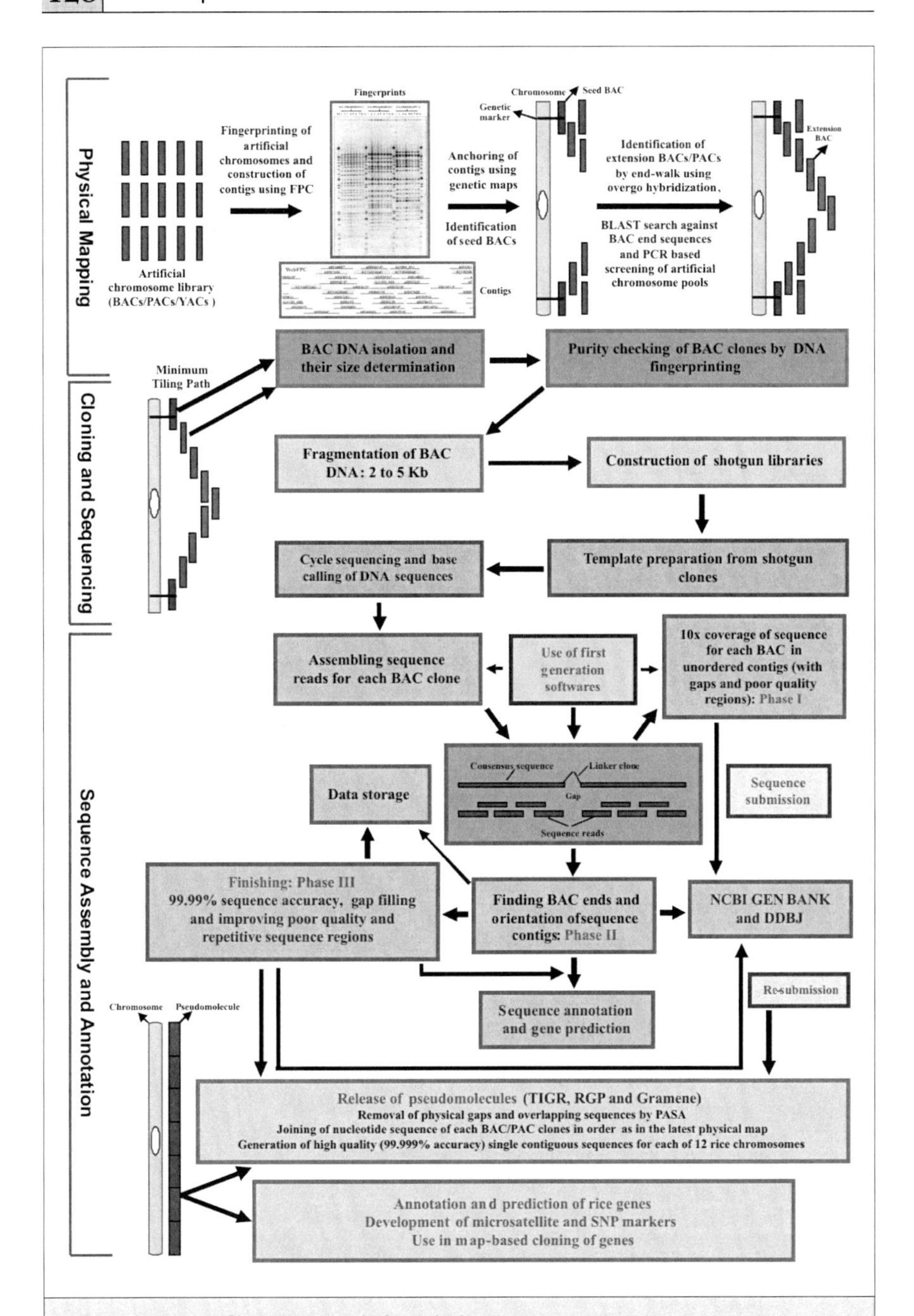

Fig. 1 Clone-by-Clone approach for whole genome sequencing in rice.
(Color image of this figure appears in the color plate section at the end of the book).

3.2 Whole-Genome Shotgun Approach

In this approach, genomic DNA is sheared to obtain 2- and 10-kb fragments by either squeezing the DNA through a pressurized syringe, nebulization or hydroshear, which are then cloned in a plasmid vector to obtain 2 kb and 10 kb libraries separately. Both the libraries are sequenced to generate 500 bp from each end of each cloned genomic DNA fragment. Sequencing both ends of each insert is critical for the assembling the entire chromosome. Sequences of a large number of such clones are then assembled to derive small contigs, which are ordered and oriented into scaffolds. The scaffolds are then mapped to chromosomal locations using known markers. The assembly process consists of five major stages: screener, overlapper, unitigger, scaffolder and repeat masker. The screener marks the sequence for microsatellites and screens out all known long interspersed repeats. The overlapper compares every read of sequence against every other in search of complete end-to-end overlaps of at least 40 bp. The unitigger identifies the uniquely assembled contigs, which are then joined by the scaffolder. The repeat masker is used to resolve repeats and thus eliminates errors in the scaffolding process. This method is more straightforward and less time consuming. It requires high level of computing expertise and facility. For resolving some of the conflicts, it requires information from the genetic and physical maps. The advantage of the whole genome shotgun approach is that it requires no prior physical mapping. It is especially valuable if there is an existing 'scaffold' of organized sequences, localized to the genome, derived from other projects. When the whole genome shotgun data are laid on the 'scaffold' sequence, it is easier to resolve ambiguities. Today, whole genome shotgun is used for most bacterial genomes and as a 'top-up' of sequence data for many other genome projects. Its disadvantage is that large genomes need vast amounts of computing power and sophisticated software to reassemble the genome from its fragments. To sequence the genome from a mammal (billions of bases long), about 60 million individual DNA sequence would be required. Reassembling these sequenced fragments requires huge investments in computational facilities. Unlike the clone-by-clone approach, assemblies can't be produced until the end of the project.

4 NEW GENOME SEQUENCING TECHNOLOGIES

Majority of DNA sequencing technology since the early 1990s has almost exclusively relied on capillary-based and semi-automated implementations of the Sanger biochemistry (Sanger 1988; Hunkapiller et al. 1991) and pyrosequencing (Ronaghi et al. 1996). With the gradual improvement and optimization of sequencing strategies, the sequence read-length up to ~1,000 bp with 99.999%—accuracies per base was achieved. Over the past three years, massively parallel DNA sequencing platforms have become available that reduce the cost of DNA sequencing by over two orders of magnitude and increase high-throughput by large-scale multiplex genome sequencing. The commercial launch of the first massively parallel pyrosequencing platform in 2005 ushered in the new era of high-throughput genomic analysis now referred to as 'Next-Generation Sequencing'. Considerable potential exists for Next-Generation DNA Sequencing technologies to bring enormous changes in genetic and biological research and to enhance our fundamental biological knowledge, by enabling the comprehensive analysis of genes, genomes, transcriptomes and interactomes in a rapid cost-effective and high-throughput manner (Shendure and Ji 2008). The Next-Generation Sequencing/New Genome Sequencing technologies have recently been realized in various commercial DNA sequencing platforms, which mainly include 454 Genome Sequencer (Roche Applied Science, Basel, USA), Solexa Illumina Genome Analyzer (San Diego, CA, USA), and SOLiD platform (Applied Biosystems, Foster City, CA, USA). Although the platforms differ in their engineering configurations and sequencing chemistries, however, all these technologies share a common technical paradigm in which sequencing is performed in a massively parallel manner with use of spatially separated clonally amplified DNA templates or single DNA molecules in a flow cell. The importance of next-generation sequencing is well understood with several useful modification of sequencing strategies in plant (Nordborg and Weigel 2008; Rounsley et al. 2009) and human genomes including whole genome and targeted genomic resequencing, metagenomic and paired-end sequencing, transcriptome and small RNA sequencing, chromatin immunoprecipitation sequencing (ChIP-Seq) for genome-wide mapping of DNA-protein interactions, large-scale analysis of DNA methylation by deep sequencing of bisulfite-treated DNA and molecular barcoding. The advantages of new sequencing technologies developed so far relative to Sanger sequencing are well demonstrated (Mardis 2008; Schuster 2008). Sequencing features like in vitro construction of a sequencing library followed by in vitro clonal amplification by new technologies circumvent several bottlenecks restricting the parallelism of conventional sequencing such as transformation and colony picking. Besides, the new technologies create much higher degree of parallelism with hundreds of millions of sequencing reads by chip-array based sequencing platforms than conventional capillary-based sequencing. Collectively, all these differences translate into dramatically lower costs. Generation of massive DNA sequence information at a considerably lesser time is an additional advantage. The principles, methods, relative strengths, and limitations of three major commercial new genome sequencing platforms such as 454, Solexa and SOLiD used extensively now for studying structural, comparative and functional genomics of prokaryotes and eukaryotes are briefly discussed below.

4.1 454 Pyrosequencing

The 454 FLX Genome Sequencer is the first next-generation commercial sequencing platform (Margulies et al. 2005) developed by Roche Applied Science, USA. The 454 sequencer works based on 'sequencing-by-synthesis' using the emulsion PCR and polymerase pyrosequencing method, which reduces the scale and complexity of the conventional pyrosequencing protocol and increases its productivity in a genomic scale. In this approach library preparation is accomplished by random fragmentation of genomic DNA followed by in vitro end-ligation of two universal common adaptor sequences. Clonal sequencing of mixture of short, adaptor-flanked fragments are generated by emulsion PCR (Dressman et al. 2003), with amplicons captured to the surface of 28 μm beads. After breaking the emulsion, beads are treated with denaturant and subjected to a hybridization-based enrichment for amplicon-bearing beads. A sequencing primer is hybridized to the universal adaptor immediately adjacent to the start of unknown sequence. Millions of copies of single clonal fragment containing beads are poured over a microfabricated array of 40,000 picoliterscale wells and sequenced by pyrosequencing method (Ronaghi et al. 1996). During the sequencing, one side of the semi-ordered array functions as a flow cell for introducing and removing sequencing reagents, whereas the other side is bonded to a fiber-optic bundle for CCD (charge coupled device)-based signal detection. Nucleotides are flowed sequentially in a fixed order across the picotiter plate device during a sequencing run. During the nucleotide flow, hundreds of thousands of beads each carrying millions of copies of a unique single-stranded DNA molecule are sequenced in parallel. If a nucleotide complementary to the template strand is flowed into a well, the polymerase extends the existing DNA strand by adding nucleotides. Incorporation of one or more nucleotide results in release of pyrophosphate through the actions of ATP sulfurylase and luciferase in a reaction and generates a burst of light signal that is detected by the

CCD camera as corresponding to the array coordinates of specific wells. The GS FLX Titanium series software automatically tracks the location of DNA carrying beads on a XY axis. Each bead corresponds to a XY-coordinate on a series of images. The signal strength/fluorescence intensity is proportional to the number of nucleotides incorporated in a single nucelotide flow. The signal intensity per nucleotide flow is recorded for each bead over-time and is plotted to generate over one million flowgrams in every 10 hours of sequencing run. Relative to other next-generation platforms, the key advantage of the 454 platform is to produce long read-length, exceptional accuracy and high-throughput. It generates more than 1,000,000 individual reads per instrument run with improved Q20 read length of 400 to 500 bp. This platform has the capability of combining long single reads and Long-Tag Paired end reads to completely assemble genomes often within a single run. However, currently, the per-base cost of sequencing with the 454 platform is much higher than that of other platforms (Solexa and SOLiD). A major limitation of the 454 technology relates to generation of homopolymers because of the absence of terminating moiety that prevents multiple consecutive incorporations in a given cycle. The dominant error type for the 454 platform is insertion-deletion, rather than substitution. In spite of disadvantages, the 454 sequencing platform could be the method of choice for certain applications where long sequence read-lengths are critical including de novo sequencing, resequencing of whole genomes and target DNA regions, metagenomics and RNA analysis.

4.2 Solexa Genome Analyzer

Illumina Genome Analyzer (*http://www.illumina.com*) is powered by Solexa sequencing technology, which uses a massively parallel sequencing-by-synthesis approach to generate billions of bases of high-quality DNA sequences per run. It is a highly robust, accurate, and scalable sequencing platform and sets a new standard for productivity, cost-effectiveness, and accuracy among next-generation sequencing technologies. It performs sequencing-by-synthesis on random blunt-end-ligated fragments to adapter primers, and mutiplexes the process with massive arrays of reactions derived from single molecules. Library construction that generates a mixture of adaptor-flanked fragments up to several hundred base pairs (bp) in length and primer hybridization with blunt-end-ligated genomic DNA is similar to the method described above for 454 sequencing platform. Illumina Sequencing technology relies on the attachment of randomly fragmented genomic DNA to a planar, optically transparent surface. Attached DNA fragments are extended and bridge PCR amplified (Fedurco et al. 2006) to create an ultra-high density sequencing flow cell with hundreds of millions of clusters, each containing about 1,000 copies of the same template/clonal amplicons. These templates are sequenced using a robust four-color DNA sequencing-by-synthesis technology that employs reversible terminators with removable fluorescent dyes. This novel approach ensures high accuracy and true base-by-base sequencing eliminating sequence-context specific errors and enabling sequencing through homopolymers and repetitive sequences. Several million clusters can be amplified to distinguishable locations within each of eight independent lanes that are on a single flow-cell. After cluster generation, the amplicons are linearized and a sequencing primer is hybridized to a universal sequence flanking the region of interest. Each cycle of sequencing consists of single-base extension with a modified DNA polymerase and a mixture of four nucleotides. Solexa incorporates the nucleotides one-at-a-time, taking a photograph of the 4-color fluorescence intensity after each cycle followed by chemical activation of the 3′OH. All four nucleotides are added simultaneously, and each is fluorescently tagged with a different color. Instead of a picotiter plate, Solexa sequencing occurs in microfluidic flow cells, which are coated with two different chemically-ligated primers. High-sensitivity fluorescence detection is achieved using laser excitation and total internal reflection optics. After completion of the first read, the templates can be regenerated in situ to enable a second 75+ bp read from the opposite end of the fragments. The Paired-End Module directs the regeneration and amplification operations to prepare the templates for the second round of sequencing. First, the newly sequenced strands are stripped off and the complementary strands are bridge amplified to form clusters. Once the original templates are cleaved and removed, the reverse strands undergo sequencing-by-synthesis. The second round of sequencing occurs at the opposite end of the templates generating 75+ bp reads for a total of >20 gigabyte of paired-end data per run. Read-lengths up to 100 bp can be generated routinely by Solexa, however; longer reads possibly incur a higher error rate. The dominant error type is substitution, rather than insertions or deletions and also homopolymers are certainly less than that obtained with other platforms such as 454. Average raw sequence error rates are in the order of 1-1.5%, but higher accuracy bases with error rates of $\leq$0.1% can be identified through quality metrics associated with each base-call. However, currently, the per-base cost of sequencing with the Solexa is much lower than that of 454 sequencing platform. The Solexa flexible system enables researchers to discover genomic variation, characterize all transcripts without any prior sequence information, identify epigenetic modifications, and study the interaction of proteins and nucleic acids across the entire genome. It generates gigabases of data quickly with walk-away automation genome-scale analysis. Besides, it opens the opportunity to redesign genotyping strategies for more effective genetic mapping and genome analysis. This includes a high-throughput method for genotyping recombinant populations utilizing whole genome resequencing data (Huang et al. 2009). With continuous advances in sequencing technologies, this genome-based method is expected to replace the conventional marker-based genotyping approach to provide a powerful tool for large-scale gene discovery and marker assisted breeding.

4.3 ABI SOLiD

ABI SOLiD platform has its origin in the system originally described by Shendure and colleagues (Shendure et al. 2005) followed by McKernan and colleagues (McKernan et al. 2006) at the Agencourt Personal Genomics (Beverly, MA, USA) that is now acquired by the Applied Biosystems (Foster City, CA, USA). The massive multiplex SOLiD platform works almost similar to 454 and Solexa sequencing technologies and amplifies fragmented DNA ligated to two generic primers in single-template emulsion PCR with primer-coated beads. Libraries may be constructed by random fragmentation/mate-paired tag that gives rise to a mixture of short, adaptor-flanked fragments. Beads without template and beads with two different templates are removed by an enrichment step before sequencing and an error-checking step during analysis. Clonal sequencing features are generated by emulsion PCR, with amplicons captured to the surface of 1 μM paramagnetic beads. After breaking the emulsion, beads bearing amplification products are selectively recovered, and then immobilized to a solid planar substrate to generate a dense, disordered array. Sequencing-by-synthesis is driven by a DNA ligase, rather than a polymerase. A universal primer, complementary to adaptor sequence, is hybridized to the array of amplicon-bearing beads. Each cycle of sequencing involves the ligation of a degenerate population of fluorescently labeled octamers. The octamer mixture is structured and the identity of specific position(s) within the octamer correlates with the identity of the fluorescent label. However, the SOLiD platform has a unique protocol, which reads every nucleotide twice in successive dinucleotide sequencing reactions and consists of an error-checking mechanism that is not offered by 454 and Solexa. The sequencing reaction is repeated five times on each template strand with four-color fluorescence of 8-mer probes. Each of the five sequencing-by-ligation runs is preceded by hybridization of a universal sequencing primer. After ligation, images are acquired in four channels, effectively collecting data for the same base positions across all template-bearing beads. The octamer is then chemically cleaved between positions five and six, removing the fluorescent label. Progressive rounds of octamer ligation enable sequencing of every fifth base (such as bases 5, 10, 15, 20). Upon completion of several such cycles, the extended primer is denatured to reset the system. Subsequent iterations of this process can be directed at a different set of positions (e.g., bases 4, 9, 14, 19) either by using a primer that is set back one or more bases from the adaptor-insert junction, or by using different mixtures of octamers where a different position (e.g., base 2) is correlated with the label. Each base position is then queried twice (once as the first base, and once as the second base, in a set of 2 bp interrogated on a given cycle) such that miscalls can be more readily identified. A disadvantage common to 454 and SOLiD is that emulsion PCR can be cumbersome and technically challenging. On the other hand, it is possible that sequencing on a high-density array of very small (1 μm) beads with sequencing-by-ligation may represent the most straightforward opportunity to achieve extremely high sequencing data densities, simply because 1 μm beads physically exclude another nucleotide at a spacing that is on the order of the diffraction limit. The SOLiD platform can sequence 50 bp short-read sequences more easily with comparable per-base sequencing cost of Solexa. It offers broad application flexibility including ultra-high throughput, scalable automation, superior accuracy and robustness and thus can be used for whole genome and targeted resequencing, sequence variant discovery, de novo sequencing gene expression analysis and ChiP based sequencing.

5 COMPARATIVE OVERVIEW OF CONVENTIONAL AND NEW SEQUENCING TECHNOLOGIES

In terms of costs, advantages, limitations and practical aspects of implementation, clear differences between conventional sequencing and the second-generation new sequencing technologies (Table 1) can be observed that might determine which sequencing platform represents the best option to serve the purposes. The applications of conventional Sanger sequencing have grown as diverse in area of genomics and for

Table 1 Comparative overview of conventional and new sequencing technologies.

Sequencing Approach	*First generation sanger sequencer*	*454 FLX/Roche*	*Solexa/Illumina*	*SOLiD/ABI*
Cost per instrument	$200,000	$500,000	$430,000	$591,000
Mode of synthesis	Polymerase (dideoxy nucleotide chain termination)	Polymerase (pyrosequencing)	Polymerase (reversible terminators)	Ligase (octamers with two base encoding)
Feature generation	Normal PCR with thermostable polymerase	Emulsion PCR	Bridge PCR	Emulsion PCR
Starting materials needed	100 ng/μl of plasmid DNA and 50 ng/μl of PCR product	3 to 5 μg DNA and 5 to 10 μg total RNA	1 to 5 μg DNA and 1 to 2 μg total RNA	1 μg DNA and 5 ng total RNA
Average sequence read length	400 to 500 bp in PCR plate	400 to 500 bp in large picotiter plate	100 bp in flow cell	50 bp in 1 slide
Sequence data file size	7,000 bp/run	12-15 Gb/run	5-10 terabyte/run (Storage image file)	15-20 terabyte/run (Storage image file)
Bar-code scanning	Not implemented	Implemented	Implemented	Implemented
Tail Paired ends	Not implemented	Implemented	Implemented	Implemented
Phred quality score	Not implemented	Implemented	Implemented	Implemented
Cost per Mb sequence	$10,000 (approx.)	$60 (approx.)	$2 (approx.)	$2 (approx.)

Contd.

Table 1 continued

Sequencing Approach	*First generation sanger sequencer*	*454 FLX/Roche*	*Solexa/Illumina*	*SOLiD/ABI*
Cost per run	$1,500 (approx.) for 384 samples (0.15 Mb)	$8,200 (approx.)	$3,976 (approx.)	$3,650 (approx.)
Abundant error type	Indels	Indels	Substitutions	Substitutions
Accuracy	High sequencing inaccuracies (about 1 to 2%)	Generation of homopolymers	Near about 99% (<1% error rate)	Near about 99.94% (<0.6% error rate)
Running time	4 hour (about 0.15 Mb) in 384 PCR plates	7.5 hour	2 to 3 days (about 1 Gb)	3 to 4 days (about 1 Gb)—
Softwares	First Generation	First Generation	Second Generation and Third-party Genome Analyzer	Second Generation and Third-party Genome Analyzer
Applications	Whole genome sequencing, automated fragment analysis for high-throughput genotyping of microsatellite markers and SNP detection	Whole genome resequencing, transcriptome analysis, CHiP sequencing, methylation pattern study and SNP/InDel detection and more specifically for de novo sequencing	Whole genome resequencing, transcriptome analysis, CHiP sequencing, methylation pattern study and SNP/InDel detection	Whole genome resequencing, transcriptome analysis, CHiP sequencing, methylation pattern study and particularly for detection of SNPs, and small and large

Contd.

Table 1 continued

Sequencing Approach	*First generation sanger sequencer*	*454 FLX/Roche*	*Solexa/Illumina*	*SOLiD/ABI*
				size InDels, and determination of small copy number variants (CNVs) for studying structure and evolution of many shall and larger complex genome species.
Advantages	Low instrument cost and generates long sequence read-length up to 1000 bp, thus suitable for long fragment sequencing	Long read-length, exceptional accuracy and high throughput and useful compared to conventional Sanger sequencer. Simpler and easier to analyze the sequence data.	Lower cost and require less material and generates lower homopolymers than 454. Simpler and easier to analyze and widely used in plant genomes.	Lower cost than 454 and require less material than Solexa and generates lower homopolymers. Quality control measures for checking sequence accuracy.
Disadvantages	Inability for GC rich and repetitive region sequencing and high error prone and low sequence accuracy.	Generation of homopolymers and InDel error types and no quality control measures for testing sequence accuracy.	Shorter reads and fewer runs per year and no quality control measures for testing sequence accuracy.	Shortest reads and fewer runs per year, and requires massive computational and data storage resources.

small-scale kilobase-to-megabase range genome sequencing projects like rice, *Arabidopsis* and *Poplar*. In contrast, large-scale genome sequencing projects from Megabase to Gigabase range now depend entirely on new genome sequencing technologies. The second generation sequencing technologies have several advantages and disadvantages with useful applications. However, only SOLiD provides error checking as a part of the sequencing process, while Solexa and 454 guarantee high quality sequences but do not implement quality control measures. On the other hand, the SOLiD platform requires massive computational and data-storage resources due to the complexity of the approach. The overall accuracy as well as the specific error distributions of individual technologies also differ. The insertion-deletion errors are observed in case of 454, whereas substitution errors are predominant for Solexa and SOLiD. The 454 sequencing platform generates long-read sequence length up to 500 bp in contrast to Solexa and SOLiD producing short-read sequence. Among these three sequencing platforms, Solexa is capable of analyzing long-tail paired end sequences while SOLiD is efficient enough to analyze mate-paired end sequences. The three new genome sequencing technologies have been used widely and extensively for several modified sequencing applications like ChIP based Sequencing/Protein-Nucleic Acid Interaction (Ingolia et al. 2009; Ketel et al. 2009; Zhu et al. 2009), de novo sequencing (Moran et al. 2009; Reinhardt et al. 2009), DNA methylation and histone analysis (Jiang and Pugh 2009; Pomraning et al. 2009), metagenomics (Nusbaum et al. 2009), targeted resequencing (Korbel et al. 2009), whole-genome resequencing (Huang et al. 2009) and transcriptome (mRNA coding and miRNA non-coding) sequencing (Wei et al. 2009; Yoder-Himes et al. 2009) both for prokaryotic and eukaryotic genomes and more than 100 articles have been published till now using these novel technologies. Among these three sequencing platforms, however, 454 pyrosequencer has been used specifically for long-read de novo sequencing, while SOLiD has been utilized widely for more precise analysis of sequence polymorphisms (SNPs and InDels) and small copy number variants, and genome evolution studies.

6 DEVELOPMENT OF SOFTWARE AND HARDWARE TO HANDLE GIGABYTE DATA POINTS

Rapid evolution of next-generation sequencing technology is posing challenges for bioinformatics in areas including sequence quality scoring, alignment, assembly and data release. For large-scale sequencing projects with new sequencing platforms, there is a need for systematic assessment of data quality, reliability and reproducibility, their biological relevance, distribution of estimated accuracies for raw base-calls (phred like quality scores), error patterns in raw or consensus sequence data, and skewing of true ratios in tag-counting applications. To meet this requirement, a variety of softwares and bioinformatic tools are now being developed, evaluated for their efficiency and used for analyzing and interpreting the vast amount of sequencing data generated by different Next-Generation Sequencing platforms. Based on their functions, the tools have been classified into several general categories including alignment of sequence reads to a reference, base-calling and/or polymorphism detection, de novo assembly creation from paired or unpaired reads, and genome browsing and annotation of long and short-read sequence data. The first generation software tools used for alignment and assembly of sequences generated from conventional Sanger sequencer can also be utilized for analyzing the long-read sequence data generated particularly by 454 platform.

6.1 First Generation Softwares

The first generation softwares are useful particularly to derive biologically meaningful results from the million base-pairs of long-read sequence data generated by conventional Sanger sequencer and next generation sequencing platform like 454 pyrosequencer quickly and affordably with the powerful suite of bioinformatic analysis tools. For instance, software tools like Phred, Phrap, Consed, PolyPhred and BLAST can be used for checking sequence quality and assembling, homology searches with reference sequences and variant detection in megabase amount of sequence data. Phred (*http://www.phrap.org*; Ewing et al. 1998) is a perl scripting program that reads the DNA sequence trace files, calls bases and finally assigns a quality value to each called base. Sequences containing at least 100 continuous nucleotides with a phred score greater than 16 are considered significant. Phrap (*http://www.phrap.org*; Ewing and Green 1998) is another perl scripting program used for assembling the sequences. It allows use of the entire sequence read and also the trimmed high quality part. Moreover, it uses a combination of user supplied data quality information both direct (from phred trace analysis) and indirect (from pair-wise read comparisons) to delineate the likely accurate base calls in each sequence read, helps to discriminate repeat sequences, permits use of the full reads in assembly and finally allows to generate a highly accurate consensus sequence. Sequences containing at least 100 continuous nucleotides clustered by phrap with a minimum score of 80 are considered significant. The high quality consensus sequences obtained after *phrap* analysis can be analyzed automatically employing an integrated software tool *polyphred* (Nickerson et al. 1997) to detect sequence variants with >99% accuracy. Consed (Gordon et al. 1998) is a graphical computer based tool for viewing, editing and finishing sequence assemblies created with Phrap. While viewing the assembly, it helps deciding where additional data or editing is necessary or identifying other anomalies, obtaining additional read data, and editing to correct errors in the assembly or consensus sequences. However, for more precise sequence data analysis and annotation, three different next generation bioinformatics tools are now integrated with 454 pyrosequencer for supporting the applications that includes: de novo assembly up to 400 megabases, resequencing of genomes of any size and amplicon variant detection by comparison with a known reference sequence.

6.2 Next Generation Softwares

To derive biologically meaningful results from the sequence data generated by Next-Generation Sequencing platforms quickly and affordably, a powerful suite of bioinformatic software analysis tools and hardwares have been implemented in different platforms. These software and hardware tools can be classified according to their suitability to analyze the sequence read length (bp) generated by various sequencing platforms. For instance, the 454 pyrosequencer mostly generates long-read sequences while Solexa and SOLiD platforms used to produce short-read sequence data. Consequently, the types of softwares required for these platforms are different, which are briefly outlined here.

6.2.1 Long-read Sequence Analysis

The 454 pyrosequencer, i.e. Genome Sequencer FLX System updated with GS FLX Titanium series contains features like user-friendly Graphical User Interface (GUI), a Laboratory Information Management System (LIMS) software, Bar-code scanning, and Phred quality scores and hardware like 'Computer Cluster' integrated with manageable data storage capacity for data management, processing and long read sequence analysis. This platform is also featured with software tools like GS Run Browser, de novo Assembler, Reference Mapper, Amplicon Variant Analyzer, and FLX Titanium Cluster softwares for easier analysis and interpretation of genome-scale large sequencing data. Interactive application of GS Run Browser Software displays the results of a sequencing run as well as raw images and graphic representations of various metrics files in 454 pyrosequencer. It assesses the quality of a run, facilitates quick evaluation of the results of titration experiments and easily exports data to an Excel spreadsheet. The specific functions of these software tools are described below.

6.2.1.1 GS De Novo Assembler Software

- Performs whole genome shotgun assembly of genomes with or without paired-end data.
- Orders contigs into scaffolds using supported paired-end reads.
- Assembles larger, more complex genomes up to 400 megabases in size with a 64-bit assembler or up to 20 megabases with a 32-bit assembler.
- Generates data outputs: fna.file (sequence of contigs, FASTA format), qual.file (corresponding Phred equivalent quality score), ace.file (alignment of the reads to contig sequence).

6.2.1.2 GS Reference Mapper Software

- Reads any reference genome maps and generates a consensus sequence.
- Easily views all differences compared to the reference sequence with automatic output to separate files: Insertions (blocks up to 50 bases), Deletions (blocks up to 50 bases) and SNPs.
- Quickly identifies high confidence difference compared to the reference genome, which is singled out in a separate file.
- Compares large, complex genomes of any size including resequencing of whole genomes of several eukaryotes and prokaryotes.
- Generates data outputs like fna.file, qual.file, and ace.files like *De Novo Assembler*.

6.2.1.3 GS Amplicon Variant Analyzer Software

- Automatically computes the alignment of reads from amplicon-based samples (ultra deep amplicon sequencing) against a reference sequence.
- Quickly identifies variants and their respective frequencies in large pools of data-screen for unknown variants, discovers unknown variants and identifies haplotypes.

- Detects low-frequency (<1%) variants.
- Discovers SNPs, insertions and deletions on a population level, detects rare somatic mutations, and performs viral subtyping.
- Analyzes DNA methylation patterns in epigenetic studies.
- Metagenomic and genetic diversity analysis.
- Generates data outputs like ace.file (alignment of the reads against a reference sequence), and png.file (graphical file format, tab delimited text file).

6.2.1.4 GS FLX Titanium Cluster

- The GS FLX Titanium cluster is a Roche-certified automated plug-and-play computing solution for the image processing and data analysis for GS FLX System without use of any other software or hardware.
- This pre-configured computer setup is a convenient, affordable and time-saving option that ensures optimal sequencing results with 454 pyrosequencer.

6.3 Short-read Sequence Analysis

The algorithm implemented under the above defined software tools is not adequate for alignment of high volume short-read sequencing data generated specifically by Solexa and SOLiD. An increasing number of bioinformatics software tools have recently been developed particularly for rapid alignment of large sets of short reads and removal of mismatches and gaps. This include sequence alignment tools (Table 2) like Cross_match (*http://www.phrap.org/phredphrapconsed.html*), ELAND (*http://www.illumina.com*), Exonerate (*http://www.ebi.ac.uk/guy/exonerate*), Mosaik (*http://bioinformatics.bc.edu/marthlab/Mosaik*), RMAP (*http://rulai.cshl.edu/rmap*), SHRiMP (*http://compbio.cs.toronto.edu/shrimp*), SOAP (*http://soap.genomics.org.cn*), SSAHA2 (*http://www.sanger.ac.uk/Software/analysis/SSAHA2*), and SXOligo Search Alignment (*http://synasite.mgrc.com.my:8080/sxog/NewSXOligoSearch.php*), alignment and sequence variant detection tool like MAQ (*http://maq.sourceforge.net*), base caller software like PyroBayes (*http://bioinformatics.bc.edu/marthlab/PyroBayes*), and variant detection softwares like PbShort (*http://bioinformatics.bc.edu/marthlab/PbShort*) and ssahaSNP (*http://www.sanger.ac.uk/Software/analysis/ssahaSNP*). Some of these tools use well-established earlier alignment algorithms, such as Smith-Waterman. Others have been significantly modified with the innovation of developing new algorithms specifically

Table 2 Second generation softwares available in commercial and public domain.

Programs	*Categories*	*Utilities*	*Websites*
CLCbio Genomics	Integrated solutions	De novo and reference assembly of sequencing data, SNP detection and CHiP-sequencing	*http://www.clcbio.com*
Galaxy	Integrated solutions	De novo and reference assembly of sequencing data, SNP detection and CHiP-sequencing	*http://g2.trac.bx.psu.edu*
Genomatix Genome Analyzer	Integrated solutions	Sequencing data analysis	*http://www.genomatix.de*
JMP Genomics	Integrated solutions	Sequencing data visualization and statistics tool from SAS	*http://www.jmp.com*
NextGENe	Integrated solutions	De novo and reference assembly of sequences. Uses a novel condensation assembly tool approach where reads are joined via "anchors" into mini-contigs before assembly. Includes SNP detection and CHiP-seq browsers.	*http://softgenetics.com*
SeqMan Genome Analyser	Integrated solutions	Software for sequence assembly of Illumina, Roche FLX and Sanger data integrating with Lasergene Sequence Analysis software for additional analysis and visualization capabilities.	*http://www.dnastar.com*
SHORE	Integrated solutions	For short read, is a mapping and analysis pipeline for short DNA sequences produced on a Illumina Genome Analyzer	*http://1001genomes.org*
ABySS	Align/Assemble	Assembly by short sequences. It is a *de novo* sequence assembler designed for very short reads. The single-processor version is useful for assembling genomes up to 40-50 Mb in size.	*http://www.bcgsc.ca/platform/bioinfo/software/abyss*

Contd.

Table 2 continued

Programs	*Categories*	*Utilities*	*Websites*
BFAST	Align/Assemble	Blat-like Fast Accurate Search Tool	*https://secure.genome.ucla.edu/index.php/BFAST*
Bowtie	Align/Assemble	Ultrafast and memory-efficient short read aligner. It aligns short DNA sequences (reads) at a rate of 25 million reads per hour on a typical workstation with 2 gigabytes of memory.	*http://bowtie-bio.sourceforge.net*
BWA	Align/Assemble	Heng Lee's BWT Alignment program: A progression from Maq. BWA is a fast light-weighted tool that aligns short sequences to a sequence database.	*http://maq.sourceforge.net*
ELAND	Align/Assemble	Efficient large-Scale alignment of nucleotide databases. Whole genome alignments to a reference genome.	*http://bioinfo.cgrb.oregonstate.edu/docs/solexa*
Exonerate	Align/Assemble	Various forms of pair-wise alignment of DNA/protein against a reference.	*http://www.ebi.ac.uk/~guy/exonerate*
GenomeMapper	Align/Assemble	GenomeMapper is a short read mapping tool designed for accurate read alignments. It quickly aligns millions of reads either with ungapped or gapped alignments.	*http://1001genomes.org/downloads/genomemapper.html*
GMAP	Align/Assemble	GMAP (Genomic Mapping and Alignment Program) for mRNA and EST Sequences.	*http://www.gene.com/share/gmap*
gnumap	Align/Assemble	The Genomic Next-generation Universal MAPper (gnumap) is a program designed to accurately map	*http://dna.cs.byu.edu/gnumap*

Contd.

Table 2 continued

Programs	*Categories*	*Utilities*	*Websites*
		sequence data obtained from next-generation sequencing machines (specifically that of Solexa/Illumina) back to a genome of any size.	
MAQ	Align/Assemble	Mapping and Assembly with Qualities (renamed from MAPASS2). Particularly designed for Illumina with preliminary functions to handle ABI SOLiD data.	*http://sourceforge.net/projects/maq*
MOSAIK	Align/Assemble	Produces gapped alignments using the Smith-Waterman algorithm	*http://bioinformatics.bc.edu/marthlab/Mosaik*
MUMmer	Align/Assemble	Modular system for the rapid whole genome alignment of finished or draft sequence. Released as a package providing an efficient suffix tree library, seed-and-extend alignment, SNP detection, repeat detection, and visualization tools.	*http://mummer.sourceforge.net*
Novocraft	Align/Assemble	Tools for reference alignment of paired-end and single-end Illumina reads.	*http://www.novocraft.com/index.html*
RMAP	Align/Assemble	Assembles 20-64 bp Illumina reads to a FASTA reference genome	*http://rulai.cshl.edu/rmap*
SeqMap	Align/Assemble	Supports up to 5 or more bp mismatches/INDELs	*http://biogibbs.stanford.edu/~jiangh/SeqMap*
SHRiMP	Align/Assemble	Assembles to a reference sequence	*http://compbio.cs.toronto.edu/shrimp*

Contd.

Table 2 continued

Programs	*Categories*	*Utilities*	*Websites*
Slider	Align/Assemble	An application for the Illumina Sequence Analyzer output that uses the probability files instead of the sequence files as an input for alignment to a reference sequence or a set of reference sequences.	*http://www.bcgsc.ca/platform/bioinfo/software/slider*
SOAP	Align/Assemble	A Short Oligonucleotide Alignment Program. A program for efficient gapped and ungapped alignment of short oligonucleotides onto reference sequences.	*http://soap.genomics.org.cn*
SSAHA	Align/Assemble	SSAHA (Sequence Search and Alignment by Hashing Algorithm) is a tool for rapidly finding near exact matches in DNA or protein databases using a hash table.	*http://www.sanger.ac.uk/Software/analysis/SSAHA*
Vmatch	Align/Assemble	A versatile software tool for efficiently solving large scale sequence matching tasks.	*http://www.vmatch.de*
Zoom	Align/Assemble	ZOOM (Zillions Of Oligos Mapped) is designed to map millions of short reads, emerged by next-generation sequencing technology, back to the reference genomes, and carry out post-analysis	*http://www.bioinformaticssolutions.com/products/zoom*
Edena	De novo Align/Assemble	Edena (Exact De Novo Assembler) is an assembler dedicated to process the millions of very short reads produced by the Illumina Genome Analyzer.	*http://www.genomic.ch/edena.php*
EULER-SR	De novo Align/Assemble	Short read de novo assembly	*http://euler-assembler.ucsd.edu/portal*

Contd.

Table 2 continued

Programs	*Categories*	*Utilities*	*Websites*
MIRA2	De novo Align/ Assemble	MIRA (Mimicking Intelligent Read Assembly) is able to perform true hybrid de-novo assemblies using reads gathered through 454 sequencing technology.	*http://chevreux.org/projects_mira.html*
SEQUAN	De novo Align/ Assemble	A Consistency-based Consensus Algorithm for De Novo and Reference-guided Sequence.	*http://www.seqan.de/projects/consensus.html*
SHARCGS	De novo Align/ Assemble	De novo assembly of short reads.	*http://sharcgs.molgen.mpg.de*
SSAKE	De novo Align/ Assemble	Short Sequence Assembly: Aggressively assembling millions of short nucleotide sequences.	*http://www.bcgsc.ca/platform/bioinfo/software/ssake*
VCAKE	De novo Align/ Assemble	De novo assembly of short reads with robust error correction.	*http://sourceforge.net/projects/vcake*
Velvet	De novo Align/ Assemble	De novo genomic assembler specially designed for short read sequencing technologies, such as Solexa or 454. Need about 20-25x coverage and paired reads.	*http://www.ebi.ac.uk/~zerbino/velvet*
ssahaSNP	SNP/InDel discovery	Polymorphism detection tool. It detects homozygous SNPs and InDels by aligning shotgun reads to the finished reference genome sequence by filtering out highly repetitive elements.	*http://www.sanger.ac.uk/Software/analysis/ssahaSNP*

Contd.

Table 2 continued

Programs	*Categories*	*Utilities*	*Websites*
PolyBayesShort	SNP/InDel discovery	A re-incarnation of the PolyBayes SNP discovery tool. This version is specifically optimized for the analysis of large numbers (millions) of high-throughput next-generation sequence reads and aligned to whole chromosomes.	*http://bioinformatics.bc.edu/marthlab/PbShort*
PyroBayes	SNP/InDel discovery	PyroBayes is a novel base caller for pyrosequences from the 454. It assigns more accurate base quality estimates to the 454 pyrosequences.	*http://bioinformatics.bc.edu/marthlab/PyroBayes*
BS-Seq	CHiP-Seq, Bis-Seq, CNV-Seq	The source code and data for the "Shotgun Bisulphite Sequencing" of the *Arabidopsis* genome reveals "DNA Methylation Patterning"	*http://epigenomics.mcdb.ucla.edu/BS-Seq/download.html*
CHiPSeq	CHiP-Seq, Bis-Seq, CNV-Seq	Chromatin immunoprecipitation assay to study in vivo protein-DNA interactions and map these comprehensively across whole genome	*http://woldlab.caltech.edu/html/chipseq_peak_finder*
CNV-Seq	CHiP-Seq, Bis-Seq, CNV-Seq	A new method to detect copy number variation using high-throughput sequencing.	*http://tiger.dbs.nus.edu.sg/cnv-seq*
FindPeaks	CHiP-Seq, Bis-Seq, CNV-Seq	Perform analysis of ChIP-Seq experiments. It uses a naive algorithm for identifying regions of high coverage, which represent Chromatin Immunoprecipitation enrichment of sequence fragments, indicating the location of a bound protein of interest.	*http://www.bcgsc.ca/platform/bioinfo/software/findpeaks*

Contd.

Table 2 continued

Programs	*Categories*	*Utilities*	*Websites*
MACS	CHiP-Seq, Bis-Seq, CNV-Seq	Model-based Analysis for ChIP-Seq. MACS empirically models the length of the sequenced ChIP fragments, which tends to be shorter than sonication or library construction size estimates, and uses it to improve the spatial resolution of predicted binding sites.	*http://liulab.dfci.harvard.edu/MACS*
PeakSeq	CHiP-Seq, Bis-Seq, CNV-Seq	Systematic scoring of ChIP-Seq experiments relative to controls. A two-pass approach for scoring ChIP-Seq data relative to controls. The first pass identifies putative binding sites and compensates for variation in the mappability of sequences across the genome. The second pass filters out sites that are not significantly enriched compared to the normalized input DNA and computes a precise enrichment and significance.	*http://www.gersteinlab.org/proj/PeakSeq*
QuEST	CHiP-Seq, Bis-Seq, CNV-Seq	Quantitative enrichment of sequence tags	*http://mendel.stanford.edu/sidowlab/downloads/quest*
SISSRs	CHiP-Seq, Bis-Seq, CNV-Seq	Site identification from short sequence reads	*http://dir.nhlbi.nih.gov/papers/lmi/epigenomes/sissrs*
Rolexa	Alternate Base Calling	R-based framework for base calling of Solexa data.	*http://svitsrv25.epfl.ch/R-doc/library/Rolexa/html/00Index.html*
Alta-cyclic	Alternate Base Calling	A novel Illumina Genome-Analyzer (Solexa) base caller.	*http://hannonlab.cshl.edu/Alta-Cyclic/main.html*

Contd.

Table 2 continued

Programs	*Categories*	*Utilities*	*Websites*
ERANGE	Transcriptomics	Mapping and quantifying mammalian transcriptomes by RNA-Seq. Supports Bowtie, BLAT and ELAND.	*http://woldlab.caltech.edu/rnaseq*
G-Mo.R-Se	Transcriptomics	A method aimed at using RNA-Seq short reads to build de novo gene models. First, candidate exons are built directly from the positions of the reads mapped on the genome (without any ab initio assembly of the reads), and all the possible splice junctions between those exons are tested against unmapped reads.	*http://www.genoscope.cns.fr/externe/gmorse/*
MapNext	Transcriptomics	A software tool for spliced and unspliced alignments and SNP detection of short sequence reads.	*http://evolution.sysu.edu.cn/english/software/mapnext.htm*
QPalma	Transcriptomics	Optimal Spliced Alignments of Short Sequence Reads.	*http://www.fml.tuebingen.mpg.de/raetsch/suppl/qpalma*
RSAT	Transcriptomics	RNA-Seq Analysis Tools.	*http://biogibbs.stanford.edu/~jiangh/rsat/*
TopHat	Transcriptomics	TopHat is a fast splice junction mapper for RNA-Seq reads. It aligns RNA-Seq reads to large genomes using the ultra high-throughput short read aligner Bowtie, and then analyzes the mapping results to identify splice junctions between exons.	*http://tophat.cbcb.umd.edu/*

tailored for short reads. For instance, SOAP (Li et al. 2008), a software package for efficient gapped or ungapped alignment, uses a memory-intensive seed and look-up table algorithm to accelerate alignment, while allowing iterative trimming of the 32 end of reads usually associated with a higher error rate. Other approaches used to accelerate processing include 'bit encoding' to compress sequence data into a computationally more manageable and efficient format (Ning et al. 2001; Li et al. 2008). Alignment software is increasingly taking into account the estimated quality of the underlying data in generating read-placements, as is the case with MAQ37, an alignment and variation discovery tool that works with data from either Solexa or SOLiD data, and SHRiMP (*http://compbio.cs.toronto.edu/shrimp/),* which includes a novel 'color-space to letter-space' Smith-Waterman algorithm compatible with two base–encoded sequence data from the SOLiD platform. Several sequence assembly tools like ALLPATHS (Butler et al. 2008), Edena (*http://www.genomic.ch/edena*), Euler-SR Assembly (Chaisson and Pevzner 2008), SHARCGS (*http://sharcgs.molgen.mpg.de*), SHRAP (Sundquist et al. 2007), SSAKE (*http://www.bcgsc.ca/platform/bioinfo/software/ssake*), VCAKE (*http://sourceforge.net/projects/vcake*), and Velvet (*http://www.ebi.ac.uk*) have recently been adopted or independently developed for generating assemblies from short and unpaired sequencing reads (Sundquist et al. 2007; Warren et al. 2007; Butler et al. 2008; Zerbino and Birney 2008). Taking the advantages of these algorithms, mate-paired reads and long-tailed end-pair sequence reads are possibly easier to analyze now and interpret with all the major sequencing platforms. It is anticipated to have a major impact on the overall success of de novo sequence assembly with short reads. The Next Generation Sequencing platforms like Illumina Solexa Genome Analyzer and ABI SOLiD have been integrated with several high-performance hardwares and softwares for analyzing gigabyte amount of sequencing data precisely with less time. This includes Sequencing Control Software (SCS), Pipeline and CASAVA Software, and different third-party genome analyzer data analysis bioinformatic tools, which are briefly discussed here.

6.3.1 Genome Analyzer SCS Software

SCS software performs image analysis and base calling and allows monitoring the progress of sequencing run by optimizing run conditions and assessing run-time quality statistics. It runs on the instrument computer, eliminating the need for image transfer across networks, which dramatically simplifies the Genome Analyzer data management. Its advantages include: (i) Compatible Design: Readily imports intensities or base calls and quality statistics into the Genome Analyzer Pipeline Software; (ii) Flexible Instrument Control: Easily chooses from a number of recipes for various sequencing applications; (iii) Flexible Data Migration and Archiving: Reduces raw sequence data into a compact base and quality format; (iv) Efficient and Fast Results: Minimizes time spent in subsequent analysis steps; base calling occurs simultaneously and is complete within hours of the sequencing run; and (v) Real-Time Feedback: Monitors run performance during the sequencing run.

6.3.2 Genome Analyzer Pipeline Software

The SCS output can be used as inputs for Genome Analyzer Pipeline software to derive quality scored alignments. This software runs on numerous Linux operating systems and is an open source program, making it entirely customizable. For ready-to-use, this software is available in Pipeline Analysis Server. Pipeline data further can be easily imported into CASAVA and Genome studio Software allowing to explore data from multiple genetic analysis applications and obtain a complete picture of the genome. It also offers several advantages including: (i) Maximization of the number of clusters used to generate sequence data with automated image calibration; (ii) Greater reliability due to filtering for high-quality reads using accurate cluster intensity scoring algorithms; (iii) Higher accuracy by minimization of the propagation of downstream sequencing errors with quality-calibrated base calls; (iv) Customization that reduces the need for elaborate computer infrastructures with highly optimized genomic alignment tools and more flexibility in respect of identification of structural variants and sequencing repetitive regions with intelligent paired-end logic.

6.3.3 Genome Analyzer System Software: CASAVA

Illumina's Consensus Assessment of Sequence and Variation (CASAVA) software captures summary information for resequencing and counting studies and places the data in a compact structure for visualization within GenomeStudio Software or several publicly available third-party genome analyzer bioinformatics tools. CASAVA can create genomic builds, call SNPs, and count reads from data generated from one or more runs of the Genome Analyzer across a broad range of sequencing applications. Its advantages include: (i) Aggregation: Condensing real results from many runs of a resequencing or counting experiment; (ii) Consensus Calling: Generating base-calls for every position in the genome where supporting evidence is found; (iii) Polymorphism Detection: Producing SNP reports for use in association studies and for correlation with microarray experiments and Counting Statistics: Generating lists of counts for reads that align to specific genomic entities such as genes, exons, and splice junction.

6.3.4 Third-party Genome Analyzer Data Analysis Tools

Many hardware and software tools (Table 2) have been designed to be used for a variety of sequencing applications such as Resequencing, ChIP based Sequencing, and Digital Gene Expression. For genome alignment, browsers like Gbrowse-Genomic Browsing: Generic Model Organism Database Project (*http://www.gmod.org/wiki/index.php/Gbrowse*), UCSC (University of California Santa Cruz) Genome Browser (*http://genome.ucsç.edu/goldenPath/help/customTrack.html*), and Staden Tools (GAP4 and TGAP): Alignment and visualization for small data sets (*http://sourceforge.net/projects/staden*), for alignment and polymorphism detection browsers like BFAST: Blat-like Fast Accurate Search Tool (*http://genome.ucla.edu/bfast*), MAQ: Mapping and Assembly with Quality (*http://maq.sourceforge.net/maq-man.shtml*), and Bowtie: an ultrafast memory-efficient short read aligner (*http://bowtie-bio.sourceforge.net*), and for genomic assembly softwares such as Velvet: De novo assembly of short reads (*http://www.ebi.ac.uk/zerbino/velvet*), SSAKE: assembly of short reads (*http://bioinformatics.oxfordjournals.org*), and Euler Genomic Assembly (*http://nbcr.sdsc.edu/euler*) are available. For ChIP Sequencing: ChIP-Seq Peak Finder (*http://woldlab.caltech.edu*), Digital Gene Expression: Comparative Count Display (*ftp://ftp.ncbi.nlm.nih.gov/pub/sage/obsolete/bin/ccd*), and SAGE DGED Tool (*http://cgap.nci.nih.gov/SAGE/SDGED*) are freely available. The Transmogrifier-4, an FPGA-based hardware development system with multi-gigabyte memory capacity, high host and memory bandwidth (Fender et al. 2005) and an 'Abstract Interface' integrated with coarse-grain SIMD operating system has been developed (Fernandez et al. 2006) for analyzing tera and gigabyte amount of large genome sequencing data generated by 454, Solexa and SOLiD. The CLC Genomics (*www.clcbio.com/genomics*) workbench is the first comprehensive analysis integrated package that has been developed recently to analyze and visualize data from all the major Next Generation Sequencing platforms. It takes the full advantage of 'paired end' data, and supports a number of features and work-tasks, such as reference assembly of genomes, de novo assembly of genomes, SNP detection using advanced models, multiplexing, and high-throughput trimming. Recently, the ABI SOLiD software community has developed a comprehensive suite of open source commercial software tools (*http://www.appliedbiosystems.com/solidsoftware*) for systematic analysis of large amount of short-read sequencing data generated by SOLiD. The goal of this community is to directly address the challenges associated with analyzing and managing the vast amounts of research data generated by the ultra-high-throughput sequencing technology. The community involves the freely accessible softwares like SOLiD™ Accuracy Enhancement Tool (SAET) to correct miscalls within raw data of sequence reads either prior to mapping or contig assembly, AB WT Analysis Pipeline tool to map transcriptome reads to a reference genome, count tags for exons and genes and view data in UCSC Genome Browser, AB Small InDel and Inversion Tool to identify small InDels and inversion from mate-pair sequence orientation analysis, De Novo Assembly Tools to create de novo assemblies from SOLiD™ colorspace reads, and SOLiD™ System Analysis Pipeline Tool (Corona Lite) to map color space reads to large or small genomes and to place and annotate paired reads. More recently, a simpler and user-friendly SOLiD™ BioScope Software has been developed by ABI which provides a validated and single framework pipeline for resequencing and whole-transcriptome off-instrument analysis. The integration of a scalable mapping engine with application specific pipelines in this software enables researchers for high-throughput and less error prone sequencing and thus facilitates the efficient conversion of large amount of short-read sequence data generated by Next Generation platforms into biologically meaningful results.

7 BRIEF OVERVIEW OF WHOLE GENOME SEQUENCING IN PLANTS

The genetic enhancement of crop species necessitates identification and implementation of innovative DNA diagnostic tools such as reliable DNA markers, and isolation and structural and functional characterization of specific genes that are predictive of specific plant characteristics such as disease resistance, stress tolerance, improved productivity and better quality. This can be realized by decoding all the genetic information encoded in the DNA. The way genes are organized within the genome and the way they interact to determine biological functions can be more precisely understood using the genome sequence information. Collecting and analyzing this information for the whole genome illustrate the key concept of genome analysis and genomics. The advancement in genome analysis supported by automation and innovative software tools has accelerated the whole genome sequencing in several plant species, especially the ones having small genome and well characterized genetic resources such as rice, *Arabidopsis*, *Sorghum*, *Poplar* and grape. A brief overview is provided here on the efforts made in different plant species to sequence the genome completely to generate a wealth of information having many different applications in basic and applied aspects of plant biology.

7.1 *Arabidopsis*

Arabidopsis is a weed of the mustard family that has a thin 6-inch-long stem, small green leaves, and tiny white flowers. Although it has no commercial, medicinal, decorative, or other practical uses, *Arabidopsis* is the powerhouse of the plant geneticists. It is easy to plant and grow, maturing in a couple of weeks producing plenty of seeds very quickly so that many seed generations can be studied in a short time. Being small, it requires little precious laboratory space to grow. It has many known biological processes and availability of many characterized genes (Meinke et al. 1998). *Arabidopsis* is the first plant species whose genome was completely sequenced. The sequencing of its genome was originally proposed in 1994 for completion in the year 2004, but finished in the year 2000 (AGI 2000). The *Arabidopsis* Genome Initiative (AGI) was launched to sequence its 125 Mb genome following the clone-by-clone strategy, with participation of seven laboratories from the US, 17 from the European Union, and one from Japan. In this approach, the integrated resources of physical maps of genome of ecotype Columbia assembled by fingerprinting analysis of BACs, TACs and cosmid clones (Marra et al. 1999) and high-density genetic maps were used to assemble sets of contigs into sequence ready tilling paths. Ten contigs representing the chromosome arms and centromeric heterochromatin were assembled from 1,596 BACs, TACs and cosmid clones (average size 100 kb) to generate the whole genome sequences including 10 Mb of centromeres (AGI 2000) with 98% sequence accuracies. The sequences of each of the five chromosomes were reported (Lin et al. 1999; Mayer et al. 1999; Salanoubat et al. 2000; Theologis et al. 2000; Tabata et al. 2000) of which chromosome 1 had the maximum physical size (29.1 Mb). Analysis of its sequence suggested that the evolution of this small genome species involved a whole genome duplication, followed by subsequent gene loss and extensive local gene duplications. A total of 25,498 genes encoding proteins from 11,000 families could be predicted, which was similar to the functional diversity reported earlier for the other sequenced multicellular eukaryotes such as *Drosophila* and *Caenorhabditis elegans*. *Arabidopsis* genome was found to have many families of new proteins and lacked several common protein families, indicating that the sets of common proteins had undergone differential expansion and contraction in these three multicellular eukaryotes. To improve the quality of sequence data including removal of sequence inaccuracies, misassembly and gaps, the *Arabidopsis* Information Resources (TAIR) has analyzed the *A. thaliana* genomic sequence data generated by AGI. They utilized sequence data of 1,604 BAC clones representing five *A. thaliana* chromosomes and generated *A. thaliana* whole genome pseudomolecule database (TAIR release 5.0) consisting of high quality non-overlapping 134.61 Mb genomic sequences with 99.999% sequence accuracies. The complete genome sequence has provided the foundations for more comprehensive comparison of conserved processes in all eukaryotes, identifying a wide range of plant-specific gene functions and establishing rapid systematic ways to identify genes for crop improvement (See for details Chapter 6 in this volume).

7.2 Rice

Rice is one of the world's most important food crops, having important syntenic relationships with the other cereal species. It is a model plant for the grasses. The International Rice Genome Sequencing Project (IRGSP), a consortium of publicly funded laboratories, was established in 1997 to obtain a high quality, map-based sequence of the rice genome using the cultivar Nipponbare of *Oryza sativa* ssp. *japonica*. The consortium comprised 10 member countries: Japan, US, China, Taiwan, Korea, India, Thailand, France, Brazil, and UK. The IRGSP adopted the clone-by-clone shotgun sequencing strategy in order to associate clones with a specific position on the genetic map. In December 2004, the IRGSP completed the sequencing of the rice genome. The consortium utilized the resources developed over decades such as high density genetic and physical maps (Sasaki 1998; Chen et al. 2002; McCouch et al. 2002) and large scale ESTs and cDNA sequence database to obtain complete finished quality sequence. The sequencing strategies adopted by IRGSP was a comprehensive genetically anchored physical map based clone by clone hierarchical shotgun approach using minimum tiling paths (MTPs) of nine large insert genomic library constructed in BAC/PAC clones (Sasaki and Burr 2000). In this method, the clones of chromosomal regions were selected from the assembly of their fingerprinted data and converted into contigs by FPC software. The contigs were mapped to chromosomes with help of clone end sequences and molecular markers, and the templates of BAC/PAC clones were selected from the MTP and used for sequencing.

In the year 2005, IRGSP reported complete sequencing that covered 95% of the 389 Mb rice genome, including virtually all of the euchromatin and two complete centromeres (IRGSP 2005). Three completely sequenced rice chromosomes (1, 4 and 10) have also been published (Feng et al. 2002; Sasaki et al. 2002) along with two completely sequenced centromeres (Nagaki et al. 2004; Wu et al. 2004; Zhang et al. 2004). Besides, the draft sequence of the *indica* cultivar 93-11 was published by the Beijing Genomics Institute (Yu et al. 2002). Prior to the IRGSP effort, Syngenta (Goff et al. 2002) tried to sequence the Nipponbare genome using whole genome shotgun approach. However, further genetic studies using these draft sequences are restricted due to ambiguities or incompleteness related to the experimental strategy, the occurrences of 62 telomeric and centromeric physical and sequence gaps remaining on the chromosomes. In the high quality sequence of the IRGSP, a total of 37,544 non-transposable-element-related protein-coding genes were identified, of which 71% had a putative homolog in *Arabidopsis*. Twenty-nine per cent of the predicted protein coding genes appeared in clustered gene families. The number and classes of transposable elements found in the rice genome were consistent with the expansion of syntenic regions in maize and sorghum genomes. Widespread and recurrent gene transfer from the organelles to the nuclear chromosomes was evident. The genome sequence has been updated with the recent release of the Build 5 pseudomolecules representing the 12 rice chromosomes. The pseudomolecule for each chromosome was constructed by joining the nucleotide sequences of each PAC/BAC clone based on the order of the clones on the latest physical map. The overlapping sequences were removed and the physical gaps were replaced by successive nucleotide sequences. The updated pseudomolecules (version 6.1) were constructed based on the sequence data fixed on June 23, 2009. The nucleotide sequences of seven new clones mapped on the euchromatin-telomere junctions were added in the new genome assembly. In addition, several clones in the centromere region of chromosome 5 were improved and one gap on chromosome 11 was closed. The map-based sequence, which is in the public domain, has proven useful for the identification of genes underlying agronomic traits. The SNPs and microsatellites identified in the genome are accelerating genetic improvement of rice (See for details Chapter 7 in this volume).

7.3 *Populus*

Trees are unique among plant species having distinct developmental, physiological and anatomical characteristics. Especially the large size and long-generation distinguish trees from other organisms. To study the cellular and molecular mechanisms underlying unique biology of woody tree plants, the black cottonwood *Populus trichocarpa* variety 'Nisqually-1' was selected as the model perennial forest tree species for genome sequencing. This species was chosen due to its modest genome size (520 Mb), ease of routine transformation systems (Meilan 2006) and availability of numerous genetic tools (Brunner et al. 2004; Tuskan et al. 2004). Interestingly, under optimal conditions, poplars can add a dozen feet of growth each year and reach maturity in as few as four years, permitting selective breeding for large-scale sustainable plantation forestry. This rapid growth coupled with conversion of the lignocellulosic portion of the plant to ethanol has the potential to provide a renewable energy resource along with a reduction of greenhouse gases. The sequencing of the poplar genome was led by the US Department of Energy's Joint Genome Institute (DOE-JGI) and Oak Ridge National Laboratory (ORNL) and involved 34 institutions from around the world, including the University of British Columbia, and Genome Canada; Umea University, Sweden; and Ghent University, Belgium. The researchers employed the whole genome shot-gun approach (Schiex et al. 2001) using the information already available including the physical maps based on 12 BAC restriction fragment fingerprints and BAC-end sequencing, and extensive genetic maps based on microsatellite polymorphisms (Cervera et al. 2001; Yin et al. 2008). The root derived genomic DNA obtained from 'Nisqually-1' was used to prepare phosmid libraries (15× clone coverage) and partially *Hind*III-digested genomic BAC DNA libraries (~10× clone coverage) and sequenced. Roughly 4.2 billion high quality base pairs representing 19 chromosomes of *Populus* genome were assembled into 2,447 major scaffolds containing an estimated 410 Mbp of genomic DNA, 157 and 803 kb DNA of chloroplast and mitochondrial genomes, respectively. The 410 Mb of assembled scaffolds were anchored to 535 sequence-tagged microsatellite genetic map (Cervera et al. 2001) and 356 microsatellite markers based chromosome-scale linkage maps (Tuskan et al. 2004) that were found to be colinear with the majority (91%) of the mapped microsatellite markers. A set of 2,460 physical map contigs constructed from the paired BAC end sequences were positioned on the genome assembly. This effort generated an approximately 6x coverage of the genome with high quality sequences

(Tuskan et al. 2006). Sequence analysis led to prediction of more than 45,000 putative protein-coding genes in the *Populus* genome and suggested a whole-genome duplication event, though only about 8,000 pairs of duplicated genes have survived. Besides, a second but older duplication event was thought to be indistinguishably coincident with the divergence of the *Populus* and *Arabidopsis* lineages. A comparative analysis of *Populus* and *Arabidopsis* genomes suggested that nucleotide substitution, tandem gene duplication, and gross chromosomal rearrangement proceeded more slowly in the tree species. *Populus* was found to have more protein-coding genes than *Arabidopsis*, ranging on average from 1.4 to 1.6 putative *Populus* homologs for each *Arabidopsis* gene. Over-represented exceptions in *Populus* include genes associated with ligno-cellulosic wall biosynthesis, meristem development, disease resistance, and metabolite transport. The genome sequence of poplar is providing new insights into the genetic programs controlling ontogeny, ecological adaptation and environmental physiology of trees (See for details Chapter 8 in this volume).

7.4 Grape

Grapes (*Vitis vinifera*) are mainly grown for producing wine. The genome of the grape is spread over 19 pairs of chromosomes and is around 500 megabases in length. The draft sequence of the grapevine genome is the fourth one to be produced for flowering plants, the second for a woody species and the first for a fruit crop cultivated for both fruit and beverage (Jaillon et al. 2007; Velasco et al. 2007). Velasco et al. (2007) used a shotgun sequencing approach, which resulted in 10.7× coverage, of which 4.2× was based on pyrosequencing and 6.5× by Sanger sequencing method. At the same time, the genome of the grape chloroplast was also sequenced. High degree of heterozygosity was observed between pairs of chromosomes with about 11.2% of the sequence differing between homologous regions. Moreover, more than two million SNPs were reported in 87% of the 29,585 identified genes. Jaillon et al. (2007) reported several large expansions of gene families with roles in aromatic features. The grapevine genome was found not to have undergone recent genome duplication, thus enabling the discovery of ancestral traits and features of the genetic organization of flowering plants. Contribution of three ancestral genomes to the grapevine haploid content was observed by this group. The ancestral arrangement was common to many dicotyledonous plants but was reported absent from the genome of rice, which is a monocotyledon. The breeding of grape vines is difficult because they take several years to grow to maturity and domesticated grapes tend to have very low fertility. For this reason, grapes are usually propagated by cuttings or graftings so that vineyards are filled with hundreds of thousands of genetically identical clones. This leaves grapes highly susceptible to the emergence of aggressive microorganisms, such as phyloxera, which devastated European grape production in the 19th and early 20th century, and powdery mildew, which continues to threaten American harvests to this day. The description of the grape genome has presented an opportunity to direct genetic improvement for disease resistance without disturbing the biochemistry of taste and grape quality.

7.5 Sorghum

Sorghum bicolor is a close relative of maize and sugarcane. It is grown for food, feed, fiber and fuel in Africa and much of the developing world because of its superior tolerance to arid growth conditions. Bedell et al. (2005) first generated sequence from the hypomethylated portion of the sorghum genome by applying methylation filtration (MF) technology. About 96% of the genes could be sequence tagged, with an average coverage of 65% across their length. This level of gene discovery was accomplished after generating a raw coverage of less than 300 megabases of the 735-megabase genome. MF preferentially captured exons and introns, promoters, microRNAs, and microsatellites, and minimized interspersed repeats, thus providing a robust view of the functional parts of the genome. More recently, Paterson et al. (2009) generated and analyzed 730-megabase sorghum genome placing approximately 98% of the genes in their chromosomal context using whole-genome shotgun sequence validated by genetic, physical and syntenic information. Genetic recombination was largely confined to about one-third of the sorghum genome with gene order and density similar to those of rice. Retrotransposon accumulation in recombinationally recalcitrant heterochromatin explained the approximately 75% larger genome size of sorghum compared with rice. Although gene and repetitive DNA distributions were found preserved since palaeopolyploidization approximately 70 million years ago, most duplicated gene sets lost one member before the sorghum-rice divergence. About 24% of the genes were reported to be grass-specific and 7% were sorghum-specific.

7.6 Tomato

The tomato (*Solanum lycopersicum* L.) is an important vegetable crop world over. Its genome is comprised of approximately 950 Mb of DNA—more than 75% of which is heterochromatin and largely devoid of genes. The majority of genes are found in long contiguous stretches of gene-dense euchromatin located on the distal portions of each chromosome arm. The genome of tomato is being sequenced by an international consortium of 10 countries (Korea, China, the UK, India, Netherlands, France, Japan, Spain, Italy, and US) as part of the larger International Solanaceae Genome Project (SOL). The tomato genome sequencing project uses an ordered bacterial artificial chromosome (BAC) approach to generate a high-quality tomato euchromatic genome sequence for use as a reference genome for the Solanaceae and Euasterids. Currently, there are around 1,000 BACs finished or in progress, representing more than one-third of the projected euchromatic portion of the genome. The expected number of genes in the euchromatin is ~40,000, based on an estimate from a preliminary annotation of 11% of finished sequence (Mueller et al. 2009). The whole-genome shotgun technique is proposed to be combined with the BAC-by-BAC approach to cover the entire tomato genome. The high-quality reference euchromatic tomato sequence is expected to be near completion by 2010 (See for details Chapter 10 in this volume).

7.7 *Brassica*

The International *Brassica* Genome Consortium involving Multinational *Brassica* Genome and *Brassica rapa* Genome Sequencing Projects (*Br*GSP, *http://brassica.bbsrc.ac.uk http://www.niab.go.kr*) started in the year 2006 for sequencing of *ca.* 500 Mb *B. rapa pekinensis* (Chinese cabbage) genome. The consortium assigned 10 *B. rapa* chromosomes to different *Br*GSP consortium participants to sequence particularly the gene-space regions of each chromosome based on clone-by-clone strategy. Of the 10 chromosomes, eight were allocated to the participating countries including Korea (R3 and R9), Canada (R2 and R10), UK and China (R1 and R8), USA (R6), and Australia (R7) while remaining two chromosomes, R4 and R5 are unassigned till now. The genomic resources and the reference mapping populations are prerequisite to undertake sequencing of such a large *Brassica* genome. Two reference mapping populations derived from two *B. rapa* ssp. *pekinensis* recombinant inbred and double haploid lines have been developed to be used for construction of reference genetic maps (Lim et al. 2006). Three *Hin*dIII, *Bam*HI, and *Sau*3AI digested BAC libraries containing 56592, 50688 and 55296 clones with an average insert size of 115 kb, 124 kb and 100 kb, respectively have also been generated. The resources include 200,017 end sequences of BAC clones, a reference genetic linkage map (*http://ukcrop.net/perl/ace/search/BrassicaDB* and *http://www.niab.go.kr*) using RAPDs, AFLPs, microsatellites and gene based markers, and recently a high resolution genetic map using 1,000 microsatellite markers derived from BAC end and EST sequences. All the 10 linkage groups of a reference genetic map of *B. rapa* have been assigned to the corresponding chromosomes through fluorescence in situ hybridization (FISH) using locus-specific BAC clones as probes. Several research groups involved in this genome sequencing project have contributed significantly including the creation of a draft physical maps (Mun et al. 2008) of the *Brassica* 'A' and 'C' genomes by fingerprinting BAC libraries and their integration with the *Arabidopsis* genome sequence by hybridization with selected gene anchor probes (*http://brassica.bbsrc.ac.uk/IGF, http://brassica.bbsrc.ac.uk/IMSORB*), development of a new and economical allele-specific amplification assay method for SNP markers in oilseed rape (*http://brassica.bbsrc.ac.uk/IMSORB*), assembling of oilseed rape genomic BAC clones into 1,429 sets of contigs by finger printed contigs (FPC) software (*http://brassica.bbsrc.ac.uk/IGF/index.htm* and *http://www.brassicagenome.org*) and construction of a deep-coverage genome-wide BAC based physical map particularly for the two biggest chromosomes A03 and A09 of *B. rapa* genome by the SNaPshot fingerprinting of 67,468 BAC clones and their assembling into 1,428 contigs spanned over 717 Mb physical region (*http://www.brassica.info*). Seed BACs for genome sequencing have been selected through *in silico* allocation of *B. rapa* BAC-end sequences onto counterpart locations of *Arabidopsis* chromosomes (Yang et al. 2005). So far, based on the physical map of *B. rapa* and the *in silico* comparative mapping of its BAC-ends onto *Arabidopsis* chromosomes, the anchoring of 1000 seed BACs unambiguously to the 125 Mb euchromatic sequenced regions of *A. thaliana* genome (*http://www.brassica-rapa.org*) has already been completed by STS mapping and FISH analysis. The seed BACs which are anchored and sequenced are being used as stepping stones for sequencing of the 10 *B. rapa* chromosomes. Considering the large genome size of *B. rapa*, the possibility of whole genome shotgun approach using various next generation sequencing platforms have also been proposed by several consortium partners. Using the whole genome shotgun sequencing approach, the *B. oleracea* Genome Sequencing project has started (*http://www.tigr.org/tdb/e2k1/bog*) in the year 2007. Six *B. oleracea* BAC contigs homeologous to a duplicated region of *A. thaliana* chromosomes 4 and 5 identified by O'Neill and Bancroft (2000) were completely sequenced and the rest are in progress. The genome sequencing of *Brassica* offers a new perspective for plant biology and evolution in the context of polyploidization. Besides, these genomic sequence information would help to understand the structural consequences of plant genome evolution and identification molecular markers and genes associated with important agricultural traits, thereby enabling improvement of *Brassica* species economically (See for details Chapter 10 in this volume).

7.8 Wheat

The first initiative towards whole genome sequencing (WGS) in wheat was organized in the form of International Wheat Genome Sequencing Consortium (IWGSC). The IWGSC (*www.wheatgenome.org*) is a collaboration focused on building the foundation for advancing agricultural research for wheat production and utilization by developing DNA-based tools and resources that result from the complete genome sequence of bread (hexaploid) wheat. The principal goal of the WGSC is to obtain a publicly available, complete sequence of common (hexaploid) wheat, since it is grown on over 95% of the wheat growing area. The complete sequence of this particular wheat holds the key to genetic improvements that will allow growers to meet the growing demands for high quality food produced in an environmentally sensitive, sustainable, and profitable manner. Further, polyploidy is a driving force for crop genome evolution and thus to understand all the dynamics and possibilities, whole genome sequencing of a hexaploid common wheat is essential. The IWGSC has selected the specific cultivar, Chinese Spring (CS) for whole genome sequencing, since this crop already has ample genetic and molecular resources (Gill et al. 2004). With the availability of new sequencing technologies provided by 454/Roche and those provided by Illumina/Solexa and ABI SOLiD (Gupta et al. 2008), sequencing of gene space of the wheat genome, which was once thought to be almost impossible, should become possible within the foreseeable future. For sequencing the complete bread wheat genome the following short and long term strategies are proposed towards reaching the final objective. This includes development of an integrated genetic and physical map for Chinese Spring, assessing several alternative genome sequencing approaches developed so far, sequencing and annotating the genic regions, obtaining full-length cDNAs for all expressed genes, finally completing the sequencing of Chinese Spring genome and high quality annotation of its genome. At this time, the precise method to sequence the ~17 Gb hexaploid wheat genome is unknown. The hexaploid coupled with the repetitive nature of the genome will present technical challenges in the current methodologies. Several methods or combinations of approaches have been proposed (Gill 2004) including whole genome shotgun sequencing, enrichment methods such as methylation filtration or high C_0t, and/or clone-by-clone sequencing. Three phases were proposed for sequencing the wheat genome: pilot, assessment, and scale up. The first phase was recommended for 5-years and is mainly focused on the short-term goal of IWGSC, involving physical and genetic mapping along with sample sequencing of the wheat genome aimed at better understanding of the wheat genome structure. The pilot phase involves the strategies like ascertaining the gene enrichment capabilities of methylation filtration and high C_0t, constructing and assessing the quality of chromosome-specific, and chromosome-arm specific BAC libraries, developing BAC fingerprint contig maps for each chromosome, assessing the distribution of genes across the genome, investigating the ability to differentiate homeologous sequences and developing bioinformatics tools for an improved semi-automated annotation of large sequences. The assessment phase will involve determining the method(s) to be used in a cost-effective manner to generate the sequence of the wheat genome. After a full assessment, the scale-up phase will involve the deployment of optimal methods on the whole genome, obtaining the genome sequence and annotation, which is the long-term goal of IWGSC. A first pilot project led by the INRA in France was initiated in 2004 to assess point 3 using the largest wheat chromosome (chr. 3B, 1GB = 2*x* the rice genome) of hexaploid wheat as a model. As many as 68,000 BAC clones from a 3B chromosome specific BAC library (Safar et al. 2004) were fingerprinted and assembled into contigs, which were then anchored to wheat bins, covering ~80% of chromosome 3B. Currently, one or more of these contigs are being sequenced, which will demonstrate the feasibility of large-scale sequencing of complete gene space of wheat genome. The first phase (2,500 contigs) has been recently achieved (Moolhuijzen et al. 2007) and the anchoring of the physical contigs to the genetic map is currently underway. This demonstrates the feasibility of the chromosome specific approach for the sequencing of the hexaploid wheat genome (See for details Chapter 10 in this volume).

7.9 Peach

The peach genome sequencing project started at the US Department of Energy's Joint Genome Institute (JGI), California in the year 2008. The peach haploid cultivar "Lovell" which had well developed genomic information like integrated high saturation genetic and physical maps and large-insert genomic BAC libraries were selected for 8-9x coverage sequencing of its genome using whole-genome shotgun approach. Construction of the peach physical map relied on a combination of BAC fingerprinting and library hybridization with molecular genetic markers and cDNAs (Sosinski et al. 2009). Previously identified microsatellite markers were also integrated with the physical map by overgo hybridization to the BAC library filters and eventually to BAC tiling path based on the physical map. Marker poor regions of the genetic/physical map were targeted for microsatellite marker development by sequencing microsatellite containing fragments of BAC clones from these regions. These efforts have created a genome-wide framework physical map (Zhebentyayeva et al. 2008) of peach containing 2,138 contigs composed of 15,655 BAC clones using high-information content fingerprinting (HICF) and FPC software. The total physical length of all contigs was estimated at 303 Mb or 104.5% of the peach genome. The framework physical map was anchored on the genetic reference map and integrated with the peach transcriptome map (*http://www.rosaceae.org*). Till now about 2,636 genetic markers including peach ESTs, gene-specific and overgo probes were incorporated into the physical framework to support the accuracy of contig assembly (Jung et al. 2008). The sequenced reads are deposited at the NCBI GenBank. Reads are aligned by using various genome assemblers to produce the primary draft assembly, which consists of contigs linked into larger scaffolds by paired-end information. The JGI and several partner institutions namely, Stanford University, Los Alamos National Laboratory, and Lawrence Livermore National Laboratory were involved in the finishing works such as gap closing, quality improvement, and assembly verification for further refinement of peach genome sequencing data. The sequencing data generated for peach genome are being finally annotated using the automated softwares like minimal automated annotation and Genome Portal to derive biological useful information for improvement of this tree species and other species belonging to Rosaceae (See for details Chapter 9 in this volume).

8 FUTURE PROSPECTS OF WHOLE GENOME SEQUENCING

The Human Genome Project, the first to be conceived to completely sequence a large eukaryotic genome, catalyzed rapid developments in whole genome sequencing of other species including plants having large and complex genomes. A consortium mode of operation involving several diverse groups of researchers with varied expertise drawn from both public and private institutions spread across continents having commitment for sharing of responsibilities as well as credits became highly successful yielding unprecedented results. Consequently, our understanding of genome structure has vastly improved. Enormous amount of information about the gene content, nature and distribution of non-genic components of the genome particularly the mobile genetic elements and microsatellites, and the extent of sequence variation especially the SNPs has become readily available in the public domain for unrestricted use. Comparative analysis of sequence within and between genomes has provided deeper insight into the pattern of genome evolution. New tools designed by using the whole genome sequence resources are greatly facilitating defining the genome function, enhancing capability in disease diagnosis and drug design, and accelerating genetic enhancement of agricultural species. Thus the conventional Sanger sequencing coupled with enabling high throughput technologies has been immensely successful both for small and large genome species although at a considerably higher cost and efforts spread over several years. Sequencing of large genomes species such as wheat and maize however, has been slow. Fortunately, approaches alternative to the conventional Sanger sequencing employing second generation sequencing technologies namely 454 Pyrosequencer, Solexa Genome Analyzer, and SOLiD have been developed. These technologies have already catapulted to prominence with increasingly widespread adoption of platforms that individually implement different flavors of massively parallel cyclic-array sequencing. The reduction in the costs of DNA sequencing and generation of high-quality low error prone giga-byte amount of large sequencing data by several orders of magnitude in less time with these new sequencing technologies have the potential to improve and revolutionize the whole genome sequencing strategies for small as well as large genome crop species in near future. Besides, these new platforms are capable of efficient resequencing of whole genome and targeted gene regions, as well as sequencing of hypermethylated genomic region, and GC-rich repetitive centromeric and transposon associated chromosomal regions more precisely and accurately in a high-throughput manner. Given the fast pace of development, it is difficult to peep even a few years into the future. It is anticipated that second generation sequencing platforms supported by appropriate bioinformatic tools would become widespread, commoditized and routine for whole genome sequencing of eukaryotes including plants to extract biologically meaningful and useful information to be used in several applications in eukaryotic structural, comparative and functional genomics.

References

Arabidopsis Genome Initiative (2000) Analysis of the genome sequence of the flowering plant *Arabidopsis thaliana*. Nature 408: 796-815

Bedell JA, Budiman MA, Nunberg A, Citek RW, Robbins D, Jones J, et al. (2005) *Sorghum* genome sequencing by methylation filtration. PLoS Biol 3: e13

Bergot JB (1994) Spectrally resolvable rhodamine dyes for nucleic acid sequence determination. US Patent 5,366,860

Brunner AM, Busov VB, Strauss SH (2004) Poplar genome sequence: functional genomics in an ecologically dominant plant species. Trend Plant Sci 9: 49-56

Butler J, MacCallum I, Kleber M, Shlyakhter IA, Belmonte MK, Lander ES, Nusbaum C, Jaffe DB (2008) ALLPATHS: *de novo* assembly of whole-genome shotgun microreads. Genome Res 18: 810-820

Cervera M-T, Storme V, Ivensa B, Gusmao J, Liu BH, Hostyn V, Slycken JV, Montagu MV, Boerjan W (2001) Dense genetic linkage maps of three *Populus* species (*Populus deltoides*, *P. nigra* and *P. trichocarpa*) based on AFLP and microsatellite markers. Genetics 158: 787-809

Chaisson MJ, Pevzner PA (2008) Short read fragment assembly of bacterial genomes. Genome Res 18: 324-330

Chen M, Presting G, Barbazuk WB, Goicoechea JL, Blackmon B, Fang G, Mohapatra T, Singh NK, Wing RA, et al. (2002) An integrated physical and genetic map of the rice genome. Plant Cell 14: 537-545

Dressman D, Yan H, Traverso G, Kinzler KW, Vogelstein B (2003) Transforming single DNA molecules into fluorescent magnetic particles for detection and enumeration of genetic variations. Proc Natl Acad Sci USA 100: 8817-8822

Duthie RS, Kalve IM, Samols SB, Hamilton S, Livshin I, Khot M, Nampalli S, Kumar S, Fuller CW (2002) Novel cyanine dye-labeled dideoxynucleoside triphosphates for DNA sequencing. Bioconjug Chem 13: 699-706

Elkin C, Kapur H, Smith T, Humphries D, Pollard M, Hammon N, Hawkins T (2002) Magnetic bead purification of labeled DNA fragments for high-throughput capillary electrophoresis sequencing. Biotechniques 32: 1296-1302

Ewing B, Green P (1998) Base calling sequencer traces using Phred II Error probabilities. Genome Res 8: 186-194

Ewing B, Hillier L, Wendl M, Green P (1998) Base calling sequencer traces using Phred I Accuracy assessment. Genome Res 8: 175-185

Fangan BM, Dahlberg OJ, Deggerdal AH, Bosnes M, Larsen F (1999) Automated system for purification of dye-terminator sequencing products eliminates up-stream purification of templates. Biotechniques 26: 980-983

Fedurco M, Romieu A, Williams S, Lawrence I, Turcatti GBTA (2006) A novel reagent for DNA attachment on glass and efficient generation of solid-phase amplified DNA colonies. Nucl Acids Res 34: e22

Fender J, Rose J, Galloway D (2005) The Transmogrifier-4: an FPGA-based hardware development system with multi-gigabyte memory capacity and high host and memory bandwidth, Field-Programmable Technology. In: Proc IEEE Intl Conf, Vol 11, pp 301-302

Feng Q, Zhang Y, Hao P, Wang S, Fu G, Huang Y, Li Y, Zhu J, et al. (2002) Sequence and analysis of rice chromosome 4. Nature 420: 259

Fernandez J, Frachtenberg E, Petrini F, Sancho J-C (2006) An abstract interface for system software on large-scale clusters. Comp J 49: 454-469

Gill BS, Appels R, Botha-Oberholster AM, Buell CR, Bennetzen JL, Sasaki T (2004) A workshop report on wheat genome sequencing: international genome research on wheat consortium. Genetics 168: 1087-1096

Gill KS (2004) Gene distribution in cereal genomes. In: Gupta PK, Varshney RK (eds) Cereal Genomics. Kluwer Academic Publ, Dordrecht, Netherlands, pp 361-385

Goff SA, Ricke D, Lan T-H, Presting G, Wang R, Dunn N, Briggs S, et al. (2002) A draft sequence of the rice genome (*Oryza sativa* L. ssp. *japonica*). Science 296: 92-100

Gordon D, Abajian C, Green P (1998) Consed: A graphical tool for sequence finishing. Genome Res 8: 195-202

Gupta PK, Mir RR, Mohan A, Kumar J (2008) Wheat Genomics: Present Status and Future Prospects. Int J Plant Genom 2008: 896451

Hawkins TL, O'Connor-Morin T, Roy A, Santillan C (1994) DNA purification and isolation using a solid-phase. Nucl Acids Res 22: 4543-4544

Hong CP, Kwon S-J, Kim JS, Yang T-J, Park B-S, Lim YP (2008) Progress in understanding and sequencing the genome of *Brassica rapa*. Int J Plant Genom 2008: 582837

Huang X, Feng Q, Qian Q, Zhao Q, Wang L, Wang A, Guan J, Fan D, Weng Q, Huang T, Dong G, Sang T, Han B (2009) High-throughput genotyping by whole-genome resequencing. Genome Res 19: 1068-1076

Hultman T, Uhlen M (1994) Solid-phase cloning to create sub-libraries suitable for DNA sequencing. J Biotechnol 35: 229-238

Hultman T, Murby M, Stahl S, Hornes E, Uhlen M (1990) Solid phase *in vitro* mutagenesis using plasmid DNA template. Nucl Acids Res 18: 5107-5112

Hunkapiller T, Kaiser RJ, Koop BF, Hood L (1991) Large-scale and automated

DNA sequence determination. Science 254: 59-67

Ingolia NT, Ghaemmaghami S, Newman JR, Weissman JS (2009) Genome-wide analysis *in vivo* of translation with nucleotide resolution using ribosome profiling. Science 324: 218-223

International Rice Genome Sequencing Project (IRGSP, 2005) The map-based sequence of the rice genome. Nature 436: 793-800

Jaillon O, Aury J-M, Noel B, Policriti A, Clepet C, et al. (2007) The grapevine genome sequence suggests ancestral hexaploidization in major angiosperm phyla. Nature 449: 463-467

Jiang C, Pugh BF (2009) Nucleosome positioning and gene regulation: advances through genomics. Nat Rev Genet 10: 161-172

Johnson AF, Wang R, Ji H, Chen D, Guilfoyle RA, Smith LM (1996) Purification of single-stranded M13 DNA by cooperative triple-helix-mediated affinity capture. Anal Biochem 234: 83-95

Ju J (2002) DNA sequencing with solid-phase-capturable dideoxynucleotides and energy transfer primers. Anal Biochem 309: 35-39

Jung S, Staton M, Lee T, Blenda A, Svancara R, Abbott A, Main D (2008) GDR (Genome Database for Rosaceae): integrated web-database for Rosaceae genomics and genetics data. Nucl Acids Res 36: D1034-D1040

Ketel C, Wang HS, McClellan M, Bouchonville K, Selmecki A, Lahav T, Gerami-Nejad M, Berman J (2009) Neocentromeres form efficiently at multiple possible loci in *Candida albicans*. PLoS Genet 5: e1000400

Korbel JO, Abyzov A, Mu XJ, Carriero N, Cayting P, Zhang Z, Snyder M, Gerstein MB (2009) PEMer: a computational framework with simulation-based error models for inferring genomic structural variants from massive paired-end sequencing data. Genome Biol 10: R23

Kumar S, Sood A, Wegener J, Finn PJ, Nampalli S, Nelson JR, Sekher A, Mitsis P, Macklin J, Fuller CW (2005) Terminal phosphate labeled nucleotides: Synthesis, applications, and linker effect on incorporation by DNA polymerases. Nucleos, Nucleot Nucl Acids 24: 401-408

Li R, Li Y, Kristiansen K, Wang J (2008) SOAP: short oligonucleotide alignment program. Bioinformatics 24: 713-714

Lim YP, Plaha P, Choi SR, Uhm T, Hong CP, Bang JW, Hur YK (2006) Towards unraveling the structure of *Brassica rapa* genome. Physiol Planta 126: 585-591

Lin X, Kaul S, Rounsley S, Shea TP, Benito MI, Town CD, et al. (1999) Sequence and analysis of chromosome 2 of the plant *Arabidopsis thaliana*. Nature 402: 761-768

Mardis ER (2008) The impact of next-generation sequencing technology on genetics. Trends Genet 24: 133-141

Margulies M, Egholm M, Altman WE, Attiya S, Rothberg JM, et al. (2005) Genome sequencing in microfabricated high-density picolitre reactors. Nature 437: 376-380

Marra M, Kucaba T, Sekhon M, Hillier L, Martienssen R, Chinwalla A (1999) zA map for sequence analysis of the *Arabidopsis thaliana* genome. Nat Genet 22: 265-270

Mayer K, Schüller C, Wambutt R, Murphy G, Volckaert G, Pohl T, Dusterhot A, et al. (1999) Sequence and analysis of chromosome 4 of the plant *Arabidopsis thaliana*. Nature 402: 769-777

McCouch SR, Teytelman L, Xu Y, Lobos KB, Clare K, Walton M, Fu B, et al. (2002) Development and mapping of 2240 new SSR markers for rice (*Oryza sativa* L.). DNA Res 9: 199-207

McKernan K, Blanchard A, Kotler L, Costa G (2006) Reagents, methods, and libraries for bead-based sequencing. US patent application 20080003571

Meilan R (2006) Challenges to commercial use of transgenic plants. J Crop Improv 18: 433-450

Meinke DW, Cherry JM, Dean C, Rounsley SD, Koornneef M, et al. (1998) *Arabidopsis thaliana*: a model plant for genome analysis. Science 282: 662-682

Millar DS, Withey SJ, Tizard ML, Ford JG, Hermon-Taylor J (1995) Solid-phase hybridization capture of low-abundance target DNA sequences: application to the polymerase chain reaction detection of *Mycobacterium paratuberculosis* and *Mycobacterium avium* subsp. *silvaticum*. Anal Biochem 226: 325-330

Moolhuijzen P, Dunn DS, Bellgard M, Carter M, Jia J, Kong X, Gill BS, Feuillet C, Breen J, Appels R (2007) Wheat genome structure and function: genome sequence data and the international wheat genome sequencing consortium. Aust J Agri Res 58: 470-475

Moran NA, McLaughlin HJ, Sorek R (2009) The dynamics and time scale of ongoing genomic erosion in symbiotic bacteria. Science 323: 379-382

Mueller LA, Lankhorst RK, Tanksley SD, Giovannoni JJ, White R, et al. (2009) A snapshot of the emerging tomato genome sequence. Plant Genome 2: 78-92

Mun J-H, Kwon S-J, Yang T-J, Kim H-S, Choi B-S, Baek S, Kim JS (2008) The first generation of a BAC-based physical map of *Brassica rapa*. BMC Genom 9: 280

Nagaki K, Cheng Z, Ouyang S, Talbert PB, Kim M, Jones KM, Henikoff S, Buell CR, Jiang J, et al. (2004) Sequencing of a rice centromere uncovers active genes. Nature 36: 138-145

Ning Z, Cox AJ, Mullikin JC (2001) SSAHA: a fast search method for large DNA databases. Genome Res 11: 1725-1729

Nordborg M, Weigel D (2008) Progress article next-generation genetics in plants. Nature 456: 720-723

Nusbaum C, Ohsumi TK, Gomez J, Aquadro J, Victor TC, Warren RM, Hung DT, Birren BW, Lander ES, Jaffe DB (2009) Sensitive, specific polymorphism discovery in bacteria using massively parallel sequencing. Nat Meth 6: 67-69

O'Meara D, Nilsson P, Nygren PA, Uhlen M, Lundeberg J (1998) Capture of single-stranded DNA assisted by oligonucleotide modules. Anal Biochem 255: 195-203

O'Neill C, Bancroft I (2000) Comparative physical mapping of segments of the genome of *Brassica oleracea* var alboglabra that are homoeologous to sequenced regions of the chromosomes 4 and 5 of *Arabidopsis thaliana*. Plant J 23: 233-243

Paterson AH, Bowers JE, Bruggmann R, Dubchak I, Grimwood J, Gundlach H, et al. (2009) The *Sorghum bicolor* genome and the diversification of grasses. Nature 457: 551-556.

Pomraning KR, Smith KM, Freitag M (2009) Genome-wide high throughput analysis of DNA methylation in eukaryotes. Methods 47: 142-150

Prober JM, Trainor GL, Dam RJ, Hobbs FW, Robertson CW, Zagursky RJ, Cocuzza AJ, Jensen MA, Baumeister K (1987) A system for rapid DNA sequencing with fluorescent chain-terminating dideoxynucleotides. Science 238: 336-341.

Reinhardt JA, Baltrus DA, Nishimura MT, Jeck WR, Jones CD, Dangl JL (2009) *De novo* assembly using low-coverage short read sequence data from the rice pathogen *Pseudomonas syringae* pv. *oryzae*. Genome Res 19: 294-305

Ronaghi M, Karamohamed S, Pettersson B, Uhlen M, Nyren P (1996) Real-time DNA sequencing using detection of pyrophosphate release. Anal Biochem 242: 84-89

Rounsley S, Marri PR, Yu Y, Wing RA (2009) *De Novo* next generation sequencing of plant genomes. Rice 2: 35-43

Safar J, Bartos J, Janda J, Bellec A, Kubalakova M, Valarik M, Pateyron S, Weiserova J, et al. (2004) Dissecting large and complex genomes: flow sorting and BAC cloning of individual chromosomes from bread wheat. Plant J 39: 960-968

Salanoubat M, Lemcke K, Rieger M, Ansorge W, Unseld M, Fartmann B, Valle G, et al. (2000) Sequence and analysis of chromosome 3 of the plant *Arabidopsis thaliana*. Nature 408: 820-822

Sanger F (1988) Sequences, sequences, and sequences. Annu Rev Biochem 57: 1-28

Sasaki T (1998) The Rice Genome Project in Japan. Proc Natl Acad Sci USA 95: 2027-2028

Sasaki T, Burr B (2000) International Rice Genome Sequencing Project: The effort to completely sequence the rice genome. Curr Opin Plant Biol 3: 138-141

Sasaki T, Matsumoto T, Yamamoto K, Sakata K, Baba T, Katayose Y, Wu J, et al. (2002) The genome sequence and structure of rice chromosome 1. Nature 420: 312-316

Schiex T, Moisan A, Rouze P (2001) In: Computational Biology: Selected Papers from OBIM'2000, number 2066 in LNCS. Springer, Heidelberg, Germany, pp 118-133

Schuster SC (2008) Next-generation sequencing transforms today's biology. Nat Meth 5: 16-18

Shendure J, Hanlee Ji (2008) Next-generation DNA sequencing. Nat Biotechnol 26: 1135-1145

Shendure J, Porreca GJ, Reppas NB, Lin X, McCutcheon JP, Rosenbaum AM, Wang MD, Zhang K, Mitra RD, Church GM (2005) Accurate multiplex polony sequencing of an evolved bacterial genome. Science 309: 1728-1732

Smith LM, Sanders JZ, Kaiser RJ, Hughes P, Dodd C, Connell CR, Heiner C, Kent SB, Hood LE (1986) Fluorescence detection in automated DNA sequence analysis. Nature 321: 674-679

Sosinski B, Shulaev V, Dhingra A, Kalyanaraman A, Bumgarner R, Rokhsar D, Ignazio V, Velasco R, Abbott AG (2009) Rosaceaous genome sequencing: Perspectives and progress. In: Folta KM, Gardiner SE (eds) Plant Genetics and Genomics: Crops and Models. Vol 6: Genetics and Genomics of Rosaceae. Springer, New York, USA, pp 601-615

Springer AL, Booth LR, Braid MD, Houde CM, Hughes KA, Kaiser RJ, Pedrak C, Spicer DA, Stolyar S (2003) A rapid method for manual or automated purification of fluorescently labeled nucleic acids for sequencing, genotyping, and microarrays. J Biomol Technol 14: 17-32

Stahl S, Hansson M, Ahlborg N, Nguyen TN, Liljeqvist S, Lundeberg J, Uhlen M (1993) Solid-phase gene assembly of constructs derived from the *Plasmodium falciparum* malaria blood-stage antigen Ag332. Biotechniques 14: 424-434

Stark M, Reizenstein E, Uhlen M, Lundeberg J (1996) Immunomagnetic separation and solid-phase detection of *Bordetella pertussis.* J Clin Microbiol 34: 778-784

Sundquist A, Ronaghi M, Tang H, Pevzner P, Batzoglou S (2007) Whole-genome sequencing and assembly with high-throughput, short-read technologies. PLoS ONE 2: e484

Tabata S, Kaneko T, Nakamura Y, Kotani H, Kato T, Asamizu E, Miyajima N, et al. (2000) Sequence and analysis of chromosome 5 of the plant *Arabidopsis thaliana*. Nature 408: 823-826

Theologis A, Ecker JR, Palm CJ, Federspiel NA, Kaul S, White O, et al. (2000) Sequence and analysis of chromosome 1 of the plant *Arabidopsis thaliana*. Nature 408: 816-820

Tillett D, Neilan BA (1999) n-butanol purification of dye terminator sequencing reactions. Biotechniques 26: 606-610

Tong X, Smith LM (1992) Solid-phase method for the purification of DNA sequencing reactions. Anal Chem 64: 2672-2677

Tong X, Smith LM (1993) Solid phase purification in automated DNA sequencing. DNA Seq 4: 151-162

Tuskan GA, DiFazio SP, Teichmann T (2004) Poplar genomics is getting popular: The impact of the poplar genome project on tree research. Plant Biol 6: 2-4

Tuskan GA, DiFazio S, Jansson S, Bohlmann J, Grigoriev I, Hellsten U et al. (2006) The Genome of Black Cottonwood, *Populus trichocarpa* (Torr. & Gray). Science 313: 1596-1604

Uhlen M, Hultman T, Wahlberg J, Lundeberg J, Bergh S, Pettersson B, Holmberg A, Stahl S, Moks T (1992) Semi-automated solid-phase DNA sequencing. Trends Biotechnol 10: 52-55

Velasco R, Zharkikh A, Troggio M, Cartwright DA, Cestaro A, et al. (2007) A high quality draft consensus sequence of the genome of a heterozygous grapevine variety. PLoS One 2: e1326

Wahlberg J, Holmberg A, Bergh S, Hultman T, Uhlen M (1992) Automated magnetic preparation of DNA templates for solid phase sequencing. Electrophoresis 13: 547-551

Warren RL, Sutton GG, Jones SJ, Holt RA (2007) Assembling millions of short DNA sequences using SSAKE. Bioinformatics 23: 500-550

Wei B, Cai T, Zhang R, Li A, Huo N, Li S, Gu YQ, Vogel J, Jia J, Qi Y, Mao L (2009) Novel microRNAs uncovered by deep sequencing of small RNA transcriptomes in bread wheat (*Triticum aestivum* L.) and *Brachypodium distachyon* (L.) Beauv. Funct Integr Genomics 9: 499-511

Yang TJ, Kim JS, Lim KB, Kwon SJ, Kim JA, Jin M, Park JY, Lim MH, Kim HI, Kim SH, Lim YP, Park BS (2005) The Korea *Brassica* genome project: a glimpse of the *Brassica* genome based on comparative genome analysis with *Arabidopsis*. Comp Funct Genom 6: 138-146

Yin TM, Zhang XY, Gunter LE, Li SX, Wullschleger SD, Huang MR, Tuskan GA (2008) Microsatellite primer resource for *Populus* developed from the mapped sequence scaffolds of the Nisqually-1 genome. New Phytol 181: 498-503

Yoder-Himes DR, Chain PS, Zhu Y, Wurtzel O, Rubin EM, Tiedje JM, Sorek R (2009) Mapping the *Burkholderia cenocepacia* niche response via high-throughput sequencing. Proc Natl Acad Sci USA 106: 3976-3981

Yu J, Hu S, Wang J, Wong GK-S, Li S, Liu B, Yang H, et al. (2002) A draft sequence of the rice genome (*Oryza sativa* L. ssp. *indica*). Science 296: 79-92

Zerbino DR, Birney E (2008) Velvet: algorithms for *de novo* short read assembly using de Bruijn graphs. Genome Res 18: 821-829

Zhang Y, Huang Y, Zhang L, Li Y, Lu T, Lu Y, Feng Q, Zhao Q, Cheng Z, Xue Y, Wing RA, Han B (2004) Structural features of the rice chromosome 4 centromere. Nucl Acids Res 32: 2023-2030

Zhebentyayeva T, Swire-Clark G, Georgi L, Garay L, Jung S, Forrest S, Blenda A, et al. (2008) A framework physical map for peach, a model Rosaceae species. Tree Genet Genomes 4: 745-756

Zhu J, Davidson TS, Wei G, Jankovic D, Cui K, Schones DE, Guo L, Zhao K, Shevach EM, Paul WE (2009) Down-regulation of *Gfi-1* expression by TGF-{beta} is important for differentiation of Th17 and CD103+ inducible regulatory T cells. J Exp Med 206: 329-341

6 *Arabidopsis* Genome Initiative

Ramesh Katam[1], Dilip Panthee*[2], Evelina Basenko*[3], Rajib Bandopadhyay*[4], Sheikh M. Basha[1], Kokiladevi Eswaran[5] and Chittaranjan Kole[6]**

[1]Plant Biotechnology Laboratory, Florida A&M University, Tallahassee FL 32317, USA

[2]Department of Horticultural Science, North Carolina State University, Mountain Horticultural Crops Research and Extension Center, Mills River NC 28759, USA

[3]Department of Genetics, Davison Life Sciences Building, University of Georgia, Athens, GA 30602-7223, USA

[4]Plant Genome Mapping Laboratory, University of Georgia, Athens, GA 30602, USA

[5]Department of Plant Molecular Biology and Biotechnology, CPMB, Tamilnadu Agricultural University, Coimbatore TN 641 003, India

[6]Department of Genetics and Biochemistry and Institute of Nutraceutical Research, Clemson University, 109 Jordan Hall, Clemson, SC 29634, USA

*Authors equally contributed; **Corresponding author: katamr@yahoo.com

1 INTRODUCTION

Arabidopsis (rockcress) is a genus in the family Brassicaceae. Representatives of Brassicaceae family are small-flowering plants, some of them are important crops such as cabbage, cauliflower, radish, and canola. *Arabidopsis* genus includes nine species and eight subspecies. The subspecies delimitation is quite recent, and is based on morphological and molecular phylogenetics (Elizabeth 2000). *Arabidopsis thaliana* commonly known as thale cress, has 10 chromosomes in diploid stage and was the first plant to have its entire genome sequenced. Most of the species in *Arabidopsis* are indigenous to Europe and only two species are found in North America and Asia.

Originally *Arabidopsis* was used as a model system for dicotyledonous crops, but more recently it became an invaluable source of information for studies in monocotyledonous plants. Established techniques, fully sequenced genome (450 Mpb) and easy maintenance of this plant allow detailed dissection of this plant's traits on molecular basis (Aubourg et al. 2007). Information gained from characterization of approximately 8,000 genes led to the discovery of valuable information about genetic regulation of important agronomic traits of crop plants (Yang et al. 2008).

2 *ARABIDOPSIS* GENOME INITIATIVE

The initiative on *Arabidopsis* genome research began in 1989 as a part of innovative approach designed to combine infinite knowledge from the Biological, Behavioral, and Social Sciences Directorate (BBS) of the National Science Foundation (NSF) into a unique database. In 1990, multiple workshops and meetings involving *Arabidopsis* research community all over the world and various national funding agencies led to the formation of a plan for the coordinated *Arabidopsis* genome research project known as "A Long-Range Plan for the Multinational Coordinated *Arabidopsis thaliana* Genome Research Project". The *Arabidopsis* Genome Initiative (AGI) began in 1996 when Japanese, European and American scientists established a plan to identify and characterize all genes of this plant. Representatives from each major *Arabidopsis* sequencing centers agreed to collaborate on decoding this plant's genome. This effort was funded by NSF, the US Department of Energy (USDoE), and the US Department of Agriculture (USDA) and resulted in systemic physical mapping using cosmids and creation of *A. thaliana* data base (AtDB). The sequencing of *Arabidopsis* genome was completed in 2001. At present NSF began the next phase of *A. thaliana* genome project, which is determination of the functions of 25,000 genes.

2.1 Specific Objectives of the AGI

The primary objective of the AGI project was to encourage a coordinated research effort to use *A. thaliana* as a model system for studies of the biology of flowering plants. More specifically the AGI project include the following sections: (1) identification and characterization of the structure, function, and regulation of genes; (2) development of technologies for plant genome studies; (3) establishment of biological resource centers; (4) establishment of an informatics program; (5) development of human resources; and (6) organization of workshops and symposia.

To facilitate and coordinate collaborative research in the US, an inter-agency agreement was signed in June, 1990 by the National Institute of Health (NIH), the DoE, and the USDA. The NSF supported the *Arabidopsis* genome research project by funding Multinational Science Steering Committee meetings, publications and distribution of the annual progress reports. NSF continues to maintain communication with agencies involved in the collaborative research in other countries. *Arabidopsis* Genome Research Project not only contributes to the advancement in plant biology but also serves as a model for multinational collaboration research project.

In 1991, the *Arabidopsis* Biological Resource Center (ABRC) was established at the Ohio State University. The role of ABRC was to acquire, preserve and distribute seed and DNA resources to the *Arabidopsis* research community. Furthermore, NSF-funded institution called The *Arabidopsis* Information Resource (TAIR) was established at the Carnegie Institution of the Washington Department of Plant Biology at Stanford, California. Extensive genetic and physical mapping of all the five chromosomes was available due to the establishment of TAIR. TAIR provided research community with necessary seed samples from *Arabidopsis* wild type and mutant strains as well those necessary for plant transformation. In addition, TAIR also accumulates, organizes and publishes information generated through multiple research efforts.

3 *ARABIDOPSIS* GENOME ANALYSIS

The *Arabidopsis* research community developed procedures for genome modification through chemical and insertion mutagenesis. Research utilizing these methods produced extensive mutant lines with diverse phenotypes and a variety of chromosomal maps and molecular markers. The preferred method of DNA modification is chemical mutagenesis. This method allows to produce millions of homozygous seeds carrying a recessive mutation upon selfing of M_1 plants. Mutants can be also created through insertion mutagenesis using transferred DNA (T-DNA) approach. This approach is a bit laborious and relies on T-DNA from *Agrobacterium tumefaciens* (Bechtold et al. 1993). T-DNA method is used for a whole plant transformation and the main advantage of this method over chemical mutagenesis is a bypass of the necessity to regenerate plants in culture. Thousands of transgenic lines carrying random T-DNA insertions throughout the genome have been deposited in public stock centers. T-DNA approach also created a new research avenue studying transposable elements in a context of plant genome evolution. Several thousand mutants defective in almost every aspect of plant development were identified over the past 20 years. Genetic mutations responsible for defects in gametogenesis, seed formation, leaf and root development, flowering, senescence, signaling, as well as response to pathogens, environmental conditions and stress were identified (Meyerowitz and Somerville 1994). A limiting factor in gene replacement is the absence of an efficient system utilizing homologous recombination, however several promising advances were reported (Kempin et al. 1997).

3.1 *Arabidopsis* Genome Organization

Arabidopsis thaliana is a genetic tool box. Researchers find genes that can be tweaked to increase nutritional content and crop yields and to coax plants to grow in salty or metallic soils, or in extreme climates. Having the entire genome sequence of any organism accelerates the pace of genomic research.

For the past 40 years *Arabidopsis* was a model for gene discovery studies and served as a reference point for examination of other species' genomes. Researchers discovered several genes in *Arabidopsis* which may be helpful in crops improvement. *Arabidopsis* is considered to be a genetic model for more than 200,000 species of flowering plants, each of which shares a basic architectural foundation and similar biochemical processes. One of the most valuable features of *Arabidopsis* is its amenability to gene knockout and gene mutagenesis using chemical mutagen such as ethylmethane sulphonate (EMS). These experimental approaches have been very useful in producing valuable information about gene function.

The small flowering plant *Arabidopsis* is an excellent model system for metabolic, genetic, and developmental studies in plants. Its haploid genome is small (~100 Mb) and contains low percentage of repetitive DNA (Meyerowitz and Somerville, 1994). Each *Arabidopsis* chromosome contains total of approximately 27,000 genes (Fig. 1). Although the small size of meiotic chromosomes and the absence of polytene chromosomes limited the progress of cytogenetic studies of chromosome structure, visualization of chromosomes was significantly improved by in situ hybridization methods. Several factors converged to catalyze a large-scale sequencing program of *Arabidopsis* genome. These include the

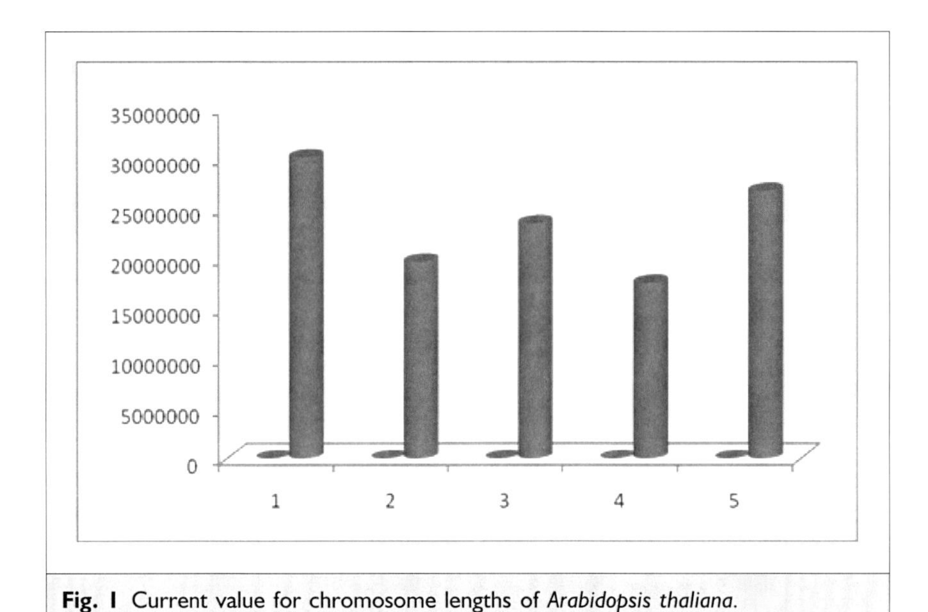

Fig. 1 Current value for chromosome lengths of *Arabidopsis thaliana*.

availability of a physical map of YACs and constant improvements in the efficiency of sequencing. Chromosomal studies produced classical genetic map, recombinant inbred (RI) map, and physical map that contained 110 cloned mutant genes among the total 27,000 genes identified. The recombinant inbred (RI) map illustrates the locations of cloned genes and molecular markers based on recombination within a defined mapping population produced through repeated selfing of plants in successive generations (Lister and Dean 1993). The recombination frequencies were estimated using phenotypic expression in F_2 generation produced by F_1 self-pollination. There are over 790 markers in the RI map, which include restriction fragment length polymorphisms (RFLPs), simple sequence length polymorphisms (SSLPs), cleaved amplified polymorphic sequences (CAPS), expressed sequence tags (EST), and the ends of bacterial (BAC) and yeast (YAC) artificial chromosomes. The length of each RI chromosome has been adjusted on the chart to match that of the classical chromosome to facilitate the comparison between equivalent regions. The genetic distances between molecular markers on the RI map will eventually become secondary to physical distances measured in base pairs (bp). Classical genetic map (Koornneef et al. 1983) and maps based on RFLPs (Chang et al. 1988; Lister and Dean 1993) are now widely available in addition to cosmid-based physical map containing ~750 contigs obtained by DNA fingerprinting (Hauge et al. 1993). Furthermore, the sequencing efficiency has been significantly increased by using advanced automation sequence technology and BAC specific hybridization in tilling assays of BAC clones. More than 98% of the identified and anchored BACs comprising 115.4 Mb net *Arabidopsis* sequences were sequenced by the year 2000 (Fig. 2). By the end of year 2000, 112,000 ESTs were produced by the AGI and other groups. As a result of this work, highly accurate contiguous sequence extending from telomeres into centromeric heterochromatin were produced. All DNA sequences were annotated and analyzed in specialized centers using EST sequences combined with computational and manual tools, and annotation. Annotated sequences were used to predict gene functions, protein structure and functions, and were extremely helpful in facilitation of scientific progress.

Genetic analysis of *Arabidopsis* has expanded in recent years to include specialized topics such as epigenetics, gene silencing, tetrad analysis, centromere mapping, and reverse genetics. Maize has been a classical model for epigenetic and paramutation studies. Epigenetic research in *Arabidopsis* also offers molecular characterization of genes important for agronomic traits. Tetrad analysis became possible in *Arabidopsis* with the isolation of the "quartet" mutant where four pollen grains derived from a single meiotic event remain attached upon release from the anther. The important implication of quartet mutant lies in the effects of fertilization by all four pollen grains (Copenhaver et al. 1998). Insertion mutagenesis approach allows discovery of molecular basis of plant's phenotype control. The precise number of insertion mutants is unknown due to differences in cataloging procedures between public and private research centers. Plans are made to improve case access to insertion mutants by scientific community (Krysan et al. 1996). Due to advances in mutant analysis and, genome sequencing *Arabidopsis* became a powerful model for studies many fundamental concepts of plant genetics.

3.2 Sequence Analysis

The sequence of chromosome 1 of *Arabidopsis* ecotype Columbia is fully annotated and consists of two contigs, the northern arm (~14.2 Mb) and the southern arm (~14.6 Mb). The sequence of metacentric chromosome 1 was derived from YAC-based physical map (Goodman 1996) and is estimated to be 31 Mb long. Sequencing of this chromosome was initiated using BACs. Three AGI groups, Genoscope, TIGR and the Plant Gene Expression Center Consortium, University of Pennsylvania deter-

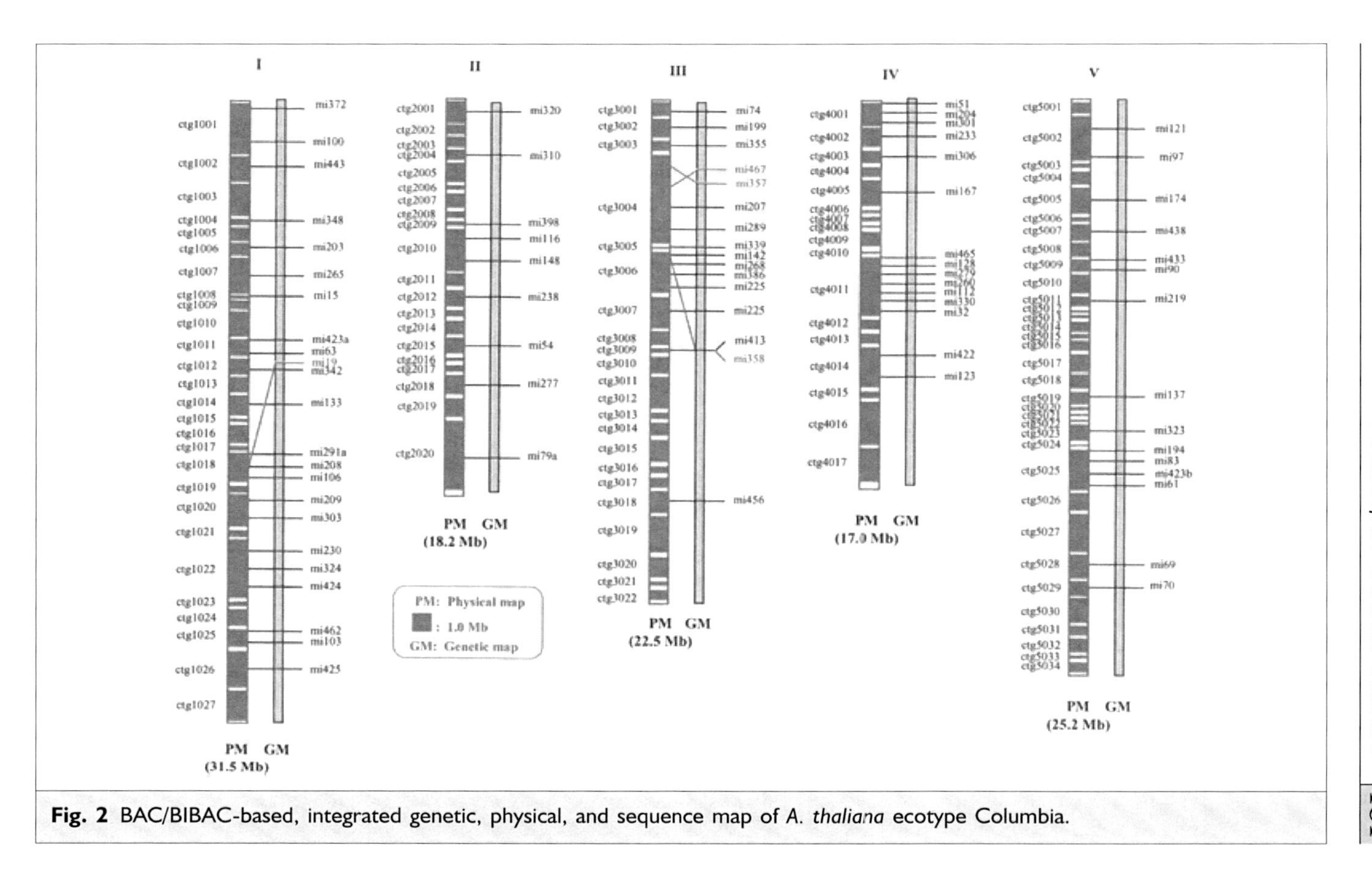

Fig. 2 BAC/BIBAC-based, integrated genetic, physical, and sequence map of *A. thaliana* ecotype Columbia.

mined the end sequences for almost all of the BACs corresponding to about 36,000 unique end sequences from approximately 18,300 BAC clones. The northern arm was sequenced completely while the southern arm has three sequencing gaps located at 15.349 Mb (BAC T18F15, C084807), 15.573 Mb (BAC T2P3, CO84820), and 20.070 Mb (BAC T4M14, AC027036). Several gene prediction programs and database searches were used to annotate chromosome 1 (Athanasios et al. 2000). It was estimated that this chromosome contain 6,850 putative genes at a density of one gene per 4.1 kb, which is more than twice the density of genes in *Drosophila* (*http://www.nature.com/nature/journal/v408/n6814/full/408816a0.html - B5#B5*). The average gene length is postulated to be approximately 2 kb and to contain five exons with an average size of 272 bp (Table 1a). Therefore, almost 50% of the chromosome is occupied by genes. Nearly 1,250 (18%) genes contain no introns and many of them have been already annotated to be hypothetical proteins. The largest gene, F5A8.4 (AC004146), contains 68 exons and encodes a 5,139-amino-acid polypeptide (relative molecular mass~500,000 Dalton), which is a ubiquitous putative membrane protein found in *Saccharomyces cerevisiae*, *Caenorhabditis elegans*, *Drosophila melanogaster* and humans. The smallest gene, T6H22.15 (AC009894), encodes L41 protein of the 60S ribosomal complex, and contains only 25 amino acids and the corresponding ESTs were discovered for this gene. At least 3,448 (50%) genes are known to be expressed, as shown by the presence of complementary DNAs or ESTs in the GenBank. Thirty-thousand (27%) of isolated 110,000 ESTs were found to be products of chromosome 1 genes, while 97% of the chromosome 1 ESTs map to annotated genes. Among the genes represented by ESTs are putative chlorophyll a/b-binding protein, small subunit of RuBP carboxylase, ferrodoxin/precursor and photosystem II 10 K polypeptide. The gene density and EST distribution are more or less the same along the chromosome, with their lowest values around the regions flanking the centromere, which is 1.3 Mb. A few EST dense regions with more than 700 ESTs per 100 kb were also discovered. These observations indicate regions undergoing transcription. *Arabidopsis* contains four pairs of megabase-scale segmental duplications located specifically within the chromosome 1. The locations of these duplications are: 3.1–3.8/19.3–19.8; 5.6–6/27.8–27.3; 6.2–7.5/25.4–26.7; and 8–8.6/25–24.3 Mb. The first and third duplications have the same orientation relative to the centromere, while the second and fourth located in opposite orientations. There are several large-scale duplications located on other *Arabidopsis* chromosomes as well. Chromosome analysis also reveals diverse repetitive elements representing 8% of the sequence and

Table 1a Genome features of Chromosome 1.

Feature	*Value*
Length	28,762,046 bp
Top arm	14,172,442 bp
Bottom arm	14,589,604 bp
Base composition (%GC)	
Overall	35.8
Coding	43.8
Non-coding	31.8
Number of genes	6.848
Gene density	4.1 kb per gene
Average gene length	2,145 bp
Average peptide length	460 amino acids
Exons	
Number	35,768
Total length	9,396,363 bp (33%)
Average per gene	5.2
Average size	272 bp
Introns	
Number	28,951
Total length	4,807,531 bp (17%)
Average size	166
Number of genes with EST	3,448 (50%)
Number of ESTs* for chromosome 1 genes	~30,000 (27%)

*Around 110,000 ESTs used for analysis.

consisting of retroelements (2.6%), DNA elements (2.4%) and a number of simple and low-complexity repeats (2.6%). Computational analysis of annotated sequences predicted 6,848 proteins, corresponding to about a quarter of the nearly 25% of *A. thaliana* putative proteins have unknown function. Predicted proteins have hypothetical function identified by gene prediction software without corresponding ESTs. Majority of the annotated proteins have some similarity to other hypothetical or putative function proteins from plants or *S. cerevisiae*, *C. elegans*, *D. melanogaster* or humans. About 1/3rd of the proteins with putative function participate in cellular metabolism, while others are involved in transcription, plant defense, signaling, and growth and development. Comparison of the predicted proteins of chromosome 1 with other organisms' databases revealed that more than 1,500 proteins have significant homology (cutoff $\leq 10^{-30}$) with proteins found in *S. cerevisiae*, *C. elegans*, *D. melanogaster* and *H. sapiens*. Proteins involved in RNA splicing and translation, tRNA biosynthesis, and cellular metabolism are highly conserved throughout the eukaryotic kingdom (Table 1b).

The chromosome 1 contains 312 families of tandemly arranged genes with 847 members comprising 12% of all the genes that have a blast score of more than 200 (Fig. 3). These families are distributed throughout the chromosome 1. One hundred and fifty-six families encode a set of distinct protein isoforms, which are not members of super families. The remaining 156 families are members of 46 super families, which vary in size from two to 23 families. For example, the *T28K15.6* (AC022522) super family contains five two-member families, which encode putative leucine-rich repeat (LRR), which is a nucleotide-binding site (NBS) of

Table 1b Proteome features of Choromosome 1.

Feature	*Value*	
Classification/function	*Number*	*(%)*
Total proteins	6,848	100
With similarities to GenBank entries	4,793	70
Unknown	1,590	23
Hypothetical	1,935	28
With putative function	3,323	49
With putative signal peptides		
Secretory pathway	1,049	15
Chloroplast	521	7.5
Mitochondria	96	1.5
Classification of proteins with putative function		
Cellular metabolism	995	30
Transcription	498	15
Plant Defense	354	11
Signaling	327	10
Growth and development	302	9
Protein fate	292	9
Intracellular transport	247	7
Ion transport	194	6
Protein synthesis	113	3
Total	**3,322**	**100**

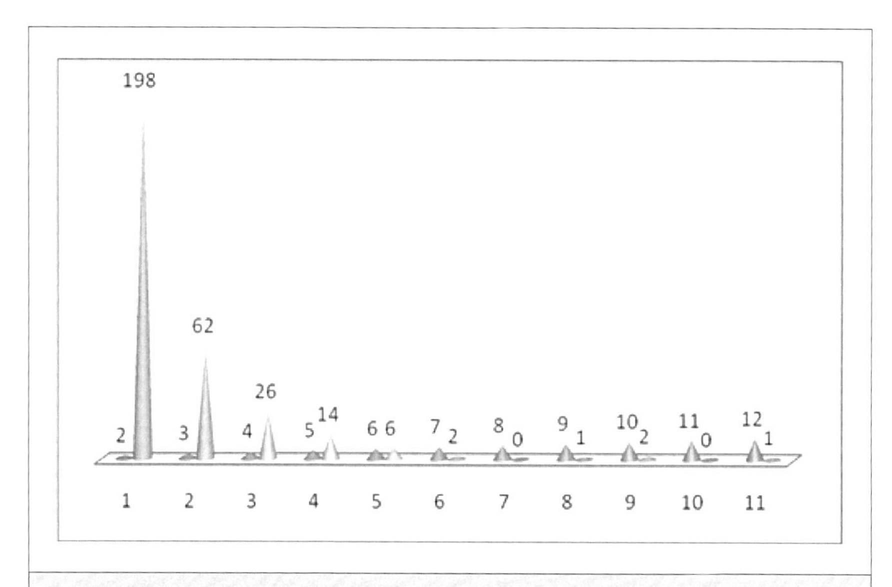

Fig. 3 Frequency of gene distribution in multigene families with tandem gene arrangements.

disease resistance protein. Two of the T28K15.6 families are located at ~4 Mb and the other three at ~20.5 Mb on the chromosome. The families of 25 super families are located within close proximity to each other (less than 2 Mb apart), whereas those of the remaining 21 are further apart (Fig. 4). In a few cases, several families can be over 20 Mb apart. For example, the T28P6.7 (AC007259) super family, which encodes *S*-locus-like receptor kinases (SRKs), has four families with two, three, twelve and two members located at 3.79, 38.1, 20.5 and 22.33 Mb, respectively. The first two SRK families are closely linked, separated by two unrelated genes. The SRK proteins have been implicated in controlling self-incompatibility in the reproductive system in Brassicaceae. It remains to be determined whether the 19 putative SRK isoforms encoded by the T28P6.7 super family and the 12 SRK isoforms encoded by the T28P6.2 super family (three families with three, seven and two members, respectively) on chromosome 1 have an *S*-locus protein kinase activity. The largest super family located on chromosome 1 is F22L4.6 (AC061957) comprised of 23 families, which encode 78 isoforms of a putative protein kinase. A large percentage of the gene families on this chromosome encode proteins that are involved in disease resistance, cell wall degradation and secondary metabolite biosynthesis.

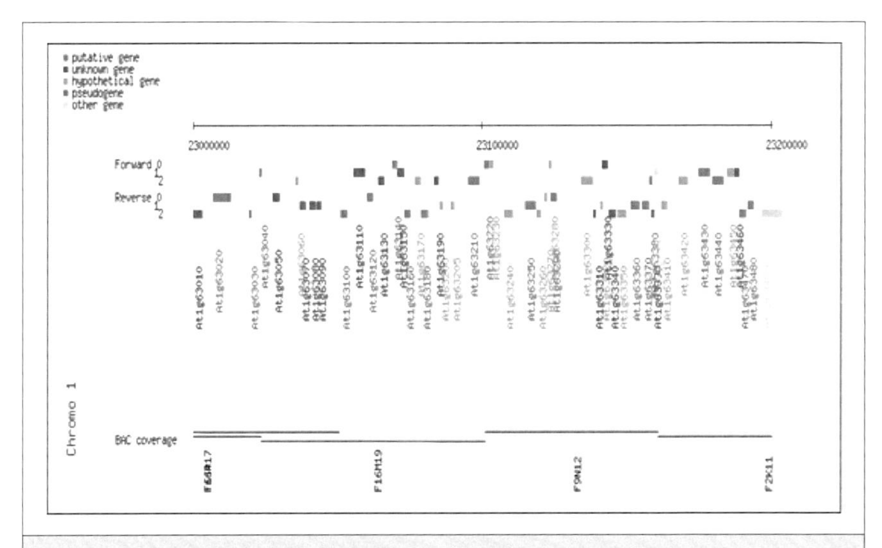

Fig. 4 Gene density on Chromosome 1.
(Color image of this figure appears in the color plate section at the end of the book).

It will be important to find out the biological significance of the multigene families. 1-aminocyclopropane-1-carboxylic acid (ACC) synthase (ACS) family has two members (ACS 2 and ACS10) on chromosome 1 and eight other members in the other four chromosomes. It has been suggested that the presence of different ACS isoforms may reflect tissue-specific expression that satisfies the biochemical properties of the cells or tissues in which each is expressed. For example, if a group of cells or tissues have low concentrations of the substrate, S-adenosyl methionine, then these cells express a high affinity ACS isoform. Distinct biological function of each isoform is defined by its biochemical properties, which in turn defines its tissue-specific expression. Such concept can explain differential expression of gene families resulting in protein variance. The sequences of chromosome 1 and the other *Arabidopsis* chromosomes already provided a wealth of information. However, more detailed experimental work is required to determine gene functions that will advance plant science. Gene predicted functions should be interpreted with caution and mapping of the transcriptional units of the *Arabidopsis* chromosomes in the future will provide experimental verification of the annotation.

Of the five *Arabidopsis* chromosomes, chromosome 2 is the smallest (16-17% of the genome) and contains one of the two genomic rDNA clusters (A1bini 1994). A detailed coverage of chromosome 2 was achieved with a minimal tiling path of 34 YACs. Furthermore, contact points between the genetic and physical maps were identified by 58 probes mapped to chromosome 2 by RFLP analysis. The order of the DNA probes in this physical map is consistent with the genetic map (Lister and Dean 1993). Given the low incidence of chimeric clones in the CIC library (Creusot et al. 1995), this map is a reliable source of information, despite five linkages that exhibit <0.25 units of fractional overlap with an adjacent YAC. Four sets of YACs, covering of approximately 90% of the chromosome were aligned to the genetic map (Lister and Dean 1993). There are three gaps estimated to be 120-1,000 kb, 340 kb, and 720 kb, respectively. These distances were identified by using the genetic distance between the markers bordering the internal gaps and a conversion factor of 200 kb/cM (Koornneef et al. 1983). A combined size of all YAC contigs (13 Mb), rDNA cluster of 3.5 Mb, gaps and telomeres suggest a minimum size of 18 Mb for chromosome 2. Out of 18 Mb, 13-15 Mb is represented by low-copy DNAs, while remaining sequence encodes rDNA (Albini 1994). Chromosome 2 thus encodes 2,500-3,000 genes (5 kb/gene) (Meyerowitz and Somerville 1994).

Chromosomes 2 and 4 have been completely sequenced and combed by gene prediction computer programs. These two chromosomes, which represent about 30% of the *Arabidopsis* genome, contain approximately 7,781 genes. The function of about half of these genes are still unknown, while other genes bear a strong resemblance to those found in plants, animals and bacteria. The Institute of Genome Research (TIGR) has classified 51% of the 4,037 genes found on Chromosome 2, based on comparison to genes with known functions. Of these, about 100 genes are suggested to directly influence plant growth and development, another 80 to encode responses to pathogens, insects, and fungi and another 40 genes to control responses to environmental stresses. The future investigation on the role of these genes will help to understand the genetic basis of disease or pest resistance and provide new avenues for genetic modification of crops. Many genes have eluded characterization because there is more than one copy in the genome. One particularly surprising finding of extensive gene duplication was observed for chromosomes 2 and 4. A similar sequence of 400–500 kb length, was found incorporated in chromosomes 2 and 4, indicating that these chromosomes share several hundred genes (Fig. 5).

Fig. 5 A sequence-based map of genes with mutant phenotypes. Gene locations are marked with horizontal lines. A single line at this scale may represent two or more neighboring genes. The length of each chromosome is proportional to its sequence. Centromeric gaps are marked by short constrictions.

4 GENETIC EVOLUTION AND GENE DUPLICATION

To improve crops, geneticists and breeders often use genetic mutations. Twenty to fifty percent of a rapeseed crop harvested for oil, can be lost because the pods open and release the seeds before the harvest time. However, when two genes called *SHATTERPROOF1* (*SHP1*) and *SHATTERPROOF2* (*SHP2*) are mutated, the seed pods fail to burst open. The seed pod is just like a small pea pod that is joined at the seam with cell-secreted chemical called pectin, which acts as a glue. When the seed pod matures, some cells secrete pectinases that degrade the 'glue', while other cells near the seam shrink, becoming dry, woody and taut, creating a 'coiled spring' type mechanism that flings the seeds out of the pod. Yanofsky et al. (1996) reported the discovery of two weakened *SHATTERPROOF* genes in *Arabidopsis* genes that control pod shattering in *Arabidopsis* are likely to be present in close relatives like cauliflower, broccoli, Brussels sprouts, peas, soybeans and other important food crops. Finding weak versions of *SHATTERPROOF* genes in rapeseed or creating transgenic lines would increase the efficiency of harvest and crop cost.

5 FUNCTION OF *ARABIDOPSIS* SMALL RNA

RNA interference (RNAi) mechanism is adopted by plants for gene suppression. Significant attention is concentrated on development of RNAi technologies for improving plant crops as well as treating human diseases. Micro-RNAs (miRNAs) and small-interfering-RNAs (siRNAs) have recently emerged as key regulators of gene expression in eukaryotes. miRNAs are small, ~21-nucleotides long, that originate from non-coding single-stranded RNAs. These RNAs are characterized by a fold-back structure, which is a subtracted for Dicer-like 1 RNase (DCL-1). The siRNAs are double-stranded, 21- to 26-nucleotide long, are derived from the processing of long double-stranded RNAs (Waterhouse et al. 2001). After processing, si- or miRNAs are incorporated into RNA-induced silencing complex (RISC) or RNA-induced transcriptional silencing (RITS) complex. These small RNAs guide RISC or RITS complexes to target mRNAs for post-transcriptional gene silencing or to homologous DNA sequences for transcriptional gene silencing. Small RNAs, particularly miRNAs have been found to play a central role in plant development, and abiotic and biotic stress responses of plants. Further identification of small RNA-guided gene regulation stress responses in plants will provide new tools for improving plant stress tolerance. The total number of miRNAs in a plant is unknown but was estimated to be 1% of the total coding sequence (250 to 300 miRNAs in *Arabidopsis*) (Sunkar et al. 2005). Three groups have independently reported miRNAs from *Arabidopsis* (Llave et al. 2002; Park et al. 2002; Reinhart et al. 2002). So far, it is estimated that 250 to 300 miRNAs are distributed into 15 families in *Arabidopsis* (Bartel 2004). Several of these miRNAs were demonstrated to play critical roles in leaf and flower development through targeting developmental regulators for mRNA cleavage or translational repression (Aukerman and Sakai 2003; Chen 2004).

siRNAs were originally discovered in plants undergoing transgene silencing or infection with viruses (Hamilton and Baulcombe 1999; Zamore et al. 2000). Endogenous small RNAs with characteristic features of siRNAs have been cloned from *Arabidopsis* (Reinhart et al. 2002). Double-stranded RNA may be derived from the transcription of inverted-repeat loci that could be converging promoters or products of polymerization events of host/viral-encoded RNA-dependent RNA polymerases (Aravin et al. 2003; Sijen et al. 2001). siRNAs target homologous RNA sequences for endonucleolytic cleavage, which is referred to as RNA interference in animals, post-transcriptional gene silencing in plants, and quelling in fungi. The biological roles of siRNA-mediated RNA silencing include protection of the genome against mobile DNA elements (Ketting et al. 2001) and resistance against viruses (Voinnet 2003). In addition to post-translational gene silencing (PTGS), siRNAs may be involved in transcriptional gene silencing by regulating DNA methylation and chromatin modification (Hamilton et al. 2002; Volpe et al. 2002; Zilberman et al. 2004). A distinguishing feature of plants is that they are sessile and thus have to cope with, rather than move to avoid, adverse environments. Plants have evolved sophisticated mechanisms to adapt to environmental stresses (Zhu 2001). Abiotic stresses, such as drought, salinity, and cold, regulate the expression of thousands of genes in plants at both transcriptional and post-transcriptional levels. Sunkar et al. (2005) reported 13 families of miRNAs from *Arabidopsis* sequences. These include miR156, miR157, miR158, miR159, miR161, miR163, miR166, miR168, miR169, miR171, miR172, miR173, and miR319/ miRJAW (Llave et al. 2002; Park et al. 2002; Palatnik et al. 2003). They have also identified many new miRNAs and putative siRNAs. The ability of RNA to fold into hairpin miRNA precursors is a distinguishing mark of miRNAs. Ambros et al. (2003) analyzed miRNA precursors of approximately 300 nucleotides for hairpin structures using the "mfold" program. The analysis revealed that they successfully identified and cloned 21 novel *Arabidopsis* miRNAs. miRNA mutants (dcl1, hen1-1, and hyl1) demonstrate severe developmental defects such as floral morphology and embryo formation (Lu and Federoff 2000). Phenotypic defects such as plant height, seed size, etc. of these mutants could be attributed to lower levels of miRNA production.

6 *ARABIDOPSIS* COMPARATIVE GENOMICS

Comparative genomics helps to understand structural and functional genetics of one species with respect to others (Koch et al. 2001). Many different features are considered when comparing genomes of different species such as sequence similarity, gene location, the length and number of exons and introns, and extent of sequence homology. Genomic comparison can be done using various computer programs (Delseny 1999). One of the most widely used programs is basic local alignment search tool (BLAST), which is available from the National Center for Biotechnology Information website (*http://www.ncbi.nlm.nih.gov*). The BLAST is a set of programs designed to perform a similarity search on all electronically available sequence data. In this chapter, we present the basic principles behind comparative genomics, how it is done and how its interpretation make readers acquainted with this relatively new scientific concept. We also present an illustration of comparative genomics study of plant species with respect to *Arabidopsis thaliana*.

The use of molecular markers made the task of comparative genomics much easier since most of these markers are PCR-based and with the use of high throughput genotyping methods by which the polymorphism can be detected with high sensitivity. RFLP marker, a hybridization-based molecular marker system (Botstein et al. 1980), also made possible comparative mapping of sexually compatible and incompatible species. Construction of linkage maps in related species is done by examining cross-hybridization results between DNA sequences from various organisms and RFLP markers' locations (Lander and Botstein 1986). Conserved genetic linkage between homologous loci normally indicates structural similarity between different genomes. Presence of this structural similarity between different plant genomes suggests their evolutionary pathways. It also helps to link positional information on genes and their functions in various species (Schranz et al. 2006). PCR-based molecular markers such as simple sequence repeat (SSR), amplified fragment length polymorphism (AFLP), single nucleotide polymorphism (SNP), cleaved amplified polymorphic sequence (CAPS) and random amplified polymorphic DNA (RAPD) are much easier to work with, while collecting genotypic data (Torjek et al. 2003). Molecular markers have been extensively used for the study of comparative genomics in crops like Brassicas (Kole et al. 2001, 2002; Lou et al. 2007; Suwabe et al. 2004) and soybean (Shultz et al. 2007).

6.1 Principles and Considerations in Comparative Genomics

It is well known that genetic variation between species is due to differential arrangements of the four basic nucleotides. These differences are introduced by mutations during the evolution of the species. The major principle behind comparative genomics is the alignment of genome sequence, its structure and function between any two or more plant species. Similar sequences can be pulled together and their function can be determined on the basis of already known information and such alignment allows to determine relative evolutionary distance between the species (Mushegian 2007). Sequence alignment could be either global or local. Global alignment spans the entire length of sequences from several species, whereas local alignment utilizes only certain regions. Global alignment is most useful when the sequences in the query are similar in sequence and size (Huang and Chao 2003). A general global alignment technique is based on the Needleman-Wunsch algorithm and dynamic programming. Local alignments are more useful in comparison of dissimilar sequences that are suspected to contain regions of similarity or similar sequence motifs within a larger sequence context. This method of genome analysis utilizes the Smith-Waterman algorithm (Saigo et al. 2006). In practice, local alignment is less time consuming, more straightforward, and therefore is the most preferred method of comparative genomics.

6.2 Basic Tools and Methods of Comparative Genomics

Powerful DNA sequencers such as ABI Prism series or CEQ series (Applied Biosystems, Inc.) produce valuable information for private and public research. Various restriction digests of genomic DNA and its subsequent probing with various markers produce a certain patterns in each species (Gebhardt et al. 2003). PCR-based molecular markers such as SSR, SNP, CAPS or AFLP can be used as well to collect data from multiple species. The genotypic data collected by these methods of genome evaluation is used to construct linkage maps (Cregan et al. 1999; Haanstra et al. 1999). Such computational tools such as MapMaker and JoinMap (Stam 1993) can be used to develop maps for positioning the quantitative trait loci (QTL). These tools can be found at TAIR (*http://www.arabidopsis.org/*), soybase (*www.soybase.org*), solanaceae genomics network (*http://www.sgn.cornell.edu/*) and others. DNA or protein sequences from any species can be compared with *Arabidopsis* ESTs or *Arabidopsis* genomic sequence. Marker sequences with similarity to retro-elements and/or having more than 30 'hits' in the *Arabidopsis* genome can be excluded from synteny (co-localization of gene or locus on the same chromosome) analysis. All *Arabidopsis* BAC, TAC, and P1 genomic clones with the *E-value* (expected value) $\leq 10^{-10}$ sequence similarity to the markers can be localized on the physical map of *Arabidopsis* using the TAIR Map Viewer tool.

6.3 Illustration with Some Examples

In this section, we present some examples of comparative genomics performed between two or more plant species. Such comparison or alignment were made at DNA or protein level. Overall genome comparison was also made on the basis of molecular linkage maps using molecular markers. For example, a genetic map of potato (*Solanum tuberosum*) was constructed based on 293 RFLP markers and 31 EST markers of *Arabidopsis* (Gebardt et al. 2003). The comparison of all marker sequences with the *Arabidopsis* genomic sequence resulted in 189 markers that were detected in 787 loci of *Arabidopsis*. Based on conserved linkage between groups of at least three different markers on the genetic map of potato and the physical map of *Arabidopsis*, 90 putative syntenic blocks were identified covering 41% of the potato genetic map and 50% of the *Arabidopsis* physical map (Gebardt et al. 2003). The existence and distribution of syntenic blocks suggested a higher degree of structural conservation in some parts of the potato genome when compared to *Arabidopsis*. Some duplicated potato syntenic blocks correlated well with ancient segmental duplications in *Arabidopsis*. Syntenic relationships between these two species indicated that potato or *Arabidopsis* genome evolution included ancient intra- and inter-chromosomal duplications (Gebardt et al. 2003).

In another study, an SSR-based linkage map was constructed in *Brassica rapa* including 113 SSR, 87 RFLP, and 62 RAPD markers (Suwabe et al. 2006). They performed synteny analysis between *B. rapa* and *Arabidopsis* which belong to the same family and they found a number of small genomic segments homologous to *Arabidopsis* sequence that were found to be scattered throughout an entire *B. rapa* linkage map. This study pointed out the complex genomic rearrangements during the course of evolution of the species in Brassicaceae. A 282.5-cM region in the *B. rapa* map was found to have synteny with *A. thaliana* map. Three QTLs (*Crr1*, *Crr2*, and *Crr4*) were identified for club root resistance and synteny analysis revealed that *Crr1* and *Crr2* overlapped in a small region of *Arabidopsis* chromosome 4. This *A. thaliana* region is known to contain disease-resistance gene clusters. These results suggest that the resistance genes for club root originated from a common ancestor and subsequently were then distributed to the different regions by the process of evolution. Similar observations were made with respect to the flowering time in *Brassica rapa* (Kole et al. 2001), and locations of resistance loci to white rust caused by *Albugo candida* (Kole et al. 2002). In a similar comparative study, 34 genomic regions of *A. thaliana* were

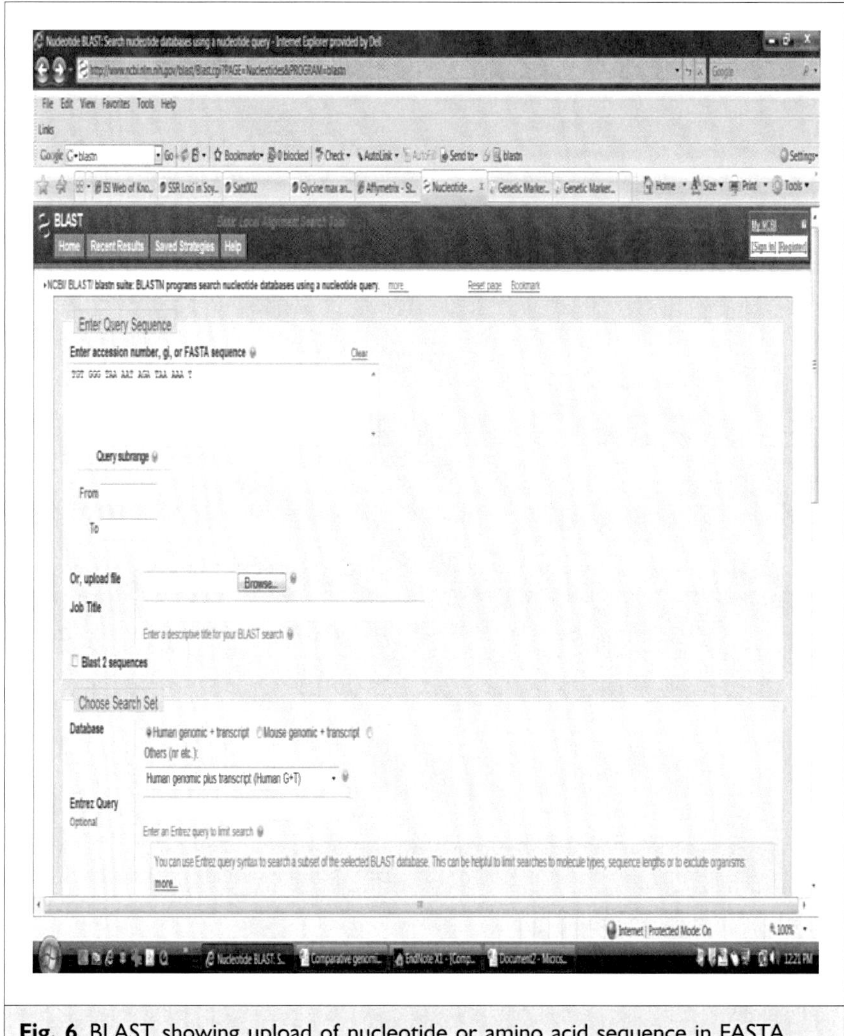

Fig. 6 BLAST showing upload of nucleotide or amino acid sequence in FASTA format.

found to be colinear with more than 28% of the *B. oleracea* genetic map (Lukens et al. 2003). These regions had a mean of 3.3 molecular markers spanning 2.1 Mbp of the *A. thaliana* genome and 2.5 centiMorgan (cM) of the *B. oleracea* genetic map. Conserved sequences between *Brassica rapa* and *Arabidopsis thaliana* was also detected using RFLP markers (Teutonico and Osborn 1994).

Arabidopsis genome is searched from the BLAST menu in NCBI

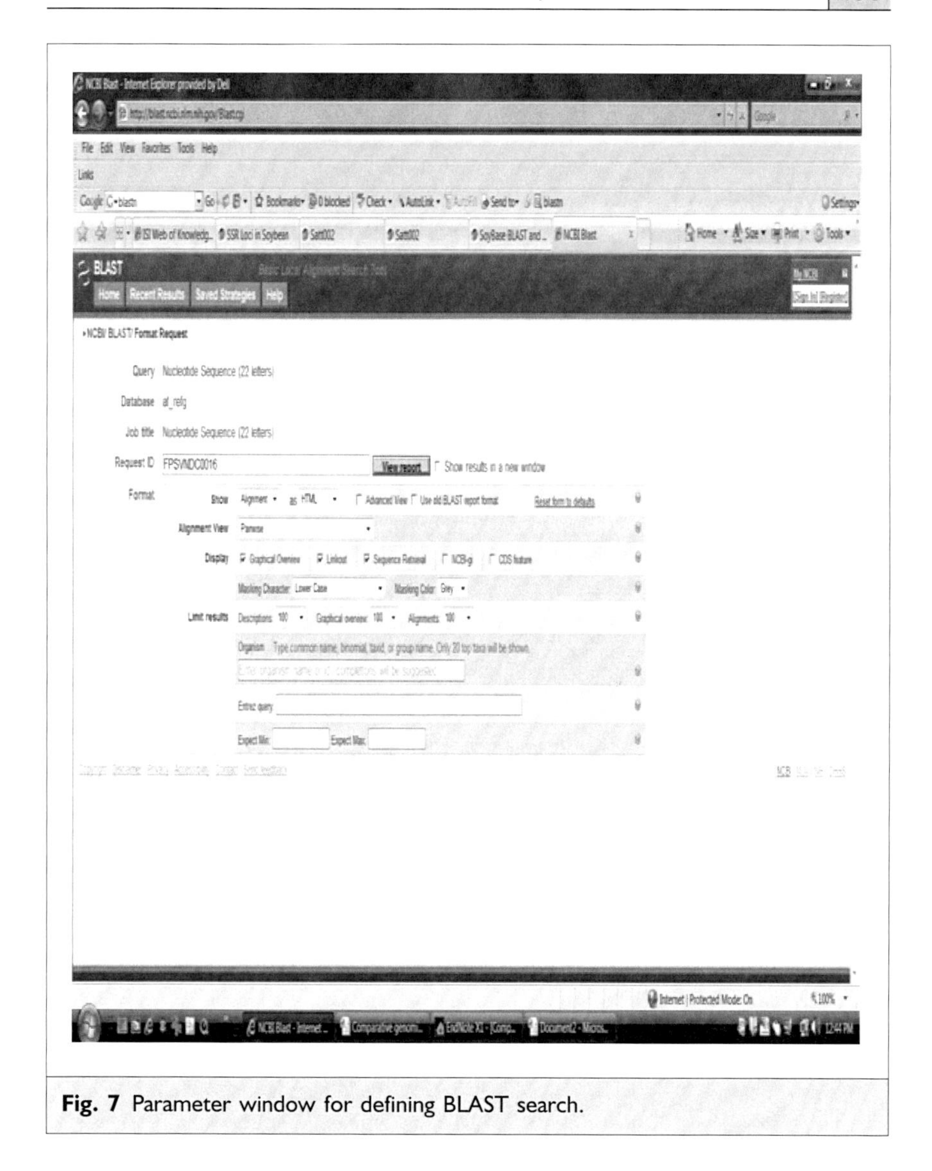

Fig. 7 Parameter window for defining BLAST search.

database and uploading a gene sequence in FAST format in space provided (Fig. 6). To obtain a gene sequence in FAST format searches for a gene name using 'Nucleotide' menu option on the NCBI. For demonstration purposes, we downloaded the sequence of the soybean marker Satt002 using default parameters. The information will be verified under 'search' options (Fig. 7). In yet another example, we selected a molecular

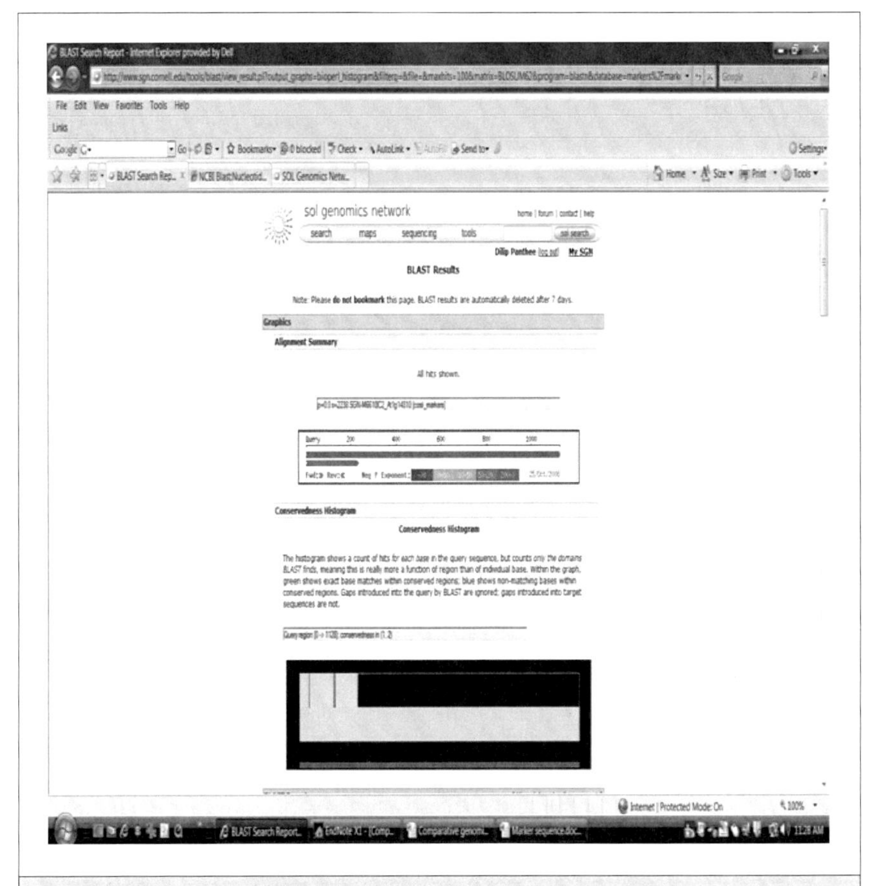

Fig. 8 BLAST results of sequence comparison between the tomato molecular and *Arabidopsis*.

marker from chromosome 1 of tomato obtained from the web site *www.sgn.cornell.edu* and blasted it against *Arabidopsis* sequence.

The BLAST result interpretation for the sequences based on discovered similarity between query and target sequences is shown in Fig. 8. Color key displayed below the results indicates the extent of homology found between two sequences. The second histogram demonstrates sequence to sequence comparison of the homologous regions found between the two species tested (Fig. 9). A red coloring of the histogram indicates that the query sequence (tomato marker) is highly similar when compared to the sequence of *Arabidopsis*. The actual BLAST report follows at the end of the analysis and provides infor-

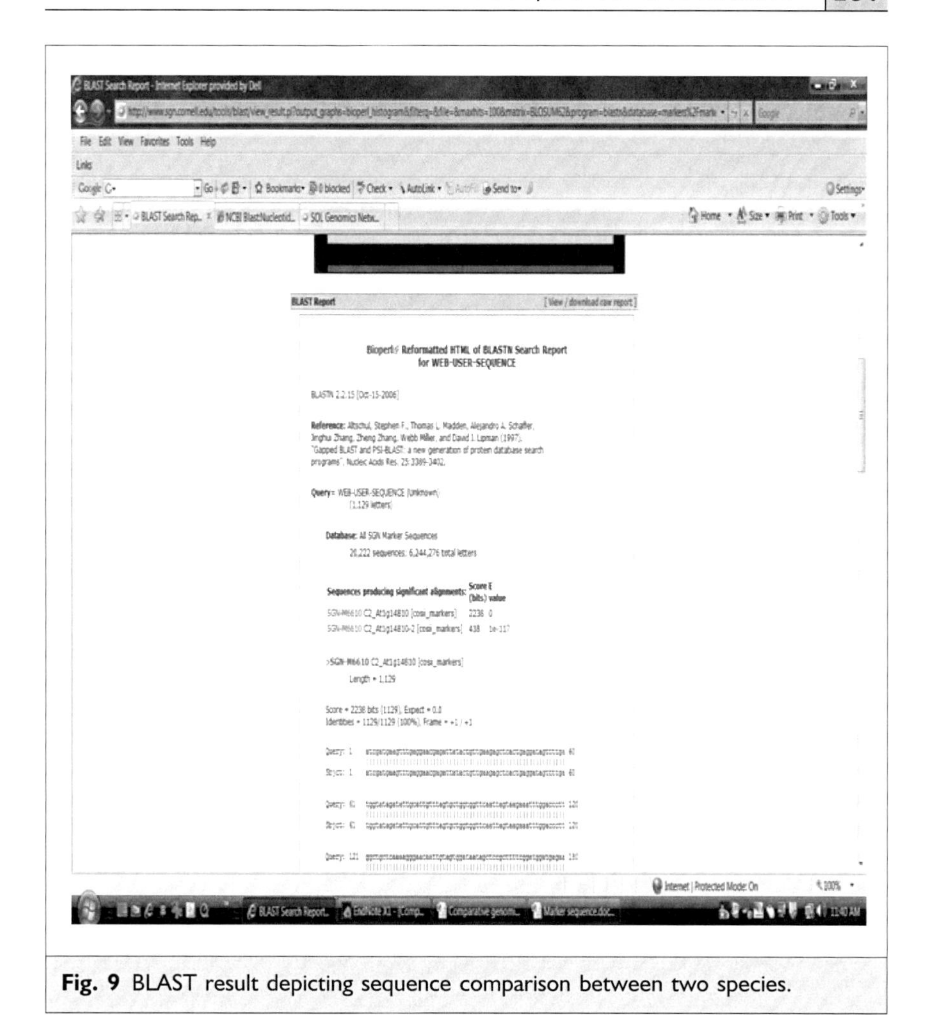

Fig. 9 BLAST result depicting sequence comparison between two species.

mation on other similar hits found in different species *E-value* length of the sequence, and percentage similarity. In our BLAST example, the sequence of tomato marker (the query sequence) length of 1,129 bp was found to be similar to three *Arabidopsis* sequences; C2_At1g14810-1 (cosii markers), C2_At1g14810-2 (cosii markers) and C2_At1g14810-3 (cosii markers). The *E-value* provides the probability of a mis-match between the two sequences, meaning that the smaller the *E-value* the more similarity exists between the sequences studied. The *E-value* of 10^{-12} is negligible when compared to *E-value* is 0.05, meaning that there is a 5% probability of getting a wrong match. In our example, the most similar hit is C2_At1g14810-1 with '0' *E-value* indicating the perfect match. This is further supported by 100% sequence similarity shown in the second half of the report.

7 OTHER IMPLICATIONS OF *ARABIDOPSIS* RESEARCH

Arabidopsis researchers developed a variety of tools to systematically dissect the *Arabidopsis* genome. Some of the tools utilize synthetic DNA markers in genome mapping, and transformation protocols for creation of mutants with compromized gene function. Also, construction of genetic maps and computing could not be possible without bioinformatics tools. All of these and many more tools allow scientists to investigate *Arabidopsis* genome in order to better understand the molecular mechanisms of a plant cell. Studies of *Arabidopsis* has already improved our understanding of plant disease resistance, root development and other important plant processes.

Certain varieties of crops are more resistant than others to viral, bacterial, or fungal pathogens. Improving disease resistance in plants is a major goal for most plant breeding programs, but production of such plants through this method is time consuming but direct genetic modification is much easier. The molecular cloning of an *Arabidopsis* disease resistance gene *RPS2* has significantly added to our understanding of how this gene and its homologs function in economically important plants (Mauricio et al. 2003).

Arabidopsis studies indicate that plants respond to light through a complex genetic network. Cloning of genes revealed previously undetected chemical nature of blue light receptor and suggested previously unknown mechanism of physiological response. The potential of this discovery may lead to production of plants able to grow with less light. To improve diet value of vegetable oils one can design crops with lesser content of polyunsaturated oils. Genes guiding oil synthesis in *Arabidopsis* are very similar to those in commercial crops. This relationship is being exploited to produce plants with healthier edible oils.

Genetic transformation caused accumulation of biodegradable plastic called polyhydroxy butyrate (PHB). Plant's dry weight can contain up to 20% PHB-rich fiber (Coats et al. 2008). Several companies have already begun programs to develop PHB producing crops in order to extract biodegradable plastic.

Agricultural industry uses ethylene to control the ripening of fruits and vegetables and the aging of flowers. Crops deficient in ethylene production or response can be a very powerful industrial tool allowing faster or slower ripening of fruits and vegetables. A gene for ethylene response was detected in *Arabidopsis* (Chang et al. 1993). A mutation in this gene could make plants completely resistant to ethylene, which opened a new era of research where plants can be kept fresh for longer.

Studies of genetics of plants' resistance to environmental stresses such as heat, cold, salty or metallic soils, or high ozone levels are of great interest. Plants capable of growing in extreme environments have economic advantages. Soil salinization from years of continuous irrigation is a widespread problem in the Midwest. Most of the crops will not grow in salty soils, but creating salt resistant plants will allow to overcome this problem. Researchers identified a mutation allowing plants to withstand four times the normal levels of aluminum in the soil (Larsen et al. 1998). A gene for aluminum resistance in *Arabidopsis* was located on chromosome 1. Rising levels of ozone is a growing agronomical concern because it weakens immune system of the plants, producing brown spots on leaves. Federoff's team is studying genes, which become suppressed in high ozone environments (Yoshida et al. 2009). With changing climate around the globe, it is vital to understand adaptive characteristics of plants to create reliable and highly efficient crops capable of supporting growing human population.

8 FUTURE PERSPECTIVE

Arabidopsis, has been proven to be an excellent model for various genetics, genomics and metabolomics studies, as well as a model system for gene discovery in other plant species. It will continue to serve as the reference species for biological researches. The purpose of comparative genomics is to gain a better understanding of how species evolved and also to determine the function of coding and non-coding regions of the genome.

Acknowledgements

RK is grateful for the financial support received from FAMU. RB is thankful to Prof. AH Paterson at PGML, UGA for hosting as a BOYSCAST Fellow and gratefully acknowledges the financial support from the Department of Science and Technology, GOI (SR/BY/L-08/2007).

References

Albini SM (1994) A karyotype of the *Arabidopsis thaliana* genome derived from synaptonemal complex analysis at prophase I of meiosis. Plant J 5: 665-672

Ambros V, Lee RC, Lavanway A, Williams PT, Jewell D (2003) MicroRNAs and other tiny endogenous RNAs in *C. elegans*. Curr Biol 13: 807-818

Aravin A, Lagos-Quintana M, Yalcin A, Zavalon M, Marks D, Snyder B, Gaasterland T, Meyer J, Tuschl T (2003) The small RNA profile during *Drosophila melanogaster* development. Dev Cell 5: 337-350

Athanasios T, Ecker JR, Palm CJ (2000) Sequence and analysis of chromosome 1 of the plant *Arabidopsis thaliana* Nature 408: 816-820

Aubourg S, Martin-Magniette ML, Brunaud V, Taconnat L, Bitton F, Balzergue S, Jullien PE, Ingouff M, Thareau V, Schiex T, Lecharny A, Renou JP (2007) Analysis of CATMA transcriptome data identities hundreds of novel functional genes and improves gene models in the *Arabidopsis* genome. BMC Genom 8: 401-409

Aukerman MJ, Sakai H (2003) Regulation of flowering time and floral organ identity by a microRNA and its APETALA2-Like target genes. Plant Cell 15: 2730-2741

Azpiroz-Leehan R, Feldmann KA (1997) T-DNA insertion mutagenesis in *Arabidopsis*: going back and forth. Trends Genet 13: 152-156

Bartel DP (2004) MicroRNAs: Genomics, biogenesis, mechanism, and function. Cell 116: 281-297

Bechtold N, Ellis J, Pelletier G (1993) *In planta Agrobacterium*-mediated gene transfer by infiltration of adult *Arabidopsis thaliana* plants. CR Acad Sci Paris Life Sci 316: 1194-1199

Botstein D, White RL, Skolnick M, Davis RW (1980) Construction of a genetic linkage map in man using restriction fragment length polymorphisms. Am J Hum Genet 32: 314-331

Chang C, Kwok SF, Bleecker AB, Meyerowitz EM (1993) *Arabidopsis* ethylene-response gene ETR1: similarity of product to two-component regulators. Science 262: 539-544

Chang C, Bowman JL, DeJohn AW, Lander ES, Meyerowitz EM (1988) Restriction fragment length polymorphism linkage map for *Arabidopsis thaliana*, Proc Natl Acad Sci USA 85 (18): 6856-6860

Chen X (2004) A microRNA as translational repressor of APETALA2 in *Arabidopsis* flower development. Science 303: 2022-2025

Coats ER, Loge FJ, Wolcott MP, Englund K, McDonald AG (2008) Production of natural fiber reinforced thermoplastic composites through the use of olyhydroxybutyrate-rich biomass Bioresour Technol 99 (7): 2680-2686

Copenhaver GP, Doelling JH, Gens JS, Pikaard CS (1995) Use of RFLPs larger than 100kbp to map the position and internal organization of the nucleolus organized region on chromosome 2 in *Arabidopsis thaliana*. Plant J 7: 273-286

Copenhaver GP, Browne WE, Preuss D (1998) Assaying genome-wide recombination and centromere functions with *Arabidopsis* tetrads. Proc Natl Acad Sci USA 95: 247-252

Cregan PB, Jarvik T, Bush AL, Shoemaker RC, Lark KG, Kahler AL, Kaya N, Vantoai TT, Lohnes DG, Chung L, Specht JE (1999) An integrated genetic linkage map of the soybean genome. Crop Sci 39: 1464-1490

Creusot F, Fouilloux E, Dron M, Lafleuriel J, Picard G, Billault A, Paslier D, Cohen D, Chaboute M, Durr A, Fleck J, Gigot C, Camilleri C, Bellini C, Caboche M, Bouchez D (1995) The CIC library: a large insert YAC library for genome mapping in *Arabidopsis thaliana*. Plant J 8: 763-770

Delseny M (1999) Genomics: methods and initial findings. Ocl-Oleagineux Corps Gras Lipides 6: 136-143

Elizabeth P (2000) Plants join the genome sequencing bandwagon. Science 290: 2054-2055

Gebhardt C, Walkemeier B, Henselewski H, Barakat A, Delseny M, Stüber K (2003) Comparative mapping between potato (*Solanum tuberosum*) and *Arabidopsis thaliana* reveals structurally conserved domains and ancient duplications in the potato genome. Plant J 34: 529-541

Goodman EA Z, Wang ML, Dewdney J, Bouchez D, Camilleri C, Belmonte S, Huang L, Dolan M, et al. (1996) Physical map of chromosome 2 of *Arabidopsis thaliana*. Genome Res 1996 6: 19-25

Hamilton AJ, Baulcombe DC (1999) A species of small antisense RNA in post transcriptional gene silencing in plants. Science 286: 950-952

Hamilton A, Voinnet O, Chappell L, Baulcombe D (2002) Two classes of short interfering RNA in RNA silencing. EMBO J 21: 4671-4679

Haanstra JPW, Wye C, Verbakel H, Meijer-Dekens F, van den Berg P, Odinot P, van Heusden AW, Tanksley S, Lindhout P, Peleman J (1999) An integrated high density RFLP-AFLP map of tomato based on two *Lycopersicon esculentum* x *L, pennellii* F2 populations. Theor Appl Genet 99: 254-271

Hauge BM, Hanley SM, Cartinhour S, Cherry JM, Goodman HM, Koornneef M, Stam P, Chang C, Kempin S, Medrano L (1993) An integrated genetic/RFLP map of the *Arabidopsis thaliana* genome. Plant J 3: 745-754

Huang XQ, Chao KM (2003) A generalized global alignment algorithm. Bioinformatics 19: 228-233

Kempin SA, Sherry A, Kempin A, Liljegren SJ, Block LM, Rounsley SD, Yanofsky MF, Lam E (1997) Targeted disruption in Arabidopsis. Nature 389: 802

Ketting RF, Fischer SEJ, Bernstein E, Sijen T, Hannon GJ, Plasterk RHA (2001) Dicer functions in RNA interference and in synthesis of small developmental timing in *C. elegans*. Gen Dev: 15: 2654-2659

Koch MA, Weisshaar B, Kroymann J, Haubold B, Mitchell-Olds T (2001) Comparative genomics and regulatory evolution: Conservation and function of the Chs and Apetala3 promoters. Mol Biol Evol 18: 1882-1891

Kole C, Quijada P, Michaels SD, Amasino RM, Osborn TC (2001) Evidence for homology of flowering-time genes *VFR2* from *Brassica rapa* and *FLC* from *Arabidopsis thaliana*. Theor Appl Genet 102: 425-430

Kole C, Williams PH, Rimmer SR, Osborn TC (2002) Linkage mapping of genes controlling resistance to white rust (*Albugo candida*) in *Brassica rapa* (syn. *campestris*) and comparative mapping to *Brassica napus* and *Arabidopsis thaliana*. Genome 45: 22-27

Koornneef M, Van Eden J, Hanhart CJ, Stam P, Braaksma FJ, Feenstra WJ (1983) Linkage map of *Arabidopsis thaliana*. J Hered 74: 265-272

Krysan PJ, Young JC, Tax F, Sussman MR (1996) Identification of transferred DNA insertions within *Arabidopsis* genes involved in signal transduction and ion transport. Proc Natl Acad Sci USA 93: 8145-8150

Lander ES, Botstein D (1986) Strategies for studying heterogeneous genetic traits in humans by using a linkage map of restriction fragment length polymorphisms. Proc Natl Acad Sci USA 83: 7353-7357

Larsen PB, Degenhardt J, Tai CY, Stenzler LM, Howell H, Kochian LV (1998) Aluminum-resistant *Arabidopsis* mutants that exhibit altered patterns of aluminum accumulation and organic Acid release from roots. Plant Physiol 117: 9-17

Leitch IJ, Bennett MD (1997) Polyploidy in angiosperms. Trends Plant Sci 2: 470-476

Lister C, Dean C (1993) Recombinant inbred lines for mapping RFLP and phenotypic markers in *Arabidopsis thaliana*. Plant J 4: 745-750

Llave C, Xie Z, Kasschau KD, Carrington JC (2002) Cleavage of Scarecrew-like mRNA targets directed by a class of Arabidopsis microRNA. Science 297: 2053-2056

Lou P, Zhao JJ, Kim JS, Shen SX, Del Carpio DP, Song XF, Jin MN, Vreugdenhil D, Wang XW, Koornneef M, Bonnema G (2007) Quantitative trait loci for flowering time and morphological traits in multiple populations of *Brassica rapa*. J Exp Bot 58: 4005-4016

Lu C, Fedoroff NA (2000) Mutation in the Arabidopsis HYL1 gene encoding a dsRNA binding protein affects responses to abscisic acid, auxin, and cytokinin. Plant Cell 12: 2351-2365

Lukens L, Zou F, Lydiate D, Parkin I, Osborn T (2003) Comparison of a *Brassica oleracea* genetic map with the genome of *Arabidopsis thaliana*. Genetics 164: 359-372

Mauricio R, Eli AS, Tonia K, Dacheng T, Martin K, Joy B (2003) Natural selection for polymorphism in the disease resistance Gene *Rps2* of *Arabidopsis thaliana*. Genetics 163: 735-746.

Meyerowitz EM, Somerville CR (eds) (1994) Arabidopsis. Cold Spring Harbor Lab Press, Cold Spring Harbor, NY, USA

Mushegian AR (2007) (ed) Foundations of Comparative Genomics. Elsevier Academic Press, Boston, USA

Palatnik JF, Allen E, Wu X, Schommer C, Schwab R, Carrington JC, Weigel D (2003) Control of leaf morphogenesis by microRNAs. Nature 425: 257-263

Park W, Li J, Song R, Messing J, Chen X (2002) CARPEL FACTORY, a Dicer homolog, and HEN1, a novel protein, act in micro-RNA metabolism in *Arabidopsis thaliana*. Curr Biol 12: 1484-1495

Preuss D, Rhee SY, Davis RW (1994) Tetrad analysis possible in Arabidopsis with mutation of the QUARTET (QRT) genes. Science 264: 1458-1460

Reinhart BJ, Weinstein EG, Jones-Rhoades MW, Bartel B, Bartel DP (2002) MicroRNAs in plants. Genet Dev 16: 1616-1626

Rodney M, Stahl EA, Korves T, Tian D, Kreitman M, Bergelson J (2003) Natural selection for polymorphism in the disease resistance gene *Rps2* of *Arabidopsis thaliana*. Genetics 163 (2): 735-746

Saigo H, Vert JP, Akutsu T (2006) Optimizing amino acid substitution matrices with a local alignment kernel. BMC Bioinformat 7: 246-257

Schranz ME, Lysak MA, Mitchell-Olds T (2006) The ABC's of comparative genomics in the Brassicaceae: building blocks of crucifer genomes. Trends Plant Sci 11: 535-542

Shultz JL, Ray JD, Lightfoot DA (2007) A sequence based synteny map between soybean and *Arabidopsis thaliana*. BMC Genom 8: 8

Sijen T, Fleenor J, Simmer F, Thijssen KL, Parrish S, Timmons L, Plasterk RHA, Fire A (2001) On the role of RNA amplification in dsRNA triggered gene silencing. Cell 107: 465-476

Somerville C, Dangl J (2000) Plant biology in 2010. Science 290: 2077-2078

Stam P (1993) Construction of integrated genetic linkage maps by means of a new computer package: JoinMap. Plant J 3: 739-744

Sunkar R, Girke T, Jain PK, Zhu JK (2005) Cloning and characterization of MicroRNAs from Rice. Plant Cell 17: 1397-1411

Suwabe K, Iketani H, Nunome T, Ohyama A, Hirai M, Fukuoka H (2004) Characteristics of microsatellites in *Brassica rapa* genome and their potential utilization for comparative genomics in Cruciferae. Breed Sci 54: 85-90

Suwabe K, Tsukazaki H, Iketani H, Hatakeyama K, Kondo M, Fujimura M, Nunome T, Fukuoka H, Hirai M, Matsumoto S (2006) Simple sequence repeat-based comparative genomics between *Brassica rapa* and *Arabidopsis thaliana*: The genetic origin of clubroot resistance. Genetics 173: 309-319

Teutonico RA, Osborn TC (1994) Mapping of RFLP and qualitative trait loci in *Brassica rapa* and comparison to the linkage maps of *B. napus, B. oleracea*, and *Arabidopsis thaliana*. Theor Appl Genet 89: 885-894

The *Arabidopsis* Initiative (2000) Analysis of the genome sequence of the flowering plant *Arabidopsis thaliana*. Nature 408: 796-815

Törjék O, Berger D, Meyer RC, Müssig C, Schmid KJ, Sörensen TR, Weisshaar B, Mitchell-Olds T, Altmann T (2003) Establishment of a high-efficiency SNP-based framework marker set for Arabidopsis. Plant J 36: 122-140

Voinnet O, Rivas S, Mestre P, Baulcombe D (2003) An enhanced transient expression system in plants based on suppression of gene silencing by the p19 protein of tomato bushy stunt virus. Plant J 33: 949-956

Volpe TA, Kidner C, Hall IM, Teng G, Grewal SI, Martienssen RA (2002) Regulation of heterochromatic silencing and histone H3 lysine-9 methylation by RNAi. Science 297: 1833-1837

Waterhouse PM, Wang MB, Lough T (2001) Gene silencing as an adaptive defense against viruses. Nature 411: 834-842

Yang ZF, Gu SL, Wang XF, Li WJ, Tang ZX, Xu CW (2008) Molecular evolution of the CPP-like gene family in plants: Insights from comparative genomics of Arabidopsis and rice. J Mol Evol 67: 266-277

Yanofsky C, Horn V, Nakamura Y (1996) Loss of overproduction of polypeptide release factor 3 influences expression of the tryptophanase operon of *Escherichia coli*. J Bacteriol 178 (13): 3755-3762

Yoshida Y, Nakano Y, Ueno S, Liu J, Fueta Y, Ishidao T, Kunugita N, Yanagihara N, Sugiura T, Hori H, Yamashita U (2009) Effects of 1-bromopropane, a substitute for chlorofluorocarbons, on BDNF expression. Int Immunopharmacol 9 (4): 433-438

Zamore PD, Tuschl T, Sharp PA, Bartel DP (2000) RNAi: double-stranded RNA directs the ATP-dependent cleavage of mRNA at 21 to 23 nucleotide intervals. Cell 101: 25-33

Zhu JK (2001) Plant salt tolerance. Trends Plant Sci 6: 66-71

Zilberman D, Cao X, Johansen LK, Xie Z, Carrington JC, Jacobsen SE (2004) Role of *Arabidopsis* ARGONAUTE 4 in RNA-directed DNA methylation triggered by inverted repeats. Curr Biol 14: 1214-1220

7 Rice Genome Initiative

Kevin L. Childs[1*] and Shu Ouyang[2]

[1]166 Plant Biology Building, Department of Plant Biology, Michigan State University, East Lansing 48824, USA

[2]CambridgeSoft Corporation, 1003 W. 7th Street, Suite 205, Frederick, MD 21701, USA

*Corresponding author: kchilds@plantbiology.msu.edu

1 INTRODUCTION

Rice (*Oryza sativa* L.) was the first crop plant to have its genome sequenced, and there were numerous reasons for this. Rice is consumed throughout the world more than any other cereal (FAO 2006). Nearly half of the people on the planet obtain nearly half of their daily calories from rice, and the demand for rice and all cereals is projected to increase with increasing global population and increasing affluence (Khush 1997; Gilland 2002). While global rice production continues to increase, global reserves of rice grain are declining (Childs 2008). Changing weather patterns are predicted to challenge the ability of rice producing nations to continue to increase production (Peng et al. 2004; Ohyanagi et al. 2006; Naylor et al. 2007). These agricultural facts made rice a candidate for genome sequencing, but biological considerations also favored choosing rice over other cereals. Based on flow cytometry, the rice genome has a calculated size of 415-439 Mb (*Oryza sativa* L. ssp. *Japonica*; Arumuganathan and Earle 1991). Of agronomically important grasses, rice has the smallest genome size. The experimentally calculated sizes of sorghum, maize, barley, oat and wheat are 772 Mb, 2.7 Gb, 4.9 Gb, 11.3 Gb and 16.0 Gb, respectively (Arumuganathan and Earle 1991). The genomes of cereals are known to be highly colinear. Even after 60 million years of evolution, it is still possible to find fine-scale conservation of chromosome organization (Gale and Devos 1998). Thus, lessons learned from the analysis of the rice genome would be immediately applicable to other grass species. These are the types of reasons that rice was the first cereal genome to be sequenced, and in fact, rice has been the target of four separate genome sequencing efforts (Butler and Pockley 2000; Dalton 2000; Sasaki and Burr 2000; Dickson and Cyranoski 2001; Yu et al. 2002; International Rice Genome Sequencing Project 2005). These different projects have approached the problem of sequencing rice using different strategies and even different biological material. While the production of the rice genome sequence was impressive, the analysis of the rice genome has yielded an equally valuable resource. The computational analyses of the rice genome sequence have facilitated many facets of rice research from breeding to molecular genetics to phylogenetics to comparative genomics. This chapter will describe the sequencing of the rice genome and some of the computational analyses that have made rice genomic resources the focus of cereal genetic research for the last half dozen years.

2 RICE GENOME INITIATIVES

2.1 Monsanto

Although the first public rice genome project began in 1998 (Sasaki and Burr 2000), two private rice genome sequencing efforts provided the first draft versions of the rice genome. In 2000, an agricultural company, Monsanto, announced the completion of a draft sequence of rice (Butler and Pockley 2000). Using a BAC-by-BAC approach (Barry 2001), Monsanto sequenced 3,391 BACs containing genomic DNA sequence from the rice cultivar *Nipponbare* to obtain 259 Mb of assembled sequence. The sequencing was contracted to the University of Washington. Sequencing was only done to 5× coverage, and this meant that a substantial number of gaps existed. The Monsanto sequence was only assembled into contigs. This sequence represented a majority of the rice genome, and it is considered "draft" sequence. The data were made available to public researchers, but they were not directly deposited within any public databases. Users were required to access the data through a Monsanto website. However, the Monsanto sequence was provided to members of the International Rice Genome Sequencing Project (IRGSP) in order to aid individual sequencing groups in the assembly of their genome sequences, and much of the Monsanto rice sequence was ultimately incorporated into the IRGSP sequencing effort (Barry 2001).

2.2 Syngenta

Shortly after the announcement of Monsanto's rice genome sequence, Syngenta, another agribusiness, announced that they too had sequenced the genome of *Nipponbare* rice. Unlike the Monsanto sequence, initially Syngenta only allowed academic researchers to analyze their sequence data after agreeing to formal collaborations (Dickson and Cyranoski 2001). Also unlike the random BAC-by-BAC sequencing strategy of Monsanto, the Syngenta effort utilized the whole-genome shotgun sequencing method (Goff et al. 2002). Sequence production was performed by Myriad Genetics for Syngenta, and the depth of sequencing was more extensive than the Monsanto sequence. Sheared nuclear DNA was cloned, and both ends of each clone were sequenced. A sufficient number of clones were sequenced to provide an overall 6× coverage of the genome, and this resulted in an estimated sequencing accuracy of 98%, although comparison to the sequence from the IRGSP genome effort suggested that the sequencing accuracy was up to 99.8% accurate. Over 389 Mb of unique assembled sequence was obtained in the final assembly. Alignment of nearly 500 full length cDNA sequences to the Syngenta sequence indicated that more than 99% of the bases from these genes were represented in the final genome sequence (Goff et al. 2002).

2.3 Beijing Genomics Institute

While the other private and public rice genome sequencing efforts all utilized the *Nipponbare* cultivar of rice, which is a member of the *japonica* subspecies, the rice genome sequencing project led by the Beijing Genomics Institute (BGI) worked with the *indica* subspecies of rice. Chinese and many other Asian rice farmers typically grow rice from the *indica* subspecies of rice. The cultivars *93-11*, a member of the *indica* subspecies, and *Pei-Ai 64s* (*PA64s*), with genetic background from *indica*, *japonica* and *javanica*, are particularly important to Chinese rice production as they are respectively the paternal and maternal parents of *Liang-You-Pei-Jiu*, a high-yielding hybrid cultivar that is grown in China. Draft genome sequences were produced for both *93-11* and *PA64s* (Yu et al. 2002). As with the Syngenta sequence, the BGI-led work also used a whole-genome shotgun approach to generate these draft sequences. The initial sequencing resulted in a 4× coverage assembly for *93-11* and in a very low coverage sequence for *PA64s*. Only the *93-11* sequences are available in public sequence databanks (Yu et al. 2002). The initial *93-11* total contig length was 361 Mb from over 103,000 scaffolds. The BGI sequencing project assumed a genome size for *indica* rice of 430 Mb. So, they estimated that they had only assembled 84% of the full genome. However, sequencing continued and a 6× draft assembly was later released (Yu et al. 2005). The more complete draft sequence was estimated to have a sequencing accuracy of 97.2%. The final draft sequence was calculated to represent 466 Mb of genomic sequence. This is a likely overestimate of the total amount of the genome that was sequenced because the assembly method used by the BGI researchers did not identify overlapping regions between contigs that ended in identical repetitive sequence.

Assembly of a shotgun sequenced whole genome is a challenge (Yu et al. 2002, 2005). The most basic idea for sequence assembly is to identify regions of individual reads that overlap and to combine overlapping reads into longer contiguous sequences. This simple strategy is complicated by the fact that the genomes of the grasses contain a high proportion of repetitive sequences (Kumar and Bennetzen 1999). Some type of strategy is necessary for dealing with repetitive sequences in any whole-genome shotgun sequencing effort because the dozens or hundreds of identical copies of a single repetitive element from multiple regions of the genome could be assembled together into a single contig. It is then impossible to confidently identify the true non-repetitive sequence that lies on each side of a single repetitive region. The BGI assembly strategy did not attempt to utilize repetitive sequences when identifying overlapping regions during contig assembly. The program that was used to generate this assembly was designed to ignore repetitive sequences when attempting to join separate sequence contigs (Yu et al. 2005). With the BGI assembly program, if two contigs each end in identical sequence, but that sequence is determined to be repetitive, the two contigs will not be combined into a larger contig. Thus, some portion of the sequence in the BGI genome assembly represents sequences that actually overlap but that cannot be recognized as overlapping.

For the initial assembly of the *93-11* genome, individual sequence reads were assembled into contigs. A pair of contigs that each contained paired-end sequence reads from a single genomic clone was assembled with a gap into a scaffold. Thus, the first release of the *93-11* genome consisted of contigs and longer scaffolds (Yu et al. 2002). For the updated version of the *93-11* genome, all individual sequences were reassembled

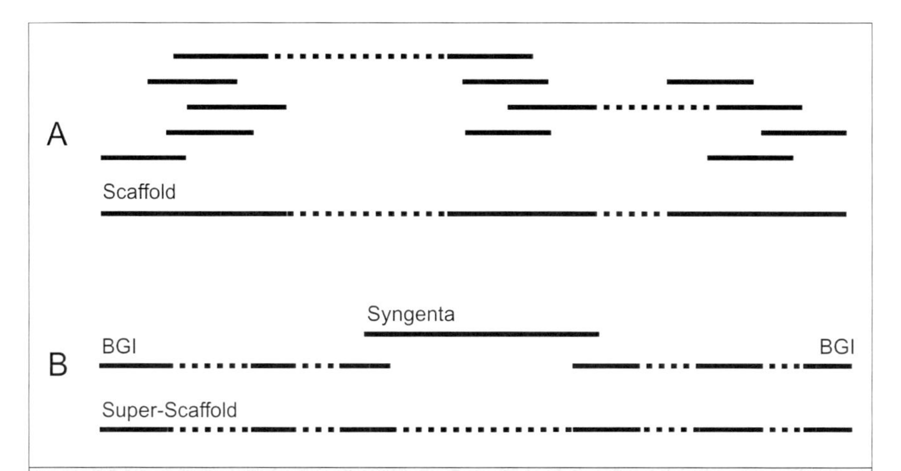

Fig. 1 Rice sequence assembly by the Beijing Genomics Institute. A. Sequences were obtained from whole genome shotgun clones (solid lines). Paired reads from opposite ends of a single clone are separated by gaps (dotted lines). Identical overlapping sequences were assembled into contigs. Different contigs that contained opposite ends of paired reads from a single clone were combined into gapped scaffolds. B. BGI scaffolds that overlapped with BGI assembled scaffolds from the Syngenta sequence were combined into gapped super-scaffolds. None of the sequence from the Syngenta scaffolds were included in the BGI super-scaffolds.

from scratch with an updated assembly program. Additionally, the original sequence reads from the Syngenta *Nipponbare* genome project were also assembled with the same assembly program. The contigs and scaffolds from the rebuilt *93-11* assembly were then compared to the contigs and scaffolds of the rebuilt Syngenta *Nipponbare* assembly. The two genome assemblies were then compared to each other to improve the assemblies. When two scaffolds from one genome aligned to a single scaffold from the other genome, the two scaffolds were linked with a gap into a super-scaffold. The median size of the super-scaffolds was 8.3 Mb. The entire BGI assembly strategy is outlined in Fig. 1. Where possible, super-scaffolds were mapped to either genetic or physical maps. Using this ordering, pseudomolecules were created from the assembled scaffolds. Strings of N's were inserted to represent the within-scaffold and between-scaffold and between-super-scaffold gaps. The sequencing error rate was estimated to be less than one error in 10,000 bp for more than 94% of the *93-11* assembly (Yu et al. 2005).

Although the calculated size of the assembled *93-11* and reassembled Syngenta *Nipponbare* sequences were very close to the expected sizes, these assemblies include numerous gaps and likely contain assembly errors. Additional measures of completeness were sought by the BGI researchers. By aligning a non-redundant set of 19,079 full-length *japonica* rice cDNAs (KOME cDNAs; Kikuchi et al. 2003) to both the *93-11* and the reassembled Syngenta *Nipponbare* pseudomolecules, they determined that 98.6% and 99.7% of the cDNAs were aligned without discrepancies to the *93-11* and reassembled *Nipponbare* pseudomolecules. The BGI scientists also used another estimate of completeness for these two assemblies (Yu et al. 2005). The percentages of KOME cDNA coding sequences (i.e. no UTRs were used) that aligned with 95% coverage to the *93-11* and reassembled *Nipponbare* sequences were calculated: (1) for fragmented cDNAs that could possibly align across gaps or to multiple locations; and (2) for cDNAs that aligned within a single gapless contig. For *93-11*, 98.1% and 91.2% of KOME cDNAs aligned with possible gaps and in a single contig, respectively. For the reassembled *Nipponbare* genome, 99.3% and 94.8% of KOME cDNAs aligned with possible gaps and in a single contig, respectively. Most importantly, these results indicate that between 1 and 9% of rice genes fall across gaps or align to multiple locations within the BGI pseudomolecules. This simply reflects the difficulty of assembling a 6× whole-genome shotgun sequence of a highly repetitive genome.

2.4 International Rice Genome Sequencing Project

In 1998, scientists from 10 countries formed the International Rice Genome Sequencing Project (IRGSP) in order to collaborate on sequencing the rice genome (International Rice Genome Sequencing Project 2005). A clone-by-clone sequencing strategy using BACs and PACs was chosen (Fig. 2), and the consortium agreed to sequence the *Nipponbare* cultivar from the *japonica* subspecies of rice. In order to ensure that the final sequence covered the entire genome, BAC and PAC libraries were generated using several different restriction enzymes. Extensive BAC end sequencing was performed before large scale sequencing began. BAC fingerprinting was used to generate physical maps, and ESTs were used to link BACs and PACs to a transcript-based genetic map. This preparatory work was executed for several reasons. Unlike the other rice genome sequencing projects, the IRGSP work would be performed in multiple sequencing centers. Participating laboratories were each responsible for a different chromosome or a different fragment of a chromosome. Data from the physical and genetic maps allowed the assignment of a subset of individual BACs/PACs to specific chromo-

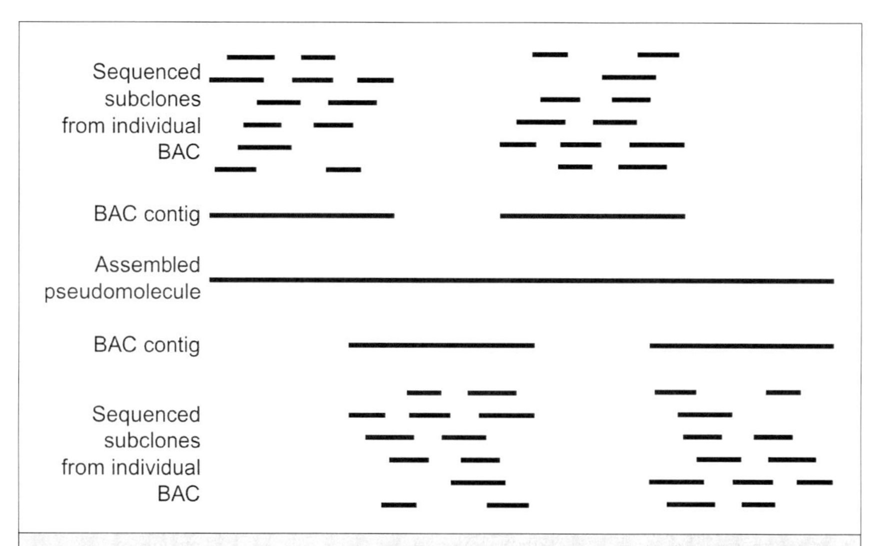

Fig 2. Rice sequence assembly by IRGSP. Individual BACs were subcloned, and the subclones were sequenced to 10X coverage. The subclone sequences from a single BAC were assembled into a BAC contig. Any gaps in a BAC sequence necessitated additional random sequencing of subclones from the BAC or directed sequence to close the gap. Overlapping BAC contigs were used to assemble pseudomolecules. Gaps in the pseudomolecules result from regions for which a BAC clone was not identified or difficult to sequence BACs that could not be completely sequenced or assembled.

somes. When a clone was sequenced, the entire set of BAC-end sequences could be aligned to the newly sequenced clone, and an overlapping clone could be identified and chosen to continue sequencing along that chromosome. BAC sequences from the Monsanto rice genome work were incorporated into IRGSP sequence throughout the sequencing project. Sequence from Monsanto BACs alone was not used to fill gaps, because the sequence was of draft quality. Instead, the sequence from Monsanto was combined with IRGSP-derived sequence reads to provide a high-quality gap-filling sequence. Similarly, sequence from the Syngenta draft genome was also used to aid in the assembly of the IRGSP sequence. If any gaps remained after 10-fold sequencing of a BAC, they were filled by completely sequencing available bridging clones, by generating and sequencing gap-filling PCR products or by directly sequencing the BAC clone using primer walking (International Rice Genome Sequencing Project 2005).

The BAC-by-BAC approach of genome sequencing lends itself to a conceptually straightforward method for sequence assembly and pseudomolecule creation. Complex assembly is generally limited to individual BACs. As with the assembly of WGS sequence, individual sequence reads are aligned to each other. However, in the BAC-by-BAC method, the reads are only aligned to other sequences from the same BAC. Thus, there is less possibility for sequences from different regions of the genome to align incorrectly. With the BAC-by-BAC method, repetitive sequence is much less problematic during assembly of a single BAC sequence. In cases, where repetitive sequence does confound the assembly process, it is possible to use primer walking to directly sequence through troublesome regions. In general, 10× sequencing will produce a completely ordered assembly in more than 50% of BACs. Once a BAC is completely sequenced, it is aligned to the library of BAC-end sequences to find overlapping BACs that would be used to pick the next BAC for sequencing. Once the adjoining BAC is fully sequenced, overlapping sequences are identified. For the IRGSP work, if there were any major discrepancies within the overlapping region, those regions were reanalyzed (International Rice Genome Sequencing Project 2005). The discrepancies were resolved by reassembling the BAC sequences or by additional sequencing. BACs/PACs were linked to genetic and physical maps before sequencing was performed. The genetic position, and thus chromosomal order, of many BACs/PACs were known before they were fully sequenced. Production of the pseudomolecule sequences was, therefore, conceptually performed as additional BAC sequence was produced.

Nonetheless, differences between the pseudomolecules from the two groups are known to exist. The first evidence of this was seen when The Institute for Genomic Research (TIGR) published their description of the TIGR Rice Genome Annotation Database (Yuan et al. 2005). The TIGR assemblies of the IRGSP *Nipponbare* pseudomolecules, which are now available from and maintained at Michigan State University (*http://rice.plantbiology.msu.edu*), used slightly different sets of BACs than the IRGSP build. This was mainly due to redundancy in the sequenced BAC set. Additionally, small differences in the assembly procedure resulted in the inclusion of different sequences from the overlap regions between BACs. Finally, for regions within the genome that represent tandem duplications or nearly-identical repetitive regions that are in close proximity to each other, it is possible for one assembly to have the region present in one orientation and the other assembly to have the same region present in the opposite orientation (KL Childs and CR Buell, unpub data).

Recently, an optical map has been created for *Nipponbare* rice, and this map shows the positions of assembly errors in both the IRGSP and TIGR/MSU rice pseudomolecules (Zhou et al. 2007). Optical mapping generates a high resolution restriction enzyme-based physical map at the genome level. Sheared genomic DNA is mounted on a glass slide support. A restriction enzyme, in this case *Swa*I, is used to digest the DNA. A fluorescent dye is used to label the DNA, and a microscope allows visualization of the digestion pattern of a large region of a single chromosome. DNA with a known restriction pattern is included in the process and acts as a size marker for calibration. The sizes of the restricted chromosomal regions can then be accurately estimated, and published pseudomolecule sequences can be virtually digested and searched for a matching pattern. For the rice optical map, approximately 97% of the *Nipponbare* genome was mapped using *Swa*I restriction digest fragments. The mapped fragments ranged in size from 300 kb to 3.6 Mb. A very large 317× coverage of the genome was obtained. The final restriction digest map published by Zhou et al. (2007) was created without the aid of the published sequences. Comparison of the resulting optical map to both the IRGSP and TIGR/MSU pseudomolecules was very informative. Regions of mis-assembly in the rice genomic sequences were identified. The known gaps in the pseudomolecules were accurately sized, and the existence of gaps that were unrecognized in the pseudomolecules was revealed. Some of the differences between the optical maps and pseudomolecules were shared by the IRGSP and TIGR/MSU assemblies. Some of the differences were only found in one or the other of the pseudomolecule versions. A total of 82 unrecognized gaps (1.4 Mb) and 53 mis-assembled regions (2.9 Mb) were identified in the IRGSP pseudomolecule assembly (build 4), and 93 new gaps (1.9 Mb) and 39 mis-assembled regions (1.5 Mb) were found in the TIGR/MSU pseudomolecules (release 4). Finally, comparison of the sizes of the IRGSP and TIGR/MSU genome assemblies to the calculated size of the optical map indicate that the IRGSP and TIGR/MSU pseudomolecules respectively represent 94.9% and 95.9% of the true genome size. After a detailed analysis, it was determined that the discordances between the rice optical map and TIGR/MSU pseudomolecules almost exclusively pre-existed at the BAC sequence level. However, using the optical mapping data, the TIGR/MSU rice pseudomolecules could be improved (release 6) in a few areas, such as anchoring previously unanchored clones, bridging gaps, and replacing BAC clones with similar clones with better sequencing quality (S Ouyang and CR Buell, unpub).

The IRGSP agreed to produce a rice genome sequence that conformed to the quality standard of the Human Genome Project (Sasaki and Burr 2000). This standard required sequence to be of high quality with less than one error per 10,000 bp. Although the final assembly was an incomplete sequence because of the existence of large gaps, 97% of the BACs/PACs that were sequenced were submitted to the public databases as finished quality sequence (i.e. with no gaps). Additionally, the IRGSP members submitted the sequences of individual BACs/PACs shortly after the sequence was assembled. Because the map-based clone-by-clone strategy is more complex, a final version of the IRGSP rice genome sequence took seven years to complete (International Rice Genome Sequencing Project 2005). The final assembled pseudomolecules represent 370 Mb of sequence with only 36 gaps within chromosome arms. Gaps also existed in the telomeric regions and in ten of twelve centromeres. The total size of the gaps was estimated to be 18 Mb, and thus, the total estimated length of the *Nipponbare* genome was calculated to be 388 Mb. Using identical sequences produced by multiple labs, the total sequencing accuracy was estimated to be better than 99.99%.

Despite being first, the private rice genome sequences are rarely used directly by the rice community. While the Monsanto and Syngenta sequences were utilized by the IRGSP project, both of the private rice sequence assemblies were never fully improved and remain draft-quality assemblies. The BGI *indica* sequence and the IRGSP and TIGR/MSU *japonica* pseudomolecules are the three rice genome assemblies that are typically used by rice genomics community. The remainder of this chapter will only discuss these public sequences.

3 RICE GENOME ANNOTATION

3.1 Initial Rice Genome Annotation Efforts

Having the complete and correct sequence of any genome is quite significant. However, a genome sequence alone is of little use to most biologists. An annotated genome is really the resource that is useful to biologists. Of course, even for so-called complete genomes, there will continue to be a need for additional sequencing and assembly refinement as the optical maps of Zhou et al. (2007) indicate, and annotation efforts can begin at any time that a significant amount of high quality sequence is obtained.

Annotating a genome is the process of describing the many features of the genome sequence. An obviously important aspect of genome annotation is the identification of the location and structure of the genes that are encoded by a given sequence. Genes that are typically identified by an annotation project include not only protein coding genes but also tRNAs, miRNAs, snRNAs and other smRNA genes. While the positional location of a gene is fundamentally useful to plant scientists, a tentative assignment of the functional identity of a gene is possibly of greater importance. Unfortunately, the functional annotation is often unsatisfying, and the process of generating functional annotation is fraught with difficulty. Annotation can also include the description of other biologically relevant features on a genome including transposable elements (TE), other repetitive elements and the regions of the genome that align with ESTs and FL-cDNAs. Researchers are interested in non-biologically-functional features of genomes too. These features include the identification of genetic markers, the location of expression array probes and oligos, the regions of the genome that have sequence similarity to gene models, proteins, ESTs and other sequence from different species. Annotation is a never-ending process because additional biological data that can be used to improve the annotation are continually being generated. Biologists will always be interested in how the latest data relate to their genome of interest. Anyone who thinks that annotation ends once the basic features of a genome are recognized should consider the fact that the importance and widespread nature of miRNAs were only recognized in 2003 (Lagos-Quintana et al. 2001; Lau et al. 2001; Lee and Ambros 2001). The work from the Human Encode project has expanded our understanding of the variety of mechanisms that are used to regulate genome function. Almost the entirety of the human genome is now expected to be transcriptionally expressed (Weinstock 2007). New biologically relevant features are undoubtedly waiting to be discovered in the rice genome. Annotation of the rice genome will be an ongoing endeavor for years to come.

While gene identification has been performed on rice BACs and chromosome sequence (Feng et al. 2002; Sakata et al. 2002; Rice Chromosome 10 Sequencing Consortium 2003; Yuan et al. 2003), gene finding on a genome scale is much more informative and satisfying. The first opportunities for genome wide annotation of genes came with the publication of the rice genome sequences. BGI analyzed the genome sequences of their *indica* assemblies, their reassembly of the Syngenta *japonica* shotgun sequence and the IRGSP *japonica* pseudomolecules. The BGI approach to gene identification began with the ab initio gene prediction program FGENESH that is produced by Softberry (Salamov and Solovyev 2000). It is widely held that FGENESH is one of the best gene finders for rice, and it is the preferred gene finder that is used by a number of rice annotation efforts (Yu et al. 2002, 2005; International Rice Genome Sequencing Project 2005; Yuan et al. 2005; Ohyanagi et al. 2006). Initial gene models identified by FGENESH were analyzed for similarity to known TEs and for alignment to 20-mers that are highly represented in the genome. Gene models that had alignment along more than 50% of the length to either transposable element (TE) sequences or to highly represented 20-mers were removed from the gene model list. BGI researchers did not want to over-estimate the number of genes in rice by including those that are TE-related. As TEs are duplicated within the genome, they can capture a true gene and carry a copy of the gene during duplication. These TE-related genes often become inactive. However, they can still be identified by gene prediction software because the most of them do contain open reading frames, supposedly for genes that are directly involved in TE excision, duplication and reintegration into the genome. In order to not inflate the total gene number with these 'uninteresting' genes, the BGI strategy was to remove them from consideration. Over ten thousand TE-related genes were eliminated from each of the three BGI analyzed genome assemblies. For the BGI *indica*, Syngenta reassembled *japonica* and IRGSP *japonica* genomes, the numbers of identified genes after filtering were 49,088, 45,824 and 43,635, respectively. These initial gene counts were suspected as being too high, and a complex correction was made that estimated a lower bound for the gene numbers as being 40,216, 37,794 and 37,581, respectively for the BGI *indica*, Syngenta reassembled *japonica* and IRGSP *japonica* assemblies. The average gene size for the BGI predicted genes was 2.5 kb. If the predicted gene models are enhanced with UTRs based on alignments to FL-cDNAs, the average size for those genes is 3.6 kb. The functional annotation pipeline, Bioverse (McDermott and Samudrala 2003), was used to provide functional assignments to the predicted gene models. Bioverse uses several different methods to provide functional annotation to genes. Importantly, the BGI researchers did not allow the use of transitive annotations. Transitive annotation occurs when automated procedures are used that do not consider the quality of the source of the evidence that is being used to support the functional assignment. For example, if gene A has sequence similarity to gene B and if gene B has been annotated as an alcohol dehydrogenase, it does not mean that gene A should also be labeled as an alcohol dehydrogenase. Perhaps genes A and B are only similar over 75% of their lengths. If the functional annotation of gene B was due to the 25% of that gene that is not similar to gene A, then gene A is almost certainly not an alcohol dehydrogenase. By carefully choosing the evidence that they used for performing their functional annotation, the BGI scientists attempted to minimize the dangers of transitive annotation (Yu et al. 2005).

The International Rice Genome Sequencing Project (2005) also performed an initial full genome annotation with the publication of the *Nipponbare* genome. Also concerned with the identification of TE-related genes, the *Nipponbare* assemblies were masked of repetitive sequence before using FGENESH (Salamov and Solovyev 2000) to create 37,544 ab initio gene models. Small gene models (CDS less than 150 bp) and gene models without proper start and stop codons were excluded from this collection. This number is nearly identical to the number of gene models (37,581) that were identified by the BGI analysis as the lower bounds gene count for the IRGSP *japonica* genome, and it suggests that the correction that was applied to BGI gene counts was warranted (Yu et al. 2005). FL-cDNA sequences (Kikuchi et al. 2003) were used to improve the initial IRGSP gene models. Gene models created by FGENESH do not include UTR sequences. Additionally, the structure of FGENESH gene models often disagree slightly with transcript evidence, and FGENESH never predicts alternative splice variants for a gene locus. The use of FL-cDNA alignments allows ab initio gene models to be improved through the addition of these features. Of the initial gene models, over 17,000 gene structures were improved through the use of FL-cDNA evidence. Evidence for these gene models was extensive. Rice FL-cDNAs and ESTs had significant alignment scores with 22,840 of the gene models. There were 330 gene models that were not supported by rice transcript evidence but that did have significant alignments with ESTs from other cereal species. Translations of the FGENESH gene models were aligned with protein sequences from Swiss-Prot. Over 19,000 models aligned significantly with Swiss-Prot sequences, and 4,500 of these models had no transcript support. Overall, more than 73% of all IRGSP *Nipponbare* gene models were supported by some type of biological evidence.

Summaries about the gene annotations for a genome are useful in a publication, but individual researchers with interest in a particular gene or a specific region of the genome will want to directly delve into the detailed annotation of a genome. The annotations from BGI and IRGSP genome analyses are available to varying degrees on the internet. Both BGI *indica* and Syngenta-based *japonica* gene models are available at the BGI website (*http://rice.genomics.org.cn*) as flat files as well as in graphical format via a MapViewer tool. Unfortunately, functional annotation of these gene models is not available. The positions of common molecular markers and SNPs that distinguish *indica* and *japonica* are also provided in files and in the MapViewer. A number of other classes of data are only available through the BGI MapViewer including FL-cDNA alignments, original FGENESH models, tiling expression data, repeat sequence locations, regions of homology between *indica* and the Syngenta *japonica* sequence and BAC-end positions. Gene models and functional annotation for those models are available for version 3 of the IRGSP derived *japonica* genome at the IRGSP website (*http://rgp.dna.affrc.go.jp/IRGSP/index.html*). However, those data are only available as flat files, and no other annotation is directly available from the IRGSP. These annotations are, of course, out of date. In the years since the publication of these genomes, researchers have generated new data that can be used to improve the annotation of the rice genome.

3.2 Other Sources for Rice Genome Annotation

There are three major sources of rice genome annotation that are used by plant researchers. The Rice Annotation Project (RAP) is a sister consortium to the IRGSP that was formed to provide annotation of the IRGSP *japonica* genome, and researchers with the RAP have annotated multiple aspects of the rice genome and provide these annotations through a website (Ohyanagi et al. 2006; Tanaka et al. 2008). The TIGR/MSU Rice Genome Annotation Project has created a database of annotation for an independently assembled version of the IRGSP BAC/PAC sequences. While the RAP and TIGR/MSU annotations are only for the IRGSP *japonica* genome, annotation for both *indica* and *japonica* genomic sequence is available through Gramene (*www.gramene.org*), which acts as a repository for genome and mapping data for all major cereal crops. Each of these three rice annotation projects provides a few similar types of data. All have sets of rice gene models and alignments of transcript and protein sequence relative to the rice genome. However, the methods used for generating gene models differ between the three. Additionally, the mandate of each of these annotation projects differs, and so, the non-gene model annotation provided by each group also varies.

The Rice Annotation Project was founded about the time that the IRGSP *japonica* genome sequence was made public, and the goal of the RAP is to provide high quality annotation of the rice genome (Ohyanagi et al. 2006). The foremost concern of any annotation project is to identify accurate gene models. RAP gene loci are identified by two methods that each use repetitive element masked pseudomolecule sequences. The first method uses high confidence alignments of mRNAs and FL-cDNAs to define loci. Overlapping transcripts define a single locus, but the transcript with the longest ORF is chosen as the representative transcript for that locus. Overlapping loci are predicted when the mRNAs from one locus fall within an intron of a second locus or when the coding sequences for the overlapping loci fall on opposite strands. Gene loci are also identified by using Combiner to generate a gene model based on comparisons of *ab initio* predictions from FGENESH, GENSCAN (Burge and Karlin 1997) and GLocate (*http://glocate.dna.affrc.go.jp/*). When the ab initio predictions are different but similarly scored, FGENESH models are preferentially chosen. The RAP provides two types of gene models, those based with transcript evidence (FL-cDNA-derived models [26,259] and models based on ab initio predictions and with EST support [5,180]) and ab initio gene models with no supporting transcript evidence (22,022; Tanaka et al. 2008). The predicted ORFs for all genes are aligned to a database of known protein sequences and the Interpro protein domain database. If an ORF has a sequence identity greater than 50% of the length of database protein, the functional description of the database protein is assigned to the gene that codes the ORF. If no significant protein alignment was found, a significant protein domain alignment was used to provide a functional description of the gene. However, all functional annotations are manually reviewed to ensure that the homologous proteins from which the functional descriptions were obtained had been originally given those functional descriptions as the result of actual experimental studies (Itoh et al. 2007). A gene with an ORF that aligned to a protein that did not have a known function was classified as a 'conserved hypothetical' gene. Those genes with ORFs that did not match any known proteins or protein domains were classified as 'hypothetical'. Ab initio gene predictions without transcript support were not classified, but are by definition 'hypothetical'. A total of 23,402 hypothetical genes were identified by the Tanaka et al. (2008).

The philosophy of the RAP functional annotation is that computationally-derived functional assignments are highly inaccurate. This was not always true. Before the movement to high-throughput sequencing, all sequences in the international sequence databases were the result of actual observational experiments. At that time, when a scientist did a BLAST alignment of an unknown sequence to a non-redundant nucleotide or protein database, the query sequence typically had context in some biological question. A new description was given to the unknown sequence using both the significant alignment results and the known biological context. With high-throughput sequencing, there often is little or no biological context. Annotations based on transitive functional descriptions are now common. By requiring manual curation of all functional annotations, the RAP researchers have attempted to reduce the likelihood of error-prone transitive annotations.

The TIGR/MSU annotation, which was originally generated at TIGR, and then moved to Michigan State University (*http://rice.plantbiology.msu.edu*; Yuan et al. 2003, 2005; Ouyang et al. 2007) in 2008, uses an in-house assembly of the *japonica* genome sequence from the IRGSP sequencing project. The aim of the TIGR/MSU rice annotation project is to provide high-quality and in-depth annotation of the rice genome. The project released the sixth version of its rice annotation in the fall of 2008.

Like the RAP, the most important products from the TIGR/MSU rice annotation are the gene models. However, the TIGR/MSU rice gene models are generated in a slightly different manner (Ouyang et al. 2007). The ab initio gene finders, FGENESH, GeneMarkHMM and Glimmer, are used to generate gene models, but the models created by FGENESH are given a preferential status. The gene model refinement program PASA (Haas et al. 2003) uses the ab initio gene models and EST, mRNA and FL-cDNA alignments to improve the initial FGENESH gene models. While the ab initio gene finders will only predict a single gene model at any given locus, transcript evidence can provide support for alternative splicing events, and PASA utilizes this evidence to generate alternatively spliced gene models at a single gene locus (Campbell et al. 2006). A small portion of gene models were manually reviewed to resolved discrepancies between PASA-generated models and the initial FGENESH gene predictions. Functional annotation of the TIGR/MSU gene models was based on an automated pipeline with manual review. Functional assignments were made based on protein alignments and similarity to PFAM protein domains. Manual review of the functional assignments was performed to increase uniformity and to replace those automated descriptions that were clearly uninformative. While the RAP annotation made use of a large number of manual reviewers during an 'annotation jamboree', the TIGR/MSU annotation project performs most manual review with a small number of in-house annotators. However, the TIGR/MSU annotation does benefit from focused community annotation that is contributed by expert rice researchers (Ouyang et al. 2007; Thibaud-Nissen et al. 2007). Scientists who have special understanding of a group of genes have submitted improved structural annotation of those genes and/or more insightful functional annotations.

With release five of the TIGR/MSU Rice Genome Annotation Project, a total of 41,046 non-TE-related genes were identified (*http://rice.plantbiology.msu.edu/riceInfo/info.shtml*). Transcript support exists for 29,098 of those genes, but 9,767 hypothetical and 2,181 conserved hypothetical genes were also identified. An additional 15,236 TE-related genes were found as well. The RAP gene set did not include any TE-related genes because the *japonica* pseudomolecules were masked before gene identification was performed (Itoh et al. 2007). The BGI gene set also contains few TE-related genes because initial gene models that matched known TEs and gene models that contain repetitive 20-mers were discarded from the initial FGENESH gene set (Yu et al. 2005). A more complete description of transposable elements and their special role in the rice genome will be given below, but it should be noted now that transposable elements are segments of DNA that can move from one genomic location to another. This movement is called transposition, and it requires several proteins that are encoded in genes that are carried by each transposon (Volff 2006). Some researchers believe that TE-related genes should not be included in 'official' gene sets (Bennetzen et al. 2004), and this is the apparent rationale used by BGI and RAP annotators when they exclude TE-related genes from their analyses. However, the TIGR/MSU annotators have taken an inclusive approach and identify TE-related genes so that rice researchers can view all genes and make their own decisions about which genes are relevant to their work.

The final popular source for rice genome annotation that will be covered here is Gramene (Jaiswal et al. 2006; Liang et al. 2008). Unlike the RAP and TIGR/MSU annotation projects, Gramene produces annotation for multiple species for which there are varying types of available geneome sequence. Both the *indica* and *japonica* rice genomes have been annotated by Gramene. The genome assemblies annotated by Gramene were the BGI scaffolds from *indica* rice (Yu et al. 2005) and the TIGR/MSU pseudomolecules from *japonica* rice (Ouyang et al. 2007). Additionally, Gramene also provides gene models that were produced in-house using the GeneBuilder gene prediction program (Milanesi et al. 1999). GeneBuilder uses homology-based evidence and functional signals to identify complete gene models that include UTRs and alternatively spliced transcripts. For the *japonica* genome, a total of 37,178 genes were identified by this method. TE-related genes were excluded from the GeneBuilder gene set. Short functional gene descriptions are only provided for the RAP and TIGR/MSU gene models. More encompassing functional descriptions can be gleaned from gene report pages.

Each of the rice annotation groups provides additional annotation for each gene locus in the form of gene report pages. Each gene report page displays the gene's identifier, a functional description of the gene, a list of any GO terms that describe the gene, a list of any InterPro or PFAM protein domains that are found within the gene, the chromosomal coordinates of the gene and the sequence of the gene. The TIGR/MSU and Gramene gene report pages also have data about putative homologous genes, and these pages describe the structural organization of the gene. The TIGR/MSU pages show the top results from BLASTP alignments to the Genbank non-redundant protein database. The Gramene pages show putative predicted orthologs to genes from a select group of species, and these pages also list putative paralogous genes from within rice.

The most common and most intuitive method for presenting annotation to researchers is through a graphical, web-based genome browser. The RAP and TIGR/MSU projects use the GMOD GBrowse genome browser (Stein et al. 2002), while Gramene presents data with the Ensembl-based genome browser (Stalker et al. 2004; Hubbard et al. 2007). Both browsers present annotation in a common manner. The user can select a particular chromosomal region to view by clicking on a graphical representation of a chromosome or by specifically designating chromosomal coordinates in a search box. Text-based searches are also available to find all annotation features that are described by a specific term. Within the browsers, the user can choose to view as many or as few annotation classes as are needed. Data from each annotation class are displayed as single tracks within the graphical view, and each annotation datum is generally shown as a single glyph. A name and often a short description are provided for each glyph, and by clicking on the glyph, the users can typically discover additional information about the annotation at a new web page that is either internal or external to the rice annotation website. A useful example of the type of additional information that can be accessed from the browser is that each of the rice annotation websites provides a gene report page that can be viewed when a user clicks on a gene model in their genome browsers.

The three popular rice annotation groups provide through their websites many additional types of annotation besides gene models. Many of these annotations types are common to all three annotation providers, but this is not always the case. Besides protein coding genes, other non-coding genes, such as tRNAs, rRNAs, miRNAs and siRNAs, have also been identified within the rice genomes. Characterized transposons, plastidic genome insertions, QTLs and genetic markers, such as

RFLPs, SSRs, SNPs, have all been localized within the rice genomes. The positions of alignments of rice ESTs, FL-cDNAs, mRNAs and rice gene models from other annotation groups are viewable from the rice genome browsers. In order to allow a genomic context for the interpretation of expression experiments, the probes and oligos that are used by common large-scale expression platforms have been localized within the genome. While just the location of these expression features will be of interest to some researchers, the actual expression values obtained by experiments reported in public sources can be found for some platforms. Besides rice-centric data, genes, proteins and other transcript evidence from numerous other species have been aligned to the rice genomes, and the locations of these alignments are available within the genome browsers. Additionally, the BAC end sequences from other *Oryza* species sequenced by the OMAP project (Wing et al. 2005) have been aligned to both *indica* and *japonica* rice. All of these types of data are available through one or more of the various rice genome browsers, but often search pages have also been created to allow a more direct access to a particular type of annotation.

3.3 Repetitive Sequence Identification and Analysis

Repetitive sequences, along with low-complexity sequences, often cause mis-assembly of genomic sequences. To prevent these errors, highly repetitive sequences sometimes are masked prior to assembly of large genomic sequences. In addition, repetitive sequences present challenges in genome annotation due to the abundance of transposable elements with variable characteristics which confounds detection of bona fide protein-coding regions. Therefore, repetitive sequence identification is a critical component of genome sequencing and annotation projects.

Transposable elements make up the majority of the dispersed repeats. According to the newly proposed TE hierarchical classification system (Wicker et al. 2007), eukaryotic TEs can be divided into retrotransposons (class I) and DNA transposons (class II) based on whether an RNA transposition intermediate is involved. Retrotransposons utilize a so-called "copy-and-paste" approach with a RNA intermediate while DNA transposons replicate without the benefit of an RNA intermediate (Finnegan 1989). DNA transposons can be further grouped into subclasses based on whether the TEs replicate themselves or are replicated by genes belonging to other active transposons.

Repetitive sequences can be identified with de novo repetitive sequence identification computer programs, which exclusively rely on sequence composition (Haas and Salzberg 2007). Different software packages including Miropeats, REPuter/REPfind, RepeatFinder, RECON, PILER, and RepeatScout have been developed to identify dispersed repetitive sequences with different algorithms and features (Parsons 1995; Kurtz et al. 2001; Volfovsky et al. 2001; Bao and Eddy 2002; Edgar and Myers 2005; Price et al. 2005). Computationally identified repetitive sequences must be further filtered to exclude low-complexity sequences and sequences from large paralogous gene families, which are readily identified along with true TE-related sequences. For organisms like rice with well-studied repetitive sequences, sequence similarity searches using pre-existing databases of known repetitive sequences can effectively detect the repetitive sequences in the genomic sequences. Programs such CENSOR, RepeatMasker and MaskerAid have been developed particularly for the repetitive sequence analysis (Jurka et al. 1996; Bedell et al. 2000). Other common DNA similarity search tools, such as BLAST, can also be used to identify this class of sequences.

Using RepeatMasker, the TIGR/MSU *Oryza* repeat database and release 6 of the TIGR/MSU rice pseudomolecules, ~37.2% of the rice genome was classified as repetitive (Ouyang and Buell 2004; Ouyang et al. 2007). TEs account for 91.6% of the entire length of the repetitive sequences identified in the rice genome. While retrotransposons dominate in the total number of bases (56.7% of total length of all TEs), miniature inverted-repeat transposable elements (MITEs) are the most abundant in absolute numbers (115,673 elements). Transposons were found to comprise 19.2% of the total of all TEs by size. Other repetitive sequence types, including centromeric repetitive sequences, telomeric repetitive sequences, and rDNAs were also identified.

3.4 Caveat Utilitor

The functional annotation for a gene is provided as a one-sentence description of a gene, and it is often considered to be the summation of the annotation for a gene. While this is generally true, it is easy to be lulled into a false sense of confidence about the completeness of the functional annotations that are provided by the rice annotation projects. Users of these data must understand how the functional annotations are generated and also the weaknesses of the annotation process. The one line description that is provided as the functional annotation of a gene is only one piece of information that can be used to more fully understand the role of the gene. Whether functional annotation is performed in an automated fashion or by manual curation, it will never be completely satisfactory. It is very difficult to write a program that will pick the best description for a gene. Even with manual curation of functional annotation, a gene may have multiple roles, but only one is likely to be used for the functional description of the gene. A few examples of the limitations of functional annotation are illustrative.

The gene names *LOC_Os01g23380* and *Os01g0335700* are synonyms for a rice gene that were identified by the TIGR/MSU and RAP annotation projects, respectively. The functional annotation for *LOC_Os01g 23380* is 'stripe rust resistance protein, Yr10, putative, expressed'. The description of *Os01g0335700* is 'Disease resistance protein family protein'. The gene models for the two annotation projects differ only in that the 5' UTR of *LOC_Os01g23380* is slightly longer than the 5' UTR of *Os01g0335700*. The ORFs for both gene models are identical. The functional annotations associated with this gene are reasonably descriptive. Both indicate that the gene is involved with disease resistance. The RAP annotation is less specific than the TIGR/MSU annotation which indicates that this gene offers resistance to stripe rust and that, in fact, this is the *Yr10* resistance gene. The TIGR/MSU annotation is based on sequence similarity between *LOC_Os01g23380* and *Yr10* from wheat (GenBank accession AAG42167). However, searching the TIGR/MSU browser with the term *Yr10* yields 27 genes on seven different chromosomes each with the name 'stripe rust resistance protein, Yr10'. Clearly, all of these genes cannot be orthologous to the wheat *Yr10* gene. At best, these genes are simply homologs to wheat *Yr10* which happens to be a stripe rust resistance gene. While less descriptive, perhaps the RAP description of 'Disease resistance protein family protein' is better because it is less likely to be misleading. However, if more information is needed about the mechanism of action of *LOC_Os01g23380*, examination of the published work about *Yr10* might be a good place to start. More general information about this gene is already available. On the TIGR/MSU gene report page for *LOC_Os01g23380*, the results of peptide sequence similarity to known PFAM domains show that this gene contains an NB-ARC domain (PFAM accession PF00931). The RAP gene report page for *Os01g0335700* indicates that this gene has amino acid sequence similarity to NB-ARC and 'disease resistance protein' InterPro domains (InterPro accessions IPR002182 and IPR000767). The PFAM and InterPro entries for these domains indicate that these domains have ATP-binding activity and contain leucine rich repeats which are often responsible for mediating protein-protein interactions. All of this information together provides a better understanding of the possible function of the rice *LOC_Os01g23380/Os01g0335700* gene. A better description of this gene could be a 'NB-ARC domain containing disease resistance gene similar to the wheat *Yr10* stripe rust resistance gene'.

The rice locus *LOC_Os03g16210* from the TIGR/MSU gene set is annotated to be a putative tropinone reductase. Tropinone is an alkaloid and a precursor to a number of additional plant alkaloids. The gene report page for *LOC_Os03g16210* shows that this gene contains three PFAM domains: oxidoreductase, short chain dehydrogenase/reductase family; NAD-binding domain 4 and NAD dependent epimerase/ dehydratase family. The reported BLASTP analysis indicates that there is similarity to other genes that are described as tropinone reductases and genes that are more generally referred to as oxidoreductases or short chain alcohol dehydrogenases, which is consistent with the PFAM domain similarity. While tropinone reductase is a more specific and perhaps more desirable description than oxidoreductase or short chain dehydrogenase, all of the tropinone BLASTP results in the gene report page are labeled as 'putative'. A little more investigation seems warranted. By using BLASTP to align the predicted amino acid sequence of *LOC_Os03g16210* to the Genbank non-redundant protein database, one plant protein with the accession number CA02390 from *Cochlearia officinalis* is found with high similarity to *LOC_Os03g16210*, and it is described as a non-putative tropinone reductase. This protein happens to have been used in the analysis of tropinone reductase activity (Brock et al. 2008). While the *C. officinalis* gene is an active tropinone reductase, the *Arabidopsis thaliana* putative tropinone reductase was not active. The active site of the *C. officinalis* protein was identified to contain a tyrosine that was not present in the *A. thaliana* protein. The homologous tyrosine is also not found in the predicted protein sequence from *LOC_Os03g 16210*. While the *A. thaliana* protein was not active in reducing tropinone, the same enzyme was able to reduce related classes of molecules: 4-methylcyclohexaone, N-propylpiperidin-4-one, N-methypiperidin-4-one, quinuclidinone and 8-thiabicycl[3.2.1]octan-3-one. This examination of the *A. thaliana* putative tropinone reductase gene also brings into question the likelihood of the rice *LOC_Os03g16210* gene producing an active tropinone reductase gene. Brock et al. (2008) performed a phylogenetic analysis of a number of putative tropinone reductases which shows that the putative tropinone reductases tend to be more similar to genes from the same species than to homologous genes from other species. They also suggest that the *A. thaliana* gene be described as a short chain dehydrogenase/reductase with low specificity. This is probably better functional annotation for *LOC_Os03g16210* as well.

The *LOC_Os03g16210* locus from the TIGR/MSU gene set will be used for one final example of the care with which a user of public annotation resources should approach these data. Immediately adjacent to *LOC_Os03g16210* are the loci *LOC_Os03g16220* and *LOC_Os03g16230*. These adjacent genes are also annotated as putative tropinone reductases. However, like *LOC_Os03g16210*, the predicted protein sequence from neither of these other putative tropinone reductases contains the tryptophan residue that Brock et al. (2008) found to be necessary for tropinone reductase activity. Both *LOC_Os03g16220* and *LOC_Os03g16230* are both better described as short chain dehydrogenase/reductase with low specificity. At this location of the IRGSP version 4 pseudomolecules, the RAP version 2 annotation shows a single 13.9 kb locus, *Os03g0268900*, instead of three separate loci. The RAP transcript map shows numerous transcripts that align to this locus. Most of the transcripts fall into three separate regions that correspond to the three TIGR/MSU loci, but there are three long transcripts that each spans almost half of the locus. The rules for RAP locus identification indicate that overlapping transcripts should be grouped into a single locus, and generally, this leads to a proper grouping of transcripts. However, in this case, it is probably more biologically sensible to consider these transcripts to come from three separate loci with the occasional aberrant transcript that is derived from more than one locus. In fact, in version 1 of the RAP annotation, three loci were identified at the current location of *Os03g0268900*. These three loci likely arose from two tandem duplication events. The high sequence similarity between the loci may make transcription across multiple loci more common than usual, but such unusual examples of transcription are not difficult to find. It is a good practice to compare between annotation resources, and to reexamine all available evidence for gene structure.

4 GENOME ORGANIZATION

Once the rice genome was sequenced, researchers began to examine the organization of the genome, but genome structure has been studied using older techniques. Mapping studies have shown that all the cereal genomes feature a high degree of conserved colinearity that has been disrupted to some degree by genome rearrangements (Gale and Devos 1998). Early molecular mapping work had suggested that genome duplication events have occurred in rice. Nagamura et al. (1995) found that a subset of their RFLP mapping probes displayed multiple bands in Southern blots, and this indicates that the probes hybridized to multiple distinct regions of genomic DNA. A subset of these probes mapped to the ends of both chromosome 11 and chromosome 12, and the order of these probes was consistent between each chromosome. Nagamura et al. (1995) concluded that the ends of chromosomes 11 and 12 were involved in an ancient duplication event.

Copying of a large region of the genome that contains multiple genes from one region to another is referred to as segmental duplication. With the finished rice genome sequences, a number of studies have extended segmental duplication research using more comprehensive techniques. Analyses of genome reorganization have been performed on both the *indica* and *japonica* sequences (Guyot and Keller 2004; Wang et al. 2005; Yu et al. 2005). In each case, all vs. all BLAST alignments of either protein or nucleotide coding sequences were performed using all non repeat-related genes in the genome. Regions of the genome that have a high proportion of genes that have significant similarity to another region of the genome are identified. Depending on the precise technique used, between 47% and 65% of the rice genome has been found to be duplicated with a small portion of the genome having been duplicated more than once. The regions of duplication are relatively few in number. Using synonymous substitution rates, the majority of the rice genome duplication was estimated to have occurred before the radiation of the Poaceae more than 50 to 70 million years ago (Wang et al. 2005; Yu et al. 2005). The small number of duplicated regions and the common age of those duplications indicate that the current rice genome is the product of an ancient polyploidy event. However, based on synonymous substitution rates, there was one duplication between chromosomes 11 and 12 that is only about 5 to 21 million years old (Wang et al. 2005; Yu et al. 2005).

While large-scale segmental duplications are rare and have apparently not occurred recently in rice, small scale genome duplications that involve tandemly repeated genes that are copied within a short distance of each other on the same chromosome are more common. The IRGSP sequencing paper found 29% of all genes in *japonica* were members of tandem duplication events (International Rice Genome Sequencing Project 2005), but only 16% of the genes in *indica* were found to be tandemly duplicated (Yu et al. 2005). Synonymous substitution rates are much lower in tandemly duplicated genes than in segmentally duplicated genes and this supports the notion that tandem duplication events occur more frequently and more recently than segmental duplications. On the other hand, non-synonymous to synonymous substitution ratios are higher in tandemly duplicated genes than in genes involved in segmental duplication. This is interpreted as indicating that after duplication one gene often endures a period of rapid evolution (Yu et al. 2005).

The combined results of segmental and tandem duplications have a significant effect on the gene content of rice. The gene pairs that result after such duplications are paralogous. If there are multiple duplication events, a large paralogous gene family can form. The members of paralogous gene families can have several fates. They can potentially become pseudogenes if their functionality does not become essential to the plant. If the coding sequence of the gene changes drastically, the gene may no longer be a member of the paralogous family. Alternatively, the functionality of the coding sequence may be unchanged, but the regulation and expression of the gene may evolve away from the ancestral form. Lin et al. (2008) have examined the functional paralogous gene families of rice. In order to ensure that only paralogous gene families that could reasonably be expected retain similar function were examined, a strict definition of paralogous gene families was used. All genes were examined for the presence of Pfam domains, and additional non-Pfam domains were identified by using all-vs-all BLASTP alignments. Genes that retained identical domain compositions (identical domain type, number and order) were grouped into paralogous gene families. Analysis of the GO annotations for the members of these paralogous gene families found that they are enriched for GO terms related to transcription regulation, nucleotide binding, transporter activity, kinase activity, protein binding and receptor activity. These are all terms that are related to the regulation of cellular activities. Lin et al. (2008) also found that paralogous family genes are slightly more likely to have alternative splice forms compared to non-paralogous genes (29% vs. 24%). Analysis of the age of paralogous gene families indicated that paralogous gene families are constantly forming but that there is a large set of older paralogous families that probably corresponds to the large-scale duplication event that is believed to have occurred around 70 million years ago (Lin et al. 2008). Expression data from 18 MPSS libraries (Nobuta et al. 2007) were used to analyze the expression patterns of the paralogous gene families where only two members had corresponding MPSS data. Only 674 families with qualifying MPSS data were identified, but interestingly, 280 of these families had divergent tissue-specific expression. Lin et al. (2008) interpreted these results to indicate that the members of these paralogous families had begun to undergo divergent functionalization where the current members of a paralogous family have specialized in the various functional roles that had formerly been performed by an ancestral gene.

A dramatic example of subfunctionalization can be seen in the *OSMADS3* and *OSMADS58* genes. These two paralogous genes are members of the C-class MADS box gene family and are responsible for the regulation of floral organ formation. It is believed that *OSMADS3* and *OSMADS58* are derived from a duplication event that involved an *AGAMOUS*-like ancestral gene. The *A. thaliana AGAMOUS* gene is responsible for floral organ development and floral meristem determinancy (Bowman et al. 1991), but the *OSMADS3* and *OSMADS58* genes have split these functions (Yamaguchi et al. 2006). *OSMADS3* T-DNA insertion mutants have flowers with ectopic development of floral organs in inappropriate whorls. *OSMADS58* disrupted rice show indeterminant growth of floral structures. So it appears that the rice *OSMADS3* and *OSMADS58* genes play the role of *AGAMOUS* in *A. thaliana*.

5 GENE STRUCTURE

Genome-wide analysis of gene structure is only possible with a fully sequenced genome or with a large number of large-scale genomic clones. Analysis of the TIGR/MSU annotation of the TIGR/MSU rice pseudomolecules found that the average gene has a length of ~2.5 kbp, an average of ~4.2 exons per gene, an average exon length of ~300 bp, and an average intron length of ~370 bp. The percent GC content of intergenic, exonic and intronic regions was found to be 41%, 54% and 39%, respectively, and the gene density within the genome was calculated to be 5.7 kbp per gene (Yuan et al. 2005). These gene structure statistics are very similar to those found by Yu et al. (2005) in the BGI *indica*, Syngenta *japonica* and IRGSP *japonica* pseudomolecule assemblies.

Another important aspect of gene structure is the fact that many rice genes have alternative splice forms. Intron removal from pre-mRNAs by the spliceosome is mediated by donor and acceptor sequences at the 5′ and 3′ ends of introns. However, alternative donor or acceptor sites can be used during the processing of pre-mRNAs, and alternatively spliced mature mRNAs can code for dramatically different proteins (Black 2003). Alternative splicing can be an important mechanism for regulating gene function, for creating a wider diversity of protein sequences and for transcriptional regulation of genes. Numerous classes of alternative splicing have been characterized in rice (Campbell et al. 2006; Wang and Brendel 2006; Chen et al. 2007). The classes involve the extension or truncation of introns on either the 5′ or 3′ ends, the omission or inclusion of whole exons and the skipping or inclusion of intron excision. Alternative splicing in rice may or may not also result in gross changes in the phase of the open reading frame of a gene. Wang and Brendel (2006) found 6,568 rice genes with evidence for multiple splice variants. Campbell et al. (2006) were able to identify 8,772 rice genes with variation in transcript splicing. In both of these studies, instances involving the skipping or inclusion of intron excision composed the majority of alternative splicing variation. The next most common types of alternative splicing found in these two studies were events that utilized alternative intron donor or intron acceptor sites. The majority of splicing variations that resulted from alternative donor or alternative acceptor events resulted in frame-shift changes in coding sequences. Approximately one third of alternative splicing events result in a premature termination codon and are believed to lead to nonsense-mediated transcript degradation (Lewis et al. 2003; Campbell et al. 2006; Wang and Brendel 2006; Chen et al. 2007). Comparisons of orthologous genes in rice and *A. thaliana* have provided numerous examples where alternative splicing has been conserved which indicates that the mechanism of regulation of alternative splicing, the variation produced by alternative splicing and the regulation provided by alternative splicing are not merely recent trials at producing gene structural and regulatory diversity (Campbell et al. 2006; Wang and Brendel 2006; Chen et al. 2007). At least in some instances, the utility of alternative splicing is ancient and has distinct adaptive advantages that have caused it to be maintained since the divergence of monocots and dicots.

Alternative splicing is also believed to be involved in more complex gene regulation. Chen et al. (2007) have identified 1,378 rice genes with alternatively spliced variants that differ in the first exon, and 90% of these were cases where the first exon of one splice variant was an internal exon of another splice variant. In the remaining cases, the first exons of the splice variants from a gene were mutually exclusive. A number of distinct classes were observed in these splice variants. Many cases of novel N-terminal amino acid sequences were predicted, and some of these cases are suspected to result in differential subcellular targeting of the alternative proteins. In other cases, novel N-terminal sequences were shown to result in the addition or omission of recognized domains to the encoded proteins. Another class of alternative splicing resulted from different first exons that are very distant from each other in the genome, and this suggested that each alternative splice form of such genes were regulated by separate promoter sequences.

6 RICE CENTROMERES

The centromeres of rice serve as a model for eukaryotic centromere structure and function. The centromeres of *Saccharomyces cerevisiae* are only 125 bp in length, but the centromeres that have been examined from most other eukaryotes are more complex (Clarke 1998; Schueler et al. 2001; Cheng et al. 2002; Sun et al. 2003; Lamb et al. 2007; Ma et al. 2007a). At the moment, for higher eukaryotes, the only examples of completely sequenced centromeres are from rice (Nagaki et al. 2004; Wu et al. 2004; Zhang et al. 2004).

Before the genome sequencing projects had provided large numbers of rice genomic BAC and PAC clones, (Dong et al. 1998) had identified and partially sequenced a single BAC clone that was shown to hybridize with centromeric regions of all rice chromosomes. The sequence from this centromere associated clone was found to contain seven families of repetitive sequences. Six of these families were shown to be present in the genome in fewer than a few hundred copies, but the seventh repeat is present in very high numbers. This high copy repeat was identified as a member of the RCS2 repeat family, and it consists of a 155-165 bp sequence that is arranged head-to-tail in long tandem repeats. The RCS2 family repeat found in *Oryza sativa* was eventually named CentO. This repeat, but not the other lower copy repeats, hybridized to the chromosomes of several additional members of the *Oryza* genus.

The CentO repeats in the centromeres of rice chromosomes 4 and 8 have been estimated to be ~182 and ~64 kbp in length, respectively, but other rice centromeres are estimated to contain CentO arrays that are as much as 1.9 Mbp in length (Cheng et al. 2002). Sequencing of the centromeres from chromosomes 4 and 8 has shown that the core of these centromeres contains multiple long arrays of CentO repeats, and these arrays are interrupted by other centromere-related repetitive sequences. With the sequencing of these centromeres, it was discovered that despite the sequence variation that was observed with the original description of the 155 bp CentO repeat, there is little variation within individual arrays of the CentO repeat (Dong et al. 1998; Nagaki et al. 2004; Wu et al. 2004; Zhang et al. 2004). However, the chromosome 4 centromere does contain a 165 bp variation of the basic 155 bp repeat unit, and this variant is found mixed with the 155 bp repeat within the same CentO arrays (Zhang et al. 2004). Several *Oryza* species that do not contain CentO, as well as pearl millet and maize which diverged from rice several million years ago, contain CentO-like repeats that have an 80 bp region that is identical to a portion of the CentO repeat. Other *Oryza* species have CentO-like repeats that have no sequence similarity to CentO from rice (Lee et al. 2005). So this key feature of centromeres can be rather variable even within a set of closely related species. Other repetitive features are enriched within rice centromeres, and at least one class of retroelement is preferentially found within centromeres (Zhang et al. 2004; Yan et al. 2006).

While the CentO repeats seem likely to have a central role in the structure of rice centromeres, these repeat features do not functionally define the centromere. The centromere associated histone, CenH3, has been shown to be necessary for kinetochore formation (Choo 2001). Chromatin immuno-precipitation experiments involving CenH3 antibodies have shown that CenH3 binds to a region that surrounds some but not all of the CentO arrays of the chromosome 8 centromere (Nagaki et al. 2004). The CenH3 binding region of the chromosome 3 centromere appears to more fully encompass all CentO sequences and it is apparently devoid of recombination (Yan et al. 2006).

Despite the notion of centromeres being heterochromatic, active gene transcription has been observed within the functional centromeres of chromosomes 3 and 8, and this has been suggested to indicate that these centromeres have arisen from euchromatic regions (Nagaki et al. 2004; Yan et al. 2006). It is believed that new centromeres can arise within the chromosome by neocentromere activation (Ventura et al. 2001). The process of neocentromere formation apparently does not affect gene transcription at the site of the neocentromere (Wong et al. 2006). Yan and Jiang (2007) have suggested that the presence of euchromatic regions within the functionally defined centromeres of chromosomes 3 and 8 mark these centromeres as recently formed genome structures. However, Ma et al. (2007a) have compared rice centromere 8 with the orthologous centromere from *Oryza brachyantha*, and have found that there has been a rearrangement in this centromere within the last 7 to 9 million years. Interestingly, *O. brachyantha* does not contain a CentO-like repeat. The core centromere repeat in *O. brachyantha*, CentO-F, shares no similarity with CentO. Thus, if the rice chromosome 8 centromere is not a completely new centromere, it it possible for centromeres to undergo radical sequence evolution 'in place'.

7 POST-GENOME RESEARCH

The utility of having a completely sequenced genome for rice cannot be understated. There are many types of research that are greatly aided by the existence of or that could not be done without a complete genome.

However, the rice genome sequence alone would not be very useful had it not also been annotated. While it is possible for individual researchers to annotate portions of the genome that they find to be interesting, it is not convenient for most researchers to do this. Additionally, there are added benefits of annotation to which the entire rice research community can refer (Zhao et al. 2004; Ouyang et al. 2007; Tanaka et al. 2008).

One area of research that has become more efficient with the release of the rice genome sequence is map-based cloning of genes and quantitative trait loci (QTLs). Prior to the existence of the finished rice genome finding mapping markers that flank and are physically close to a gene or QTL was a prolonged process that involved numerous cycles of marker mapping, physical map extension and new marker development. At some point, the mapping resolution of the segregating population would be reached. If the physical distance was judged to be too large, it would then be necessary to increase the size of the mapping population. Eventually, the final mapping interval would be sequenced and analyzed for its gene content. With luck, a gene candidate would be found and the cause of the genetic difference between the mapping genotypes could be identified (Zeng et al. 2002, 2004; Hong et al. 2003; Komori et al. 2004). With the help of the rice genome sequence, this entire process can be drastically shortened. While it is still necessary to have a large mapping population, identifying markers for fine-mapping is much easier. The genome sequence can be directly searched for potential SSRs. Alternatively, primers can be designed to amplify sequence within the mapping interval, and the sequence of these regions within the mapping parents can be analyzed for small exploitable polymorphisms. Unlike marker discovery before the genome sequence was available, marker discovery can now be performed in parallel for an entire mapping interval. Physical map creation is also no longer necessary, and once a final mapping interval is obtained, the annotation databases can be consulted to identify candidate genes with their functional descriptions (Konishi et al. 2006; Li et al. 2006; Qu et al. 2006; Ma et al. 2007b).

References

Arumuganathan K, Earle E (1991) Nuclear DNA content of some important plant species. Plant Mol Biol Rep 9: 208-218

Bao Z, Eddy SR (2002) Automated de novo identification of repeat sequence families in sequenced genomes. Genome Res 12: 1269-1276

Barry GF (2001) The use of the Monsanto draft rice genome sequence in research. Plant Physiol 125: 1164-1165

Bedell JA, Korf I, Gish W (2000) MaskerAid: a performance enhancement to RepeatMasker. Bioinformatics 16: 1040-1041

Bennetzen JL, Coleman C, Liu R, Ma J, Ramakrishna W (2004) Consistent over-estimation of gene number in complex plant genomes. Curr Opin Plant Biol 7: 732-736

Black DL (2003) Mechanisms of alternative pre-messenger RNA splicing. Annu Rev Biochem 72: 291-336

Bowman JL, Drews GN, Meyerowitz EM (1991) Expression of the Arabidopsis floral homeotic gene AGAMOUS is restricted to specific cell types late in flower development. Plant Cell 3: 749-758

Brock A, Brandt W, Drager B (2008) The functional divergence of short-chain dehydrogenases involved in tropinone reduction. Plant J 54: 388-401

Burge C, Karlin S (1997) Prediction of complete gene structures in human genomic DNA. J Mol Biol 268: 78-94

Butler D, Pockley P (2000) Monsanto makes rice genome public. Nature 404: 534

Campbell MA, Haas BJ, Hamilton JP, Mount SM, Buell CR (2006) Comprehensive analysis of alternative splicing in rice and comparative analyses with Arabidopsis. BMC Genom 7:327

Chen FC, Wang SS, Chaw SM, Huang YT, Chuang TJ (2007) Plant Gene and Alternatively Spliced Variant Annotator. A plant genome annotation pipeline for rice gene and alternatively spliced variant identification with cross-species expressed sequence tag conservation from seven plant species. Plant Physiol 143: 1086-1095

Cheng Z, Dong F, Langdon T, Ouyang S, Buell CR, Gu M, Blattner FR, Jiang J (2002) Functional rice centromeres are marked by a satellite repeat and a centromere-specific retrotransposon. Plant Cell 14: 1691-1704

Childs N (2008) Rice Outlook. RCS-08g, USDA, Economic Research Service, July 14, 2008

Choo KH (2001) Domain organization at the centromere and neocentromere. Dev Cell 1: 165-177

Clarke L (1998) Centromeres: proteins, protein complexes, and repeated domains at centromeres of simple eukaryotes. Curr Opin Genet Dev 8: 212-218

Dalton R (2000) Cereal gene bank accepts need for patents. Nature 404:534

Dickson D, Cyranoski D (2001) Commercial sector scores success with whole rice genome. Nature 409: 551

Dong F, Miller JT, Jackson SA, Wang GL, Ronald PC, Jiang J (1998) Rice (Oryza sativa) centromeric regions consist of complex DNA. Proc Natl Acad Sci USA 95: 8135-8140

Edgar RC, Myers EW (2005) PILER: identification and classification of genomic repeats. Bioinformatics 21 (Suppl 1): 152-158

FAO (2006) FAO Statistical Yearbook 2005/2006. Food and Agricultural Organization, Rome, Italy

Finnegan DJ (1989) Eukaryotic transposable elements and genome evolution. Trends Genet 5: 103-107

Feng Q, Zhang Y, Hao P, Wang S, Fu G, Huang Y, Li Y, Zhu J, Liu Y, Hu X, Jia P, Zhao Q, Ying K, Yu S, Tang Y, Weng Q, Zhang L, Lu Y, Mu J, Zhang LS, Yu Z, Fan D, Liu X, Lu T, Li C, Wu Y, Sun T, Lei H, Li T, Hu H, Guan J, Wu M, Zhang R, Zhou B, Chen Z, Chen L, Jin Z, Wang R, Yin H, Cai Z, Ren S, Lv G, Gu W, Zhu G, Tu Y, Jia J, Chen J, Kang H, Chen X, Shao C, Sun Y, Hu Q, Zhang X, Zhang W, Wang L, Ding C, Sheng H, Gu J, Chen S, Ni L, Zhu F, Chen W, Lan L, Lai Y, Cheng Z, Gu M, Jiang J, Li J, Hong G, Xue Y, Han B (2002) Sequence and analysis of rice chromosome 4. Nature 420: 316-320

Gale MD, Devos KM (1998) Comparative genetics in the grasses. Proc Natl Acad Sci USA 95: 1971-1974

Goff SA, Ricke D, Lan TH, Presting G, Wang R, Dunn M, Glazebrook J, Sessions A, Oeller P, Varma H, Hadley D, Hutchison D, Martin C, Katagiri F, Lange BM, Moughamer T, Xia Y, Budworth P, Zhong J, Miguel T, Paszkowski U, Zhang S, Colbert M, Sun WL, Chen L, Cooper B, Park S, Wood TC, Mao L, Quail P, Wing R, Dean R, Yu Y, Zharkikh A, Shen R, Sahasrabudhe S, Thomas A, Cannings R, Gutin A, Pruss D, Reid J, Tavtigian S, Mitchell J, Eldredge G, Scholl T, Miller RM, Bhatnagar S, Adey N, Rubano T, Tusneem N, Robinson R, Feldhaus J, Macalma T, Oliphant A, Briggs S (2002) A draft sequence of the rice genome (*Oryza sativa* L. ssp. *japonica*). Science 296: 92-100

Gilland B (2002) World population and food supply. Can food production keep pace with population growth in the next half century. Food Pol 27: 47-63

Guyot R, Keller B (2004) Ancestral genome duplication in rice. Genome 47: 610-614

Haas BJ, Salzberg SL (2007) Finding repeats in genome sequences. In: Lengauer,T (ed) Bioinformatics - From Genomes to Therapies. Vol 1. Wiley, Weinheim, Germany, pp 197-233

Haas BJ, Delcher AL, Mount SM, Wortman JR, Smith RK, Jr., Hannick LI, Maiti R, Ronning CM, Rusch DB, Town CD, Salzberg SL, White O (2003) Improving the Arabidopsis genome annotation using maximal transcript alignment assemblies. Nucl Acids Res 31: 5654-5666

Hong Z, Ueguchi-Tanaka M, Umemura K, Uozu S, Fujioka S, Takatsuto S, Yoshida S, Ashikari M, Kitano H, Matsuoka M (2003) A rice brassinosteroid-deficient mutant, ebisu dwarf (d2), is caused by a loss of function of a new member of cytochrome P450. Plant Cell 15: 2900-2910

Hubbard TJ, Aken BL, Beal K, Ballester B, Caccamo M, Chen Y, Clarke L, Coates G, Cunningham F, Cutts T, Down T, Dyer SC, Fitzgerald S, Fernandez-Banet J, Graf S, Haider S, Hammond M, Herrero J, Holland R, Howe K, Johnson N, Kahari A, Keefe D, Kokocinski F, Kulesha E, Lawson D, Longden I, Melsopp C, Megy K, Meidl P, Ouverdin B, Parker A, Prlic A, Rice S, Rios D, Schuster M, Sealy I, Severin J, Slater G, Smedley D, Spudich

G, Trevanion S, Vilella A, Vogel J, White S, Wood M, Cox T, Curwen V, Durbin R, Fernandez-Suarez XM, Flicek P, Kasprzyk A, Proctor G, Searle S, Smith J, Ureta-Vidal A, Birney E (2007) Ensembl 2007. Nucl Acids Res 35: D610-617

International Rice Genome Sequencing Project (2005) The map-based sequence of the rice genome. Nature 436: 793-800

Itoh T, Tanaka T, Barrero RA, Yamasaki C, Fujii Y, Hilton PB, Antonio BA, Aono H, Apweiler R, Bruskiewich R, Bureau T, Burr F, Costa de Oliveira A, Fuks G, Habara T, Haberer G, Han B, Harada E, Hiraki AT, Hirochika H, Hoen D, Hokari H, Hosokawa S, Hsing YI, Ikawa H, Ikeo K, Imanishi T, Ito Y, Jaiswal P, Kanno M, Kawahara Y, Kawamura T, Kawashima H, Khurana JP, Kikuchi S, Komatsu S, Koyanagi KO, Kubooka H, Lieberherr D, Lin YC, Lonsdale D, Matsumoto T, Matsuya A, McCombie WR, Messing J, Miyao A, Mulder N, Nagamura Y, Nam J, Namiki N, Numa H, Nurimoto S, O'Donovan C, Ohyanagi H, Okido T, Oota S, Osato N, Palmer LE, Quetier F, Raghuvanshi S, Saichi N, Sakai H, Sakai Y, Sakata K, Sakurai T, Sato F, Sato Y, Schoof H, Seki M, Shibata M, Shimizu Y, Shinozaki K, Shinso Y, Singh NK, Smith-White B, Takeda J, Tanino M, Tatusova T, Thongjuea S, Todokoro F, Tsugane M, Tyagi AK, Vanavichit A, Wang A, Wing RA, Yamaguchi K, Yamamoto M, Yamamoto N, Yu Y, Zhang H, Zhao Q, Higo K, Burr B, Gojobori T, Sasaki T (2007) Curated genome annotation of Oryza sativa ssp. japonica and comparative genome analysis with Arabidopsis thaliana. Genome Res 17: 175-183

Jaiswal P, Ni J, Yap I, Ware D, Spooner W, Youens-Clark K, Ren L, Liang C, Zhao W, Ratnapu K, Faga B, Canaran P, Fogleman M, Hebbard C, Avraham S, Schmidt S, Casstevens TM, Buckler ES, Stein L, McCouch S (2006) Gramene: a bird's eye view of cereal genomes. Nucl Acids Res 34: D717-723

Jurka J, Klonowski P, Dagman V, Pelton P (1996) CENSOR—a program for identification and elimination of repetitive elements from DNA sequences. Comput Chem 20: 119-121

Khush GS (1997) Origin, dispersal, cultivation and variation of rice. Plant Mol Biol 35: 25-34

Kikuchi S, Satoh K, Nagata T, Kawagashira N, Doi K, Kishimoto N, Yazaki J, Ishikawa M, Yamada H, Ooka H, Hotta I, Kojima K, Namiki T, Ohneda E, Yahagi W, Suzuki K, Li CJ, Ohtsuki K, Shishiki T, Otomo Y, Murakami K, Iida Y, Sugano S, Fujimura T, Suzuki Y, Tsunoda Y, Kurosaki T, Kodama T, Masuda H, Kobayashi M, Xie Q, Lu M, Narikawa R, Sugiyama A, Mizuno K, Yokomizo S, Niikura J, Ikeda R, Ishibiki J, Kawamata M, Yoshimura A, Miura J, Kusumegi T, Oka M, Ryu R, Ueda M, Matsubara K, Kawai J, Carninci P, Adachi J, Aizawa K, Arakawa T, Fukuda S, Hara A, Hashizume W, Hayatsu N, Imotani K, Ishii Y, Itoh M, Kagawa I, Kondo S, Konno H, Miyazaki A, Osato N, Ota Y, Saito R, Sasaki D, Sato K, Shibata K, Shinagawa A, Shiraki T, Yoshino M, Hayashizaki Y, Yasunishi A (2003) Collection, mapping, and annotation of over 28,000 cDNA clones from japonica rice. Science 301: 376-379

Komori T, Ohta S, Murai N, Takakura Y, Kuraya Y, Suzuki S, Hiei Y, Imaseki H, Nitta N (2004) Map-based cloning of a fertility restorer gene, *Rf-1*, in rice (*Oryza sativa* L.). Plant J 37: 315-325

Konishi S, Izawa T, Lin SY, Ebana K, Fukuta Y, Sasaki T, Yano M (2006) An SNP caused loss of seed shattering during rice domestication. Science 312: 1392-1396

Kumar A, Bennetzen JL (1999) Plant retrotransposons. Annu Rev Genet 33: 479-532

Kurtz S, Choudhuri JV, Ohlebusch E, Schleiermacher C, Stoye J, Giegerich R (2001) REPuter: the manifold applications of repeat analysis on a genomic scale. Nucl Acids Res 29: 4633-4642

Lagos-Quintana M, Rauhut R, Lendeckel W, Tuschl T (2001) Identification of novel genes coding for small expressed RNAs. Science 294: 853-858

Lamb JC, Yu W, Han F, Birchler JA (2007) Plant chromosomes from end to end: telomeres, heterochromatin and centromeres. Curr Opin Plant Biol 10: 116-122

Lau NC, Lim LP, Weinstein EG, Bartel DP (2001) An abundant class of tiny RNAs with probable regulatory roles in Caenorhabditis elegans. Science 294: 858-862

Lee HR, Zhang W, Langdon T, Jin W, Yan H, Cheng Z, Jiang J (2005) Chromatin immunoprecipitation cloning reveals rapid evolutionary patterns of centromeric DNA in *Oryza* species. Proc Natl Acad Sci USA 102: 11793-11798

Lee RC, Ambros V (2001) An extensive class of small RNAs in Caenorhabditis elegans. Science 294: 862-864

Lewis BP, Green RE, Brenner SE (2003) Evidence for the widespread coupling of alternative splicing and nonsense-mediated mRNA decay in humans. Proc Natl Acad Sci USA 100: 189-192

Li C, Zhou A, Sang T (2006) Rice domestication by reducing shattering. Science 311: 1936-1939

Liang C, Jaiswal P, Hebbard C, Avraham S, Buckler ES, Casstevens T, Hurwitz B, McCouch S, Ni J, Pujar A, Ravenscroft D, Ren L, Spooner W, Tecle I, Thomason J, Tung CW, Wei X, Yap I, Youens-Clark K, Ware D, Stein L (2008) Gramene: a growing plant comparative genomics resource. Nucl Acids Res 36: D947-953

Lin H, Ouyang S, Egan A, Nobuta K, Haas BJ, Zhu W, Gu X, Silva JC, Meyers BC, Buell CR (2008) Characterization of paralogous protein families in rice. BMC Plant Biol 8:18

Ma J, Wing RA, Bennetzen JL, Jackson SA (2007a) Plant centromere organization: a dynamic structure with conserved functions. Trends Genet 23: 134-139

Ma JF, Yamaji N, Mitani N, Tamai K, Konishi S, Fujiwara T, Katsuhara M, Yano M (2007b) An efflux transporter of silicon in rice. Nature 448: 209-212

McDermott J, Samudrala R (2003) Bioverse: Functional, structural and contextual annotation of proteins and proteomes. Nucl Acids Res 31: 3736-3737

Milanesi L, D'Angelo D, Rogozin IB (1999) GeneBuilder: interactive *in silico* prediction of gene structure. Bioinformatics 15: 612-621

Nagaki K, Cheng Z, Ouyang S, Talbert PB, Kim M, Jones KM, Henikoff S, Buell CR, Jiang J (2004) Sequencing of a rice centromere uncovers active genes. Nat Genet 36: 138-145

Nagamura Y, Inoue T, Antonio BA, Shimano T, Kajiya H, Shomura A, Lin SY, Kuboki Y, Harushima Y, Kurata N, Minobe Y, Yano M, Sasaki T (1995) Conservation of duplicated segments between rice chromosomes 11 and 12. Breed Sci 45: 373-376

Naylor RL, Battisti DS, Vimont DJ, Falcon WP, Burke MB (2007) Assessing risks of climate variability and climate change for Indonesian rice agriculture. Proc Natl Acad Sci USA 104: 7752-7757

Nobuta K, Venu RC, Lu C, Belo A, Vemaraju K, Kulkarni K, Wang W, Pillay M, Green PJ, Wang GL, Meyers BC (2007) An expression atlas of rice mRNAs and small RNAs. Nat Biotechnol 25: 473-477

Ohyanagi H, Tanaka T, Sakai H, Shigemoto Y, Yamaguchi K, Habara T, Fujii Y, Antonio BA, Nagamura Y, Imanishi T, Ikeo K, Itoh T, Gojobori T, Sasaki T (2006) The Rice Annotation Project Database (RAP-DB): hub for *Oryza sativa* ssp. *japonica* genome information. Nucl Acids Res 34: D741-744

Ouyang S, Buell CR (2004) The TIGR Plant Repeat Databases: a collective resource for the identification of repetitive sequences in plants. Nucl Acids Res 32: D360-363

Ouyang S, Zhu W, Hamilton J, Lin H, Campbell M, Childs K, Thibaud-Nissen F, Malek RL, Lee Y, Zheng L, Orvis J, Haas B, Wortman J, Buell CR (2007) The TIGR Rice Genome Annotation Resource: improvements and new features. Nucl Acids Res 35: D883-887

Parsons JD (1995) Miropeats: graphical DNA sequence comparisons. Comput Appl Biosci 11: 615-619

Peng JH, Zadeh H, Lazo GR, Gustafson JP, Chao S, Anderson OD, Qi LL, Echalier B, Gill BS, Dilbirligi M, Sandhu D, Gill KS, Greene RA, Sorrells ME, Akhunov ED, Dvorak J, Linkiewicz AM, Dubcovsky J, Hossain KG, Kalavacharla V, Kianian SF, Mahmoud AA, Miftahudin, Conley EJ, Anderson JA, Pathan MS, Nguyen HT, McGuire PE, Qualset CO, Lapitan NL (2004) Chromosome bin map of expressed sequence tags in homoeologous group 1 of hexaploid wheat and homoeology with rice and Arabidopsis. Genetics 168: 609-623

Price AL, Jones NC, Pevzner PA (2005) De novo identification of repeat families in large genomes. Bioinformatics 21 (Suppl 1): 351-358

Qu S, Liu G, Zhou B, Bellizzi M, Zeng L, Dai L, Han B, Wang GL (2006) The broad-spectrum blast resistance gene Pi9 encodes a nucleotide-binding site-leucine-rich repeat protein and is a member of a multigene family in rice. Genetics 172: 1901-1914

Rice Chromosome 10 Sequencing Consortium (2003) In-depth view of structure, activity, and evolution of rice chromosome 10. Science 300: 1566-1569

Sakata K, Nagamura Y, Numa H, Antonio BA, Nagasaki H, Idonuma A, Watanabe W, Shimizu Y, Horiuchi I, Matsumoto T, Sasaki T, Higo K (2002) RiceGAAS: an automated annotation system and database for rice genome sequence. Nucl Acids Res 30: 98-102

Salamov AA, Solovyev VV (2000) *Ab initio* gene finding in Drosophila genomic DNA. Genom Res 10: 516-522

Sasaki T, Burr B (2000) International Rice Genome Sequencing Project: the effort to completely sequence the rice genome. Curr Opin Plant Biol 3: 138-141

Schueler MG, Higgins AW, Rudd MK, Gustashaw K, Willard HF (2001) Genomic and genetic definition of a functional human centromere. Science 294: 109-115

Stalker J, Gibbins B, Meidl P, Smith J, Spooner W, Hotz HR, Cox AV (2004) The Ensembl Web site: mechanics of a genome browser. Genom Res 14: 951-955

Stein LD, Mungall C, Shu S, Caudy M, Mangone M, Day A, Nickerson E, Stajich JE, Harris TW, Arva A, Lewis S (2002) The generic genome browser: a building block for a model organism system database. Genom Res 12: 1599-1610

Sun X, Le HD, Wahlstrom JM, Karpen GH (2003) Sequence analysis of a functional Drosophila centromere. Genom Res 13: 182-194

Tanaka T, Antonio BA, Kikuchi S, Matsumoto T, Nagamura Y, Numa H, Sakai H, Wu J, Itoh T, Sasaki T, Aono R, Fujii Y, Habara T, Harada E, Kanno M, Kawahara Y, Kawashima H, Kubooka H, Matsuya A, Nakaoka H, Saichi N, Sanbonmatsu R, Sato Y, Shinso Y, Suzuki M, Takeda J, Tanino M, Todokoro F, Yamaguchi K, Yamamoto N, Yamasaki C, Imanishi T, Okido T, Tada M, Ikeo K, Tateno Y, Gojobori T, Lin YC, Wei FJ, Hsing YI, Zhao Q, Han B, Kramer MR, McCombie RW, Lonsdale D, O'Donovan CC, Whitfield EJ, Apweiler R, Koyanagi KO, Khurana JP, Raghuvanshi S, Singh NK, Tyagi AK, Haberer G, Fujisawa M, Hosokawa S, Ito Y, Ikawa H, Shibata M, Yamamoto M, Bruskiewich RM, Hoen DR, Bureau TE, Namiki N, Ohyanagi H, Sakai Y, Nobushima S, Sakata K, Barrero RA, Souvorov A, Smith-White B, Tatusova T, An S, An G, Satoshi OO, Fuks G, Messing J, Christie KR, Lieberherr D, Kim H, Zuccolo A, Wing RA, Nobuta K, Green PJ, Lu C, Meyers BC, Chaparro C, Piegu B, Panaud O, Echeverria M (2008) The Rice Annotation Project Database (RAP-DB): 2008 update. Nucl Acids Res 36: D1028-1033

Thibaud-Nissen F, Campbell M, Hamilton JP, Zhu W, Buell CR (2007) EuCAP, a Eukaryotic Community Annotation Package, and its application to the rice genome. BMC Genom 8:388

Ventura M, Archidiacono N, Rocchi M (2001) Centromere emergence in evolution. Genom Res 11: 595-599

Volff JN (2006) Turning junk into gold: domestication of transposable elements and the creation of new genes in eukaryotes. Bioessays 28: 913-922

Volfovsky N, Haas BJ, Salzberg SL (2001) A clustering method for repeat analysis in DNA sequences. Genom Biol 2:RESEARCH0027

Wang BB, Brendel V (2006) Genomewide comparative analysis of alternative splicing in plants. Proc Natl Acad Sci USA 103: 7175-7180

Wang X, Shi X, Hao B, Ge S, Luo J (2005) Duplication and DNA segmental loss in the rice genome: implications for diploidization. New Phytol 165: 937-946

Weinstock GM (2007) ENCODE: more genomic empowerment. Genom Res 17: 667-668

Wicker T, Sabot F, Hua-Van A, Bennetzen JL, Capy P, Chalhoub B, Flavell A, Leroy P, Morgante M, Panaud O, Paux E, SanMiguel P, Schulman AH (2007) A unified classification system for eukaryotic transposable elements. Nat Rev Genet 8: 973-982

Wing RA, Ammiraju JS, Luo M, Kim H, Yu Y, Kudrna D, Goicoechea JL, Wang W, Nelson W, Rao K, Brar D, Mackill DJ, Han B, Soderlund C, Stein L, SanMiguel P, Jackson S (2005) The oryza map alignment project: the golden path to unlocking the genetic potential of wild rice species. Plant Mol Biol 59: 53-62

Wong NC, Wong LH, Quach JM, Canham P, Craig JM, Song JZ, Clark SJ, Choo KH (2006) Permissive transcriptional activity at the centromere through pockets of DNA hypomethylation. PLoS Genet 2:e17

Wu J, Yamagata H, Hayashi-Tsugane M, Hijishita S, Fujisawa M, Shibata M, Ito Y, Nakamura M, Sakaguchi M, Yoshihara R, Kobayashi H, Ito K, Karasawa W, Yamamoto M, Saji S, Katagiri S, Kanamori H, Namiki N, Katayose Y, Matsumoto T, Sasaki T (2004) Composition and structure of the centromeric region of rice chromosome 8. Plant Cell 16: 967-976

Yamaguchi T, Lee DY, Miyao A, Hirochika H, An G, Hirano HY (2006) Functional diversification of the two C-class MADS box genes *OSMADS3* and *OSMADS58* in *Oryza sativa*. Plant Cell 18: 15-28

Yan H, Jiang J (2007) Rice as a model for centromere and heterochromatin research. Chrom Res 15: 77-84

Yan H, Ito H, Nobuta K, Ouyang S, Jin W, Tian S, Lu C, Venu RC, Wang GL, Green PJ, Wing RA, Buell CR, Meyers BC, Jiang J (2006) Genomic and genetic characterization of rice Cen3 reveals extensive transcription and evolutionary implications of a complex centromere. Plant Cell 18: 2123-2133

Yu J, Hu S, Wang J, Wong GK, Li S, Liu B, Deng Y, Dai L, Zhou Y, Zhang X, Cao M, Liu J, Sun J, Tang J, Chen Y, Huang X, Lin W, Ye C, Tong W, Cong L, Geng J, Han Y, Li L, Li W, Hu G, Huang X, Li W, Li J, Liu Z, Li L, Liu J, Qi Q, Liu J, Li L, Li T, Wang X, Lu H, Wu T, Zhu M, Ni P, Han H, Dong W, Ren X, Feng X, Cui P, Li X, Wang H, Xu X, Zhai W, Xu Z, Zhang J, He S, Zhang J, Xu J, Zhang K, Zheng X, Dong J, Zeng W, Tao L, Ye J, Tan J, Ren X, Chen X, He J, Liu D, Tian W, Tian C, Xia H, Bao Q, Li G, Gao H, Cao T, Wang J, Zhao W, Li P, Chen W, Wang X, Zhang Y, Hu J, Wang J, Liu S, Yang J, Zhang G, Xiong Y, Li Z, Mao L, Zhou C, Zhu Z, Chen R, Hao B, Zheng

W, Chen S, Guo W, Li G, Liu S, Tao M, Wang J, Zhu L, Yuan L, Yang H (2002) A draft sequence of the rice genome (*Oryza sativa* L. ssp. *indica*). Science 296: 79-92

Yu J, Wang J, Lin W, Li S, Li H, Zhou J, Ni P, Dong W, Hu S, Zeng C, Zhang J, Zhang Y, Li R, Xu Z, Li S, Li X, Zheng H, Cong L, Lin L, Yin J, Geng J, Li G, Shi J, Liu J, Lv H, Li J, Wang J, Deng Y, Ran L, Shi X, Wang X, Wu Q, Li C, Ren X, Wang J, Wang X, Li D, Liu D, Zhang X, Ji Z, Zhao W, Sun Y, Zhang Z, Bao J, Han Y, Dong L, Ji J, Chen P, Wu S, Liu J, Xiao Y, Bu D, Tan J, Yang L, Ye C, Zhang J, Xu J, Zhou Y, Yu Y, Zhang B, Zhuang S, Wei H, Liu B, Lei M, Yu H, Li Y, Xu H, Wei S, He X, Fang L, Zhang Z, Zhang Y, Huang X, Su Z, Tong W, Li J, Tong Z, Li S, Ye J, Wang L, Fang L, Lei T, Chen C, Chen H, Xu Z, Li H, Huang H, Zhang F, Xu H, Li N, Zhao C, Li S, Dong L, Huang Y, Li L, Xi Y, Qi Q, Li W, Zhang B, Hu W, Zhang Y, Tian X, Jiao Y, Liang X, Jin J, Gao L, Zheng W, Hao B, Liu S, Wang W, Yuan L, Cao M, McDermott J, Samudrala R, Wang J, Wong GK, Yang H (2005) The Genomes of *Oryza sativa*: a history of duplications. PLoS Biol 3:e38

Yuan Q, Ouyang S, Liu J, Suh B, Cheung F, Sultana R, Lee D, Quackenbush J, Buell CR (2003) The TIGR rice genome annotation resource: annotating the rice genome and creating resources for plant biologists. Nucl Acids Res 31: 229-233

Yuan Q, Ouyang S, Wang A, Zhu W, Maiti R, Lin H, Hamilton J, Haas B, Sultana R, Cheung F, Wortman J, Buell CR (2005) The institute for genomic research osa1 rice genome annotation database. Plant Physiol 138: 18-26

Zeng L, Yin Z, Chen J, Leung H, Wang GL (2002) Fine genetic mapping and physical delimitation of the lesion mimic gene Spl11 to a 160-kb DNA segment of the rice genome. Mol Genet Genom 268: 253-261

Zeng LR, Qu S, Bordeos A, Yang C, Baraoidan M, Yan H, Xie Q, Nahm BH, Leung H, Wang GL (2004) Spotted leaf11, a negative regulator of plant cell death and defense, encodes a U-box/armadillo repeat protein endowed with E3 ubiquitin ligase activity. Plant Cell 16: 2795-2808

Zhang Y, Huang Y, Zhang L, Li Y, Lu T, Lu Y, Feng Q, Zhao Q, Cheng Z, Xue Y, Wing RA, Han B (2004) Structural features of the rice chromosome 4 centromere. Nucl Acids Res 32: 2023-2030

Zhao W, Wang J, He X, Huang X, Jiao Y, Dai M, Wei S, Fu J, Chen Y, Ren X, Zhang Y, Ni PX, Zhang JG, Lil SG, Wang J, Wong GKS, Zhao HY, Yu J, Yang HM, Wang J (2004) BGI-RIS: an integrated information resource and comparative analysis workbench for rice genomics. Nucleic Acids Res 32: D377–D382

Zhou S, Bechner MC, Place M, Churas CP, Pape L, Leong SA, Runnheim R, Forrest DK, Goldstein S, Livny M, Schwartz DC (2007) Validation of rice genome sequence by optical mapping. BMC Genom 8: 278

8 The *Populus* Genome Initiative

Stephen DiFazio
Department of Biology, West Virginia University,
Morgantown, West Virginia 26508-6057 USA
Email: spdifazio@mail.wvu.edu

1 INTRODUCTION

1.1 Background on *Populus* Biology

The genus *Populus* consists of about 29 species organized into six major sections that occur primarily in the Northern Hemisphere (Eckenwalder 1996). Species from the different sections of the genus have diverse ecological characteristics. Two of the most economically important sections (Aigeiros and Tacamahaca) contain species collectively known as cottonwoods. These occur mostly in riparian zones and are characterized by primarily ruderal life history, dominating early successional stages and thriving on flood-mediated disturbance (Braatne et al. 1996; Karrenberg et al. 2002). The other major section of the genus (section Populus, also known as Leuce), contains species commonly known as aspens, which are characterized by extensive clonal growth, and which can occur in very diverse sites, from mixed upland forests to boreal regions (Barnes 1969; Peterson and Peterson 1992; Romme et al. 2001). The cottonwood sections of the genus are highly interfertile and readily hybridize, and the aspen species are also highly interfertile, but the cottonwoods are reproductively isolated from the aspens. This isolation is reflected by the strong ecological, morphological, and genetic distinctions between these groups of species (Stettler et al. 1980). All *Populus* species have 38 chromosomes in their diploid genomes (Blackburn and Heslop Harrison 1924), with exceptions apparently occurring regularly within species due to meiotic nondisjunction (Einspahr et al. 1963; Bradshaw and Stettler 1993).

1.2 Selection of *Populus* for Sequencing

Populus was the first tree selected for whole-genome sequencing, and the third plant overall. At the time of its selection in 2003, *Populus* was already well-established as a model organism because of its experimental tractability, potential economic importance, and its central role in many ecosystems (Wullschleger et al. 2002). One of the primary advantages of *Populus* is the ease with which it can be vegetatively propagated using simple stem cuttings in many species without the necessity of hormonal treatments (Heilman 1999). Another key factor is that the genus consists of ecologically contrasting species that readily hybridize. This occurs naturally between sympatric *Populus* species around the world (Eckenwalder 1984; Zsuffa et al. 1996; Lexer et al. 2004), and has been a key feature of *Populus* breeding programs for decades (Stettler et al. 1996). In fact, the combination of hybridization and vegetative propagation has dominated *Populus* genetic improvement efforts thus far, and most major breeding programs have focused on making a large number of interspecific crosses and choosing clones that show a high degree of heterosis, and display characteristics that are particularly amenable to plantation culture (Stettler et al. 1996). This trait has also made *Populus* one of the premier models for genetic engineering (Taylor 2002; Bhalerao et al. 2003), because transformed lines with desirable characteristics and low somaclonal variation can be rapidly and cheaply disseminated (Strauss et al. 1995). Hybridization has also been a key feature in ecological studies, as *Populus* hybrid zones have provided a fertile area for ecological and evolutionary genetics, and have been a favorite subject in the burgeoning new field of Community Genetics (Whitham et al. 2006).

The rationale for selecting *Populus* was based in part on its importance, and in part on its tractability as a model organism for genomics research. The *Populus* whole-genome sequencing project was initially proposed by Roger Dahlman, Program Manager of the US Department of Energy's (DOE) Terrestrial Carbon Program, in January 2001. The DOE had made major infrastructure and personnel investments for the Human Genome Project, including the creation of the largest public sequencing facility in the world, the Joint Genome Institute Production

Genomics Facility (JGI) in Walnut Creek, CA. Due to continuous improvements in the efficiency of their sequencing pipeline, JGI was capable of producing in excess of 1 Gb of high-quality sequence per month at the time (Detter et al. 2002), which was orders of magnitude higher than was expected at the time of its establishment. As the DOE portion of the Human Genome Project neared completion in 2000, DOE had begun to identify organisms worthy of sequencing that would be relevant to the DOE missions of research related to energy production and its consequences. The choice of *Populus* therefore made perfect sense, because there was a long history of DOE-funded research on *Populus* as a potential bioenergy crop (Dinus et al. 2001), and a surge of interest in *Populus* as a possible solution for carbon sequestration to counter anthropogenic climate change (Tuskan and Walsh 2001). Following Dahlman's initial suggestion, Stan Wullschleger and Jerry Tuskan of DOE's Oak Ridge National Laboratory led to efforts in organizing the *Populus* genetics and genomics community, leading to the formation of the International *Populus* Genome Consortium (*http://www.ornl.gov/sci/ipgc/*). Following substantial grass-roots lobbying and the production of a series of white papers, the sequencing project was approved by the DOE's Office of Science in October 2001.

The importance of community organization and international cooperation in the establishment and success of this project cannot be overestimated. The *Populus* Genome Sequencing Project was a gargantuan effort involving 40 laboratories in eight countries, with 108 researchers directly contributing to the sequencing and subsequent analysis, with particularly notable contributions from the United States, Canada, Sweden, and Belgium. The sequencing itself took place primarily at the DOE's Joint Genome Institute (JGI) under the direction of Daniel Rokhsar, and this certainly represented the bulk of the expense of the project. However, major contributions were made by cooperating institutions around the world. Some of this work was funded by the US National Science Foundation's Plant Genome Research Program, which supported efforts to create a genome portal and to enhance *Populus* bioinformatics tools. Genome Canada supported the involvement of the University of British Columbia, which contributed full-length cDNA sequences and BAC end-sequencing, and a BAC physical map to enhance assembly. The Swedish government supported efforts by the Umea Plant Sciences Center, which spearheaded the collection and analysis of approximately 350,000 EST sequences provided by laboratories around the world. Finally, the government of Belgium and the

European Union supported efforts by the University of Ghent, which customized software for gene prediction and annotation. Countless other individual researchers around the world provided in-kind contributions to the project, in the form of physical resources like template DNA, ESTs or BAC libraries, or analytical expertise related to some aspect of the genome analysis.

1.3 Tools in Place at the Time of Selection

From a technical standpoint, many of the required features for a successful genome sequencing project were already in place for *Populus* at the time it was selected. First, enough cytogenetics and genetic mapping work had been done to establish that *Populus* had a relatively tractable genome. Genetic maps were available for multiple species (Bradshaw et al. 1994b; Cervera et al. 2001; Yin et al. 2001), establishing that segregation occurs as expected in interspecific diploid crosses (with some notable exceptions: Bradshaw and Stettler 1993, 1994; Yin et al. 2004b). Furthermore, information from genetic mapping and flow cytometry indicated that the haploid genome size of the major *Populus* species was approximately 550 Mb (Bradshaw and Stettler 1993), about four times larger than *Arabidopsis*, but a fraction of the size of most conifers (Bradshaw et al. 2000). This was one of the primary factors in the choice of *Populus* as the first sequenced tree over more commercially-important species like loblolly pine (*Pinus taeda*) and Douglas-fir (*Pseudotsuga menziesii*), which had dominated tree improvement research funding prior to the genomics era. Another major factor was the phylogenetic similarity between *Populus* and *Arabidopsis*. Prior to sequencing, approximately 150,000 ESTs were available for *Populus*, and these showed high sequence conservation with known *Arabidopsis* genes (Sterky et al. 1998, 2004b), thereby facilitating provisional annotation of predicted genes based on similarity to intensively studied model annual species. Finally, several BAC libraries were available for *Populus*, and shotgun sequencing of a handful of these BACs had revealed microsynteny between *Arabidopsis* and *Populus*, and a moderate density of repeat elements in *Populus* (Stirling et al. 2003; Lescot et al. 2004). This initial sequencing suggested that assembly of a whole-genome shotgun sequence was feasible, and that existing algorithms for gene prediction in Angiosperms would function reasonably well for *Populus* (Lescot et al. 2004).

2 THE GENOME SEQUENCING PROJECT

The results of the genome sequencing project have previously been described by Tuskan and 108 co-authors in a publication in Science in 2006 (Tuskan et al. 2006), as well as in many subsequent publications by members of the International *Populus* Genome Sequencing Consortium. Unless otherwise specified, the original citation for my descriptions of this project below is the *Science* paper.

2.1 Sequencing Strategy

At the initiation of the project, there was substantial debate within the sequencing consortium about the overall approach to the project. The primary question was whether the genome should be approached as a true whole-genome shotgun project, or whether a BAC-by-BAC approach would be more prudent. The whole-genome shotgun approach, in which the genome would be randomly sheared into small pieces, which would then be sequenced separately and subsequently assembled computationally, had previously been used successfully for a number of large whole-genome sequencing projects at JGI and elsewhere (Aparicio et al. 2002; Dehal et al. 2002; see also Chapter 5 of this volume). The alternative BAC-by-BAC approach requires a physical map of tiled BACs, which is produced with restriction fragment length polymorphism (RFLP) fingerprinting, followed by shotgun sequencing of individual BACs of known position in the physical map (Marra et al. 1997; see also Chapter 2 of this volume). A similar debate had taken place within the Human Genome Sequencing Projects, with the public sequencing consortium initially attempting a BAC-by-BAC approach (Lander et al. 2001), and a private effort led by Celera and J. Craig Venter attempting a whole-genome shotgun approach (Venter et al. 2001). The primary advantage of the whole-genome shotgun approach is speed, and when it became clear that the private sequencing effort would be completed well ahead of the public effort, the public Human Genome Sequencing project was forced to partially switch to the shotgun strategy as well (Green 2001). In the case of *Populus*, it was unclear if a whole-genome shotgun would be feasible for two reasons: (1) the repeat composition and high heterozygosity of the genome could prevent coalescence of sequence contigs into coherent scaffolds; and (2) there was some evidence of a recent genome duplication in *Populus* (Bradshaw et al. 1994a), and it was feared that this would complicate the assembly.

In the end a hybrid strategy was agreed upon. The group at the

University of British Columbia and the BC Genome Sciences Center would produce a physical map by fingerprinting and end-sequencing a *Populus* BAC library, while JGI would proceed with a whole-genome shotgun for this same genotype. The end sequences were to be used to integrate the physical map with the sequence, and gaps and problematic areas of the assembled sequence were to be filled by shotgun sequencing of selected BAC clones. The whole-genome shotgun was to proceed using paired clone-end sequencing of random genomic fragments in independently-prepared libraries of three main sizes: 3 kb inserts, 8 kb inserts in standard plasmid libraries, and 40 kb inserts in phosmid libraries. The rationale for the different-sized libraries is that the bulk of the sequencing would be performed on 3 kb fragments, which would minimize cloning bias that is commonly observed in large-fragment libraries, while the larger libraries would enhance contiguity of scaffolds by bridging small repetitive regions of the genome.

2.2 Selection of Genotype for Sequencing

Another controversial decision was the selection of genotype to be sequenced. There was a strong push to select an aspen (*Populus tremula*, *P. tremuloides*, or *P. alba*), because most of the model transformation clones were derived from these species, and the majority of ESTs were also from that section of the genus. However, the cottonwoods are much more important commercially in the United States, and cottonwood hybrids were the leading candidates for high-yield plantations for bioenergy and carbon sequestration (Tuskan 1998; Perlack et al. 2005). Furthermore, most genetic maps and pedigrees were for cottonwoods (Bradshaw et al. 1994b; Cervera et al. 2001; Yin et al. 2004b, 2008), and the existing BAC libraries were also from cottonwoods (Stirling et al. 2001a; Lescot et al. 2004), so the most relevant resources for genome sequencing and assembly were already in hand. Therefore, for strategic reasons, and to accelerate the production of template for the sequencing pipeline, it was decided to sequence a black cottonwood tree (*Populus trichocarpa* Torr. & Gray). The selected genotype, clone number 383-2499, was originally collected along the Nisqually River in Washington State by Toby Bradshaw, one of the pioneers of *Populus* genomics. This clone, commonly called 'Nisqually-1', was the maternal parent for the largest pedigree produced for *Populus*, a cross with *P. deltoides* clone ILL-101 to produce a family of 2,028 F_1 progeny. The purpose of this pedigree was to isolate a major gene conferring resistance to a hybrid leaf pathogen, *Melampsora x columbiana* 3, which was segregating at an 1:1 ratio in this pedigree (Stirling et al. 2001a). For this purpose, a 9.5× BAC library was also prepared for this pedigree by partially digesting high molecular weight genomic DNA with *Hind*III (Stirling et al. 2001a). The existence of the large pedigree and the BAC library, coupled with the availability of abundant material in clone banks at the University of Washington and elsewhere, was enough to tip the balance in favor of this genotype. There are, however, several ironies about Nisqually-1: (1) the disease resistance gene that originally piqued interest in this genotype still has not been isolated, due in part to high complexity of this genomic region, which may be linked to suppression of recombination in the large pedigree, thereby making map-based cloning nearly impossible (Stirling et al. 2001b; Yin et al. 2004a); (2) the original ortet has since been destroyed by flooding in its native habitat; and (3) Nisqually-1 has proven to be somewhat difficult to handle in tissue culture. Even though transformation protocols have been successfully developed (Ma et al. 2004; Song et al. 2006), this clone is unlikely to supplant aspen hybrids as the model of choice for functional genomics in *Populus* (Busov et al. 2005). However, there is enough sequence conservation on a genome-wide scale between the model aspens and *P. trichocarpa* that for most purposes genomic resources from one species can be used informatively for other species in the genus (Sterky et al. 2004).

2.3 Preparation of Sequencing Template

The initial sequencing template was prepared from surface-sterilized leaves of Nisqually-1 using a CTAB-based protocol. This template was used to construct the 3 kb and 8 kb libraries that form the basis for most of the sequence data (Table 1). A second set of templates was also prepared from root tissue grown in hydroponics and tissue culture. The DNA extraction protocol involved a nucleii isolation step using a sucrose gradient followed by a cesium chloride gradient centrifugation step. This DNA was expected to be virtually free of plastid contamination, and was used to construct the fosmid libraries.

2.4 Shotgun Sequencing

A total of 7.65 million sequence reads were generated from these libraries, with 4.4 million reads coming from 3 kb libraries and 2.5 million reads from 8 kb libraries, and 650,000 reads from fosmid libraries (Table 1). In addition, 81,904 end sequences were obtained from BAC clones that averaged 100 kb in size (Kelleher et al. 2007). This resulted in a theoretical sequence coverage of the genome of nearly 10× (i.e., an

Table 1 Description of sequencing libraries generated for the *Populus trichocarpa* genome sequencing project by JGI as of January 2004. The difference between theoretical and actual coverage of the genome is based on the cumulative length of sequence actually assembled into contigs.

Insert size (kb)	*Libraries*	*Sequences*	*Theoretical sequence coverage*[a]	*Assembled sequence coverage*	*Clone coverage*[b]
3	4	4,427,983	5.48	3.57	13.69
8	4	2,570,799	3.18	2.14	21.20
36	3	651,211	0.81	0.62	24.17
100	1	81,904	0.11	0.10	9.49
Total	11	7,649,993	9.46	6.33	6

[a] Sequence coverage is calculated based on the total amount of sequence in each library divided by the estimated genome size of 485 Mb.
[b] Clone coverage is the total insert size of the clones (assuming the averages given in the insert size column) divided by the estimated genome size of 485 Mb.

Table 2 Kingdoms represented among unassembled sequence reads and small scaffolds from the *Populus* shotgun sequence dataset, based on WU-BLASTN searches versus the NCBI non-redundant nucleotide database.

	Unassembled reads		*Small scaffolds (<10 Kb)*	
Kingdom	*Taxa*	*Sequences*	*Taxa*	*Sequences*
Fungi	78	540	1	1
Metazoa	175	10,638	6	45
Archaea	9	54	0	0
Bacteria	291	13,656	40	231
Eukaryota	40	477	2	2
Viruses	27	407	0	0
Vector	67	1,996	5	7
Viridiplantae	723	577,511	35	2,900
Total	1,410	605,279	89	3,186

average of 10 sequences representing each nucleotide position), and an expected clone coverage of nearly 70× (i.e., the average number of clone inserts covering each position in the genome, though only the ends of the clones are represented by actual sequence). Therefore, a highly contiguous assembly was expected.

2.5 Sequence Assembly

The shotgun sequences were initially assembled based on homology and paired end read information using the JAZZ assembler (Aparicio et al. 2002). The assembly process began with identification and masking of reads derived from repetitive regions of the genome. This was accomplished by counting the number of occurrences of all 16-mer 'words' in the entire set of 7.65 million sequences, and then masking of 16-mers that occurred more than 32 times. This resulted in removal of entire reads from the assembly process and mitigated the confounding effects of repetitive DNA on shotgun assembly (Green 2001). Pairwise alignments of all sequences were then performed, and contigs were constructed by converting pairwise relationships to a graph topology and finding the most direct route through the graph. Sequence contigs were joined using similar methodology, taking advantage of linkage information contained in paired end-read information.

The initial assembly utilized 4.8 million of the sequence reads to form approximately 45,970 sequence contigs of at least 1 kb in length, resulting in approximately 427 Mb of assembled genome sequence contigs, excluding gaps. These contigs were grouped together using paired clone end information into 22,136 sequence scaffolds that covered 464 Mb of assembled sequence and 'captured' sequence gaps (the size of which was estimated based on average clone insert size). Half of this scaffold sequence was contained in 62 major scaffolds, each of which was at least 2 Mb in size. The maximum contig size was 1.7 Mb, and the maximum scaffold size was nearly 12.5 Mb.

2.6 Contamination of the Sequencing Template

Nearly 2.85 million of the original sequence reads could not be assembled into meaningful sequence contigs in the whole-genome shotgun assembly. Approximately 750,000 of these were simply low-quality or chimeric sequences that were excluded by the assembler. However, 2.1 million were high quality sequences that otherwise should have assembled. The leaf-derived sequences failed to assemble at a much higher rate (25%) than the root-derived sequences (18%), suggesting that the DNA extraction method was related to this problem. Two sets of sequences with uniform sequence depth (954× and 60× respectively) were removed and assembled into putative chloroplast and mitochondrial genomes, respectively. These accounted for approximately 300,000 of the unassembled sequences. Another 613,000 corresponded to *Populus* repeat elements, as determined by high 16-mer composition and comparison to *Populus* repeat libraries using WU-BLASTN (see below). The remaining 1.1 million unassembled sequences were compared to the NCBI nonredundant nucleotide database using WU-BLASTN searches. Approximately 600,000 of these sequences showed no homology to known sequences, and are therefore of unknown origin. An additional 482,199 had significant hits to known, non-*Populus* sequences. Of these, the vast majority had hits ($E < 1e^{-10}$) to other plants, and likely represent inexplicably unassembled portions of the *Populus* genome. However, nearly 25,000 of the remaining sequences had hits to fungi, bacteria, and viruses that were likely endophytic or pathogenic contaminants of the sequencing template, despite the fact that the leaves and roots were surface-sterilized prior to extraction. Similar trends were seen for small scaffolds from the sequencing dataset, where nearly 300 of the scaffolds <10 kb in size were apparently of microbial origin. This provides a potentially-interesting window into the invisible and largely unknown microbial associates of *Populus* (Germaine et al. 2004).

2.7 BAC Physical Map

A physical map was constructed using a 10× Nisqually-1 BAC library that had previously been constructed by the BAC Center at the Texas A&M University. The library was constructed from high molecular weight DNA extracted from leaf tissue (Stirling et al. 2001a) and partially digested with *Hind*III. Restriction enzyme fingerprinting (Marra et al. 1997) using *Hind*III was performed on 46,025 clones from this library with an average insert size of 100 kb, providing 9.4-fold coverage of the physical map (Kelleher et al. 2007). This resulted in production of 3,471 contigs containing 11 BAC clones each on average. One exceptional contig consisted of over 1,200 clones, which was ultimately discovered to represent chloroplast contamination. The relative lack of contiguity in this library appears to be the result of complex haplotype structure in the *Populus* genome (Kelleher et al. 2007). In particular, heterozygous polymorphisms at *Hind*III sites prevented haplotypes from converging in the assembly, leading to complex forking patterns in the tiling path. This is likely a problem for the sequence-based assembly as well. Another complicating factor is the apparent existence of *Hind*III 'deserts' in the *Populus* genome: large regions entirely lacking *Hind*III sites. These regions would not be represented in this BAC library, since it was constructed with a single restriction enzyme. Such complexity greatly mitigates the advantages of a BAC-by-BAC sequencing approach in a highly heterozygous organism like *Populus*, and tips the balance strongly in favor of the more efficient whole-genome shotgun approach. Nevertheless, BAC end-sequences and the BAC physical map were extremely useful in enhancing the contiguity of the shotgun assembly, so BAC fingerprinting and mapping still plays a vital role in genome sequencing projects.

2.8 Map-Based Assembly

The large number of scaffolds in both the sequence assembly and the physical map posed substantial challenges for the analysis and application of the genome sequence. We, therefore, sought to anchor sequence contigs onto a genetic map representing the 19 *Populus* chromosomes (Yin et al. 2004b, 2008). This was accomplished by using sequence tags provided by PCR primers for 356 simple sequence repeat (SSR) markers that could be placed unambiguously in the sequence as well as on the genetic map. Location of the primer sequences in the assembled genome sequence was accomplished by performing BLASTN searches and requiring that both primers match the putative SSR locus in inverse orientation relative to each other, and at a distance that was consistent with the known size of the SSR. This resulted in linking 155 major sequence scaffolds and 335 Mb of sequence into chromosomal linkage groups (LGs) (Fig. 1). The smallest chromosome, LG_IX, was covered by two scaffolds containing 12.5 Mb of sequence. In contrast, the largest chromosome, LG_I, contained 21 scaffolds representing 35.5 Mb of sequence.

Some caveats are in order regarding the map-based assembly. First, the vast majority of the markers used for genome assembly were mapped with only 44 progeny, so the resolution of the map is quite low, and small sequence scaffolds are not always positioned or oriented accurately. This is also a problem for scaffolds that are only mapped with one marker, which is true for 75 out of 155 scaffolds. Second, the low resolution of the map can also lead to tandem assembly of scaffolds representing different haplotypes that should actually be assembled to the same position. This type of misassembly can easily be misinterpreted as large-scale tandem or segmental duplications. It is extremely difficult to distinguish between mis-assemblies of this type and true duplication events. Examples of problematic regions of the assembly are the peri-telomeric (top) portions of LG_X and LG_XIX (Fig. 1). LG_XIX has been investigated in some detail, and demonstrated by intensive mapping to have very strong haplotypic divergence in this peritelomeric region (Yin

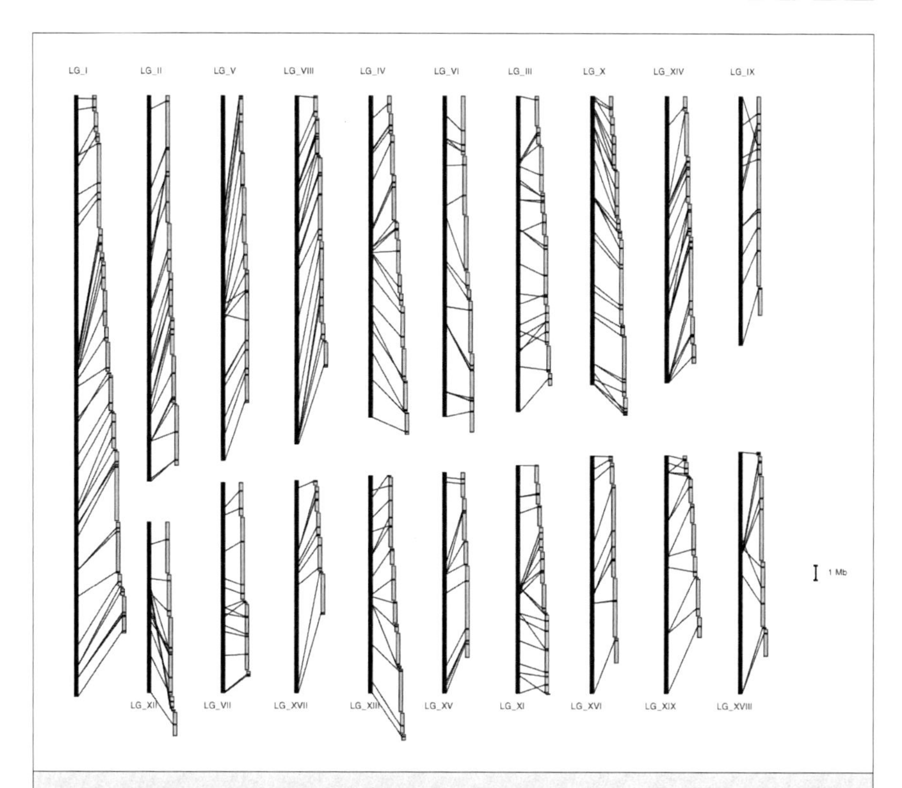

Fig. 1 Representation of a recent map-based assembly of the *Populus* genome. Black bars represent chromosomal linkage groups derived from genetic mapping, and white bars represent sequence scaffolds from the sequence-based assembly. Positions of microsatellite markers are indicated by lines connecting the linkage groups to the sequence scaffolds.

et al. 2008). This, coupled with strong suppression of recombination and the mapping of sex determination to this region (Markussen et al. 2007; Gaudet et al. 2008), has led to the speculation that this chromosome might be in the early stages of evolving into a sex chromosome (Yin et al. 2008).

Integration of the physical and genetic maps afforded the opportunity to examine the ratio of physical to genetic distance in *Populus*. The median ratio of physical to genetic distance was 118.5 kb/centi-Morgan, based on 54 'framework' SSR markers (mapped with at least 150 progeny) located on the same sequence scaffold. There is, as expected, a substantial amount of variation in these estimates (Fig. 2), reflecting real differences in recombination frequency across the genome, as well as

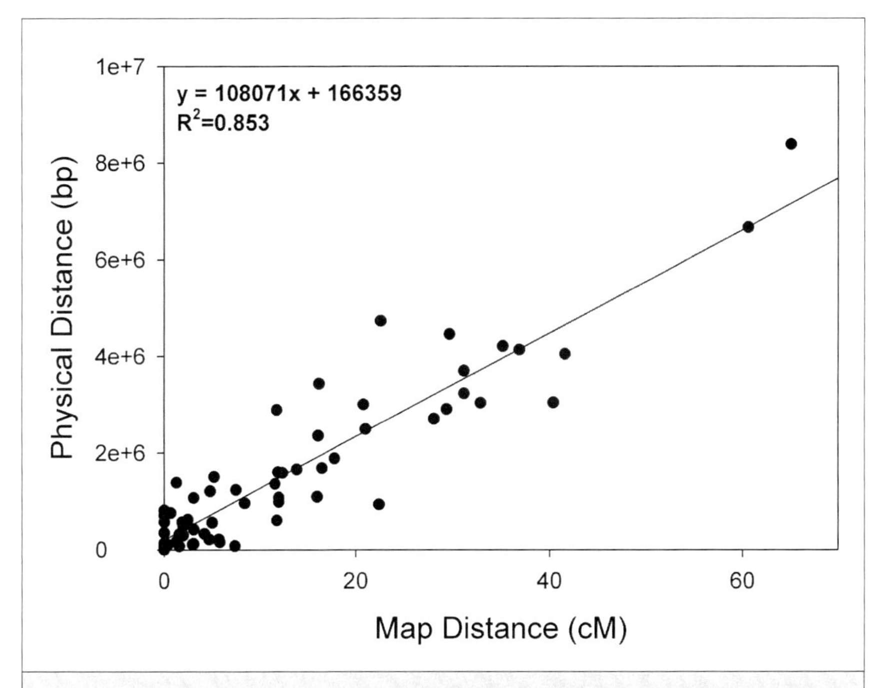

Fig. 2 Relationship between genetic distance and physical distance for *Populus* sequence scaffolds for Populus framework markers, which were mapped with over 150 progeny (Yin et al. 2004b).

errors in estimation of physical genome size caused by sequence gaps, imprecision in mapping positions, and actual differences in genome composition between Nisqually-1 and the clone used in the mapping pedigree, 93-968 (Yin et al. 2004b).

3 *POPULUS* GENOME CHARACTERISTICS

3.1 Gene Content

Gene content prediction was carried out using four different approaches, and the results were merged to provide a consensus list of gene models. The four main gene-calling algorithms were ab initio FgenesH, homology-based FgenesH (which uses EST evidence) (Solovyev et al. 2006), Genewise, GrailExp6, and EuGène (Foissac et al. 2008), all of which were trained with a set of over 4,664 full-length cDNA sequences (Ralph et al. 2008). These gene predictions were carried out by three independent groups (JGI, ORNL, and the University of Ghent), and then merged by the JGI to produce consensus predictions. Most gene prediction programs provide markedly different results, and each has its own strengths and weaknesses, with major tradeoffs between specificity and sensitivity, depending on the weight given to different evidence sources (ESTs, full-length cDNAs, alignments to other genomes) in the training and analysis phases (Foissac et al. 2008). As expected, the gene finding algorithms produced quite different results for *Populus* as well (Fig. 3), and it was challenging to identify the best model at each locus, and derive a consensus set of predicted genes that are likely to be true, protein coding genes. The initial consensus set of contained 58,036 putative genes, only 25% of which were predicted by two or more algorithms. This set was quickly discovered to contain many pseudogenes and transposable elements, and was gradually reduced to the publicly-released set of 45,555. However, this set still contains approximately 370 genes with strong hits to known transposable elements.

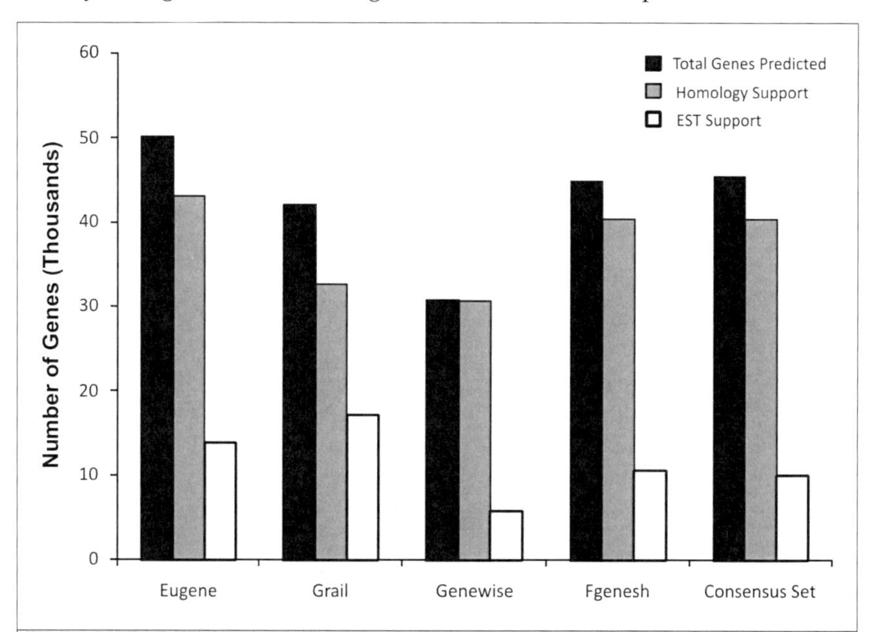

Fig. 3 Number of genes predicted by each set of gene prediction programs, as well as the consensus set of gene models released by the genome consortium in 2004. 'Homology support' indicates the number of gene models with significant BLASTN ($E < 1e^{-10}$) alignments versus know proteins in the NCBI nonredundant database, and 'EST Support' is the number of gene models with significant BLASTN alignments to one of the ~250,000 Populus ESTs that were available at the time of prediction.

Furthermore, it appears that many bona fide genes from the initial set are not included in the final set of released genes. BLASTP searches revealed that 4,520 of these excluded genes have significant hits to plant proteins in the NCBI non-redundant database. Furthermore, 3,675 of these excluded genes showed some evidence of expression in whole-genome microarray experiments using diverse *Populus* tissues (Brunner et al. in prep.). Therefore, the gene content of *Populus* is still poorly determined, and efforts to improve the selection and annotation of gene models are continuing. Nevertheless, it seems clear that the final number of *Populus* genes will exceed 40,000, based on expression evidence and homology to known genes.

3.2 Comparison of Gene Content with *Arabidopsis* and Grape

In one sense, the gene content of *Populus* is surprisingly similar to that of *Arabidopsis*, despite the considerable phylogenetic distance between these taxa and their obviously contrasting biological characteristics. *Populus* is a member of the Eurosid I subclass, while *Arabidopsis* is in the Eurosid II subclass. Furthermore, *Arabidopsis* is a diminutive herbaceous annual with perfect flowers and largely selfing mating system (see also Chapter 6 of this volume), while *Populus trichocarpa* is the tallest perennial angiosperm in the northern hemisphere, with a dioecious, completely outcrossing breeding system (DeBell 1990). Nevertheless, 72% of the *Populus* genes had significant BLASTP hits to at least one *Arabidopsis* gene, with an average expectation score of 7.3×10^{-13} and an average of 59% (+/-1.6%) amino acid identity over their aligned lengths. The comparison looks even more favorable in the opposite direction, with 87.4% of *Arabidopsis* genes showing significant hits to *Populus* proteins, with an average amino acid identity of 61% (+/-1.6%) over their aligned length, with 17% of gene models having 80% or greater amino acid identity (Fig. 4). The discrepancy in the reciprocal comparisons probably reflects a higher frequency of incorrectly annotated pseudogenes in *Populus*, since the *Arabidopsis* annotation has received substantially more attention and resources than the *Populus* annotation (Swarbreck et al. 2008).

The complete genome of the wine grape (*Vitis vinifera*) was also recently published (Jaillon et al. 2007), and this genome makes for some interesting comparisons with *Populus*. First, one might expect more similarity in coding sequences because grape is also a perennial woody plant. However, the phylogenetic position of grape relative to the rosids

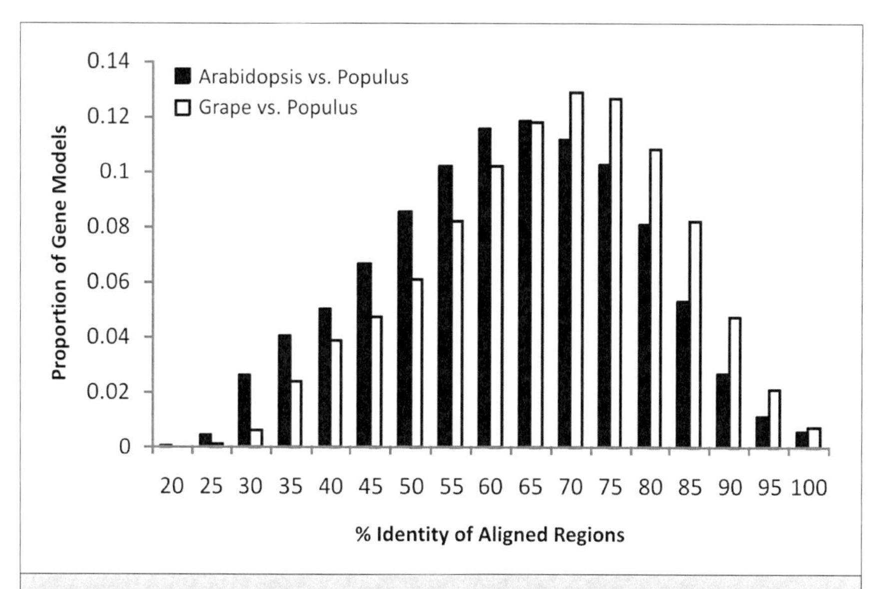

Fig. 4 Comparison of *Populus* predicted proteins to those in the *Arabidopsis* and grape genomes. Comparison only includes genes with significant BLASTP alignments (*E* score <1e-10). The % identity is the weighted average of all High Scoring Pairs from the BLASTP alignments. The *Arabidopsis* comparison is based on 26,994 significant alignments out of 30,900 proteins compared (87.4%), while the grape comparison is based on 46,288 significant alignments out of 55,990 proteins compared (82.7%).

has been the subject of some controversy, and recent analyses seem to place the order Vitales as a sister group to the Rosids (Jansen et al. 2006). Therefore, based on phylogeny one might expect grape proteins to be more divergent from *Populus* than *Arabidopsis* proteins. Surprisingly, the percentage of proteins showing significant hits to *Populus* proteins was somewhat higher in grape (84.9% of 30,442 proteins) compared to *Arabidopsis*, using the criteria described above. Furthermore, the amino acid identity was 65.1% (+/-1.49%), with 26.5% of models showing 80% or greater amino acid identity to *Populus* genes (Fig. 4). Therefore, grape genes have substantially higher identity to *Populus* genes than to *Arabidopsis* genes, despite the fact that *Arabidopsis* is closer evolutionarily (Velasco et al. 2007). This is probably due to the higher rate of evolution in *Arabidopsis* and other herbaceous annuals, which have many more generations per time interval than long-lived woody perennials like *Populus* and grape (Tuskan et al. 2006; Semon and Wolfe 2007).

The relative abundance of genes in functional categories provides a more sensitive and informative measure than gross gene content differences between genomes. The Gene Ontology classification system provides a convenient means of doing this (Harris et al. 2006). Because GO classifications have not yet been performed for *Populus*, we assigned provisional GO classifications using *Arabidopsis* best hits, utilizing the simpler GO-Slim terms. There were significant differences between *Populus* and *Arabidopsis* for almost all GO-Slim categories, but only five categories were different between *Populus* and grape (Fig. 5). Interestingly, the classes that showed significant differences between grape and *Populus* were transcription factor activity and kinase activity, which were over-represented in *Populus*, and 'response to stress' and 'cell wall', which were over-represented in grape. Both of these latter classes were in turn strongly over-represented in *Populus* compared to *Arabidopsis*, so perhaps both of these classes are related to the woody, perennial habit. However, in the case of the cell wall class, a large portion of those over-represented in grape are related to flavonoid and polyphenol production, which is one of the aspects of grape chemistry that makes this species so desirable for wine production (Jaillon et al. 2007a; Velasco et al. 2007).

3.3 Genome Structure

Assembly of the genome and corresponding gene content to linkage groups made it possible to investigate the gross structure of the genome at a chromosomal scale. This revealed the striking existence of two whole genome duplication events. This was accomplished by making pairwise comparisons among all *Populus* genes using double-affine Smith-Waterman alignments. This revealed the presence of large syntenic blocks of genes on different linkage groups that had approximately concordant genetic distances (Fig. 6). Blocks of these syntenic genes were defined based on the existence of two or more genes aligning on different chromosomes, with fewer than 10 intervening, nonaligning genes. The genetic distance between these aligning genes was calculated based on the number of transversion substitutions at four-fold degenerate nucleotide sites (4DTV), which is a conservative estimate of genetic divergence that should be less susceptible to multiple substitutions than more commonly-occurring synonymous substitutions (Comeron 1995). Comparison of the size of the syntenic blocks versus the mean 4DTV value for those blocks revealed two clear groups of blocks that were of approximately uniform age (Fig. 6). The group of larger blocks centered at 4DTV = 0.068 represents the most recent whole genome duplication

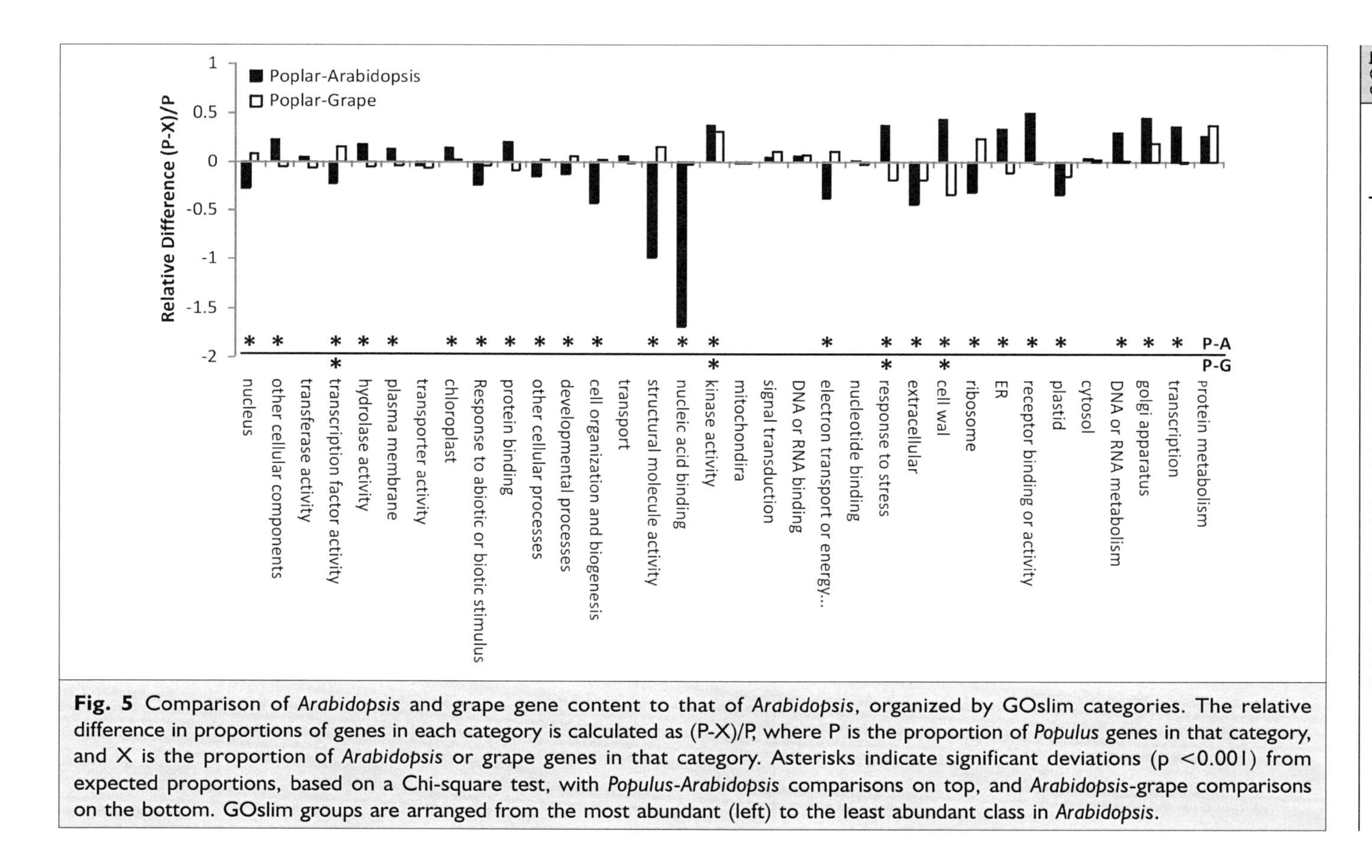

Fig. 5 Comparison of *Arabidopsis* and grape gene content to that of *Arabidopsis*, organized by GOslim categories. The relative difference in proportions of genes in each category is calculated as (P-X)/P, where P is the proportion of *Populus* genes in that category, and X is the proportion of *Arabidopsis* or grape genes in that category. Asterisks indicate significant deviations ($p < 0.001$) from expected proportions, based on a Chi-square test, with *Populus-Arabidopsis* comparisons on top, and *Arabidopsis*-grape comparisons on the bottom. GOslim groups are arranged from the most abundant (left) to the least abundant class in *Arabidopsis*.

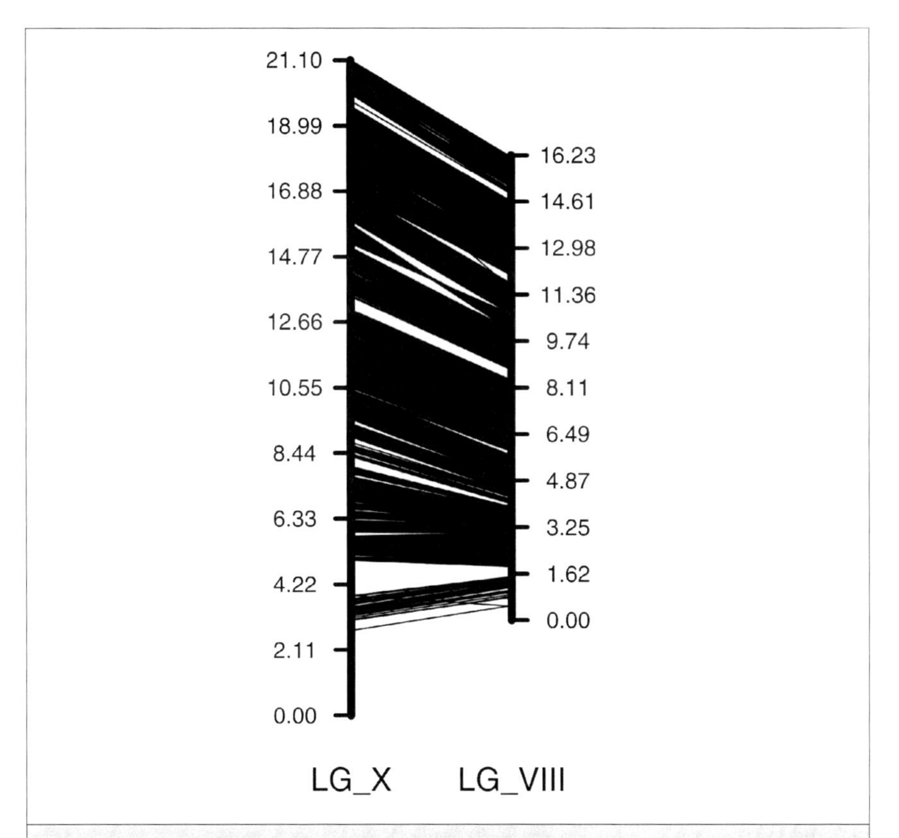

Fig. 6 Comparison of genes with significant alignments between two linkage groups, chromosome X (left) and Chromosome VIII (right). Genes are represented by black lines connecting the heavy lines representing the chromosomes. Position on the linkage group is given in megabases to the side of each group. Genetic distances between genes in these large syntenic blocks are highly concordant, indicating that these syntenic chromosome blocks originated from whole genome duplication events.

in *Populus*, while the group of smaller blocks at 4DTV = 0.31 represents a more ancient duplication event (Sterck et al. 2005). Dating of these events is difficult, because the *Populus* genome is evolving considerably slower than genomes that have previously been used to calibrate the angiosperm molecular clock. Using the molecular clock as calibrated by fossil records for the Brassicaceae, for example, the most recent duplication dates to 8 million years ago (Sterck et al. 2005). However, the *Populus* genus has been in existence for at least 50 million years (Eckenwalder 1996), and the genome duplication is shared by many species in the genus (Sterck et al. 2005), so the *Populus* genome is clearly evolving at a much slower rate than herbaceous angiosperms, which is to be expected based on generation time (Bell et al. 2005).

The *Arabidopsis* genome also shows evidence of at least two whole-genome duplication events (Blanc et al. 2003; see also Chapter 6 of this volume), but following these events the genome has become substantially rearranged, making it difficult to reconstruct the older events (Blanc et al. 2003). A similar rearrangement has occurred in *Populus*, but much less severe (see figure in Tuskan et al. 2006). Extensive rearrangements following genome doubling is a common component of the diploidization process (Adams and Wendel 2005; Semon and Wolfe 2007). The structural complexity of these two genomes, coupled with the high rates of gene evolution in *Arabidopsis*, make it particularly difficult to establish orthology and determine whether the ancient duplication event in *Populus* is shared with *Arabidopsis*. The timing of the event is similar to the timing of the split of the *Arabidopsis* and *Populus* lineages, as determined by pairwise comparisons of genetic distances between *Populus* duplicated genes, *Arabidopsis* duplicated genes, and between putative *Arabidopsis* and *Populus* orthologs. Given the close timing of these events, it is tempting to speculate that the genome duplication was a primary driver of the diversification of the rosids (Lynch and Conery 2003).

In contrast to *Populus* and *Arabidopsis*, the grape genome has been comparatively quiescent, with minimal rearrangements, and equivocal evidence of a single whole-genome duplication that could be shared with *Populus* and *Arabidopsis* (Jaillon et al. 2007b; Velasco et al. 2007). This structural simplicity has allowed reconstruction of the truly ancient whole genome duplication event that is shared by all angiosperms. It appears that this event resulted in hexaploidy in the ancient angiosperm progenitor, as suggested by the presence of three syntenic blocks in rice, *Populus*, and *Arabidopsis* for every one block in grape (Jaillon et al. 2007a). However, evidence for this event is still weak, because the genetic distance is too great to allow relative dating with nucleotide substitution rates, and it is possible to confound two different duplication events that occurred at very different times, followed by massive rearrangement and gene loss. This same analysis (Jaillon et al. 2007a) suggested that only one duplication event had occurred in the *Populus* genome, despite the existence of virtually unequivocal evidence for two events when relative levels of divergence are taken into account (Fig. 7). Part of the problem is the confusion about the proper relationship of

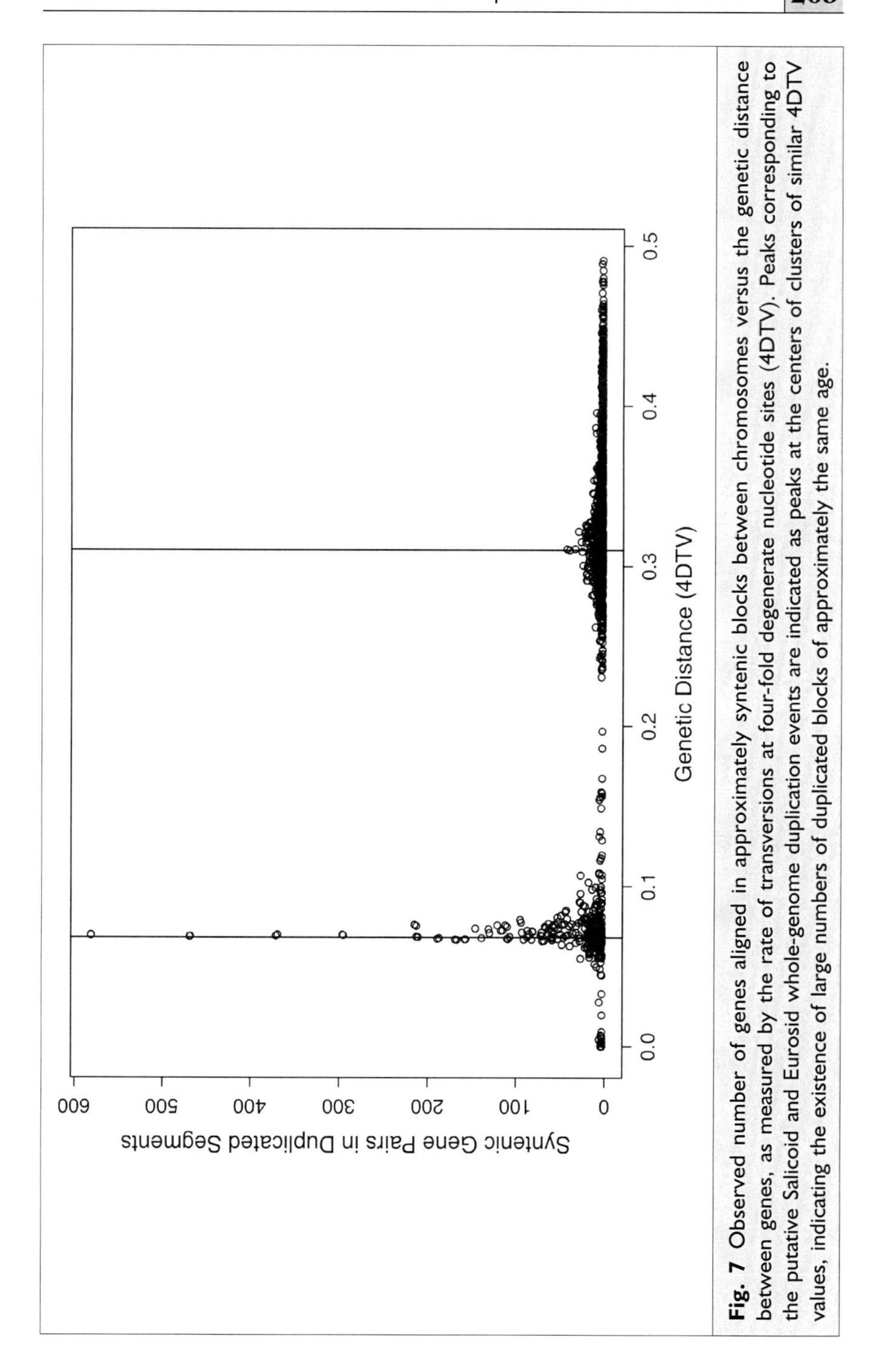

Fig. 7 Observed number of genes aligned in approximately syntenic blocks between chromosomes versus the genetic distance between genes, as measured by the rate of transversions at four-fold degenerate nucleotide sites (4DTV). Peaks corresponding to the putative Salicoid and Eurosid whole-genome duplication events are indicated as peaks at the centers of clusters of similar 4DTV values, indicating the existence of large numbers of duplicated blocks of approximately the same age.

grape to the Rosids. If grape is taken as on outgroup, then a clear model that incorporates both *Populus* duplications can be accommodated (Velasco et al. 2007). However, if grape is assumed to be closer to *Populus* than *Arabidopsis*, as sequence similarity suggests, then grape would have to share the duplication event that occurred near the time of the split with *Populus*. However, this event was not detected in grape or *Populus*, based on an analysis using reciprocal BLASTP hits, without regard to degree of divergence of putative orthologs (Jaillon et al. 2007b). Fortunately, many more plant genomes are currently in sequencing pipelines, so the duplication history of the angiosperms will soon become much clearer.

4 IMPLICATIONS AND APPLICATIONS OF THE GENOME SEQUENCE

The genome sequence was truly a watershed event for the tree genetics community, and the impacts have reverberated throughout forest science and even into other parts of plant science. The genome sequence has provided a nearly complete catalog of all genes and regulatory elements in this model tree, thus opening up a whole realm of research that was not possible before the sequencing project. One index of the impact of the sequence is the number of citations of journal articles related to *Populus* has more than doubled since 2004, the year that the sequence was first publicly released, and the number of *Populus* publications has nearly doubled since 2000. The main article describing the genome sequence (Tuskan et al. 2006) has been cited over 250 times since it was published in September, 2006.

The genome sequence has already had extensive applications in applied science. For example, *Populus* is currently the focus of three major bioenergy projects, two in the US and one in Canada, with a total committed funding of more than $20 million over the next few years. One of these projects, the DOE Bioenergy Science Center headed by Oak Ridge National Lab, is resequencing 18 *Populus* genotypes using next-generation sequencing technology. The project will characterize single nucleotide polymorphisms (SNPs) across the genome for the purpose of genetic association studies to identify genes underlying cell wall biosynthesis, with the ultimate goal of reducing the recalcitrance of lignocellulosic feedstocks to cellulose extraction (Rubin 2008). The project will then use Illumina Bead Arrays to assay 20,000 SNPs for over 1,000 trees collected across the range of *Populus trichocarpa*. These trees will be established in three different common gardens encompassing most of the range of the species (California, Oregon, and British Columbia), and phenotyping will be performed for a large number of traits. This project, and others like it, will therefore propel *Populus* from the realm of comparative genomics, and almost complete reliance on gene homology to herbaceous models for functional annotation, to direct functional characterization of a large fraction of the genes in the genome. *Populus* will thus be solidly established as a premier model organism for functional genomics.

The impact in areas outside of genomics has been equally profound. The fields of community genetics and ecological genomics are flourishing, with *Populus* as one of the primary model organisms, driven by the availability of the genome sequence and the central importance of *Populus* in many ecosystems (Whitham et al. 2006, 2008). Multiple large-scale ecological genomics studies have been funded in *Populus* since the publication of the genome sequence, including an NSF-FIBR project, two Plant Genome Research Program projects and several large projects in Europe. The genome sequence is allowing exploration of diverse questions, such as exploration of the genetic architecture of species barriers, based on patterns of introgression across hybrid zones (Lexer et al. 2007), or the genetic basis of sexual selection (Yin et al. 2008). Furthermore, additional species associated with *Populus* have also been sequenced (Martin et al. 2004, 2008), and many more are in progress. We are truly on the threshold of a brave new era in which genome sequencing of entire communities will become entirely plausible, potentially allowing elucidation of fundamental truths about the mechanisms of the assemblage and persistence of ecological communities (Whitham et al. 2008). This is likely to fundamentally change the way we approach ecological and evolutionary research.

Acknowledgements

The *Populus* Genome consortium, led by Jerry Tuskan, Dan Rokhsar, Carl Douglas, Stefan Jansson, Goran Sandberg, and Yves Van de Peer, made all of the work reported in this chapter possible. In particular, I have directly co-opted analyses performed by Uffe Hellsten, Nik Putnam, and Igor Grigoriev. This work was supported by the US Department of Energy, Office of Science, Biological and Environmental Research, NSF-FIBR, and the NSF Plant Genome Research Program.

References

Adams KL, Wendel JF (2005) Polyploidy and genome evolution in plants. Curr Opin Plant Biol 8: 135-141

Aparicio S, Chapman J, Stupka E, Putnam N, Chia J, Dehal P, Christoffels A, Rash S, Hoon S, Smit A, Gelpke MDS, Roach J, Oh T, Ho IY, Wong M, Detter C, Verhoef F, Predki P, Tay A, Lucas S, Richardson P, Smith SF, Clark MS, Edwards YJK, Doggett N, Zharkikh A, Tavtigian SV, Pruss D, Barnstead M, Evans C, Baden H, Powell J, Glusman G, Rowen L, Hood L, Tan YH, Elgar G, Hawkins T, Venkatesh B, Rokhsar D, Brenner S (2002) Whole-genome shotgun assembly and analysis of the genome of Fugu rubripes. Science 297: 1301-1310

Barnes BV (1969) Indications of possible mid-Cenozoic hybridization in aspens of the Columbia Plateau. Rhodora 69: 70-81

Bell CD, Soltis DE, Soltis PS (2005) The age of the angiosperms: A molecular timescale without a clock. Evolution 59: 1245-1258

Bhalerao R, Nilsson O, Sandberg G (2003) Out of the woods: forest biotechnology enters the genomic era. Curr Opin Biotechnol 14: 206-213

Blackburn KB, Heslop Harrison JW (1924) A preliminary account of the chromosomes and chromosome behaviour in the Salicaceae. Ann Bot 38: 361-378

Blanc G, Hokamp K, Wolfe KH (2003) A recent polyploidy superimposed on older large-scale duplications in the Arabidopsis genome. Genom Res 13: 137-144

Braatne JH, Rood SB, Heilman PE (1996) Life history, ecology, and reproduction of riparian cottonwoods in North America. In: Stettler RF, Bradshaw HD Jr, Heilman PE, Hinckley TM (eds) Biology of Populus and its Implications for Management and Conservation. NRC Res Press, Ottawa, Canada, pp 57-85

Bradshaw HD, Jr, Stettler RF (1993) Molecular genetics of growth and development in Populus. I. Triploidy in hybrid poplars. Theor Appl Genet 86: 301-307

Bradshaw HD, Stettler RF (1994) Molecular genetics of growth and development in Populus. II. Segregation distortion due to genetic load. Theor Appl Genet 89: 551-558

Bradshaw HD, Villar M, Watson BD, Otto KG, Stewart S, Stettler RF (1994a) Molecular-genetics of growth and development in Populus. 3. A genetic-linkage map of a hybrid Poplar composed of RFLP, STS, and RAPD Markers. Theor Appl Genet 89: 167-178

Bradshaw HDJ, Villar M, Watson BD, Otto KG, Stewart S, Stettler RF (1994b) Molecular genetics of growth and development in Populus. III. A genetic linkage map of a hybrid poplar composed of RFLP, STS, and RAPD markers. Theor Appl Genet 89: 167-178

Bradshaw HD, Ceulemans RE, Davis J, Stettler RF (2000) Emerging model systems in plant biology: poplar (*Populus*) as a model forest tree. J Plant Growth Reg 19: 306-313

Busov VB, Brunner AM, Meilan R, Filichkin S, Ganio L, Gandhi S, Strauss SH (2005) Genetic transformation: a powerful tool for dissection of adaptive traits in trees. New Phytol 167: 9-18

Cervera MT, Storme V, Ivens B, Gusmao J, Liu BH, Hostyn V, Slycken JV, Montagu MV, Boerjan W (2001) Dense genetic linkage maps of three *Populus* species (*Populus deltoides, P. nigra* and *P. trichocarpa*) based on AFLP and microsatellite markers. Genetics 158: 787-809

Comeron JM (1995) A method for estimating the numbers of synonymous and nonsynonymous substitutions per site. J Mol Evol 41: 1152-1159

DeBell DS (1990) *Populus trichocarpa* Torr. & Gray, Black Cottonwood. In: Burns RM, Honkala BH (eds) Silvics of North America. Vol 2. Hardwoods. USDA For Serv Agri Handbook USDA For Serv, Washington DC, USA, pp 570-576

Dehal P, Satou Y, Campbell RK, Chapman J, Degnan B, De Tomaso A, Davidson B, Di Gregorio A, Gelpke M, Goodstein DM, Harafuji N, Hastings KEM, Ho I, Hotta K, Huang W, Kawashima T, Lemaire P, Martinez D, Meinertzhagen IA, Necula S, Nonaka M, Putnam N, Rash S, Saiga H, Satake M, Terry A, Yamada L, Wang HG, Awazu S, Azumi K, Boore J, Branno M, Chin-bow S, DeSantis R, Doyle S, Francino P, Keys DN, Haga S, Hayashi H, Hino K, Imai KS, Inaba K, Kano S, Kobayashi K, Kobayashi M, Lee BI, Makabe KW, Manohar C, Matassi G, Medina M, Mochizuki Y, Mount S, Morishita T, Miura S, Nakayama A, Nishizaka S, Nomoto H, Ohta F, Oishi K, Rigoutsos I, Sano M, Sasaki A, Sasakura Y, Shoguchi E, Shin-i T, Spagnuolo A, Stainier D, Suzuki MM, Tassy O, Takatori N, Tokuoka M, Yagi K, Yoshizaki F, Wada S, Zhang C, Hyatt PD, Larimer F, Detter C, Doggett N, Glavina T, Hawkins T, Richardson P, Lucas S, Kohara Y, Levine M, Satoh N, Rokhsar DS (2002) The draft genome of *Ciona intestinalis*: Insights into chordate and vertebrate origins. Science 298: 2157-2167

Detter JC, Jett JM, Lucas SM, Dalin E, Arellano AR, Wang M, Nelson JR, Chapman J, Lou YI, Rokhsar D, Hawkins TL, Richardson PM (2002) Isothermal strand-displacement amplification applications for high-throughput genomics. Genomics 80: 691-698

Dinus RJ, Payne P, Sewell MM, Chiang VL, Tuskan GA (2001) Genetic modification of short rotation poplar wood: properties for ethanol fuel and fiber productions. Crit Rev Plant Sci 20: 51-69

Eckenwalder JE (1984) Natural intersectional hybridization between North American species of *Populus* (*Salicaceae*) in sections *Aigeiros* and *Tacamahaca*. II. Taxonomy. Can J Bot 62: 336-342

Eckenwalder JE (1996) Systematics and evolution of *Populus*. In: Stettler RF, Bradshaw HD Jr, Heilman PE, Hinckley TM (eds) Biology of Populus and its Implications for Management and Conservation. NRC Res Press, Ottawa, Canada, pp 7-32

Einspahr D, Benson MK, Peckham JR (1963) Natural variation and heritability in triploid aspen. Silvae Genet 12: 51-58

Foissac S, Gouzy J, Rombauts S, Mathe C, Amselem J, Sterck L, Van de Peer Y,

Rouze P, Schiex T (2008) Genome annotation in plants and fungi: EuGene as a model platform. Curr Bioinform 3: 87-97

Gaudet M, Jorge V, Paolucci I, Beritognolo I, Mugnozza GS, Sabatti M (2008) Genetic linkage maps of *Populus nigra* L. including AFLPs, SSRs, SNPs, and sex trait. Tree Genet Genom 4: 25-36

Germaine K, Keogh E, Garcia-Cabellos G, Borremans B, van der Lelie D, Barac T, Oeyen L, Vangronsveld J, Moore FP, Moore ERB, Campbell CD, Ryan D, Dowling DN (2004) Colonisation of poplar trees by gfp expressing bacterial endophytes. Fems Microbiol Ecology 48: 109-118

Green ED (2001) Strategies for the systematic sequencing of complex genomes. Nat Rev Genet 2: 573-583

Harris MA, Clark JI, Ireland A, Lomax J, Ashburner M, Collins R, Eilbeck K, Lewis S, Mungall C, Richter J, Rubin GM, Shu SQ, Blake JA, Bult CJ, Diehl AD, Dolan ME, Drabkin HJ, Eppig JT, Hill DP, Ni L, Ringwald M, Balakrishnan R, Binkley G, Cherry JM, Christie KR, Costanzo MC, Dong Q, Engel SR, Fisk DG, Hirschman JE, Hitz BC, Hong EL, Lane C, Miyasato S, Nash R, Sethuraman A, Skrzypek M, Theesfeld CL, Weng SA, Botstein D, Dolinski K, Oughtred R, Berardini T, Mundodi S, Rhee SY, Apweiler R, Barrell D, Camon E, Dimmer E, Mulder N, Chisholm R, Fey P, Gaudet P, Kibbe W, Pilcher K, Bastiani CA, Kishore R, Schwarz EM, Sternberg P, Van Auken K, Gwinn M, Hannick L, Wortman J, Aslett M, Berriman M, Wood V, Bromberg S, Foote C, Jacob H, Pasko D, Petri V, Reilly D, Seiler K, Shimoyama M, Smith J, Twigger S, Jaiswal P, Seigfried T, Collmer C, Howe D, Westerfield M (2006) The Gene Ontology (GO) project in 2006. Nucl Acids Res 34: D322-D326

Heilman PE (1999) Planted forests: poplars. New for 17/18: 89-93

Jaillon O, Aury JM, Noel B, Policriti A, Clepet C, Casagrande A, Choisne N, Aubourg S, Vitulo N, Jubin C, Vezzi A, Legeai F, Hugueney P, Dasilva C, Horner D, Mica E, Jublot D, Poulain J, Bruyere C, Billault A, Segurens B, Gouyvenoux M, Ugarte E, Cattonaro F, Anthouard V, Vico V, Del Fabbro C, Alaux M, Di Gaspero G, Dumas V, Felice N, Paillard S, Juman I, Moroldo M, Scalabrin S, Canaguier A, Le Clainche I, Malacrida G, Durand E, Pesole G, Laucou V, Chatelet P, Merdinoglu D, Delledonne M, Pezzotti M, Lecharny A, Scarpelli C, Artiguenave F, Pe ME, Valle G, Morgante M, Caboche M, Adam-Blondon AF, Weissenbach J, Quetier F, Wincker P (2007) The grapevine genome sequence suggests ancestral hexaploidization in major angiosperm phyla. Nature 449: 463-467

Jansen RK, Kaittanis C, Saski C, Lee SB, Tomkins J, Alverson AJ, Daniell H (2006) Phylogenetic analyses of Vitis (Vitaceae) based on complete chloroplast genome sequences: effects of taxon sampling and phylogenetic methods on resolving relationships among rosids. BMC Evol Biol 6:

Karrenberg S, Edwards PJ, Kollmann J (2002) The life history of Salicaceae living in the active zone of floodplains. Freshwater Biol 47: 733-748

Kelleher CT, Chiu R, Shin H, Bosdet IE, Krzywinski MI, Fjell CD, Wilkin J, Yin

TM, DiFazio SP, Ali J, Asano JK, Chan S, Cloutier A, Girn N, Leach S, Lee D, Mathewson CA, Olson T, O'Connor K, Prabhu AL, Smailus DE, Stott JM, Tsai M, Wye NH, Yang GS, Zhuang J, Holt RA, Putnam NH, Vrebalov J, Giovannoni JJ, Grimwood J, Schmutz J, Rokhsar D, Jones SJM, Marra MA, Tuskan GA, Bohlmann J, Ellis BE, Ritland K, Douglas CJ, Schein JE (2007) A physical map of the highly heterozygous Populus genome: integration with the genome sequence and genetic map and analysis of haplotype variation. Plant J 50: 1063-1078

Lander ES, Linton LM, Birren B, Nusbaum C, Zody MC, Baldwin J, Devon K, Dewar K, Doyle M, FitzHugh W, Funke R, Gage D, Harris K, Heaford A, Howland J, Kann L, Lehoczky J, LeVine R, McEwan P, McKernan K, Meldrim J, Mesirov JP, Miranda C, Morris W, Naylor J, Raymond C, Rosetti M, Santos R, Sheridan A, Sougnez C, Stange-Thomann N, Stojanovic N, Subramanian A, Wyman D, Rogers J, Sulston J, Ainscough R, Beck S, Bentley D, Burton J, Clee C, Carter N, Coulson A, Deadman R, Deloukas P, Dunham A, Dunham I, Durbin R, French L, Grafham D, Gregory S, Hubbard T, Humphray S, Hunt A, Jones M, Lloyd C, McMurray A, Matthews L, Mercer S, Milne S, Mullikin JC, Mungall A, Plumb R, Ross M, Shownkeen R, Sims S, Waterston RH, Wilson RK, Hillier LW, McPherson JD, Marra MA, Mardis ER, Fulton LA, Chinwalla AT, Pepin KH, Gish WR, Chissoe SL, Wendl MC, Delehaunty KD, Miner TL, Delehaunty A, Kramer JB, Cook LL, Fulton RS, Johnson DL, Minx PJ, Clifton SW, Hawkins T, Branscomb E, Predki P, Richardson P, Wenning S, Slezak T, Doggett N, Cheng JF, Olsen A, Lucas S, Elkin C, Uberbacher E, Frazier M, Gibbs RA, Muzny DM, Scherer SE, Bouck JB, Sodergren EJ, Worley KC, Rives CM, Gorrell JH, Metzker ML, Naylor SL, Kucherlapati RS, Nelson DL, Weinstock GM, Sakaki Y, Fujiyama A, Hattori M, Yada T, Toyoda A, Itoh T, Kawagoe C, Watanabe H, Totoki Y, Taylor T, Weissenbach J, Heilig R, Saurin W, Artiguenave F, Brottier P, Bruls T, Pelletier E, Robert C, Wincker P, Rosenthal A, Platzer M, Nyakatura G, Taudien S, Rump A, Yang HM, Yu J, Wang J, Huang GY, Gu J, Hood L, Rowen L, Madan A, Qin SZ, Davis RW, Federspiel NA, Abola AP, Proctor MJ, Myers RM, Schmutz J, Dickson M, Grimwood J, Cox DR, Olson MV, Kaul R, Raymond C, Shimizu N, Kawasaki K, Minoshima S, Evans GA, Athanasiou M, Schultz R, Roe BA, Chen F, Pan HQ, Ramser J, Lehrach H, Reinhardt R, McCombie WR, de la Bastide M, Dedhia N, Blocker H, Hornischer K, Nordsiek G, Agarwala R, Aravind L, Bailey JA, Bateman A, Batzoglou S, Birney E, Bork P, Brown DG, Burge CB, Cerutti L, Chen HC, Church D, Clamp M, Copley RR, Doerks T, Eddy SR, Eichler EE, Furey TS, Galagan J, Gilbert JGR, Harmon C, Hayashizaki Y, Haussler D, Hermjakob H, Hokamp K, Jang WH, Johnson LS, Jones TA, Kasif S, Kaspryzk A, Kennedy S, Kent WJ, Kitts P, Koonin EV, Korf I, Kulp D, Lancet D, Lowe TM, McLysaght A, Mikkelsen T, Moran JV, Mulder N, Pollara VJ, Ponting CP, Schuler G, Schultz JR, Slater G, Smit AFA, Stupka E, Szustakowki J, Thierry-Mieg D, Thierry-Mieg J, Wagner L, Wallis J, Wheeler R, Williams A, Wolf YI, Wolfe KH, Yang SP, Yeh RF, Collins F, Guyer MS, Peterson J, Felsenfeld A, Wetterstrand KA,

Patrinos A, Morgan MJ (2001) Initial sequencing and analysis of the human genome. Nature 409: 860-921

Lescot M, Rombauts S, Zhang J, Aubourg S, Mathe C, Jansson S, Rouze P, Boerjan W (2004) Annotation of a 95-kb Populus deltoides genomic sequence reveals a disease resistance gene cluster and novel class I and class II transposable elements. Theor Appl Genet 109: 10-22

Lexer C, Heinze B, Alia R, Rieseberg LH (2004) Hybrid zones as a tool for identifying adaptive genetic variation in outbreeding forest trees: lessons from wild annual sunflowers (Helianthus spp.). For Ecol Manag 197: 49-64

Lexer C, Buerkle CA, Joseph JA, Heinze B, Fay MF (2007) Admixture in European Populus hybrid zones makes feasible the mapping of loci that contribute to reproductive isolation and trait differences. Heredity 98: 74-84

Lynch M, Conery JS (2003) The evolutionary demography of duplicate genes. J Struct Funct Genom 3: 35-44

Ma C, Strauss SH, Meilan R (2004) *Agrobacterium* -mediated transformation of the genome-sequenced poplar clone, Nisqually-1 (*Populus trichocarpa*). Plant Mol Biol Rep 22: 311-312

Markussen T, Pakull B, Fladung M (2007) Positioning of sex-correlated markers for Populus in a AFLP- and SSR-marker based genetic map of *Populus tremula* x *tremuloides*. Silvae Genet 56: 180-184

Marra MA, Kucaba TA, Dietrich NL, Green ED, Brownstein B, Wilson RK, McDonald KM, Hillier LW, McPherson JD, Waterston RH (1997) High throughput fingerprint analysis of large-insert clones. Genom Res 7: 1072-1084

Martin F, Tuskan GA, DiFazio SP, Lammers P, Newcombe G, Podila GK (2004) Symbiotic sequencing for the *Populus mesocosm*. New Phytol 161: 330-335i

Martin F, Aerts A, Ahren D, Brun A, Danchin EGJ, Duchaussoy F, Gibon J, Kohler A, Lindquist E, Pereda V, Salamov A, Shapiro HJ, Wuyts J, Blaudez D, Buee M, Brokstein P, Canback B, Cohen D, Courty PE, Coutinho PM, Delaruelle C, Detter JC, Deveau A, DiFazio S, Duplessis S, Fraissinet-Tachet L, Lucic E, Frey-Klett P, Fourrey C, Feussner I, Gay G, Grimwood J, Hoegger PJ, Jain P, Kilaru S, Labbe J, Lin YC, Legue V, Le Tacon F, Marmeisse R, Melayah D, Montanini B, Muratet M, Nehls U, Niculita-Hirzel H, Oudot-Le Secq MP, Peter M, Quesneville H, Rajashekar B, Reich M, Rouhier N, Schmutz J, Yin T, Chalot M, Henrissat B, Kues U, Lucas S, de Peer YV, Podila GK, Polle A, Pukkila PJ, Richardson PM, Rouze P, Sanders IR, Stajich JE, Tunlid A, Tuskan G, Grigoriev IV (2008) The genome of *Laccaria bicolor* provides insights into mycorrhizal symbiosis. Nature 452: 88-92

Perlack RD, Wright LL, Turhollow AF, Graham RL, Stokes BJ, Erbach DC (2005) Biomass as Feedstock for a Bioenergy and Bioproducts Industry: The Technical Feasibility of a Billion-Ton Annual Supply. Oak Ridge Natl Lab, Oak Ridge, TN, USA

Peterson EB, Peterson NM (1992) Ecology, Management, and Use of Aspen and Balsam Poplar in the Prairie Provinces, Canada. Forestry Canada, Victoria, British Columbia, Canada

Ralph SG, Chun HJE, Cooper D, Kirkpatrick R, Kolosova N, Gunter L, Tuskan GA, Douglas CJ, Holt RA, Jones SJM, Marra MA, Bohlmann J (2008) Analysis of 4,664 high-quality sequence-finished poplar full-length cDNA clones and their utility for the discovery of genes responding to insect feeding. BMC Genom 9: 57

Romme WH, Floyd-Hanna L, Hanna DD, Bartlett E (2001) Aspen's ecological role in the West. In: Shepperd WD, Binkley D, Bartos DL, Stohlgren TJ, Eskew LG (eds) Sustaining Aspen in Western Landscapes: Symposium Proceedings; 13-15 June 2000. RMRS-P-18., US Department of Agriculture, Forest Service, Rocky Mountain Research Station, Fort Collins, CO, USA, pp 243-259

Rubin EM (2008) Genomics of cellulosic biofuels. Nature 454: 841-845

Semon M, Wolfe KH (2007) Consequences of genome duplication. Curr Opin Genet Dev 17: 505-512

Solovyev V, Kosarev P, Seledsov I, Vorobyev D (2006) Automatic annotation of eukaryotic genes, pseudogenes and promoters. Genome Biol 7: S10

Song JY, Lu SF, Chen ZZ, Lourenco R, Chiang VL (2006) Genetic transformation of *Populus trichocarpa* genotype Nisqually-1: A functional genomic tool for woody plants. Plant Cell Physiol 47: 1582-1589

Sterck L, Rombauts S, Jansson S, Sterky F, Rouze P, Van de Peer Y (2005) EST data suggest that poplar is an ancient polyploid. New Phytol 167: 165-170

Sterky F, Regan S, Karlsson J, Hertzberg M, Rohde A, Holmberg A, Amini B, Bhalerao R, Larsson M, Villarroel R (1998) Gene discovery in the wood-forming tissues of poplar: analysis of 5,692 expressed sequence tags. Proc Natl Acad Sci USA 95: 13330-13335

Sterky F, Bhalerao RR, Unneberg P, Segerman B, Nilsson P, Brunner AM, Charbonnel-Campaa L, Lindvall JJ, Tandre K, Strauss SH, Sundberg B, Gustafsson P, Uhlen M, Bhalerao RP, Nilsson O, Sandberg G, Karlsson J, Lundeberg J, Jansson S (2004) A Populus EST resource for plant functional genomics. Proc Natl Acad Sci USA 101: 13951-13956

Stettler RF, Koster R, Steenackers V (1980) Interspecific crossability studies in poplars. Theor Appl Genet 58: 273-282

Stettler RF, Zsuffa L, Wu R (1996) The role of hybridization in the genetic manipulation of Populus. In: Stettler RF, Bradshaw HD, Heilman PE, Hinckley TM (eds) Biology of Populus and its Implications for Management and Conservation. NRC Research Press, Ottowa, Canada, pp 87-112

Stirling B, Newcombe G, Vrebalov J, Bosdet I, Bradshaw HD (2001a) Suppressed recombination around the MXC3 locus, a major gene for resistance to poplar leaf rust. Theor Appl Genet 103: 1129-1137

Stirling B, Newcombe G, Vrebalov J, Bosdet I, Bradshaw HDJr (2001b) Suppressed recombination around the *MXC3* locus, a major gene for resistance to poplar leaf rust. Theor Appl Genet 103: 1129-1137

Stirling B, Yang ZK, Gunter LE, Tuskan GA, Bradshaw HD (2003) Comparative sequence analysis between orthologous regions of the Arabidopsis and Populus genomes reveals substantial synteny and microcollinearity. Can J For Res 33: 2245-2251

Strauss SH, Rottmann WH, Brunner AM, Sheppard LA (1995) Genetic engineering of reproductive sterility in forest trees. Mol Breed 1: 5-26

Swarbreck D, Wilks C, Lamesch P, Berardini TZ, Garcia-Hernandez M, Foerster H, Li D, Meyer T, Muller R, Ploetz L, Radenbaugh A, Singh S, Swing V, Tissier C, Zhang P, Huala E (2008) The Arabidopsis Information Resource (TAIR): gene structure and function annotation. Nucl Acids Res 36: D1009-D1014

Taylor G (2002) Populus: Arabidopsis for forestry. Do we need a model tree? Ann Bot 90: 681-689

Tuskan GA (1998) Short-rotation woody crop supply systems in the United States: What do we know and what do we need to know? Biomass Bioenerg 14: 307-315

Tuskan GA, Walsh ME (2001) Short-rotation woody crop systems, atmospheric carbon dioxide and carbon management: A US case study. For Chron 77: 259-264

Tuskan GA, DiFazio S, Jansson S, Bohlmann J, Grigoriev I, Hellsten U, Putnam N, Ralph S, Rombauts S, Salamov A, Schein J, Sterck L, Aerts A, Bhalerao RR, Bhalerao RP, Blaudez D, Boerjan W, Brun A, Brunner A, Busov V, Campbell M, Carlson J, Chalot M, Chapman J, Chen GL, Cooper D, Coutinho PM, Couturier J, Covert S, Cronk Q, Cunningham R, Davis J, Degroeve S, Dejardin A, dePamphilis C, Detter J, Dirks B, Dubchak I, Duplessis S, Ehlting J, Ellis B, Gendler K, Goodstein D, Gribskov M, Grimwood J, Groover A, Gunter L, Hamberger B, Heinze B, Helariutta Y, Henrissat B, Holligan D, Holt R, Huang W, Islam-Faridi N, Jones S, Jones-Rhoades M, Jorgensen R, Joshi C, Kangasjarvi J, Karlsson J, Kelleher C, Kirkpatrick R, Kirst M, Kohler A, Kalluri U, Larimer F, Leebens-Mack J, Leple JC, Locascio P, Lou Y, Lucas S, Martin F, Montanini B, Napoli C, Nelson DR, Nelson C, Nieminen K, Nilsson O, Pereda V, Peter G, Philippe R, Pilate G, Poliakov A, Razumovskaya J, Richardson P, Rinaldi C, Ritland K, Rouze P, Ryaboy D, Schmutz J, Schrader J, Segerman B, Shin H, Siddiqui A, Sterky F, Terry A, Tsai CJ, Uberbacher E, Unneberg P, Vahala J, Wall K, Wessler S, Yang G, Yin T, Douglas C, Marra M, Sandberg G, Van de Peer Y, Rokhsar D (2006) The genome of black cottonwood, *Populus trichocarpa* (Torr. & Gray). Science 313: 1596-1604

Velasco R, Zharkikh A, Troggio M, Cartwright DA, Cestaro A, Pruss D, Pindo M, FitzGerald LM, Vezzulli S, Reid J, Malacarne G, Iliev D, Coppola G, Wardell B, Micheletti D, Macalma T, Facci M, Mitchell JT, Perazzolli M, Eldredge G, Gatto P, Oyzerski R, Moretto M, Gutin N, Stefanini M, Chen Y, Segala C, Davenport C, Dematt+¿ L, Mraz A, Battilana J, Stormo K, Costa F, Tao Q, Si-Ammour A, Harkins T, Lackey A, Perbost C, Taillon B, Stella

A, Solovyev V, Fawcett JA, Sterck L, Vandepoele K, Grando SM, Toppo S, Moser C, Lanchbury J, Bogden R, Skolnick M, Sgaramella V, Bhatnagar SK, Fontana P, Gutin A, Van de Peer Y, Salamini F, Viola R (2007) A high quality draft consensus sequence of the genome of a heterozygous grapevine variety. PLoS ONE 2: e1326

Venter JC, Adams MD, Myers EW, Li PW, Mural RJ, Sutton GG, Smith HO, Yandell M, Evans CA, Holt RA, Gocayne JD, Amanatides P, Ballew RM, Huson DH, Wortman JR, Zhang Q, Kodira CD, Zheng XH, Chen L, Skupski M, Subramanian G, Thomas PD, Zhang J, Gabor Miklos GL, Nelson C, Broder S, Clark AG, Nadeau J, McKusick VA, Zinder N, Levine AJ, Roberts RJ, Simon M, Slayman C, Hunkapiller M, Bolanos R, Delcher A, Dew I, Fasulo D, Flanigan M, Florea L, Halpern A, Hannenhalli S, Kravitz S, Levy S, Mobarry C, Reinert K, Remington K, Abu-Threideh J, Beasley E, Biddick K, Bonazzi V, Brandon R, Cargill M, Chandramouliswaran I, Charlab R, Chaturvedi K, Deng Z, Di F, V, Dunn P, Eilbeck K, Evangelista C, Gabrielian AE, Gan W, Ge W, Gong F, Gu Z, Guan P, Heiman TJ, Higgins ME, Ji RR, Ke Z, Ketchum KA, Lai Z, Lei Y, Li Z, Li J, Liang Y, Lin X, Lu F, Merkulov GV, Milshina N, Moore HM, Naik AK, Narayan VA, Neelam B, Nusskern D, Rusch DB, Salzberg S, Shao W, Shue B, Sun J, Wang Z, Wang A, Wang X, Wang J, Wei M, Wides R, Xiao C, Yan C, Yao A, Ye J, Zhan M, Zhang W, Zhang H, Zhao Q, Zheng L, Zhong F, Zhong W, Zhu S, Zhao S, Gilbert D, Baumhueter S, Spier G, Carter C, Cravchik A, Woodage T, Ali F, An H, Awe A, Baldwin D, Baden H, Barnstead M, Barrow I, Beeson K, Busam D, Carver A, Center A, Cheng ML, Curry L, Danaher S, Davenport L, Desilets R, Dietz S, Dodson K, Doup L, Ferriera S, Garg N, Gluecksmann A, Hart B, Haynes J, Haynes C, Heiner C, Hladun S, Hostin D, Houck J, Howland T, Ibegwam C, Johnson J, Kalush F, Kline L, Koduru S, Love A, Mann F, May D, McCawley S, McIntosh T, McMullen I, Moy M, Moy L, Murphy B, Nelson K, Pfannkoch C, Pratts E, Puri V, Qureshi H, Reardon M, Rodriguez R, Rogers YH, Romblad D, Ruhfel B, Scott R, Sitter C, Smallwood M, Stewart E, Strong R, Suh E, Thomas R, Tint NN, Tse S, Vech C, Wang G, Wetter J, Williams S, Williams M, Windsor S, Winn-Deen E, Wolfe K, Zaveri J, Zaveri K, Abril JF, Guigo R, Campbell MJ, Sjolander KV, Karlak B, Kejariwal A, Mi H, Lazareva B, Hatton T, Narechania A, Diemer K, Muruganujan A, Guo N, Sato S, Bafna V, Istrail S, Lippert R, Schwartz R, Walenz B, Yooseph S, Allen D, Basu A, Baxendale J, Blick L, Caminha M, Carnes-Stine J, Caulk P, Chiang YH, Coyne M, Dahlke C, Mays A, Dombroski M, Donnelly M, Ely D, Esparham S, Fosler C, Gire H, Glanowski S, Glasser K, Glodek A, Gorokhov M, Graham K, Gropman B, Harris M, Heil J, Henderson S, Hoover J, Jennings D, Jordan C, Jordan J, Kasha J, Kagan L, Kraft C, Levitsky A, Lewis M, Liu X, Lopez J, Ma D, Majoros W, McDaniel J, Murphy S, Newman M, Nguyen T, Nguyen N, Nodell M (2001) The sequence of the human genome. Science 291: 1304-1351

Whitham TG, Bailey JK, Schweitzer JA, Shuster SM, Bangert RK, LeRoy CJ, Lonsdorf EV, Allan GJ, DiFazio SP, Potts BM, Fischer DG, Gehring CA,

Lindroth RL, Marks JC, Hart SC, Wimp GM, Wooley SC (2006) A framework for community and ecosystem genetics: from genes to ecosystems. Nat Rev Genet 7: 510-523

Whitham TG, DiFazio SP, Schweitzer JA, Shuster SM, Allan GJ, Bailey JK, Woolbright SA (2008) Perspective—Extending genomics to natural communities and ecosystems. Science 320: 492-495

Wullschleger SD, Janssson S, Taylor G (2002) Genomics and forest biology: *Populus* Emerges as the Perennial Favortie. Plant Cell 14: 2651-2655

Yin TM, Huang MR, Wang MX, Zhu LH, Zeng ZB, Wu RL (2001) Preliminary interspecific genetic maps of the Populus genome constructed from RAPD markers. Genome 44: 602-609

Yin TM, DiFazio SP, Gunter LE, Jawdy SS, Boerjan W, Tuskan GA (2004a) Genetic and physical mapping of Melampsora rust resistance genes in Populus and characterization of linkage disequilibrium and flanking genomic sequence. New Phytol 164: 95-105

Yin TM, DiFazio SP, Gunter LE, Riemenschneider D, Tuskan GA (2004b) Large-scale heterospecific segregation distortion in Populus revealed by a dense genetic map. Theor Appl Genet 109: 451-463

Yin TM, DiFazio SP, Gunter LE, Zhang X, Sewell MM, Woolbright SA, Allan GJ, Kelleher CT, Douglas CJ, Wang M, Tuskan GA (2008) Genome structure and emerging evidence of an incipient sex chromosome in Populus. Genom Res 18: 422-430

Zsuffa L, Giordano E, Pryor LD, Stettler RF (1996) Trends in poplar culture: som global and regional perspectives. In: Stettler RF, Bradshaw HD Jr, Heilman PE, Hinckley TM (eds) Biology of Populus and its implications for management and conservation. NRC Res Press, Ottawa, Canada, pp 515-539

9 Peach Genome Initiative

Albert G. Abbott
Clemson University, Department of Genetics and Biochemistry, 116 Jordan Hall, Clemson, SC 29634, USA
Email: aalbert@clemson.edu

1 INTRODUCTION

Plants of the Rosaceae comprise many of the most important specialty crops worldwide. They are grown both for their fruits (peaches, plums, apples, cherries, strawberries, raspberries and others), lumber (black cherry) and ornamental value (roses). Collectively, they comprise one of the most important temperate region plant families. In the genus *Prunus,* the species that produce drupe fruits (peaches, apricots, almonds, plums, cherries) are significant agriculture crops in many local economies worldwide and provide important components of healthy diets. However, like most crop species, their growth, fruit production and sustainability are influenced by many different biotic and abiotic factors. Rosaceous breeding programs are continually confronted with the need to find genetic solutions to everchanging problems posed by disease and pests (e.g., viruses, fungi, insects) and the everchanging environmental landscape (e.g., drought, global warming, cold temperatures, marginal lands).

Until recently, the genetic understanding of the characters important to stone fruit agriculture lagged far behind that of the large commodity crops such as maize, rice, soybean, etc. This was primarily due to the lack of significant public investment and the refractory nature of the plants, for example juvenility periods in years, requirements for large amounts of labor and space, for achieving high-resolution genetic manipulations. However, with the shift in the market-place focus in many countries from large-scale commodity crops to smaller specialty crops of increased nutritional value, the development of genetic resources in the stone fruit crops came to the forefront. In the last 8-10 years, significant investment by the United States Department of Agriculture (USDA) in the United States and the European Union (EU) granting agencies in Europe and the state of the art in molecular genetics technologies, have substantially contributed to the progress in the development of genetic and genomic resources for key *Prunus* species. In *Prunus*, the peach [*Prunus persica* (L.) Batsch] is currently the most highly genetically characterized species and serves as a reference genome for identification and characterization of genes important to *Prunus* agriculture and more broadly to other species in the Rosaceae as well.

Within the context presented above, it is the intent of this chapter, to provide the most current picture of our understanding of the peach genome; its structural organization, sequence composition, its expression and its relationships to other genomes within and outside the Rosaceae.

2 PEACH STRUCTURAL GENOMICS

In the structural genomics era, merger of high-throughput genomics technologies with traditional genetic approaches has led to an unprecedented discovery rate for genes controlling important characters in model genetic systems. Fig. 1 depicts the central input technologies essential for this gene discovery process.

In peach, through the efforts of numerous laboratories worldwide, we are currently in a position to understand the basic structure and composition of the genome, thus providing the gene information for improvement of the species through modern breeding technologies. These efforts are outlined below.

The peach belongs to the Rosaceae, subfamily Prunoideae, genus *Prunus* and subgenus *Amygdalus*. The peach karyotype consists of one clearly identifiable large submetacentric chromosome, and seven chromosomes of smaller size, two of them being acrocentric (Jelenkovic and Harrington 1972; Salesses and Mouras 1977). Due to the small size of peach chromosomes, little work has been done to characterize the gene content of individual chromosomes, however in almond use of fluorescence in situ hybridization (FISH) has enabled assignment of the rDNA to a specific chromosome and fluorescence microscopy has aided in distinguishing the chromosomes by their size (Corredor et al. 2004).

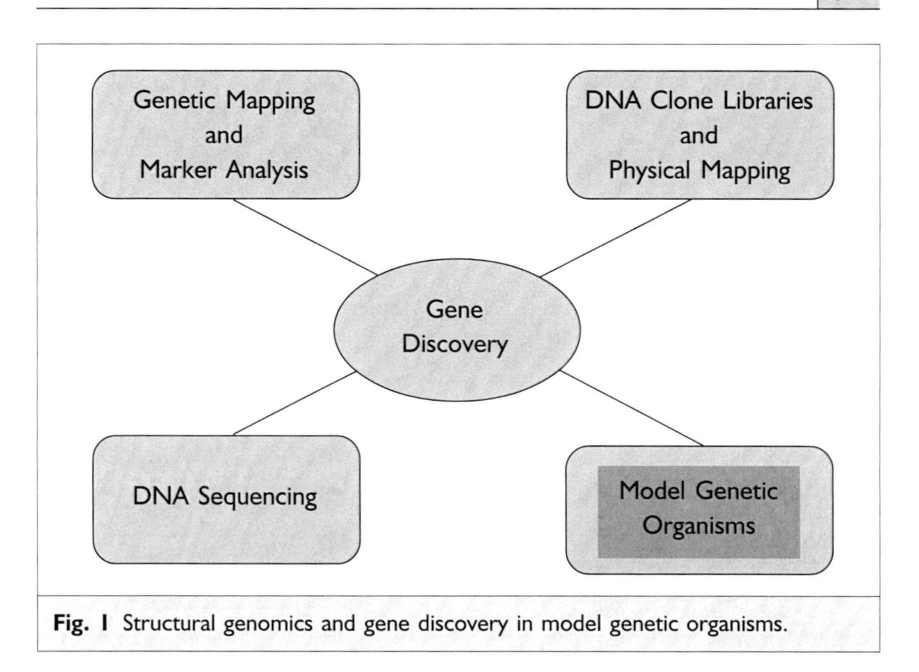

Fig. 1 Structural genomics and gene discovery in model genetic organisms.

As peach crosses with other species in *Amytgdalus* and produce fertile hybrids, it is likely that the peach karyotype is similar to that of almond. In contrast, crosses with species in the other subgenera (*Prunophora* and *Cerasus*) are possible, but fertile hybrids are rare (Scorza and Sherman 1996) suggesting that chromosomal structural organization differs significantly from that of the more distant relatives such as apricot, plum, and cherry.

Peach is predominantly a self-fertilizing species, unlike the majority of its congeneric species that exhibit varying levels of gametophytic self-incompatibility. The genetic base for peach in the US is bottlenecked predominantly due to its history of import from China and its self-compatibility (Scorza et al. 1985; Miller et al. 1989). This has resulted in a relatively low level of genetic variability as evidenced from the difficulties in identifying polymorphism in intraspecific genetic mapping families (see Table 1 below for references) in contrast to other species such as apricot and almond.

A total of 42 morphological characters of simple Mendelian inheritance were discovered during the last century (Dirlewanger and Arús 2004), however until the development of molecular marker maps, only a few linkage relationships had been determined. Five linkage groups involving 11 major genes were reported by Monet et al. (1996).

Table 1 Peach inter- and intra-specific maps (adapted from P. Arus et al. 2006).

Population	*Species*	*Type*	*Marker #*	*TxE anchors*	*LG[1] #*	*Total map distance (cM)*	*References[2]*
'Texas' x 'Earlygold'	almond x peach	F_2	817	817	8	519	Joobeur et al. 1998; Aranzana et al. 2003; Dirlewanger et al. 2004; Howad et al. 2005
NC174RL x 'Pillar'	peach	F_2	88	0	15	396	Chaparro et al. 1994
'N J Pillar' x KV77119	peach	F_2	47	2	8	332	Rajapakse et al. 1995
'Padre' x'54P455'	almond x peach	F_2	161	23	8	1,144	Foolad et al. 1995; Bliss et al. 2002
'Ferjalou Jalousia®' x 'Fantasia'	peach	F_2	124	49	7	518	Dirlewanger et al. 1998; Etienne et al. 2002
'Lovell' x 'Nemared'	peach	F_2	153	1	15	1,297	Lu et al. 1998
'Garfi' x 'Nemared'	almond x peach	F_2	51	51	7*	474	Jáuregui et al. 2001
IF7310828 x *P. ferganensis*	peach x *P. ferganensis*	BC_1	216	71	8	665	Dettori et al. 2001; Verde et al. 2005
'Akame'x 'Juseitou'	peach	F_2	178	45	7*	571	Yamamoto et al. 2001, pers. comm.
'Summergrand' x P1908	peach x *P. davidiana*	F_2	153	57	8	874	Foulongne et al. 2003a
'Contender' x Fla92-2c	peach	F_2	126	96	8	535	Fan et al. 2008

[1] LG = Linkage groups; *Linkage groups 6 and 8 of these maps were mapped as a single group due to the effects of a reciprocal translocation.

[2] When more than one reference is given, the data presented are either from the most recent publication or from the combination of the data from all publications.

2.1 Molecular Genetic Mapping in Peach

Chaparro et al. (1994) constructed the first molecular genetic map in fruit trees consisting of 83 random amplified polymorphic DNA (RAPD) markers, one isozyme marker and four morphological characters in a peach intraspecific F_2 progeny. Two more maps based on restriction fragment length polymorphism (RFLP) markers were published shortly thereafter; the first constructed in a peach x peach F_2 progeny (Rajapakse et al. 1995) and the second in a peach x almond F_2 progeny (Foolad et al. 1995). Later, peach maps utilized amplified fragment length polymorphism (AFLP) markers (Lu et al. 1998) or combinations of AFLPs and RAPDs (dominant markers) with codominant (RFLPs) and morphological markers (Dirlewanger et al. 1998). These maps were considered low level saturated maps (average marker density 4.5-8.5 cM/marker).

Joobeur et al. (1998) published a core genetic map for *Prunus* constructed with transferable markers (11 isozymes and 226 RFLPs) in a 'Texas' almond x 'Earlygold' peach F_2 population. This map was composed of eight linkage groups with a total distance of 491 cM, representing an average density of 2.0 cM/marker, and maximal gap size of 12 cM. This map was improved by the addition of 185 simple sequence repeat (SSR) markers, and 126 RFLPs, most of them obtained with *Arabidopsis* DNA probes, and five sequence-tagged sites (Aranzana et al. 2003; Dirlewanger et al. 2004). Recently, 264 additional SSRs have been mapped to the TxE map using the 'bin mapping' approach (Howad et al. 2005). From the 817 markers currently placed on the TxE map, 756 (92%) are based on known publicly available DNA sequences, with at least 198 (24%) of these sequences corresponding to a putative protein. Recent expressed sequenced tag (EST) mapping through implementation of the peach physical map (see below) has placed an additional 600 EST sequences on this map (Horn et al. 2005).

The *Prunus* scientific community has adopted the TxE map as the reference map for the genus and this map has been used as the genetic anchor for the peach physical map (Zhebentyayeva et al. 2008). It provides a set of transferable markers that can be used as anchors for map construction in other progenies, a common linkage group terminology and marker order within each linkage group, and a highly polymorphic population that allows mapping markers that would not segregate in most peach intraspecific crosses. Table 1 presents a compilation of the inter- and intra-specific peach maps that have been published. Those anchored on the *Prunus* general map are highlighted.

Since peach exhibits a low level of intraspecific variation due to historical genetic bottlenecks in its dispersion and its self-compatibility, a very densely marked 'consensus' map with highly polymorphic markers well distributed in all the genomic regions would insure that segregating markers are available in regions of interest in other peach crosses. To reach this goal, researchers utilized the peach physical map resources to increase the number of SSRs mapped in parallel with targeted strategies to fill regions with low SSR density around the traits of importance (Wang et al. 2001, 2002; Georgi et al. 2002). Through this core reference map, it is now possible to locate the major genes and quantitative trait loci (QTL) that segregate in different populations (Table 2).

In total, 23 loci controlling simple-inherited characters were assigned to specific positions on the TxE map, 19 of these loci were mapped in intraspecific peach crosses and three that segregated in interspecific almond x peach crosses. For complex characters 35 QTLs for bloom and maturity time, fruit quality, tree architecture, disease resistance and chilling requirement were also placed on the map (Abbott et al. 1998; Viruel et al. 1998; Dirlewanger et al. 1999; Etienne et al. 2002; Verde et al. 2002; Foulongne et al. 2003b; Fan et al. 2008; Olukolu et al. 2008). Additionally, using the physical map resources anchored on the *Prunus* general map (see below), a resistance gene map was integrated into this *Prunus* genetic framework (Lalli et al. 2005) serving as a resource for identification of resistance gene containing regions of the peach genome.

2.2 Physical Mapping

Integrated physical/genetic maps are of critical importance for high-throughput EST mapping, QTL fine-mapping and effective positional cloning of genes (Zhang and Wing 1997; Green 2001; Zhang and Wu 2001). Fig. 2 below depicts the strategy whereby an integrated complete physical map and a high density genetic map provide cloned DNAs containing genes of importance. If one has identified genetic markers that flank a character, then direct alignment of the physically mapped clones provide the cloned genomic window where the gene controlling the trait lies. Sequencing this interval provides the candidate gene information for final identification of the gene or genes of importance.

In peach, a consortium of laboratories in the United States and abroad have developed a substantially complete physical map for peach employing essentially the strategies used to develop the physical maps for *Arabidopsis thaliana* and *Drosophila melanogaster* (Marra et al. 1999; Hoskins et al. 2000). The approach combined hybridization of the geneti-

Table 2 Major genes and QTL placed on the *Prunus* reference map (adapted from Arus et al. 2006).

Characters	*LG*[1]	*Symbol*[2]	*Populations*	*References*
Flesh color (white / yellow)	G1	*Y*	'Padre' x '54P455'	Warburton et al. 1996; Bliss et al. 2002
Evergrowing	G1	*Evg*	'Empress op dwarf' x PI442380	Wang et al. 2002
Internode length	G1	QTL	(*P. ferganensis* x 'IF310828')BC1	Verde et al. 2002
Powdery mildew resistance	G1	QTL	'Summergrand' x P1908	Foulongne et al. 2003b
Flower color	G1	*B*	'Garf'l' x 'Nemared'	Jauregui 1998
Chilling requirement	G1	*Cr* QTL	'Contender' x Fla 92-2c, 'Perfection' x A.1740	Fan et al. 2008; Olukolu et al. 2009
Root-knot nematode resistance	G2	*Mi*[(3)]	'Akame' x 'Juseitou', 'Garfi' x Nemared', 'Lowell' x 'Nemared', 'P.2175' x 'GN22', 'Padre' x '54P455'	Yamamoto et al. 2001; Jáuregui 1998; Lu et al. 1998; Claverie et al. 2004; Bliss et al. 2002
Ripening time, fruit skin color, soluble-solids content	G2	QTL	(*P. ferganensis* x 'IF310828') BC1	Verde et al. 2002
Double flower	G2	*Dl*	'NC174RL' x 'PI'	Chaparro et al. 1994
Broomy (or pillar) growth habit	G2	*Br*	Various progenies	Scorza et al. 2002
Chilling requirement	G2	*Cr* QTL	'perfection' x A.1740	Olukolu et al. 2009
Flesh color around the stone	G3	*Cs*	'Akame' x 'Jusetou'	Yamamoto et al. 2001
Anther color (yellow/anthocyanic)	G3	*Ag*	'Texas' x 'Earlygold'	Joobeur 1998

Contd.

Table 2 continued

Characters	*LG*[1]	*Symbol*[2]	*Populations*	*References*
Leaf curl resistance	G3	QTL	'Summergrand' x P1908	Viruel et al. 1998
Fruit weight, fruit diameter, glucose content	G3	QTL	'Suncrest' x 'Bailey'	Abbott et al. 1998
Polycarpel	G3	*Pcp*	'Padre' x '54P455'	Bliss et al. 2002
Flower color	G3	*Fc*	'Akame' x 'Jusetou'	Yamamoto et al. 2001
Chilling requirement	G3	*Cr* QTL	'Perfection' x A.1740	Olukolu et al. 2009
Blooming time, ripening time, fruit development period	G4	QTL	'Ferjalou Jalousia®' x 'Fantasia'; (*P. ferganensis* x 'IF310828')BC1	Etienne et al. 2002 Verde et al. 2002
Soluble-solids content, fructose, glucose	G4	QTL	'Ferjalou Jalousia®' x 'Fantasia'	Etienne et al. 2002
Flesh adhesion (clingstone / freestone)	G4	*F*	(*P. ferganensis* x 'IF310828') BC1; 'Akame'x 'Juseitou'	Verde et al. 2002; Dettori et al. 2001; Yamamoto et al. 2001
Flesh texture (melting/ non-melting)	G4	*M*	'Dr. Davis' x 'Georgia Belle' and ' Georgia Belle' ⊗	Peace et al. 2005
Chilling requirement	G4	*Cr* QTL	'Contender' x Fla 92-2c,	Fan et al. 2008
Non-acid fruit	G5	*D*	'Ferjalou Jalousia'® x 'Fantasia'	Dirlewanger et al. 1998, 1999; Etienne et al. 2002
Sucrose, malate, titrable acidity, pH, sucrose	G5	QTL	'Ferjalou Jalousia®' x 'Fantasia'	Etienne et al. 2002

Contd.

Table 2 continued

Characters	*LG*[1]	*Symbol*[2]	*Populations*	*References*
Skin hairiness (nectarine/peach)	G5	*G*	'Ferjalou Jalousia®' x ' Fantasia'; 'Padre' x '54P455'	Dirlewanger et al. 1998, 1999 Bliss et al. 2002
Kernel taste (bitter/sweet)	G5	*Sk*	'Padre' x 54P455'	Bliss et al. 2002
Chilling requirement	G5	*Cr* QTL	'Contender' x Fla 92-2c, 'Perfection' x A.1740	Fan et al. 2008; Olukolu et al. 2009
Ripening time, fruit skin color, soluble-solids content	G6	QTL	(*P. ferganensis* x 'IF310828')BC1	Verde et al. 2002
Plant height (normal / dwarf)	G6	*Dw*	'Akame' x 'Juseitou'	Yamamoto et al. 2001
Leaf shape (narrow / wide)	G6	*Nl*	'Akame' x 'Juseitou'	Yamamoto et al. 2001
Male sterility	G6	*Ps*	'Ferjalou Jalousia®' x 'Fantasia'	Dirlewanger et al. 1998
Powdery mildew resistance	G6	QTL	'Summergrand' x P1908	Foulongne et al. 2003b
Leaf curl resistance	G6	QTL	'Summergrand' x P1908	Viruel et al. 1998
Fruit shape (flat/round)	G6	*S**	'Ferjalou Jalousia®' x 'Fantasia'	Dirlewanger et al. 1998, 1999
Chilling requirement	G6	*Cr* QTL	'Contender' x Fla 92-2c, 'Perfection' x A.1740	Fan et al. 2008; Olukolu et al. 2009
Leaf color (red/yellow)	G6-G8	*Gr*	'Garfi' x Nemared'; 'Akame' x 'Juseitou'	Jauregui 1998; Yamamoto et al. 2001
Fruit skin color	G6-G8	*Sc*	'A kame' x 'Juseitou'	Yamamoto et al. 2001
Leaf gland (reniform/globose/ eglandular)	G7	*E*	(*P. ferganensis* x 'IF310828')BC1	Dettori et al. 2001

Contd.

Table 2 continued

Characters	*LG*[1]	*Symbol*[2]	*Populations*	*References*
Resistance to mildew	G7	QTL	(*P. ferganensis* x 'IF310828')BC1	Verde et al. 2002
Chilling requirement	G7	*Cr* QTL	'Contender' x Fla 92-2c, 'Perfection' x A.1740	Fan et al. 2008; Olukolu et al. 2009
Powdery mildew resistance	G8	QTL	'Summergrand' x P1908	Foulongne et al. 2003b
Quinase	G8	QTL	'Ferjalou Jalousia®' x 'Fantasia'	Etienne et al. 2002
Showy bloom	G8	Sh	'Contender' x Fla 96-2c	Fan et al. 2008
Chilling requirement	G8	*Cr* QTL	'Contender x Fla 92-2c, 'Perfection' x A.1740	Fan et al. 2008; Olukolu et al. 2009

[1] LG = Linkage group; G6-G8 genes located close to the translocation breakpoint between these two linkage groups.
[2] QTLs are included if they have been consistently found (at least in two independent measurements) in the indicated populations.
[3] One or two genes of nematode resistance with different notations and one QTL which have been described in this linkage group.

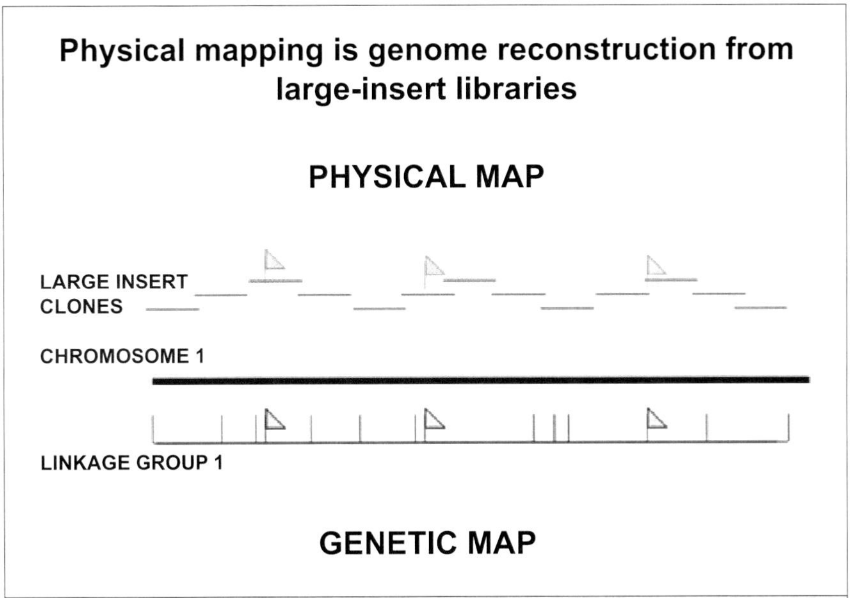

Physical mapping is genome reconstruction from large-insert libraries

PHYSICAL MAP

LARGE INSERT CLONES

CHROMOSOME 1

LINKAGE GROUP 1

GENETIC MAP

Fig. 2 Illustration of the integration of physical map and genetic maps for gene discovery. Vertical lines on the linkage group represent genetically mapped markers, flags represent genetically mapped phenotypic characters. Horizontal lines on the physical maps represent cloned tiling paths (individual cloned DNAs overlapping and spanning the chromosome (linkage group). By projecting the mapped DNA markers onto the physical map one can infer which cloned pieces of DNA must contain the sequence of the gene specifying the flagged characters.

cally mapped markers with bacterial artificial chromosome (BAC) DNA fingerprinting and, in this case, hybridization of ESTs and whole cDNA sequences, as well. An initial acrylamide gel-based physical framework for peach was established and released in 2006 (Zhebentyayeva et al. 2006). This framework, based on random fingerprinting of 3x peach genome equivalents, covered at least 50% of the genome and included hybridization data for 673 of 3,384 ESTs of the peach unigene set (PP_LEa, i.e., *Prunus persica* 'Loring' fruit ESTs). Genetically anchored BAC contigs containing EST sequences provided the substrate for a *Prunus-Arabidopsis* microsynteny study (Jung et al. 2006) and for further development of the *Prunus* transcript map (Horn et al. 2005). Since this initial map was released, we have produced a second generation map integrating global hybridization data that include additional ESTs, sequenced AFLP fragments, SSRs (microsatellites), gene-specific genomic probes and 'overgo' probes derived from BAC-end sequences. The total number of markers in this physical framework was increased to 2,636 markers. Selectively, all the hybridization positive BACs omitted during random fingerprinting were fingerprinted. As a result, an additional 1x peach genome equivalent composed of marker-positive BACs was incorporated. Finally, advantage of the high information content fingerprinting (HICF) technique along with improved FPC v8.5.2 software facilitated increasing the average number of bands per BAC clone and improving accuracy in contig assembly (Nelson et al. 2007). This HICF physical map for the peach consists of 2,138 contigs of which 252 contigs are anchored to eight linkage groups of the *Prunus* reference map (Zhebentyayeva et al. 2008). Length of the physical map contigs approximates 303 Mb, which exceeds the estimated size of the peach genome (Baird et al. 1994). Due to the abundance of hybridization data, the HICF physical map for peach is biased to the expressed genomic regions, and thus substantially covers the euchromatic portion of the peach genome (for more details, see the Genome Database for Rosaceae (GDR) at *http://www.rosaceae.org*.

2.3 Genomic Sequencing

The peach genome has been targeted by the Joint Genome Institute (JGI) for whole-genome sequencing as it is a logical reference genome for the *Prunus* species and a comparative genome for the Rosaceae. Due to the current state of structural genomics in peach, it holds the promise for the most rapid complete genome assembly. To date JGI has completed a 4-5x coverage of the genome using whole-genome shotgun Sanger sequencing. Additional short-read sequencing has been completed and a least a 40-60x coverage of the genome has been done using Solexa sequencing technologies (D. McCombie, pers. comm.). Work is currently underway to annotate and assemble the sequences that will be housed and publicly available in the GDR and NCBI databases (B. Sosinski, pers. comm.).

Several laboratories have reported pilot-scale genomic sequences of several regions of the peach genome chosen for their association with particular phenotypic characters. Bielenberg et al. (2007) published the sequences from the lower end of LG 1, the *EVG* (evergrowing) region, which contains genes controlling the process of endodormancy. The analysis of this region reveals an endoduplication of MADS box genes of the short vegetative phase (SVP) class which control vegetative to floral meristem transitions (Bielenberg et al. 2008). Currently this region is under further sequence characterization as it is associated with QTL that influence chilling requirement for bud-break in the spring (Olukolu et al. 2009; Fan et al. 2009).

Another region of the peach genome currently being sequenced contains genes important to Plum Pox Virus (PPV) resistance in apricot. BAC-scale sequencing in this region has been done to identify markers and genes candidates for PPV resistance in both the peach and apricot genomes (V Decroocq and AG Abbott, unpublished results; ML Badenes and AG Abbott, unpublished results). SSR markers developed from these BAC sequences are currently being tested for marker-assisted selection (MAS) of PPV resistance in apricot (Badenes, pers. comm.).

Additional regions of the peach genome have undergone limited BAC scale sequencing. These regions were chosen for genome comparative studies and details are provided below in the section on comparative genomics.

3 PEACH FUNCTIONAL GENOMICS, PROTEOMICS, AND METABOLOMICS

Structural approaches, as outlined above, provide gene sequences as candidates controlling characters of importance, however many characters are quantitatively inherited and may rely on systems of genes operating in concert to specify the final character. Systems approaches to genetic studies often follow a route that takes advantage of several high throughput technologies all directed at merging gene sequences and their various alleles with final metabolic products in the organism. Fig. 3 depicts the 'omics' systems and their interplay.

The status of this systems picture in peach is outlined below.

3.1 EST Sequencing in Peach

The critical substrate for functional analysis of any genome is the gene sequences. In this regard, efforts worldwide have provide a number of EST sequences for peach as part of a growing set of ESTs for the Rosaceae. Peach ESTs are discussed here in the context of the broader set of gene sequences for the family.

A recent analysis of all Rosaceae ESTs has resulted in a combined unigene set for the family, as well as for each individual species with EST representation in the database (Table 3). This set was derived from a total of 369,106 ESTs representing 151 cDNA libraries and 17 species. Further curation of the data revealed that 20 tissue types were repre-

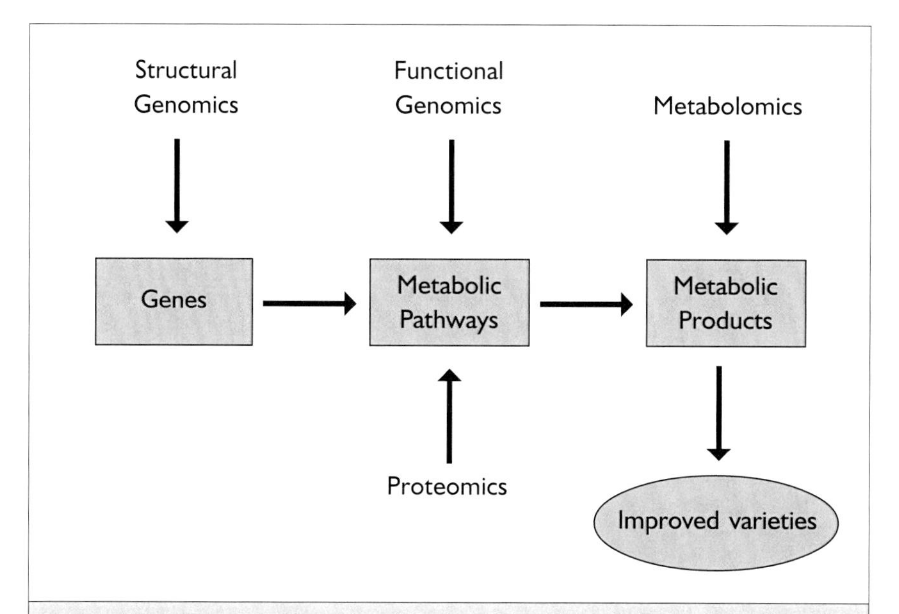

Fig. 3 Illustration depicting the informational system of organisms and the direction of information flow. The "omics" technologies are focused at specific levels of the information hierarchy. Integration of "omics" data provides a systems approach to understanding and genetically manipulating phenotype.

Table 3 Summary of Rosaceae and species specific unigenes available through GDR

Organism	*EST Number*	*Unigene Number*	*Contig Number*	*Singlet Number*
Fragaria	18,729	10,012	2,939	7,073
Malus	250,907	82,850	23,868	58,982
Prunus	83,751	23,721	8,818	14,903
Pyrus	330	271	35	236
Rosa	5,284	2,963	705	2,258
Rosaceae	359,001	90,337	13,764	76,573

sented, and that over 40% of the tissues were from fruit. This unigene set was also compared to other available plant unigenes from *Medicago, Arabidopsis,* grape, soybean, etc. and a strong sequence similarity was shown. Additional screening of the contigs revealed in excess of 33,000 SSRs from the genera specific contigs and 27,260 of them were found in the Rosaceae contigs (Staton 2007). To date, only a handful of published functional genomics studies employing microarrays have been conducted in the Rosaceaous species. These include two studies in apple (Lee et al. 2007; Schaffer et al. 2007), three in strawberry (Aharoni et al. 2000, 2002, 2004), and one from raspberry (Mazzitelli et al. 2007), but none of these employed the GDR unigenes. It is anticipated that future microarray projects within the Rosaceae will employ this useful resource thereby allowing interesting cross-species comparisons.

Given the scope and size of sequence conservation of ortholgous genes between Rosaceous species, as well as with other plant species, we believe that we will be able to generate a set of 60-mer oligonucleotides representing the Rosaceae unigene set that will be useful for expression studies for any species within the Rosaceae. This microarray resource will be of great value for expression studies for the major crop plants within the family, but perhaps more importantly, this Rosaceae-wide microarray will be an invaluable tool for use in the under-represented and under-funded species including *Rubus* (Brambles e.g., raspberry), rose, cherry and apricot. For many of these under-represented species, where EST resources have not been developed, and may not be developed in the near future, a Rosaceae microarray will facilitate functional genomics analyses that would be otherwise unavailable.

3.2 EST Sequences and Physical Mapping

Concomitant with BAC fingerprinting, in excess of 2,300 EST probes have been hybridized on the physical map. This combined physical/genetic/EST map positions a significant number of genes along each linkage group and thus provides the substrate for candidate gene approaches to search in genetically mapped trait containing intervals as depicted in Fig. 4 (for details of the peach transcriptome map, see the GDR database).

3.3 Transcript Profiling Studies

With the increase in EST, cDNA, and genomic sequences available for peach, research laboratories have focused on a number of studies on global gene expression analyses of peach tissues in a variety of conditions. Some studies have focused on analyzing the expression changes of peach genes during fruit ripening (e.g., Trainotti et al. 2006) as the quality of fruit is a major target for genetic manipulation, while others have focused attention on peach tissues challenged with environmental stresses such as cold temperatures (e.g. Gonzalez-Aguero et al. 2008).

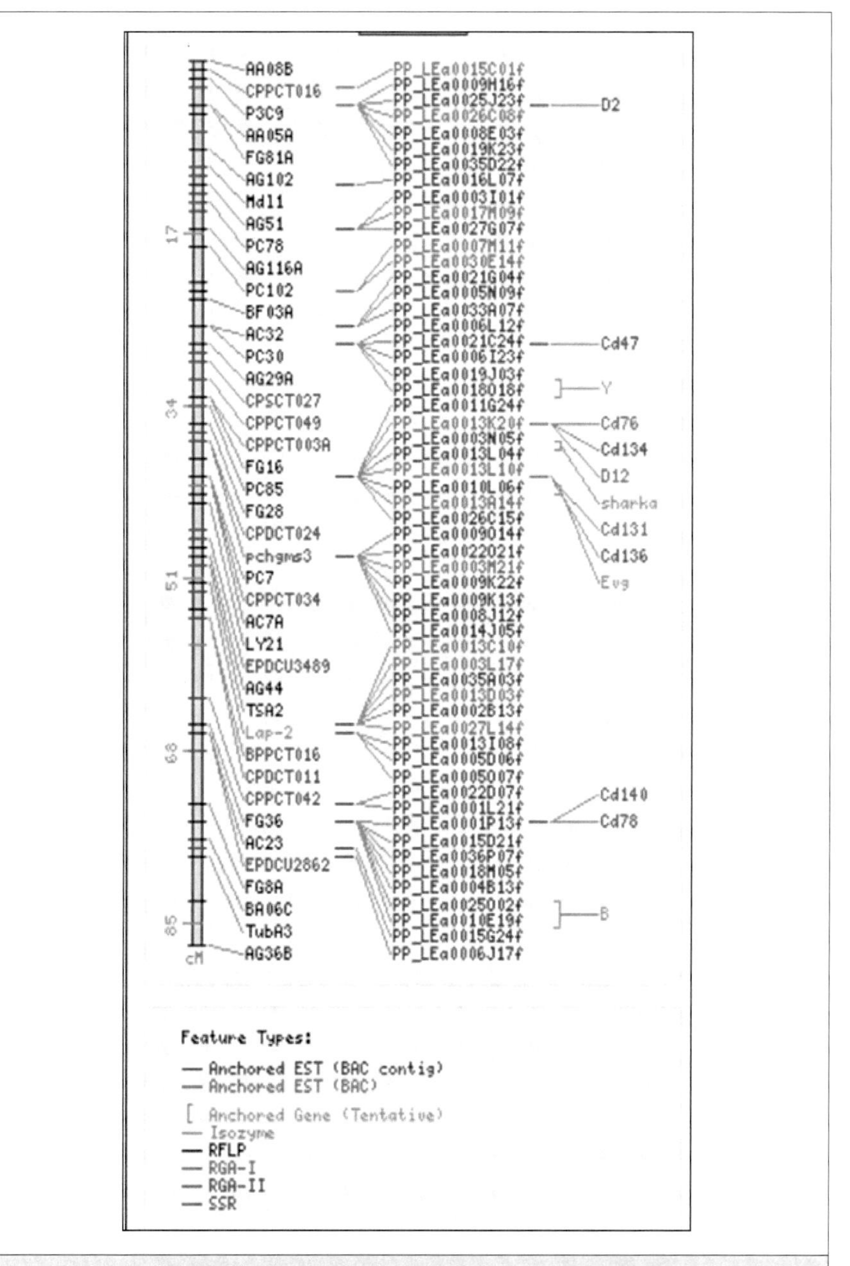

Fig. 4 LG I of the Peach transcriptome map in GDR. The PPV resistance region has been defined in the mapping analysis of crosses shown in Table 1.

(Color image of this figure appears in the color plate section at the end of the book).

The purpose of such studies is to identify and characterize expression of key genes involved in processes effecting fruit quality, stress resistance (biotic and abiotic), and other important life history characters (e.g., endodormancy, chill response) as these genes are targets for manipulation either through breeding or transgenic technologies. Results of a few of these studies are summarized below.

Peach is currently being examined by a number of laboratories as a model for the development and ripening of drupe fruits. Thus, a number of recent reports have applied transcriptomics to the study of developing fruit. For developmental characterization of gene expression in fruit ripening, Trainotti et al. (2006) first published the use of EST chips for characterizing gene expression in ripening peach fruit. In their work, they employed the peach microarray (μPEACH1.0) that contained 4,800 oligonucleotide sequences from a set of unigenes expressed during late stages of peach development. Hybridizations using probes from pre-climateric to climateric stages identified 267 up- regulated genes and 109 down-regulated genes. Using the TAIR Gene Ontology these genes were classified into three main categories based on cellular localization, molecular function and biological process. Results of this work suggested a significant increase in expression of transcription factors and enzymes involved in ethylene biosynthesis and action. Some of these include a new member of the peach ETR family (Pp-ETR) and 19 genes encoding transcription factors of the MADS-box, AUX/IAA, bZIP, bHLH, HD, and Myb families. Genes controlling fruit quality were also seen to vary in expression including cell wall-related genes, a new pectin-methyl esterase, two expansins, genes encoding enzymes in the isoprenoid pathway and genes in the carotenoid pathway.

Others have applied transcript profiling in peach or nectarine fruits challenged with specific treatments. Zillioto et al. (2008) examined nectarine fruits challenged with 1-MCP, a known ripening inhibitor of fruits. Their results indicate that after treatment there is an initial limited inhibition of genes involved in ethylene synthesis. However unlike other fruits, other mechanisms apparently modulate ethylene-dependent genes involved in ripening of nectarines. Some genes for example show a marked recovery in expression post treatment (e.g., endo-PG), while others appear to be more inhibited by the treatment (e.g., PpEXP2, PpEXP3).

Ziosi et al. (2008) examined the effects of exogenous application of jasmonates (JAs) to the ripening transcriptome in peach fruits. Besides their role in wound response and plant defenses, JAs are also involved in root growth, seed germination, pollen development, and fruit development and ripening (Pena-Corte´s et al. 2005; Wasternack 2007). Results of this work demonstrated that in peach fruit, exogenous JAs delayed ripening through interference with ripening and stress/defense-related genes. Some examples of down-regulated genes included, 1-aminocyclopropane-1-carboxylic acid oxidase (PpACO1), polygalacturonase (PG), and IAA7, a transcriptional modulator. JAs also affected the expression of cell wall-related genes (e.g., pectate lyase and expansions and additionally up-regulated several stress-related genes including those involved in the synthesis if endogenous JA.

Finally, one transcriptomics study has been reported examining the changes in peach fruit gene expression during post harvest storage (Gonzalez-Aguero et al. 2008). This study was directed specifically at identifying important gene expression differences that might lead to an understanding of the wooliness character in peaches, which is a substantial post-harvest problem leading to mealy inedible fruit. The results of this study reveal that 106 genes were differentially expressed between normal and mealy fruit. In the mealy fruit most of these genes were repressed. Further studies on several cultivars of a subset of these genes (cobra, endopolygalacturonase, cinnamoyl-CoAreductase and rab11) demonstrated that their expression was lower in mealy fruit and remained low through the ripening of process. These results indicate that woolliness involves significant changes in expression genes involved with cell wall metabolism and endomembrane trafficking and that these genes are good candidates for manipulation of this character to ameliorate the negative effects of cold storage on peach fruit.

3.4 Proteomics Studies

One of the realities of transcriptomics is that although one can access very quickly the large-scale steady-state levels of transcripts for numerous genes, there are significant questions concerning the relevance of transcript level variation to actual protein presense (the proteome) or activity. Proteomics approaches have come to the recent forefront as an essential step in the integration of genes to physiological states to final metabolic products. Arguably these data may be more critical in the long run to give us a picture of the essential controls of many of the processes key to improving plant fruits. In peach, there have been a few reports of the application of proteomics approaches of which one is highlighted one below.

Of the characters that most influence the geographic range and success of temperate fruit agriculture, cold acclimation, dormancy and cold resistance are perhaps the most important. Renaut et al. (2008) recently reported a study of cold acclimation and short-day effects on the bark proteome of peaches. Using current proteomics technologies (DiGE and mass spectrometry) differentially expressed proteins were assigned to either 'temperature responsive' or 'photoperiod responsive' categories. The most dramatic effect on the proteome was linked to low temperature; however, short photoperiod and low temperature appeared to act synergistically on the expression of some proteins. This study identified 57 proteins with altered presense. These proteins were involved in basic metabolic processes such as that of carbohydrates, (e.g., enolase, male dehydrogenase), energy production (e.g., ATP synthases and lyases), defense or protective responses (e.g., dehydrins, HSPs and PR-proteins), cytoskeletal organization (e.g., tubulins and actins), transcription regulation and others. This work when combined with other 'omics' levels of investigation should lead to a better understanding of gene activities in temperate woody plant species during acclimation and resistance to environmental stresses.

3.5 Metabolomic Studies

There are currently no published reports of the application of metabolomics technologies in peach, however, it is certain that this arena promises to be the most 'fruitful' for identifying those genes most important to the direct genetic manipulation of peach trees to improve the nutritional value and quality of the fruit. Directly identifying those compounds that promote high quality of the fruit (sugars, acids, polyphenols, and others) and linking their presence to the enzymes and thus the genes that control their manufacture will enable breeding for alleles that increase or decrease such traits. Additionally, these genes will be excellent targets for direct transgenic approaches when such technologies will become routine in peach. For the moment, we must wait for the application of the current technologies to this most important crop species however, there have been some initial reports of metabolomic approaches in other Rosaceous crops and for further details see the recent review of Shuleav et al. (2008).

4 PEACH COMPARATIVE GENOMICS

4.1 Intrafamily Comparative Genomics

The availability of extensive genomics resources in peach, in particular, a complete physical/genetic map populated with extensive ESTs, is currently allowing the use of comparative mapping strategies to identify and characterize the genetic composition of regions of the *Prunus* genome that are known to control important traits. This is achieved by cross-referencing markers of the peach physical/genetic map to the regions mapped in crosses in other species that carry the traits of interest. This immediately provides physically mapped contigs of sequences, some with ESTs to be explored as candidate genes for the traits under study. Below, two important international collaborative projects implementing this comparative strategy are outlined.

4.1.1 Plum Pox Virus (PPV)

PPV is the most devastating disease problem of fruit trees worldwide. Having its origin in Bulgaria, it has spread westward through Europe and now into the Americas (Nemeth 1994).

Currently, with the discovery of PPV in Pennsylvania, in Niagara on the Lake, Canada and the bordering regions of the New York State (2008), it is obvious that our fruit tree industry is now perched in an extremely precarious position. Control of PPV is completely limited to eradication of trees infected with the disease. For the grower and consumer, this could mean severe loss of production and of crops of fruiting trees for many years. Because of the potential disastrous effects on the American fruit tree industry, PPV was listed earlier by the federal government as one of the top ten most significant threats to agriculture in the United States (see Federal Government ACT: Public Health Security and Bioterrorism Act of 2002, described at *www.aphis.usda.gov/ppq/permits/agr_bioterrorism*). Though it is no longer on this list, PPV remains one of the greatest threats to the American stone fruit industry.

Driven by the need to formulate strategies for developing fruit trees resistant to this virus, an international consortium of laboratories devoted to the integrated study of genetic resistance to PPV was formed. Laboratories in this consortium are currently employing three strategies. These are: (1) comparative molecular mapping with several independently identified genetic sources of resistance in apricot varieties; (2) a candidate gene approach to potentially identify gene products directly related to resistance or capable of enhancing resistance; (3) *Prunus* transgenics to test individual candidate resistance genes and pyramid natural resistance genes with pathogen-derived resistance approaches.

All of this work relies heavily on integrating mapping results from resistant apricots onto the physical/genetic map of peach as a means to position in high resolution the location of regions conferring genetic resistance to this disease. Initial results have enabled a major resistance gene containing region of apricot on linkage group one to be significantly genetically resolved (Vera-Ruiz et al. 2009) and candidate genes conferring resistance to be identified (Marandel et al. 2008). Further work in this project will be expedited by the complete peach genome sequences as the complete genetic nature of the intervals where the resistance is located will be known in the very near future.

4.1.2 Chilling Requirement

Chilling requirement in peach and apricot trees as with most temperate fruiting trees is a very important trait governing the production of fruit in regions, where late frosts in the spring can severely limit production of fruits due to destruction of flower blossoms. Thus, identification of candidate genes that control chilling requirement (CR) in fruiting trees of the family Rosaceae provides an avenue to manipulate this trait in the breeding program to accommodate everchanging year to year environmental conditions. Results of studies on the genetics of this trait will provide insight into the fundamental control of this very important character in perennial tree species while providing robust markers for selection of this character in breeding programs utilizing modern translational breeding strategies. For this reason, a collaborative project designed to map CR QTLs in two different species of Rosaceae (apricot in Israel, peach in the US) has proven critical in the definition of genomic regions controlling the important character in these fruiting trees. It is anticipated that results from these studies will likely translate to other important perennial species in Rosaceae and to perennial species in other plant families as well.

The work to-date utilizing the peach physical/genetic map resources has demonstrated: (1) CR in peach and apricot is predominantly controlled by a limited number of QTL loci, seven detected in a peach F_2 derived map comprising 65% of the character and seven in an apricot F_1 map comprising 71.6% and 55.6% of the trait in the Perfection and A 1740 parental maps, respectively (Fan et al. 2007; Olukolu et al. 2008), and that peach and apricot initial maps appear to share five genomic intervals containing potentially common QTL. (2) Application of common anchor markers of the *Prunus*/peach, physical/genetic map resources has allowed the identification of the shared intervals but also to have immediately available some putative candidate gene information from these intervals, the *EVG* region on LG 1 in peach (Bielenberg et al. 2004) and the *TALY 1* region in apricot on LG 2 in peach (Olukolu et al. 2008). (3) Mapped BAC contigs are easily defined from the complete physical map resources in peach through the common SSR markers that anchor the CR maps in the two species. (4) Sequences of BACs in these regions can be easily mined for additional polymorphic markers to use in marker-assisted selection (MAS) applications.

Thus by using common markers and genomic information available in peach, comparative map integration quickly translates into potential candidate genes and markers for selection of variation in this very important trait.

4.2 Interfamily Comparative Genomics

4.2.1 Peach-Arabidopsis Comparisons

In order to examine the evolution of the plant genome, it is extremely valuable to compare structural organization of relatively similar sized genomes of plants that have diverged over significant evolutionary time. Thus, identification of significantly conserved regions potentially identifies functional chromosomal units. The *Prunus* map and the *Arabidopsis* genome sequence have been compared using a set of RFLP markers mapped in the TxE map obtained either with probes of different species (mainly *Prunus* and apple) that had a high level of sequence conservation with *Arabidopsis* (TBLASTX values lower than 10^{-15}) or with *Arabidopsis* probes that hybridized well to *Prunus* DNA (Dominguez et al. 2003). The position of 227 *Prunus* loci (map average density of 2.6 cM/marker) could be compared to that of 703 *Arabidopsis* homologous sequences. The criterion for declaring a syntenic region was that three or more homologous markers had to be located within 1% of the *Prunus* map distance (6 cM) and within a 1% of the *Arabidopsis* genome (1.2 Mb). In addition, blocks with gaps longer than 1% of either genome were rejected. With these stringent criteria, it was possible to detect 37 syntenic regions, covering 23% and 17% of the *Prunus* and *Arabidopsis* genomes, respectively. The longest of these regions included 13 markers for a distance of 25 cM in linkage group 2 of *Prunus* and 16 homologous sequences spanning 5.4 Mb in chromosome 5 of *Arabidopsis*.

Similarly, higher resolution studies have not supported extensive preservation of localized genome structure between the two genomes. The sequence of peach BACs and BAC ends located in several locations in the peach genome was compared with that of *Arabidopsis* (Georgi et al. 2003; Jung et al. 2006). Predicted genes in these sequences were homologous to genes scattered along the five chromosomes of *Arabidopsis*, with limited preservation to several genes in any one location. In summary, macro- and micro-synteny results concur in detecting a fragmentary preservation between these two genomes putatively separated for more than 90 million years.

4.2.2 Peach-Populus-Medicago Comparison

A recent study (Jung et al. 2008) highlights microsyntenic and macro-syntenic comparisons among the genomes of peach, poplar and *Medicago*. Due to extensive physically mapped EST sequences in peach and the availability of complete BAC sequences for some regions of the peach genome, it was possible to compare these regions with the complete sequences of poplar and partial sequences of *Medicago*. For this work seven *Prunus* BAC sequences, 1,093 *Prunus* genetic map-anchored sequences, and 2,140 peach physical map-anchored sequences were analyzed with the conclusion that there is well conserved microsynteny across the *Prunus* species—peach, plum, and apricot—and the *Populus* genome, however this was not true for *Arabidopsis*. The syntenic regions detected covered 477 kb in the *Populus* genome and 133 kb in the peach BACs that were completely sequenced. Two syntenic regions between *Arabidopsis* and these BACs were much shorter, around 10 kb, with only four or five gene pairs. Similarly, there were syntenic regions that are conserved between the *Prunus* BACs and the partially sequenced genome of *Medicago*, which also belong to Rosid I with *Prunus* and *Populus*. Using genetic or physical map-anchored *Prunus* sequences showed no significant macrosynteny between *Prunus* and the completely sequenced plant models *Arabidopsis* and *Populus*. Longer microsyntenic blocks of gene content were detected, which span mega base pairs, however the gene order was not necessarily conserved in those syntenic regions.

Further resolution of the evolution of the genomes of *Prunus* will await more extensive sequencing efforts. In this regard, the peach genome sequence is nearing completion through the efforts of the Joint Genome Institute (United States Department of Energy), an Italian Consortium (headed by Dr I Verde), Cold Spring Harbor Laboratory (carried out by Dr R McCombie) and Dr B Sosinski at North Carolina State University (currently doing the assembly).

5 CONCLUSIONS AND PERSPECTIVES

With significant investment in research on specialty crop genomes, such as that of the peach, major advances in our understanding of the genes that control important characters of these species is leading rapidly to improvement of breeding programs and breeding germplasm. Indeed to face the challenges presented by everchanging environmental landscape, both biotic and abiotic stress, and rapid adaptation of cultivars for special needs of the growers will continually confront breeding programs. Thus, the genomic tools developed for crops, such as, peach will play a major role in future fruit tree agriculture in the world by providing the markers and genes to meet these challenges. It is through the melding of current genomic technologies with traditional breeding programs that fruit tree agriculture will sustain production in the face of shrinking field space, various pests, diseases, and global climate change. With the current peach genomics infrastructure, we are currently well on the road to realizing this translational breeding future.

References

Abbott AG, Rajapakse S, Sosinski B, Lu ZX, Sossey-Alaoui K, Gannavarapu M, Reighard G, Ballard RE, Baird, WV, Scorza, R, Callahan A (1998) Construction of saturated linkage maps of peach crosses segregating for characters controlling fruit quality, tree architecture and pest resistance. Acta Hort 465: 41-49

Aharoni A, O'Connell AP (2002) Gene expression analysis of strawberry achene and receptacle maturation using DNA microarrays. J Exp Bot 53: 2073-2087

Aharoni A, Keizer LCP, Bouwmeester HJ, Sun ZK, Alvarez-Huerta M, Verhoeven HA, Blaas J, van Houwelingen A, De Vos RCH, van der Voet H (2000) Identification of the *SAAT* gene involved in strawberry flavor biogenesis by use of DNA microarrays. Plant Cell 12: 647-61

Aharoni A, Keizer LCP, Van den Broeck HC, Blanco-Portales R, Munoz-Blanco J, Bois G, Smit P, De Vos RCH, O'Connell AP (2002) Novel insight into vascular, stress, and auxin-dependent and -independent gene expression programs in strawberry, a non-climacteric fruit. Plant Physiol 129: 1019-1031

Aharoni A, Giri, AP, Verstappen FWA, Bertea CM, Sevenier R, Sun ZK, Jongsma MA, Schwab W, Bouwmeester HJ (2004) Gain and loss of fruit flavor compounds produced by wild and cultivated strawberry species. Plant Cell 16: 3110-3131

Aranzana MJ, Pineda A, Cosson P, Dirlewanger E, Ascasibar J, Cipriani G, Ryder CD, Testolin R, Abbott AG, King GJ, Iezzoni AF, Arús P (2003) A set of simple-sequence repeat (SSR) markers covering the *Prunus* genome. Theor Appl Genet 106: 819-825

Baird WV, Estager AS, Wells J (1994) Estimating nuclear DNA content in peach and related diploid species using laser flow cytometry and DNA hybridization. J Am Soc Hort Sci 199: 1312-1316

Bielenberg DG, Wang Y, Fan S, Reighard GL, Scorza R, Abbott AG (2004) A deletion affecting several gene candidates is present in the peach *Evergrowing* mutant. J Hered 95(5): 436-444

Bielenberg DG, Wang Y, Li Z, Zhebentyayeva T, Fan S, Reighard GL, Scorza R, Abbott AG (2008) Sequencing and annotation of the evergrowing locus in peach [*Prunus persica* (L.) Batsch] reveals a cluster of six MADS-box transcription factors as candidate genes for regulation of terminal bud formation. Tree Genet Genomes 4: 495-507

Bliss FA, Arulsekar S, Foolad MR, Becerra V, Gillen AM, Warburton ML. Dandekar AM, Kocsisne GM, Mydin KK (2002) An expanded genetic linkage map of *Prunus* based on an interspecific cross between almond and peach. Genome 45: 520-529

Brasileiro AC, Leple JC, Muzzin J, Ounnoughi D, Michael MF, Jouanin L (1991) An alternative approach for gene transfer in trees using wild-type *Agrobacterium* strains. Plant Mol Biol 17: 441-452

Chaparro JX, Werner DJ, O'Malley D, Sederoff RR (1994) Targeted mapping and linkage analysis of morphological isozyme, and RAPD markers in peach. Theor Appl Genet 87: 805-815

Claverie M, Bosselut N, Lecouls AC, Voisin R, Lafargue B, Poizat C, Kleinhentz M, Laigret F, Dirlewanger E, Esmenjaud D (2004) Location of independent root-knot nematode resistance genes in plum and peach. Theor Appl Genet: 108: 765-773

Corredor E, Román M, García E, Perera E, Arús P, Naranjo T (2004) Physical mapping of rDNA genes enables to establish the karyotype of almond. Ann Appl Biol 144: 219-222

da Câmara Machado A, Puschmann M, Puringer H, Kremen R, Katinger H, Laimer da Câmara Machado M (1995) Somatic embryogenesis of *Prunus subhirtella autumno rosa* and regeneration of transgenic plants after *Agrobacterium*-mediated transformation. Plant Cell Rep 14: 335-340

Dettori MT, Quarta R, Verde I (2001) A peach linkage map integrating RFLPs, SSRs, RAPDs, and morphological markers. Genome 44: 783-790

Dirlewanger E, Arús P (2004) Markers in tree breeding: Improvement of peach. In: Lörz H, Wenzel G (eds) Molecular Marker Systems in Plant Breeding and Crop Improvement. Biotechnology in Agriculture and Forestry. Vol 5. Springer, Heidelberg, Germany, pp 279-302

Dirlewanger E, Pronier V, Parvery C, Rothan C, Guye A, Monet R (1998) Genetic linkage map of peach (*Prunus persica* (L.) Batsch) using morphological and molecular markers. Theor Appl Genet 97: 888-895

Dirlewanger E, Moing A, Rothan C, Svanella L, Pronier V, Guye A, Plomion C, Monet R (1999) Mapping QTL controlling fruit quality in peach (*Prunus persica* (L.) Batsch). Theor Appl Genet 98: 18-31

Dirlewanger E, Cosson P, Poizat C, Laigret F, Aranzana MJ, Arús P, Dettori MT, Verde I, Quarta R (2004) Synteny within the *Prunus* genomes detected by molecular markers. Acta Hort 622: 177-187

Dolgov SV, Firsov AP (1999) Regeneration and agrobacterial transformation of sour cherry leaf disks. Acta Hort 484: 577-580

Dominguez I, Graziano E, Gebhardt C, Barakat A, Berry S, Arús P, Delseny M, Barnes S (2003) Plant genome archaeology: evidence for conserved ancestral chromosome segments in dicotyledonous plant species. Plant Biotechnol J 1: 91-99

Etienne C, Rothan C, Moing A, Plomion C, Bodenes C, Svanella-Dumas L, Cosson P, Pronier V, Monet R, Dirlewanger E (2002) Candidate genes and QTLs for sugar and organic acid content in peach [*Prunus persica* (L.) Batsch]. Theor Appl Genet 105: 145-159

Fan S, Bielenberg D, Zhebentyayeva T, Reighard G, Abbott AG (2008) The developmentof a genetic linkage map for chilling requirements in peach. In: Plant Anim Genome XV Conf, San Diego, CA, USA P132

Fan S, Bielenberg D, Zhebentyayeva T, Reighard G, Okie W, Holland D, Abbott AG (2009) Mapping quantitative trait loci associated with chilling requirement, heat requirement and bloom date in peach [*Prunus persica* (L.) Batsch]. New Phytol doi: 10.1111/j1469-8137.2009.03119.x

Foolad MR, Arulsekar S, Becerra V, Bliss FA (1995) A genetic map of *Prunus* based on an interspecific cross between peach and almond. Theor Appl Genet 91: 262-629

Foulongne M (2002) Introduction d'une résistence polygénique à l'oïdium chez le pêcher *Prunus persica* à partir d'une espèce sauvage *Prunus davidiana*. PhD Thesis, Université de la Mediterranée-Faculté de Sciences de Marseille-Luminy, France

Foulongne M, Pascal T, Arús P, Kervella J (2003a) The potential of *Prunus davidiana* for introgression into peach [*Prunus persica* (L.) Batsch] assessed by comparative mapping. Theor Appl Genet 107: 227-238

Foulongne M, Pascal T, Pfeiffer F, Kervella J (2003b) QTL for powdery mildew resistance in peach x *Prunus davidiana* crosses: consistency across generations and environments. Mol Breed 12: 33-50

Georgi LL, Wang Y, Yvergniaux D, Ormsbee T, Iñigo M, Reighard G, Abbott AG (2002) Construction of a BAC library and its application to the identification of simple sequence repeats in peach (*Prunus persica* (L.) Batsch). Theor Appl Genet 105: 1151-1158

Georgi LL, Wang L, Reighard GL, Mao L, Wing RA, Abbott AG (2003) Comparison of peach and *Arabidopsis* genomic sequences: fragmentary conservation of gene neigborhoods. Genome 46: 268-276

Gonza´lez-Aguero M, Pavez L, Ibanez F, Pacheco I, Campos-Vargas R, Meisel LA, Orellana A, Retamales J, Silva H, Gonzalez M, Cambiazo V (2008) Identification of woolliness response genes in peach fruit after post-harvest treatments. J Exp Bot 59(8): 1973-1986

Green ED (2001) Strategies for the systematic sequencing of complex genomes. Nat Rev Genet 2: 573-583

Horn R, Lecouls A-C, Callahan A, Dandekar A, Garay L, McCord P, Howad W, Chan H,Verde I, Ramaswamy K, Main D, Jung S, Georgi L, Forrest S, Mook J, Zhebentyayeva TN, Yu Y, Kim HR, Jesudurai C, Sosinski BA, Arus P, Baird V, Parfitt D, Reighard G, Scorza R, Tomkins J, Wing R, Abbott AG (2005) Candidate gene database and transcript map for peach, a model species for fruit trees. Theor Appl Genet 110: 1419-1428

Hoskins RA, Nelson CR, Berman BP, Laverty TR, George RA, Ciesiolka L, Naeemuddin M, Arenson AD, Durbin J, David RG, Tabor PE, Bailey MR, DeShazo DR, Catanese J, Mammoser A, Osoegawa K, de Jong PJ, Celniker SE, Gibbs RA, Rubin GM, Scherer SE (2000) BAC-based physical map of the major autosomes of *Drosophila melanogaster*. Science 287: 2271-2274

Howad W, Yamamoto T, Dirlewanger E, Testolin R, Cosson P, Cipriani G, Monforte AJ, Georgi L, AbbottAG, Arus P (2005) Mapping with a few plants: Using selective mapping for microsatellitte saturation of the *Prunus* reference map. Genetics 171: 1305-1309

Jáuregui B, de Vicente MC, Messeguer R, Felipe A, Bonnet A, Salesses G, Arús P (2001) A reciprocal translocation between 'Garfi' almond and 'Nemared' peach. Theor Appl Genet 102: 1169-1176

Jelenkovic G, Harrington E (1972) Morphology of the pachytene chromosomes in *Prunus persica*. Can J Genet Cytol 14: 317-324

Joobeur T, Viruel MA, de Vicente MC, Jáuregui B, Ballester J, Dettori MT, Verde I, Truco MJ, Messeguer R, Batlle I, Quarta R, Dirlewanger E, Arús P (1998) Construction of a saturated linkage map for *Prunus* using an almond x peach F_2 progeny. Theor Appl Genet 97: 1034-1041

Joobeur T, Periam N, de Vicente MC, King G, Arús P (2000) Development of a second generation linkage map for almond using RAPD and SSR markers. Genome 43: 649-655

Jung S, Main D, Staton M, Cho I, Zhebentyayeva T, Arús P Abbott AG (2006) Synteny conservation between the *Prunus* genome and both the present and ancestral *Arabidopsis* genomes. BMC Genom 7:81

Jung S, Jiwan D, Cho I, Lee T, Abbott A, Sosinski B. Main D (2009) Synteny of *Prunus* and other model plant species. BMC Genom (in press)

Laimer da Camara Machado M, da Camara Machado A, Hanzer V, Weiss H, Regner F, Steinkellner H, Mattanovich D, Plail R, Knap E, Kalthoff B, Katinger H (1992) Regeneration of transgenic plants of *Prunus armeniaca* containing the coat protein gene of plum pox virus. Plant Cell Rep 11: 25-29

Lalli DA, Decroocq V, Blenda AV, Schurdi-levraud V, Garay L, Le Gall O,

Damsteegt V, Reighard GL, Abbott AG (2005) Identification and mapping of resistance gene analogs (RGAs) in *Prunus:* A resistance map for *Prunus.* Theor Appl Genet 111: 1504-1513

Lee YP, Yu GH, Seo YS, Han SE, Choi YO, Kim D, Mok IG, Kim WT, Sung SK (2007) Microarray analysis of apple gene expression engaged in early fruit development. Plant Cell Rep 26: 917-926

Lu ZX, Sosinski B, Reighard GL, Baird WV, Abbott AG (1998) Construction of a genetic linkage map and identification of AFLP markers for resistance to root-knot nematodes in peach rootstocks. Genome 41: 199-207

Mante S, Scorza R, Cordts J (1989) Plant regeneration from cotyledons of *Prunus persica, Prunus domestica,* and *Prunus cerasus.* Plant Cell Tiss Org Cult 19: 1-11

Mante S, Morgens P, Scorza R, Cordts J, Callahan A (1991) *Agrobacterium* mediated transformation of plum (*Prunus domestica*) hypocotyl slices and regeneration of transgenic plants. Biotechnology 9: 853 857

Marandel G, Salava J, Abbott AG, Candresse T, Decroocq V (2009) Quantitative trait loci meta-analysis of *Plum Pox virus* resistance in apricot (*Prunus armeniaca* L.): New insights on the organization and the identification of resistance genomic factors. Mol Plant Pathol 10(3): 347-360

Marra M, Kucaba T, Sekhon M, Hillier L, Martienssen R, Chinwalla A, Crokett J, Fedele J, Grover H, Gund C, McCombie WR, McDonald K, McPherson J, Mudd N, Parnell L, Schein J, Seim R, Shelby P, Waterston R, Wilson R (1999) A map for sequence analysis of the *Arabidopsis thaliana* genome. Nat Genet 22: 269-270

Mazzitelli L, Hancock RD, Haupt S, Walker PG, Pont SDA, McNicol J, Cardle L, Morris J, Viola R, Brennan R, Hedley PE, Taylor MA (2007) Co-ordinated gene expression during phases of dormancy release in raspberry (*Rubus idaeus* L.) buds. J Exp Bot 58(5): 1035-1045

Miguel C, Oliveira, MM (1999) Transgenic almond (*Prunus dulcis* Mill.) plants obtained by *Agrobacterium*-mediated transformation of leaf explants. Plant Cell Rep 18: 387-393

Miller PJ, Parfitt DE, Weinbaum SA (1989) Outcrossing in peach. HortScience 24: 359-360

Nelson WM, Dvorak J, Luo MC, Messing J, Wing RA. Soderlund C (2007) Efficacy of clone fingerprinting methodologies. Genomics 89: 160-165

Nemeth M (1994) History and importance of plum pox in stone-fruit production. EPPO Bull 24: 525-536

Olukolu B, Trainin T, Kole C, Fan S, Bielenberg D, Reighard G, Abbott A, Holland D (2008) Construction of a high-density genetic linkage map and detection of QTLs controlling chilling requirement in apricot (*Prunus armeniaca* L.). Modern Variety Breeding for Present and Future Needs. Proc 18th EUCARPIA General Congr, 9-12 Sept 2008, Valencia, Spain

Olukolu BA, Trainin T, Fan S, Kole C, Bielenberg DG, Reighard GL, Abbott AG,

Holland D (2009) Molecular genetic dissection of chilling requirement in apricot (*Prunus armeniaca* L.). Genome 52(10): 819-828

Padilla IMG, Webb K, Scorza R (2003) Early antibiotic selection and efficient rooting and acclimatization improve the production of transgenic plum plants (*Prunus domestica* L.). Plant Cell Rep 22: 38-45

Peace CP, Crisosto CH, Gradziel TM (2005) Endopolygalacturonase: a candidate gene for freestone and melting flesh in peach. Mol Breed 16: 21-31

Pen˜a-Corte´s H, Barrios P, Dorta F, Polanco V, Sa´nchez C, Sa´nchez E, Rami´rez I (2005) Involvement of jasmonic acid and derivatives in plant responses to pathogens and insects and in fruit ripening. J Plant Growth Regul 23: 246-260

Rajapakse S, Bethoff LE, He G, Estager AE, Scorza R, Verde I, Ballard RE, Baird WV, Callahan A, Monet R, Abbott AG (1995) Genetic Linkage mapping in peach using morphological, RFLP and RAPD markers. Theor Appl Genet 90: 503-510

Renaut J, Hausman J, Bassett C, Artlip T, Cauchie H, Witters E, Wisniewski M (2008) Quantitative proteomic analysis of short photoperiod and low-temperature responses in bark tissues of peach (*Prunus persica* L. Batsch). Tree Genet Genom 4: 589-600

Salesses G, Mouras A (1977) Tentative d'utilisation des protoplastes pour l'etude des chromosomes chez les *Prunus*. Ann Amelior Plant 27 : 363-368

Scorza R, Sherman WB (1996) Peaches. In: Janick J, Moore JN (eds) Fruit Breeding. Vol 1: Tree and Tropical Fruits. John Wiley, New York, USA, pp 325-440

Scorza R, Mehlenbacher SA, Lightner GW (1985) Inbreeding and coancestry of freestone peach cultivars of the eastern United States and implications for peach germplasm improvement. J Am Soc Hort Sci 110: 547-552

Scorza R, Ravelonandro M, Callahan A, Cordts JM, Fuchs M, Dunez J, Gonsalves D (1994) Transgenic plums (*Prunus domestica* L.) express the plum pox virus coat protein gene. Plant Cell Rep 14: 18-22

Scorza R, Melnicenco L, Dang P, Abbott AG (2002) Testing a microsatellite marker for selection of columnar growth habit in peach (*Prunus persica* (L.) Batsch). Acta Hort 592: 285-289

Schaffer RJ, Friel EN, Souleyre EJF, Bolitho K, Thodey K, Ledger S, Bowen JH, Ma JH, Nain B, Cohen D (2007) A genomics approach reveals that aroma production in apple is controlled by ethylene predominantly at the final step in each biosynthetic pathway. Plant Physiol 144: 1899-1912

Shulaev V, Korban S, Sosinski B, Abbott AG, Aldwinckle HS, Folta KM, Iezzoni A, Main D, Arús P, Dandekar AM, Lewers K, Brown SK, Davis TM, Gardiner SE, Potter D, Veilleux RE (2008) Multiple models for Rosaceae genomics. Plant Physiol 147(3): 985-1003

Trainotti L, Bonghi C, Ziliotto F, Zanin D, RasoriA, Casadoro G, Ramina A, Tonutti P (2006) The use of microarray μPEACH1.0 to investigate transcriptome changes during transition from pre-climacteric to climacteric phase in peach fruit. Plant Sci 170 (3): 606-613

Verde I, Quarta R, Cerdrola C, Dettori MT (2002) QTL analysis of agronomic traits in a BC_1 peach population. Acta Hort 592: 291-297

Viruel MA, Madur D, Dirlewanger E, Pascal T, Kervella J (1998) Mapping quantitative trait loci controlling peach leaf curl resistance. Acta Hort 465: 79-87

Wang Y, Georgi LL, Zhebentyayeva TN, Reighard GL, Scorza R, Abbott AG (2001) High throughput targeted SSR marker development in peach (*Prunus persica* (L.) Batsch). Genome 45: 319-328

Wang Y, Georgi LL, Reighard GL, Scorza R, Abbott AG (2002) Genetic mapping of the evergrowing gene in peach (*Prunus persica* (L.) Batsch). J Hered 93: 352-358

Warburton, ML, Becerra-Velasquez VL, Goffreda JC, Bliss FA (1996) Utility of RAPD markers in identifying genetic linkages to genes of economic interest in peach. Theor Appl Genet 93: 920-925

Wasternack C (2007) Jasmonates, an update on biosynthesis, signal transduction and action in plant stress response, growth and development. Ann Bot 100: 681-697

Yamamoto T, Shimada T, Imai T, Yaegaki H, Haji T, Matsuta N, Yamaguchi M, Hayashi T (2001) Characterization of morphological traits based on a genetic linkage map in peach. Breed Sci 51: 271-278

Zhang HB, Wing RA (1997) Physical mapping of the rice genome with BACs. Plant Mol Biol 35: 115-127

Zhang HB, Wu C (2001) BAC as tools for genome sequencing. Plant Physiol Biochem 39: 195-209

Zhebentyayeva TN, Horn R, Mook J, Lecouls A, Georgi L, Abbott AG, Reighard GL, Swire-Clark G, Baird WV (2006) A physical framework for the peach genome. Acta Hort 713: 83-88

Zhebentyayeva TN, Swire-Clark G, Georgi LL, Garay L, Jung S, Forrest S, Blenda AV, Blackmon B, Mook J, Horn R, Howad W, Arús P, Main D, Tomkins JP, Sosinski B, Baird WV, Reighard GL, Abbott AG (2008) A framework physical map for peach, a model Rosaceae species. Tree Genet Genom: DOI: 10.1007/s11295-008-0147-z

Ziosi V, Bonghi C, Bregoli AM, Trainotti L, Biondi S, Sutthiwal S, Kondo S, Costa G, Torrigiani P (2008) Jasmonate-induced transcriptional changes suggest a negative interference with the ripening syndrome in peach fruit. J Exp Bot: doi:10.1093/jxb/erm331, 1-11

10 Current Status of On-going Genome Initiatives

Jaya R. Soneji[1], Madhugiri Nageswara Rao[1*], Padmini Sudarshana[2], Jogeswar Panigrahi[3] and Chittaranjan Kole[4]

[1]University of Florida, IFAS, Citrus Research & Education Center, 700 Experiment Station Road, Lake Alfred, FL 33850, USA

[2]Monsanto Research Center, #44/2A, Vasant's Business Park, Bellary Road, NH-7, Hebbal, Bangalore 560092, India

[3]Biotechnology Unit, Sambalpur University, Jyoti Vihar, Burla 768019, Orissa, India

[4]Department of Genetics and Biochemistry and Institute of Nutraceutical Research, Clemson University, 109 Jordan Hall, Clemson, SC 29634, USA

*Corresponding author: mnrao@crec.ifas.ufl.edu/ mnrbhav@yahoo.com

1 INTRODUCTION

The ever-growing demand for more quantity and higher quality of food may be met by the emerging methods of genetic improvement of crop plants. As plant biology research depends fundamentally on plant genome research, it is necessary to lay a firm foundation for an entirely new level of efficiency and success in the application of genetics and breeding to crop plants (Cook 1998). Genomics will accelerate the application of gene technology to agriculture (Briggs 1998). Whole-genome sequencing provides a bounty of information to understand the biology of our complex world. It can be applied to identify gene function and regulation, which will provide a direct access to all genes of an organism. It can also be used to study evolutionary relationships among organisms and will represent an essential step towards a systematic understanding of genome organization as well as plant biology (Yu and Wing 2005; Stein 2007).

Extensive work has been carried out to sequence the model plants, *Arabidopsis* and rice. The studies conducted on these model plants will provide significant insight for comparative genomics. The complete genome sequences of poplar and draft sequence of sorghum are also available. However, these genomes scarcely represent the rich botanical diversity that dominates our ecosystems (Pryer et al. 2002) and do not encompass all of the diverse physiological, developmental and environmental processes seen throughout the plant kingdom (VandenBosch and Stacey 2003). They may not be sufficient for understanding the mechanisms underlying economically important traits in crop plants.

Understanding of the genome of one model plant may explain the similarities but not the differences among all the crop plants (Cook 1998). Hence, there is an urgent need to focus on the genome sequencing initiatives in other crop species having traits of economic and/or agronomic importance. To bring the genomic revolution to crop species, additional genomic resources should be developed (VandenBosch and Stacey 2003). For this, it is necessary to establish genetic linkage and physical maps, bacterial artificial chromosome (BAC) libraries, expressed sequence tag (EST) collections, transformation systems, and DNA chips, etc.

Efforts are being made worldwide to sequence the genomes of various economically important crop species (Table 1). In this chapter, we will give a brief overview of the current status of genome sequencing initiatives taken in cereals, solanaceous members, legumes, and fruit crops (Table 2). We will also be explaining in details the efforts taken for whole-genome sequencing in wheat, tomato, and *Medicago* as examples.

2 CEREALS

The cereals include the world's three most important sources of staple food, viz. wheat, maize and rice. To gain maximum information and knowledge on cereals, the research should proceed simultaneously and coordinately on wheat and barley as one group, maize and sorghum as a second group, and rice as a third group. This will provide maximal information on each of these unique and economically most important representatives of this plant group (Cook 1998). The genome initiatives in rice have been discussed in details in Chapter 7 of this volume. The genetic information of rice will be useful for comparative genomics with other cereals such as maize, wheat (which has been discussed in detail later in this chapter), barley, rye and sorghum, all of which are known

Table 1 Resources on genome sequencing efforts.

Species	*Consortium*	*Website*
Cereals		
Oryza sativa	International Rice Genome Sequencing Project (IRGSP)	*http://rgp.dna.affrc.go.jp/IRGSP/*
Triticum aestivum	International Wheat Genome Sequencing Consortium (IWGSC)	*http://www.wheatgenome.org/*
Hordeum vulgare	International Barley Genome Sequencing Consortium (IBSC)	*http://barleygenome.org/*
Sorghum bicolor	-	*www.phytozome.net/sorghum*
Zea mays	Maize Genome Consortium (MGC)	*http://www.maizegenome.org/*
Brachypodium distachyon	International *Brachypodium* Initiative (IBI)	*http://www.brachypodium.org/*
Solanaceae		
Solanum lycoperscium	International Tomato Genome Sequencing Project (ITGSP)	*http://www.sgn.cornell.edu/about/tomato_sequencing.pl*
Nicotiana tabacum	Tobacco Genome Initiative (TGI)	*http://www.tobaccogenome.org/*
Solanum tuberosum	Potato Genome Sequencing Consortium (PGSC)	*http://www.potatogenome.net/*
Leguminosae		
Lotus japonicus	Lotus (Miyakogusa) Consortium	*http://www.kazusa.or.jp/lotus/*
Glycine max	-	*http://www.phytozome.net/soybean*
Phaseolus vulgaris	Phaseomics Global Initiative	*http://www.phaseolus.net/*
Medicago trunculata	*Medicago trunculata* Sequencing Consortium	*http://www.medicago.org/genome/*

Contd.

Table I continued

Species	*Consortium*	*Website*
Fruits		
Carica papaya	Papaya Genome Project	*http://asgpb.mhpcc.hawaii.edu/papaya/*
Vitis vinifera	International Grape Genome Program (IGGP)	*http://www.vitaceae.org/index.php/International_Grape_Genome_Program*
Prunus persica	Rosaceae International Genomics Initiative (RosIGI)	*http://www.bioinfo.wsu.edu/gdr/genus/prunus/peach/index.php*
Malus x domestica	Rosaceae International Genomics Initiative (RosIGI)	*http://www.bioinfo.wsu.edu/gdr/genus/malus/apple/index.php*
Fragaria vesca	Rosaceae International Genomics Initiative (RosIGI)	*http://www.bioinfo.wsu.edu/gdr/genus/fragaria/strawberry/index.php*
Citrus sinensis	International Citrus Genome Consortium (ICGC)	*http://int-citrusgenomics.org/*
Musa	Global Musa Genomics Consortium (GMGC)	*http://www.musagenomics.org/index.php?id=50*

Table 2 Plant genomes sequenced or being sequenced.

Botanical name	*Common name*	*Size (in Mb)*	*Strategy for genome sequencing*	*Group/Institute involved in sequencing*	*Sequencing status*
Cereals					
Oryza sativa	Rice	430	BAC-end sequencing	International Consortium	Completed
Triticum aestivum	Wheat	16,000	Whole-genome shotgun sequencing, methyl filtration, high C_ot and/or clone-by-clone sequencing	International Consortium	In progress
Hordeum vulgare	Barley	5,500	BAC-end sequencing	International Consortium	In progress
Sorghum bicolor	Sorghum	700-772	Whole-genome shotgun sequencing	Joint Genome Institute	Draft
Zea mays	Maize	2,500	Methyl filtration, high C_ot, BAC-end sequencing and whole-genome shotgun sequencing	Consortium (led by US researchers)	In progress
Brachypodium distachyon	Purple false brome	355	Whole-genome shotgun sequencing	Joint Genome Institute	Checkpoint assembly
Solanaceae					
Solanum lycoperscium	Tomato	950	BAC-end sequencing	International Consortium	In progress
Nicotiana tabacum	Tobacco	4,434	Methyl filtration and BAC-end sequencing	North Carolina State University	In progress
Solanum tuberosum	Potato	840	BAC-by-BAC sequencing	International Consortium	In progress
Leguminosae					
Lotus japonicus	Trefoil	472	Whole-genome shotgun sequencing	Kazusa DNA Research Institute	In progress

Contd.

Table 2 continued

Botanical name	*Common name*	*Size (in Mb)*	*Strategy for genome sequencing*	*Group/Institute involved in sequencing*	*Sequencing status*
Glycine max	Soybean	1,100	Whole-genome shotgun sequencing	Joint Genome Institute	Preliminary assembly
Phaseolus vulgaris	Common bean		Large-scale random genome sequencing	International Consortium	In progress
Medicago trunculata	Medic barrel	550	BAC-by-BAC sequencing	International Consortium	In progress
Fruits					
Carica papaya	Papaya	372	Whole-genome shotgun sequencing	International Consortium	Draft genome assembly
Vitis vinifera	Grape	475	Whole-genome shotgun sequencing	International Consortium	Draft genome assembly
Prunus persica	Peach	270	Whole-genome shotgun sequencing	Joint Genome Institute	In progress
Malus x domestica	Apple	750	Whole-genome shotgun sequencing	International Consortium	In progress
Fragaria vesca	Strawberry	200	Whole-genome shotgun sequencing	Consortium (led by US researchers)	In progress
Citrus sinensis	Sweet orange	367	Whole-genome shotgun sequencing	Joint Genome Institute	In progress
Musa	Banana	550-600	BAC-end sequencing, whole BAC sequencing and reduced representation sequencing	International Consortium	In progress

to share a common ancestor with rice (IRGSP 2005).

Barley ($2n = 14$) represents the basic genome of species belonging to the Triticeae tribe and exhibits strong colinearity to the model grass genomes of rice and *Brachypodium* (Keller and Feuillet 2000). To facilitate the sequencing of barley genome, International Barley Sequencing Consortium was founded in 2006 involving members from countries including Australia, Finland, Germany, Japan, Scotland, and the United States. It will be carried out in three phases over a period of nine years. The first phase will focus on the development of a physical map and identification of the complete gene repertoire, the second on targeted sequencing of prioritized regions of the genome, while the final phase will be devoted to gap filling and completion of the genome (Barley white paper; *www.public.iastate.edu/~imagefpc/IBSC%20Webpage/publications/whitepaper_IBSC_061110.pdf*). Germany will be providing the whole-genome physical map integrated with the current physical map developed in the USA. Australia will provide new BAC libraries for physical mapping in addition to targeted sequencing efforts on 7H (to be linked with wheat group 7). Scottish Crop Research Institute-led effort will provide BAC-end sequences (BES) anchored to the genetic map. Finland will evaluate 454 sequencing technology for analysis and annotation of repetitive DNA (Barley white paper; *www.public.iastate.edu/~imagefpc/IBSC%20Webpage/publications/whitepaper_IBSC_061110.pdf*). Japan will provide additional gene discovery and targeted BAC sequencing from Haruna Nijo, a second haplotype (Saisho et al. 2002). BAC clones have been sequenced using 454 technology (Wicker et al. 2006).

Maize ($2n = 20$) is both a classical genetic model for plant research and an economically important crop. It has a segmental allotetraploid origin (Swigonova et al. 2004). High-resolution genetic maps have been constructed using thousands of SSRs, RFLPs and SNPs and will be useful for the whole-genome sequencing of maize (Martienssen et al. 2004). The Donald Danforth Plant Science Center, The Institute for Genomic Research (TIDR), Purdue University and Orion Genomics constitute the Consortium for Maize Genomics. Methyl-filtration and high C_0t selection have been proposed for the enrichment of gene rich regions. The evaluation of the gene-hit rates, estimation of gene and maize genome coverage, comparison between methyl-filtration and high C_0t selection, and correlation of the assembled sequences to the maize genetic/physical map were to be carried out by The Donald Danforth Plant Science Center. Nearly one million gene enrichment sequences have been produced (Palmer et al. 2003; Whitelaw et al. 2003). High C_0t

libraries have been constructed by Purdue University while methyl filtration libraries were constructed by Orion Genomics. Two additional tools for genome analysis, hypomethylated partial restriction (HMPR) libraries (Emberton et al. 2005) and methylation spanning linker libraries (MSLL; Yuan et al. 2002), have also been developed in maize (Rabinowicz and Bennetzen 2006). TIGR is focusing on the sequencing of the methyl filtration and high C_0t libraries and on sequence assembly, integration and annotation. Over 500,000 BES and one million whole-genome shotgun (WGS) reads have also been deposited in the public databases (Messing et al. 2004).

Sorghum ($2n = 40$) provides food, feed, fiber, fuel and chemical/biofuel across a range of environments and production systems (Kresovich et al. 2005). DOE-JGI initiated the Sorghum genome project through their Community Sequencing program and the Consortium was led by Andrew Paterson, John Bowers, Steve Kresovich, C. Thomas Hash, Jo Messing, Daniel Peterson, Jeremy Schmutz, and Dan Rokhsar. The US sorghum inbred, BTx623, was sequenced using shotgun sequencing. A total of 10,717,203 shotgun reads were collected giving ~8× coverage (Paterson 2008). Methylation filtration was used to generate sequence from the hypomethylated portion of the sorghum genome. Around 550,000 of these reads were sequenced successfully and revealed that approximately 96% of the gene set of sorghum had been sequence-tagged with an average coverage of 65% across their length providing an 1x coverage of the methyl filteration-estimated gene space has been assembled in contigs (Bedell et al. 2005; Paterson 2008).

Due to its small haploid genome (~355 Mbp) and availability of a polyploid series with a basic chromosome number of $x = 5$, ($2n = 2x = 10$), the diploid race of *Brachypodium distachyon* can be used as a model for much larger polyploidy genomes (Ozdemir et al. 2008). To establish the *B. distachyon* as a model species, it would be necessary to assemble genetic resources (genetic stocks, segregating populations), molecular resources (traditional genomic and cDNA libraries, large insert libraries, etc.), obtaining sequence information (ESTs, etc.), optimizing transformation systems, and developing mutagenesis protocols and mutant pools (Garvin 2007). The first meeting of the International *Brachypodium* Initiative was held in 2006 and was attended by members from Australia, Denmark, Finland, France, Poland, the UK and US. John Innes Center (UK) is developing a BAC-based physical map of *B. distachyon* genotype, Bd3-1 using BAC end sequencing and fingerprinting (Bevan 2006). Another physical map of *B. distachyon* genotype, Bd-21 using snapshot-based fingerprinting is being developed at the University of

California, Davis and US Department of Agriculture (USDA) and will be used for sequence assembly, comparative genome analysis, gene isolation, and functional genomics analysis (Huo et al. 2006; Luo et al. 2007). It was being sequenced at DOE-JGI and recently, a 4x depth of the Bd-21 genotype has been covered (Ozdemir et al. 2008). Another project aimed at generating ESTs is also being carried out.

2.1 Wheat

Wheat, belonging to the tribe *Triticeae* of the family Poaceae, is the second most produced food crop among cereals and is the staple food of the world. It is adapted to the temperate regions and occupies one-sixth of the world's crop acreage (Gupta et al. 2008). It was the first domesticated crop and is the youngest polyploid species among the agricultural crops (Gill et al. 2004). An allohexaploid ($2n = 6x = 42$), it has a large genome (16,000 Mb) consisting of seven groups of chromosomes, each group containing a set of three homeologous chromosomes belonging to the A, B and D sub-genomes (Gill et al. 2004; Gupta et al. 2005, 2008). Molecular tools have been used to study the cytogenetics of wheat. Significant advances in the understanding of the wheat plant and grain biology need to be achieved to increase the yield and protect the crop from biotic and abiotic stresses (Gill et al. 2004). To accelerate the achievement of these objectives and for genetic improvement of wheat, genome sequencing is necessary. A number of initiatives have been taken to develop new tools for wheat genomics research.

2.1.1 Linkage Mapping

Extensive linkage maps have been developed in wheat using restriction fragment length polymorphism (RFLP), amplified fragment length polymorphism (AFLP), simple sequence repeat (SSR), sequence tagged site (STS), sequence tagged microsatellite (STM), target region amplified polymorphism (TRAP), EST-SSR, sequence-related amplified polymorphism (SRAP), inter-simple sequence repeat (ISSR), single nucleotide polymorphism (SNP) and diversity array technology (DArT) markers. Though a large number of molecular markers have been used for map construction, the density of wheat genetic maps was improved with the development of SSR markers (Roder et al. 1998). A high-density SSR consensus map was developed by Somers et al. (2004). Consensus maps have also been developed for wheat by pooling the information generated by various maps into a single comprehensive map (Appels 2003; Somers et al. 2004; Gupta et al. 2008). This has also led to the mapping of a number of genes controlling qualitative as well as quantitative traits.

Previously, 5,537 RFLP loci, 2,049 protein loci and genes controlling phenotypic traits were mapped in wheat (McIntosh et al. 2003). Around 2,500 mapped genomic SSR markers are available in wheat, which will help in identifying key recombination events in breeding populations and fine-map genes. More than 300 EST-SSRs could be placed on the genetic map of wheat genome. Around 1,240,000 wheat ESTs have been developed (Gupta et al. 2008). These are being used for development of functional molecular markers, preparation of transcript maps and construction of cDNA arrays. A significant number of unigene ESTs have also been mapped in 'Chinese Spring' deletion set (Sorrells et al. 2003). However, more markers are still needed for the preparation of high-density physical maps for gene cloning (Snape and Moore 2007).

Functional markers have also been developed in wheat (Bagge et al. 2007) for use in construction of molecular function maps. These functional markers will assist in the identification of genes responsible for individual traits for use in marker-assisted selection (MAS) and thereby, greatly facilitating their use in wheat breeding programs. Along with this, microarray-based DArT markers have also been developed. The performance of DArTs in allele-calling efficiency and accuracy have been evaluated and subsequently incorporated into the genetic maps (Akbari et al. 2006). Despite the large resource of ESTs in wheat, very few SNPs have been developed (Snape and Moore 2007). Hence, several high-density platforms such as Illumina's GoldenGate and ABI's SNaPshot are being used for large-scale genotyping for dozens to thousands of SNPs.

2.1.2 Physical Mapping

For the construction of physical maps of wheat, a number of methods have been employed such as deletion mapping, *in silico* physical mapping, radiation-hybrid mapping and BAC based physical mapping.

2.1.2.1 Deletion Mapping

Deletion mapping provides a simple and rapid method to construct cytogenetically based physical maps (Werner et al. 1992). Even though the three genomes in wheat differ in chromosome size and in the amount and distribution of heterochromatin, the relative gene position has been largely conserved. Using deletion stocks (Werner et al. 1992; Endo and Gill 1996), genes for morphological characters were mapped to physical segments of the wheat chromosomes (Gupta et al. 2008). This also allows the construction of consensus physical maps leading to allocation of loci across the genomes (Werner et al. 1992). Along with this, recently, genomic SSRs, ESTs and EST-SSRs have also been physically mapped using deletion stocks (Sourdille et al. 2004; Goyal et al. 2005; Peng and Lapitan 2005; Mohan et al. 2007). Physical mapping of around 16,000 EST loci, using deletion stocks, has been achieved by collaborative research between the members of 'The International Wheat Genome Sequencing Consortium' (Qi et al. 2004). Consensus genetic maps would permit the allocation of markers to specific chromosome regions across the three genomes. Deletion mapping would confirm gene allocation such that any mapped gene may be integrated into the consensus cytogenetic maps.

2.1.2.2 *In silico* Physical Mapping

Unigene derived microsatellite (UGMS) markers have been developed in wheat. Of the 429 class I UGMS markers developed, only 157 had homology with the bin-mapped ESTs and could be mapped to the wheat chromosome bins (Parida et al. 2006). Except 56 wheat UGMS markers, which mapped to unique chromosome bins, the remaining revealed intra- and inter-chromosomal duplications and could be mapped to two or more bins. These bin-mapped wheat UGMS markers provide invaluable information for a targeted mapping of genes for useful traits, comparative genomics and sequencing of gene-rich regions of the wheat genome. Mohan et al. (2007) used two approaches, the wet-lab approach and the *in silico* approach, to physically map a set of 672 loci belonging to 275 EST-SSRs (93 wheat EST-SSRs and 182 rye EST-SSRs) on 21 wheat chromosomes. The wet-lab approach involved the use of deletion stocks, while the *in silico* approach involved matching with ESTs that were previously mapped. The *in silico* approach allowed mapping of twice the number of loci (per EST-SSR) mapped using wet-lab analysis. Southern hybridizations and *in silico* mapping were also used in cDNA clones to determine whether the induced genes were associated with quantitative trait locus (QTL) for Fusarium head blight resistance (Hill-Ambroz et al. 2006). *In silico* mapping will also be done on the 16,000 wheat ESTs assigned to the deletion bins (Qi et al. 2004). This will lead to the mapping of markers with known sequences.

2.1.2.3 Radiation-hybrid Mapping

For the first time, radiation hybrid (RH) mapping was attempted in wheat to construct a high-resolution map of wheat chromosome 1D (D genome) in a tetraploid durum wheat (*T. turgidum* L., AB genomes) background (Kalavacharla et al. 2006). For this purpose, durum wheat alien substitution line for chromosome 1D, harboring nuclear-cytoplasmic compatibility gene scs^{ae} was used (Hossain et al. 2004a). To analyze the area around scs^{ae}, seed from this line was irradiated with 35-krad γ-rays and used as the basis for generating a durum wheat radiation hybrid (DWRH) panel of 87 RH lines containing various fragments of chromosome 1D (Hossain et al. 2004b). These DWRH-1D lines initially allowed detection of 88 radiation induced breaks involving 39 1D specific markers such as RFLPs, SSRs and ESTs (Kalavacharla et al. 2006; Gupta et al. 2008). This DWRH-1D map was further expanded by using 378 molecular markers, which detected 2,312 chromosome breaks. The total map distance ranged from around 3,341 $cR_{35,000}$ for five major linkage groups to 11,773 $cR_{35,000}$ for a comprehensive map (Kalavacharla et al. 2006). This gave a resolution of one break every 199 kb of DNA. To date, this is the highest resolution that has been obtained by plant RH mapping and serves as a first step for the development of RH resources in wheat.

2.1.2.4 BAC-based Physical Mapping

For developing a physical map of the wheat genome, a BAC library covering the whole wheat genome in the model variety 'Chinese Spring' has been constructed (Allouis et al. 2003). Techniques have also been developed for constructing a high quality BAC library from flow-sorted wheat chromosome 3B. This BAC library consisted of 67,968 clones with an average insert size of 103 kb. It represented 6.2 equivalents of chromosome 3B with 100% coverage and 90% specificity as confirmed by genetic markers (Safar et al. 2004a). BAC library has also been constructed from the A genome of wheat (*T. monococcum* accession DV92). The library consists of 276,480 clones with an average insert size of 115 kb (Lijavetzky et al. 1999). BAC based physical map of wheat D genome is being constructed using the diploid species, *Aegilops tauschii* (Gupta et al. 2008). To achieve this, a large number of BAC libraries have been constructed. The BAC DNA is being sequenced and arranged in to contigs. This will be beneficial in the identification, mapping and sequencing of gene-rich regions in wheat.

2.1.3 Sequencing

The first initiative towards the whole-genome sequencing of wheat was taken by the International Triticeae Mapping Initiative (ITMI) in 2003 by launching IGROW (International Genome Research of Wheat) which developed into an International Wheat Genome Sequencing Consortium (IWGSC). The IWGSC was established to facilitate and co-ordinate international efforts toward obtaining the complete sequence of the common wheat genome. IWGSC selected 'Chinese Spring', a cultivar of common wheat, for whole-genome sequencing as it already has ample genetic and molecular resources (Gill et al. 2004). The sequencing of a large genome like wheat represents a challenging endeavor. Hence, IWGSC decided to pursue short- and mid-term goals with the final objective of sequencing the complete wheat genome (Appels et al. 2005).

The wheat genome sequencing was to be conducted in three phases: pilot, assessment, and scale up. The pilot phase was recommended to be a 5-year wheat genome project and had several clear goals. They were to focus on physical and genetic mapping along with sample sequencing of the wheat genome aimed at better understanding of wheat genome structure (Gill et al. 2004). This was to be achieved by ascertaining the gene enrichment capabilities of methylation filtration and high C_0t cloning, construction of specific BAC libraries, development of BAC contigs, assessment of the distribution of genes, investigation of the ability to differentiate homeologous sequences and development of bioinformatics tools for annotation of sequences (Appels et al. 2005)

The assessment phase will involve an analysis of all available wheat sequences and annotation data to determine which method(s) will be well-developed, scientifically supported, and economically feasible approach for sequencing the entire wheat genome (Gill et al. 2004; Appels et al. 2005). After a full assessment, the scale-up phase will involve the deployment of optimal methods on the whole genome, obtaining the genome sequence and annotation, which is the long-term goal of IWGSC. Thus, the scale-up phase will not be initiated until completion of the assessment phase (Gill et al. 2004). With the availability of new sequencing technologies provided by 454/Roche and those provided by Illumina/Solexa and ABI SOLiD (Gupta 2008), sequencing of gene space of the wheat genome, which was once thought to be almost impossible, should become possible within the foreseeable future (Gupta et al. 2008).

The first pilot project was initiated in 2004 and was led by the INRA (France) using the largest wheat chromosome 3B of hexaploid wheat as a model. From a 3B chromosome specific BAC library, 68,000 BAC clones have been fingerprinted using SNAPshot protocol (Safar et al. 2004a). The fingerprints allowed 57,329 BACs to be assembled into contigs. Of these, 104 contigs were then anchored to wheat bins, covering approximately 80% of the genome of chromosome 3B. Currently, one or more of these contigs are being sequenced (Moolhuijzen et al. 2007), which will demonstrate the feasibility of large-scale sequencing of complete gene space of the wheat genome.

2.1.4 Functional Genomics

Functional genomics offers new opportunities for the rapid identification of the genes underlying traits of agronomic importance and provides novel strategies for the manipulation of these traits (Leader 2005). But the major challenge in the post-genome era of plant biology will be the determination of the functions of all the genes in a plant genome (Travella et al. 2006). At the level of functional gene analysis and the isolation of agronomically important genes, wheat (*Triticum aestivum*) is lagging behind due to the lack of efficient tools to study gene function in polyploid species. The estimation of mRNA abundance for large number of genes can be simultaneously carried out using open systems such as serial analysis of gene expression (SAGE), massively parallel signature sequencing (MPSS) and/or closed systems such as micro- and macro-arrays (Leader 2005). Microarrays have been widely used in wheat for understanding alterations in the transcriptome of hexaploid wheat during grain development, germination and plant development under abiotic stresses (Wilson et al. 2004, 2005), for introgression with cDNA (Skinner et al. 2005), to study differential gene expression between leaves of well-watered wheat plants and plants subjected to water deficit stress (Way et al. 2005), for mapping translocation breakpoints (Bhat et al. 2007) and for the comparison of the Affymetrix GeneChip® Wheat Genome Array with an in-house custom-spotted cDNA array and a quantitative reverse transcription-polymerase chain reaction (RT-PCR, Poole et al. 2007). Jordan et al. (2007) identified regions of wheat genome controlling seed development by mapping 542 expression QTLs (eQTLs), using a doubled haploid (DH) mapping population that was earlier used for mapping of SSR markers and QTL analysis of agronomic and seed quality traits (McCartney et al. 2005). Expression analysis using mRNA from developing seeds from the same mapping population was also conducted using Affymetrix GeneChip® Wheat Genome Array (McCartney et al. 2006). SAGE data platform has also been developed in wheat to investigate the developing wheat caryopsis (McIntosh et al. 2007).

2.1.4.1 RNA Interference

Due to the polyploid nature of the wheat, many wheat genes have homoeologous copies that are functionally redundant. Hence, efforts to produce loss-of-function mutants through phenotypic screenings (chemical or insertional mutagenesis) are less effective in polyploid wheat. The discovery of RNA-mediated gene silencing creates a viable alternative strategy for gene functional analysis through the simultaneous knock-down of expression of multiple related gene copies (Fu et al. 2007). Schweizer et al. (2000) reported the co-transformation of in vitro synthesized dsRNA with either the *gus*A (GUS) reporter gene or with the fusion of a germin like protein (GLP) and green fluorescence protein (GFP) (TaGLP2a::GFP) as the reporter gene using particle bombardment. A hairpin construct was used to target *TaGLP2a* and the effect of this gene on plant defense responses was determined (Christensen et al. 2004). Both these studies suggested that the delivery of specific dsRNA into single epidermal cells in wheat transiently interfered with their gene function. For stable RNAi transformation in wheat, a candidate for the vernalization gene *VRN2* (Yan et al. 2004) and a MADS-box transcription factor, *VRN1*, that promotes flowering and is up-regulated by vernalization (Loukoianov et al. 2005) were targeted. To reduce the amylase content in wheat, Li et al. (2005) used RNAi to silence the *Granule Bound Starch Synthase I* (*GBSSI*) gene. RNAi was also used to down-regulate the two different isoforms of starch-branching enzyme (SBE) II (SBEIIa and SBEIIb) in wheat endosperm to augment its amylose content (Regina et al. 2006). Recently, new studies have shed light on the potential of RNAi silencing to simultaneously knock-down transcripts of homoeologous genes in polyploid wheat (Travella et al. 2006; Uauy et al. 2006; Yue et al. 2007). Travella et al. (2006) targeted two genes, *phytoene desaturase* (*PDS*) and *ethylene insensitive 2* (*EIN2*), which have three homoeologous copies that are expressed in wheat. Silencing of these genes generates phenotypes that are easy to visualize making them ideal targets for evaluating silencing of individual homoeologous genes. Uauy et al. (2006) were able to characterize the effect of RNAi on individual homologous gene copies of wheat NAM genes, which encode NAC transcription factors affecting senescence and grain protein, zinc and iron content. Yue et al. (2007) targeted a gene encoding the high molecular weight (HMW) glutenin subunit 1Dx5 which is a seed storage protein associated with dough strength and elasticity in wheat flours and were able to obtain two transgenic lines, one with complete and one with partial 1Dx5 silencing. An alternative to large RNA fragments, synthetic microRNA (miRNA) constructs can also be used for gene silencing and was demonstrated for the first time in wheat by Yao et al. (2007). They were able to discover and predict targets for 58 miRNAs, belonging to 43 miRNA families, of which 20 were conserved and 23 were novel to wheat.

2.1.4.2 TILLING

DNA sequence information facilitates the development of molecular markers and also provides targets for transgenic alteration of gene expression as well as introduction of new genes (Slade and Knauf 2005). Targeting induced local lesions in genomes (TILLING) was developed to take advantage of this new DNA sequence information and to investigate the functions of specific genes. TILLING is a reverse genetics approach for mutation generation and discovery that does not rely on transgenic technology (Henikoff et al. 2004). Slade et al. (2005) created a TILLING library in both bread and durum wheat to determine its utility in a complex genome like wheat. The *waxy* locus encoding the *GBSSI* was targeted. Using locus-specific PCR primers, they were able to identify 246 alleles of the *waxy* genes by TILLING each homeolog in 1,920 allohexaploid and allotetraploid wheat individuals. This made available novel genetic diversity at *waxy* loci and provided a way for allele mining in important germplasm of wheat. A line of bread wheat containing homozygous mutations in two *waxy* homoeologs created through TILLING and a pre-existing deletion of the third *waxy* homoeolog displayed a near-null *waxy* phenotype. Research is also being carried out to develop TILLING technique for both diploid and hexaploid wheat to identify variant alleles of specific genes either arising naturally or via induced mutation. The *Rht* dwarfing genes, the gibberellin (GA) biosynthetic genes and three global regulators of disease resistance will initially be examined for variant allele-novel trait associations. Once other genes of high value to sustainable agriculture are identified, the same approach will be pursued to provide breeders with a wide repertoire of novel trait variants (WGIN 2008).

2.1.5 Association Mapping

Association mapping is a high-resolution method for mapping QTL based on linkage disequilibrium (LD) and holds great promise for genetic dissection of complex traits. Due to the availability of genetic and genotyping resources and technology in wheat, LD and association mapping studies are emerging and will facilitate the trait-genotype associations and fine-mapping of QTLs (Somers 2005).

Mapped DArT markers were used to find associations with resistance to stem rust, leaf rust, yellow rust, and powdery mildew, plus grain yield in five wheat international multi-environment trials from the International Maize and Wheat Improvement Center (CIMMYT). Two linear mixed models were used to assess marker-trait associations incorporating information on population structure and covariance between relatives and an integrated map containing 813 DArT markers and 831 other markers was constructed (Crossa et al. 2007). Ravel et al. (2006) were the first to report the use of association study to discriminate between two candidate genes, *Glu-B1-1*, the structural gene coding for *Glu1Bx*, and the *B* homoeologous gene coding for *SPA* (*spa-B*), a seed storage protein activator, controlling a QTL for high-molecular-weight glutenin subunit (HMW-GS) *GluBx* in wheat. Associations between markers in the region of *QSng.sfr-3BS*, a major QTL for resistance to *Stagonospora nodorum* glume blotch (SNG), and SNG resistance have also been investigated by linkage and association analyses using a recombinant inbred line (RIL) mapping population in wheat and reported that association mapping had a marker resolution which was 390-fold more powerful than QTL analysis conducted (Tommasini et al. 2007). Association mapping was performed on a selected sample of 95 cultivars of soft winter wheat for kernel morphology and milling quality (Bresghello and Sorrells 2006). In agreement with previous QTL analysis, they were able to identify markers for kernel size on the three chromosomes tested, and alleles potentially useful for selection thus demonstrating that association mapping could complement and enhance previous QTL information for marker-assisted selection (MAS).

3 SOLANACEAE

The Solanaceae is the third most economically important plant taxon and comprise more than 3,000 species, including the tuber-bearing potato, a number of fruit-bearing vegetables (tomato, eggplant, peppers), ornamental plants (petunias), narcotic plants (tobacco), plants with edible leaves (*Solanum aethiopicum, S. macrocarpon*) and medicinal plants (*Datura, Capsicum, Solanum xanthocarpum*). Genome sequencing of solanaceous species will improve our understanding on two aspects: (1) how a common set of genes/proteins can give rise to a wide range of morphologically and ecologically distinct organisms, and (2) how a deeper understanding of the genetic basis of plant diversity can be harnessed to better meet the needs of society in an environment-friendly and sustainable manner (Mueller et al. 2005).

Tobacco, *Nicotiana tabacum* L. ($2n = 48$) is an amphiploid species (Khan and Narayan 2007) with great commercial importance. It has a large genome with little information on its genome structure and organization. Tobacco Genome Initiative (TGI) is led by the scientists at the North Carolina State University at Raleigh, USA, with the goal to sequence and annotate the open reading frames (ORFs) in the genome of tobacco. TGI is supported by Phillip Morris USA, Inc. (TGI, *http://www.tobaccogenome.org/*). Methyl filtration and BAC-end sequencing have been used for physical mapping. To date more than 1,400,000 lanes of a methyl-filtered library have been sequenced and more than 75,000 BAC clones have been fingerprinted using the four-dye protocol. 80,000 ESTs have already been sequenced and six newly constructed cDNA libraries are being sequenced to increase the EST acquisitions (Opperman et al. 2006).

The potato genome consists of 12 chromosomes and is about 840 Mb in size (Jacobs et al. 2006). Ultra-dense genetic map of potato consisting of 10,000 unique AFLP markers is available for potato (van Os 2005). To facilitate the sequencing of potato genome, Potato Genome Sequencing Consortium was founded involving members from countries including Argentina (chromosome 2), Brazil (chromosome 3), Chile (chromosome 3), China (chromosome 2, 10 and 11), Great Britain (chromosome 4), India (chromosome 2), Ireland (chromosome 4), Netherlands (chromosome 1 and 5), New Zealand (chromosome 9), Peru (chromosome 3), Poland (chromosome 8), Russia (chromosome 12), and the Unites States (chromosome 6). A DNA genomic library consisting of 85,000 BAC clones have been constructed. The average insert size of these BAC clones is about 120 kb. Of these 73,000 BAC clones have been fingerprinted and assembled into contigs. A minimal tiling path will be established, which will allow a chromosome specific, BAC-by-BAC approach towards sequencing the entire potato genome (Jacobs et al. 2006).

3.1 Tomato

Tomato (*Solanum lycopersicum* L., formerly *Lycopersicon esculentum* Miller), belonging to the family Solanaceae, is widely grown around the world and constitutes a major agricultural industry. It is a fruit that is often treated as a vegetable and is the second most consumed vegetable after potato (FAOSTAT 2005: *http://faostat.fao.org*). It has a diploid genome with 12 chromosome pairs and a genome size of 950 Mb encoding approximately 35,000 genes that are largely sequestered in contiguous euchromatic regions (Michaelson et al. 1991; van der Hoeven et al. 2002; Barone et al. 2008). Due to its short generation time, photoperiod insensitivity, high self-fertility and homozygosity, great reproductive potential, ease of controlled pollination and hybridization, lack of gene duplication, amenability to asexual propagation and whole-plant regeneration, and availability of a wide array of mutants and genetic stocks, it is the most intensively investigated Solanaceous species (Lee et al. 2007). Markers have been extensively used for identification and mapping for various traits and for use in MAS. Currently, 10 countries (Korea, China, United Kingdom, India, Netherlands, France, Japan, Spain, Italy and the United States) are involved in sequencing the tomato genome (Mueller et al. 2005). The information generated by large-scale genome sequencing can lead to a major revolution in the understanding of tomato biology (Barone et al. 2008). The availability of homozygous inbred lines, characterized genetic and genomic resources as well as its economic value led to the selection of tomato for genome sequencing. Also many DNA markers distributed almost evenly on all chromosomes are available facilitating the use of fluorescence in situ hybridization (FISH) to anchor BAC clones as starting points of genome sequencing (Shibata 2005).

3.1.1 Linkage Mapping

Around 1,300 morphological, physiological and disease resistance genes have been identified in tomato of which only around 400 have been mapped onto the 12 chromosomes (Kalloo 1991; Tanksley 1993). These maps have been upgraded by the addition of molecular markers. The first high-density molecular linkage map of tomato was constructed based on 67 F_2 plants of *S. lycopersicum* cv. VF36-*Tm2a* × *S. pennellii* LA716 cross. It consisted of 1,030 markers and had a marker density of approximately one per 1.2 cM (Tanksley et al. 1992). This map has been upgraded and presently consists of 2,222 mapped molecular markers with an average marker distance of <1 cM, an average of 750 kb per cM and an average estimated total length of the tomato linkage map of 1,300 cM (Zhang et al. 2002). More than 1,000 RFLP markers have been mapped on to the 12 tomato chromosomes while more than 2,600 SSRs have been mined (Frary et al. 2005). RAPD, CAPS, AFLP, SCAR and RGA markers have also been developed and mapped in tomato (Saliba-Colombani et al. 2000; Suliman-Pollatschek et al. 2002; Zhang et al. 2002; Frary et al. 2005; Villalta et al. 2005).

Around 250,000 ESTs have been derived but only a small number has been mapped onto the tomato chromosomes (Barone et al. 2008). However, most of the available DNA markers do not detect polymorphism within the cultivated species or between the cultivated species and closely related species (Saliba-Colombani et al. 2000; Labate et al. 2007). This limited resolution restricts the use of markers in many tomato genetics and breeding programs that attempt to exploit intra-specific genetic variation. Efforts are being made to develop high-throughput markers with greater resolution, including SNPs and InDels (Labate and Baldo 2005). Such markers would allow detection of polymorphism among closely related individuals within species or between *S. lycopersicum* and closely related species (Labate et al. 2007). More recently, to allow high-throughput development of markers, oligonucleotide-based arrays have been used to identify DNA sequence polymorphisms in tomato. To identify single feature polymorphisms (SFPs), total genomic DNA hybridization methods are being exploited (Barone et al. 2008).

3.1.2 Physical Mapping and Sequencing

Of the total 950 Mb of the tomato genome, 730 Mb exists as heterochromatin and 220 Mb as euchromatin. The DNA in the heterochromatin is rich in repetitive sequences and poor in genes whereas euchromatin is thought to contain single copy sequences which include more than 90% of the genes (Mueller et al. 2005). The strategy is to sequence only the euchromatin portion of the tomato genome. For the ease of sequencing the tomato genome, the 12 chromosomes have been split up between the 10 countries (Table 3). This effort is part of a larger initiative known as the International Solanaceae Genome Project (SOL): Systems Approach to Diversity and Adaptation. BAC approach has been chosen to sequence the tomato genome. A total of 402,012 BAC libraries have been constructed of which 188,130 are *Hind*III-digested, 112,507 are *Mbo*I-digested and 101,375 are *Eco*RI-digested BAC libraries. These libraries were prepared from *S. lycopersicum* var. Heinz 1706. Sequencing will be based on the F2-2000 map (*http://sgn.cornell.edu/)* used for anchoring 1,500 markers by overgo (overlapping oligo) hybridization (Cai et al. 1998). Currently, more than 650 unambiguous anchor points are being used as 'seed BAC' for sequencing (Lee et al. 2007), but further analysis will increase this number to an estimated 800-1,000 anchor points.

In addition, the F2-2000 map is being combined with an AFLP map,

Table 3 Tomato sequencing efforts by International Tomato Sequencing Consortium.

Country	*Chromosome*	*BACs*	*BACs sequenced (%)*	*Total sequenced (%)*
United States	1	391	3	4
Korea	2	268	47	42
China	3	274	4	5
United Kingdom	4	193	56	54
India	5	111	35	39
Netherlands	6	213	55	67
France	7	277	28	36
Japan	8	175	69	76
Spain	9	164	35	34
United States	10	186	2	2
China	11	135	13	17
Italy	12	113	34	45

Source: Sol genomics network (*http://www.sgn.cornell.edu/about/tomato_sequencing.pl*)

containing more than 1,200 markers, generated at Keygene in Netherlands, providing even more anchor points (Mueller et al. 2005). After a low-coverage sequencing of each seed BAC, the construction of a minimal tiling path of BAC clones was performed by BLASTing the sequence of each 'seed' BAC against the BES Tagged Connector (STC) database to identify BAC with minimal overlap in either directions. The BAC-end database consisting of 200,000 clones (from *Eco*RI, *Hind*III, and *Mbo*I libraries) was used both to confirm and extend the euchromatin minimal tiling path (e-MTP). Each BES was subjected to automated annotation to determine the proportion of ends that are likely to correspond to genic regions (Barone et al. 2008). The tiling path will then be generated by walking out from the seed BAC sequence in both directions using the deep BES and fingerprint contig (FPC) maps available from the *Hind*III and *Mbo*I libraries. FISH analysis is additionally employed to confirm chromosome mapping and delineate the euchromatin/heterochromatin boundaries (Peterson et al. 1999). About 29% of the target regions have been sequenced (*http://sgn.cornell.edu*). So far about 10,258 BACs and 3,451 contigs have been assigned to the map. In Japan, selected BAC Mixture (SBM) shotgun sequencing has been set up. In this method, BAC clones whose end sequences do not contain repeat sequences are selected, then the selected BACs are mixed and shotgun sequencing is performed (Asamizu 2007).

As the euchromatic portion of the genome is about 220 Mb, the average physical distance between two adjacent seed BAC may be as little as 200 kb. The density and resolution of the available map may not be adequate to provide a template for complete sequencing. Hence, new strategies have been undertaken to complement the ongoing sequencing. The release of new markers by Syngenta® has helped in the identification of new candidate seed BAC. Being distributed throughout the full genome, these may be useful for filling in gene spaces (Barone et al. 2008). Other methods available for full genome sequencing have also been assessed. Full genome shotgun sequencing is not a cost-effective way to sequence a fraction of the genome. Methyl and C_0t filtering are both methods providing a bias for coding sequence and therefore euchromatic sequence. All these methods do not by themselves provide gene order, which will be critical for a reference genome (Mueller et al. 2005). Whole-genome shotgun sequencing and the availability of new generation sequencing technologies, including 454/Roche's sequencer FLX, Solexa's sequencing system, and ABI's SOLiD, may also prove useful in completing the whole-genome sequencing of the tomato genome (Barone et al. 2008). In addition, chloroplast genome sequence is available (Kahlau et al. 2006) and mitochondrial genome is to be sequenced by Argentina. The organellar genome sequences will be important for distinguishing genomic insertions of organellar sequences from organellar contamination contained in the BAC libraries (Mueller et al. 2005).

3.1.3 Functional Genomics

Designing the tools necessary for high-throughput expression studies is the major task of the tomato functional genomics program. This can be achieved by unraveling the changes in gene expression associated with the developmental processes and will be a step towards understanding the complex regulation mechanisms underlying these processes (Delalande et al. 2007).

3.1.3.1 Mutagenesis

In tomato, both classical and insertional mutagenesis have been used. More than 600 characterized monogenic mutations are available in a variety of genetic backgrounds at the Tomato Genetics Resource Center at the University of California, Davis, USA (Barone et al. 2008). Menda et al. (2004) induced mutations and obtained an extensive mutant population comprising of 6,000 ethyl-methane sulfonate (EMS)-induced and 7,000 fast neutron-induced mutant lines. For insertional mutagenesis, transposon tagging has been exploited in tomato. Several genes, *Cf-9*, *Cf-4*, *Dwarf*, *Defective Chloroplasts and Leaves* (*DCL*), *Feebly*, and *Defective embryo and meristems* (*Dem*), have been isolated by transposon tagging (Emmanuel and Levy 2002). Around 200,000-300,000 *Ds* insertions, derived from several unlinked T-DNAs, were roughly estimated to be sufficient to achieve a high probability of insertion into any specific target gene. *Ac/Ds* has also been used to deliver the firefly *luciferase* reporter gene for promoter trapping and the *β-glucuronidase* (*GUS*) gene for enhancer trapping in tomato (Meissner et al. 2000). The systems utilizing the 'Micro-Tom' have not yielded saturating mutant collections and hence have not been used much (Barone et al. 2008). High efficiency transformation protocols developed for 'Micro-Tom' may serve as a tool for extensive T-DNA mutagenesis programs (Sun et al. 2006). The chromosomal location of 405 individual inserts with a modified *Ds* transposable element has been determined in tomato. These insertion lines may facilitate cloning of gene closely linked to one of the *Ds* loci (Gidoni et al. 2003).

3.1.3.2 TILLING

For targeting local mutations in the genome and induce point mutations in genes of interest, TILLING platforms for tomato are under development. EMS has been used to generate mutants for TILLING. Using TILLING, Arcadia Biosciences has generated 220 induced mutations in 19 tomato genes implicated in fruit shelf-life. EcoTILLING of 182 tomato cultivars uncovered one non-coding SNP in the expansin (*EXP1*) gene. TILLING was used to identify nine SNPs in the mutagenized library, including two nonsense and seven missense mutations. Early results showed increased fruit firmness in a knock-out expansin line, similar to the results seen in the transgenic antisense expansin line. Because TILLING is a non-transgenic technology, the nonsense line provides tomato breeders with a novel, non-transgenic source for fruit firmness (Loeffler et al. 2008). In another study, loss of functional alleles in the *polygalacturonase* gene was identified using the TILLING reverse genetics approach (McCallum et al. 2008).

3.1.3.3 RNA Interference

Strategies for gene silencing have also been widely used as a tool for functional genomics research in tomato. Tomato fruit ripening was one of the early systems in which both sense and antisense silencing were found to be effective (Gray et al. 1992). Majority of the work on RNAi has been restricted to the fruit leading to the studies involving the silencing of fruit-specific genes (Davuluri et al. 2005). Xiong et al. (2005) reported a significant inhibition in the rate of ethylene production and prolonged shelf-life of ripened fruits and leaves of transgenic plants of tomato cv. Hezuo 906 in which a unit of tomato ACC oxidase dsRNA was introduced by *Agrobacterium tumefaciens*-mediated transformation. Hairpin RNA-mediated strategies (such as siRNAs) have been used to silence the *AC1* and *AC4* genes of the tomato leaf curl virus for effective resistance in transgenic tomato plants (Ramesh et al. 2007). A novel strategy to obtain parthenocarpic tomatoes by down-regulation of the flavonoid biosynthesis pathway using RNAi-mediated suppression of *chalcone synthase* (*CHS*), the first gene in the flavonoid pathway, has also been achieved (Schijlen et al. 2007).

3.1.3.4 Virus-induced Gene Silencing

Virus-induced Gene Silencing (VIGS) offers an attractive and quick alternative for knocking out expression of a gene without the need to genetically transform the plant. VIGS has been successfully used as functional genomics tools in tomato. It has been described in tomato roots (Valentine et al. 2004) and fruits (Fu et al. 2005) although the extent to which silencing remains confined to these organs has not been extensively investigated. A recombinant tobacco rattle virus (TRV)-based vector was used to infect the tomato plants and was able to induce efficient gene silencing by suppressing *phytoene desaturase* (*PDS*) gene, constitutive triple response 1 and 2 (*LeCTR1* and *LeCTR2*) genes in tomato (Liu et al. 2002). The suppression of *LeCTR1* gene promotes the ripening of the tomato fruit. Fu et al. (2005) demonstrated that VIGs of *LeEILs* could also inhibit fruit ripening in tomato and suggested that VIGS could co-suppress several members of a gene family and overcome possible functional redundancy by choosing regions that are conserved between genes. VIGS has been used to demonstrate the importance of tomato abscission-related polygalacturonases (TAPGs) in the abscission of leaf petioles (Jiang et al. 2008). Using a single set of primers, tomato ESTs have been cloned using a modified TRV vector, which was based on the GATEWAY recombination system allowing restriction- and ligation-free cloning. *Ribulose bisphosphate carboxylase* (*RbcS*) and an endogenous gene homologous to the tomato EST cLED3L14 were silenced using this vector. This modified vector system may facilitate large-scale functional analysis of tomato ESTs (Liu et al. 2002). VIGS has also been used for silencing abscission related expansins (*LeEXP11* and *LeEXP12*) and endoglucanases (*LeCEL1* and *LeCEL2*) in tomato but had no discernible effect on break strength, even when the two endo-glucanase genes were silenced concurrently (Jiang et al. 2008). Silencing of the *LeEIN2* gene using VIGs also resulted in the suppression of tomato fruit ripening (Fu et al. 2005) and development of parthenocarpic fruits (Zhu et al. 2006). The use of a new vector system modified from DNAβ (DNAmβ) of tomato yellow leaf curl China virus (TYLCCNV) has also been reported for VIGS analysis in tomato (Cai et al. 2007). Delay of post-harvest ripening and senescence of tomato fruit could also be achieved through virus-induced *LeACS2* gene silencing (Xie et al. 2006).

3.1.3.5 Transcriptional Profiling

Transcriptional profiling is being widely explored since the extensive EST collection available in tomato (Mueller et al. 2005) has allowed designing of several microarray platforms. The most widely used to date have been Tom1, a cDNA-based microarray containing probes for approximately 8,000 independent genes, and Tom2, a long oligonucleo-tide-based microarray containing probes for approximately 11,000 independent genes. Both these microarrays are already available from Boyce Thompson Institute for Plant Research (*http://bti.cornell.edu/CGEP/CGEP.html*) and soon Tom2 will also be available from the EU-SOL project (*http://www.eu-sol.net*). The third array is an Affymetrix Genechip, which contains probe sets for approximately 9,000 independent genes (*http://www.affymetrix.com/products/arrays/specific/tomato.affxspecific/tomato.affx*). As the tomato genome project progresses, a comprehensive, public tomato microarray platform will become indispensable.

4 LEGUMINOSAE

With more than 650 genera and 20,000 species, legumes are second only to grasses in economic importance in world agriculture (Medicago Whitepaper: http://*medicago.toulouse.inra.fr/ATS/documents/whitepaper gensequ.pdf*). The legumes have very diverse characteristics and exhibit interesting differences in secondary metabolism, pod development and other processes. They also develop important and interesting symbioses with nitrogen (N)-fixing rhizobia and with mycorrhizal fungi (Vanden-

Bosch and Stacey 2003). Genomic efforts in legumes will help in advancing the breeding programs aimed at understanding the basic plant processes and unique aspects of legume plant physiology and development.

The genus, *Lotus*, consists of more than one hundred diploid ($2n$ = 10, 12, 14), tetraploid ($2n$ = 24, 28) and hexaploid ($2n$ = 36) species including *L. japonicus* ($2n$ = 12) (Kawakami 1930; Ito et al. 2000). Lotus (Miyakogusa) Consortium was established in 1999 and around 30 laboratories from all over Japan are involved in developing linkage maps, expression arrays (of both plant and endosymbiont genes), transformation techniques, etc. (VandenBosch and Stacey 2003). The genome sequencing is being carried out at the Kazusa DNA Research Institute, Japan (Stougaard et al. 2007) and is focused on gene-rich regions and an approach using seed points anchoring sequences onto the genetic map has been developed. Transformation-competent artificial chromosome (TAC) and BAC vectors have been used to construct genomic libraries (Liu et al. 1999; Sato and Tabata 2005) and sequenced by shotgun strategy. A total of 1,351 clones have been located on the genetic linkage map using 720 microsatellite markers and 80 derived cleaved amplified polymorphic sequence (dCAPS) markers, and by overlaps with the genetically mapped clones (Sato et al. 2001; Nakamura et al. 2002; Asamizu et al. 2003; Kaneko et al. 2003). Nucleotide sequences covering 174 Mb regions, which include 129 Mb phase I sequences, have been determined and released to the public (Sato and Tabata 2005).

Soybean (*Glycine max*, $2n$ = 40) has a genome size of ~1,100 Mbp (Arumuganathan and Earle 1991). The soybean genome project was initiated through the Department of Energy-Joint Genome Institute's Community Sequencing Program (CSP). The consortium has members from Stanford University, University of Missouri-Columbia, Iowa State University, Purdue University and DOE-JGI and was supported by the USDA and the National Science Foundation (NSF). The physical mapping efforts were funded by the United Soybean Board and the NSF. Whole- genome shotgun sequencing was used and a total of about 13 million shotgun reads have been produced and deposited in the National Center for Biotechnology Information (*http://www.phytozome.net/soybean*). A preliminary assembly (representing 7.23x coverage) and annotation of the soybean genome has also been made available. In a collaborative project, funded by NSF, between the University of Missouri, Washington University Genome Center and Orion Genomics, sample sequencing of the soybean genome was also done to test methyl filtration as a means to enrich for gene-rich segments of the genome (*http://www.agbionetwork.org/~soybeangenome/GSA.php*).

Phaseolus spp. ($2n = 22$, Karpechenko 1925), consisting of about 50 species commonly known as beans, many of which are economically important, are one of the world's most ancient crops (VandenBosch and Stacey 2003) having a high nutritional value. The International Consortium (Phaseomics Global Initiative) has members from Argentina, Australia, Belgium, Brazil, Colombia, Italy, Malaysia, Mexico, Spain, Switzerland and the United States. The main goal of the consortium is to increase the genetic resources and tools available for the crop, especially large insert and cDNA libraries, genomic sequences, ESTs, and genetic markers. A large collection of ESTs and BAC libraries has been established for *Phaseolus* and will be beneficial for its sequencing. A large-scale random-sequencing of the *P. vulgaris* genome has been initiated (Lariguet et al. 2005).

4.1 Medicago

Medicago truncatula ($2n = 16$), belonging to the family Leguminosae is an important forage crop with a small genome of about 550 Mb, annual habit and rapid life cycle. The *M. truncatula* Sequencing Consortium has been formed with researchers from Belgium, France, Germany, Netherlands, United Kingdom and United States being its members. *Medicago* forms symbiotic associations with a wide array of arbuscular mycorrhizal fungi and develops root nodules with *Sinorhizobium meliloti,* which is one of the best-characterized *Rhizobium* species at the genetic level (Galibert et al. 2001). Jemalong A17 has been selected by the research community as a reference line for most genetic and genomic approaches (Ane et al. 2008).

4.1.1 Linkage Mapping

Extensive genetic maps have been developed for *M. truncatula* using F_2 populations. A number of molecular markers have been used such as RAPDs, AFLPs, CAPS, SSRs, RFLPs and isozymes (Thouquet et al. 2002; Mun et al. 2006). A core genetic map has also been established using the segregation of sequence-characterized genetic markers (such as ESTs, BAC-end sequence tags and resistance gene analogs) in an F_2 population (Choi et al. 2004). The map based on a Jemalong A17 × A20 F_2 population, is the one currently used as a reference for the genome sequencing project (Ane et al. 2008). Maps based on RILs and SSRs have also been developed. These maps will provide sustainable tools to the community.

4.1.2 Physical Mapping and Sequencing

To assemble the physical map of *M. truncatula,* a number of BAC libraries have been constructed (Nam et al. 1999). Contigs have been developed that cover approximately 480 Mbp (95%) of the genome. BAC clones have also been correlated with ESTs. The physical map has been linked to a dense genetic map using ESTs and SSR markers on BAC contigs (VandenBosch and Stacey 2003). This physical map with approximately 11x coverage, and deep BES data with more than 160,000 sequences from two BAC libraries provided tools for sequencing (Cannon et al. 2005).

The *M. truncatula* Sequencing Consortium began in 2001 with a seed grant from the Samuel Roberts Noble Foundation. In 2003, the NSF and the European Union 6th Framework Program began providing most of the funding. Among the eight chromosomes in *M. truncatula*, six are being sequenced by the United States [Nevin Young (University of Minnesota), Bruce Roe (ACGT, University of Oklahoma; chromosomes 1, 4, 6, 8), and Chris Town (TIGR; chromosomes 2, 7)] under a NSF project 'Sequencing the Gene Space of the Model Legume, *M. truncatula*'. The remaining two are being sequenced by partners in Europe with Giles Oldroyd (John Innes Center) co-ordinating sequencing of chromosome 3 at the Sanger Center, and Frederic Deballe (INRA-CNRS) coordinating sequencing of chromosome 5 at Genoscope (*http://medicago.org/genome/downloads/Mt2/*). As earlier research on *M. truncatula* had shown that most genes are found in relatively gene-rich euchromatic arms, with few found in or around centromeres, a BAC-by-BAC strategy for sequencing was selected by the international steering committee (Kulikova et al. 2001, 2004; Pedrosa et al. 2002; Young et al. 2005). This strategy to sequence the euchromatin had produced high quality, though still incomplete, pseudomolecules for all the eight chromosomes, representing the first assembly of the *M. truncatula* genome sequence comprising approximately 63% of the euchromatin and capturing around 58% of all *M. truncatula* genes (Young et al. 2007). This BAC-by-BAC approach will produce contiguous, marker-anchored sequence that will reach chromosome-arm scale in size. These arm-length sequence contigs will be important, as *M. truncatula* is intended to serve as a reference for crop legumes with genomes too large or complex to be sequenced efficiently at this time (Cannon et al. 2005).

4.1.3 Functional Genomics

In the past few years, the *M. truncatula* research community has generated approximately 226,923 ESTs from different cDNA libraries and separated them into around 200,000 tentative consensus sequences and 19,000 singletons, in the process identifying greater than 36,878 unigenes (Tadege et al. 2005). An Affymetrix chip with bioinformatically optimized oligonucleotides representing 48,000 genes is available (*http://www.affymetrix.com/support/technical/datasheets/medicagodatasheet.pdf*). Microarray and DNA chips have been used for the analysis of genes on filters during the arbuscular mycorrhizal symbiosis (Liu et al. 2003) and for comparison between wild type and non-nodulating mutants or between fix-mutants (Mitra et al. 2004; Starker et al. 2006). A dual-genome Symbiosis Chip containing 10,000 *M. truncatula* genes and the entire *S. meliloti* prokaryotic genome allows for coordinate study of signal exchange and development in a symbiotic interaction (Barnett et al. 2004). Suppressive subtractive hybridization (SSH) and SAGE have also been utilized in *M. truncatula* (Godiard et al. 2007; Ane et al. 2008). RNAi-induced gene silencing has been documented (Limpens et al. 2004). The combination of RNAi constructs and hairy root transformation is useful for large-scale screening projects to identify genes of interest for further analysis (Ane et al. 2008).

4.1.4 Reverse Genetics

Attempts have been made to develop three reverse genetic platforms for *M. truncatula* based on: (i) *Tnt1*-mutagenized population, (ii) fast-neutron mutagenized population, and (iii) EMS-mutagenized populations (GLIP 2004).

4.1.4.1 Tnt1 Mutant Collection

T-DNA tagging is not feasible in *M. truncatula* because of its large genome size and the absence of an efficient in planta transformation system (Somers et al. 2003). Transposable elements represent an attractive alternative for generating a large number of insertions in a relatively fewer number of mutant lines. On insertion, these mobile elements can disrupt genes and inactivate them. In *M. truncatula*, the tobacco retrotransposon *Tnt1* has been used to construct insertion mutant collections by in vitro somatic embryogenesis. Over 7,600 independent lines representing an estimated 190,000 insertion events were generated by tagging *Tnt1* to *M. truncatula* (Tadege et al. 2008). On an average, the insertions were at 25 different locations and were also found to be stable during subsequent generations in soil. This mutant collection could serve for the massive sequencing of insertion borders, providing a great tool for *in silico* reverse genetics (GLIP 2004).

4.1.4.2 Deletion Mutagenesis using Fast-neutron Bombardment

Fast-neutrons induce deletions in genomes which may inactivate genes. These deletions can be identified by analyzingly large mutagenized populations. A collection of 250,000 M_2 mutants has been developed and used to identify knockout mutants in important genes (GLIP 2004). Forward genetic screens using fast-neutron mutated *M. truncatula* lines have already been successfully used to identify several developmentally altered mutants and also mutants that are deficient in nodule formation upon infection with *S. meliloti* (Colbert et al. 2001; Tadege et al. 2005).

4.1.4.3 Mutagenesis using EMS and TILLING

In *M. truncatula*, EMS-mutated seeds have been used to identify several mutants that are altered in various stages of their symbiotic interactions or in general aspects of plant morphology and physiology (Ane et al. 2004; Mitra et al. 2004; Tadege et al. 2005). Mutagenesis has been performed on 4,000 lines of *M. truncatula*. Using large EMS-mutant collections, an allelic series of mutants of the gene of interest have been identified using TILLING and an interactive database was developed for the morphological phenotypes of the mutants and, when available, sequences corresponding to the altered gene sequence responsible of the mutant phenotype (GLIP 2004). Mutants for 72 genes involved in a wide variety of traits have been identified.

5 FRUIT CROPS

Fruits are an important component of our daily diet as they supply high-value nutrients and antioxidants. They are consumed in various forms such as fresh, dried, juice and processed products. The variety of flavors, textures, and levels of sweetness and acidity offered by the fruits satisfies diverse consumer taste and choice (Shulaev et al. 2008). Conventional fruit breeding, in many cases, requires a long generation time, high heterozygosity, inability to produce inbred lines, and/or apomixis. To overcome these limitations genome sequencing of fruits is a necessity.

Papaya, a fruit crop cultivated in the tropical and subtropical regions, is ranked first on nutritional scores among 38 common fruits, based on the percentage of the United States Recommended Daily Allowance for vitamin A, vitamin C, potassium, folate, niacin, thiamine, riboflavin, iron and calcium, plus fiber (Ming et al. 2008). In 2004, the University of Hawaii's Center for Genomics, Proteomics and Bioinformatics Research Initiative (CGPBRI) formed an integrative multi-institutional consortium involving other laboratories from United States as well as China. A female plant of 'SunUp', a transgenic variety of papaya developed through transformation of Sunset, was selected for sequencing. A total of 2.8 million whole-genome shotgun (WGS) sequencing reads were generated by the traditional Sanger method of which 1.6 million high-quality reads were assembled into contigs containing 271 Mb and scaffolds spanning 370 Mb including embedded gaps. Approximately 90% of the euchromatin has been covered, containing 92.1% of the unigenes and 92.4% of the genetic markers (Wei and Wing 2008). With coverage of 3x (Stokstad 2008), 'SunUp' papaya is the first commercial virus-resistant transgenic fruit tree (Gonsalves 1998) to be sequenced (Ming et al. 2008).

Grapes ($2n = 38$; Mullins et al. 1992) are consumed not only as fresh or dried fruit but also as juice and processed into wine. An international consortium, the International Grape Genome Program (IGGP) was formed for the co-ordination of the development of the genomic resources and has members from Australia, Canada, Chile, France, Germany, Italy, South Africa, Spain, and the Unites States. Grapes are highly heterozygous with around 13% sequence divergence between alleles. Hence, PN40024 genotype, originally derived from 'Pinot Noir' (Bronner and Oliveira 1990) and has been bred close to full homozygosity (estimated at about 93%) by successive selfings, was selected for sequencing. A total of 6.2 million end-reads were produced by the French-Italian consortium 'Vigna/Vigne' representing an 8.4-fold coverage of the genome (Jaillon et al. 2007). Another genome of 'Pinot Noir' clone ENTAV 115 was also sequenced using Sanger sequencing method to generate 6.5x coverage and was integrated with sequence reads generated by a scalable, highly parallel sequencing by synthesis (SBS) method which gave a 4.2x coverage (Velasco et al. 2007).

The family Rosaceae contains a number of important fruit crops (apple, pear, raspberries, strawberries, blackberries, peach/nectarine, apricot, plum, sweet and sour cherry, and almond) as well as ornamental plants (roses, flowering cherry, crabapple, quince and pear; Jung et al. 2008). Genomic studies are being carried out by various research groups to understand and improve the factors that control various traits of economic importance in these crops. For this purpose, a Genome Database for Rosaceae, an integrated web-based database containing genetic and genomic data for the Rosaceae, has been created (Jung et al. 2008). For whole-genome sequencing, peach was selected as the model plant for Rosaceae. Peach ($2n$ = 16, Jelenkovic and Harrington 1972) has a genome size of 2C = 580 Mb (Baird et al. 1994). 'Drupomics' was founded by the Italian Ministry of Agriculture for the sequencing of Peach in cooperation with the United States partners (Clemson University and the Joint Genome Institute). An initiative to sequence the peach genome was announced by the Joint Genome Institute in 2007 and 'Lovell' a doubled haploid line was selected for sequencing. An integrated approach involving Sanger (6x coverage) and 454 or Solexa (15x coverage) will be carried out. Efforts are also underway, by two groups, to sequence the genome of apple. The first group has teamed with Myriad Genetics, Inc. and 454 Life Sciences to sequence a Golden delicious x Scarlet cultivar, for a total of 4x Sanger and 10x by 454 sequencing. Currently, a 14x coverage has been sequenced and is undergoing assembly. The second group has initiated sequencing a Golden delicious doubled haploid and currently 1.5x coverage has been sequenced using 454 technology. In the spring of 2008, the Strawberry Genome Sequencing Consortium was formed with the aim to sequence the full genome of the *Fragaria vesca,* which has the smallest genome among the economically important plant species, using 454 technology (*http://strawberry.vbi.vt.edu/tiki-index.php?page=Strawberry+Genome+Sequencing+Consortium*).

Citrus ($2n$ = 18) has a genome size of around 367 Mb (Arumuganathan and Earle 1991). The International Citrus Genomics Consortium was formed by members from Australia, Brazil, China, France, Italy, Israel, Japan, Spain and the United States to sequence the genome of sweet orange (*Citrus sinensis*). Around 1.2x coverage of the whole-genome shotgun sequence of sweet orange was produced by the Joint Genome Institute. This was achieved by sequencing ends of about 126,000 phosmid and 257,000 plasmid clones containing 40 kb and 8 kb inserts respectively (Roose et al. 2007). A quality assembly of the highly heterozygous sweet orange genome could not be obtained using the available total sequence coverage of about 473 Mb. Hence, the focus has shifted to the genome sequencing of a haploid (or di- or tri-haploid) genome (Talon and Gmitter 2008). The chloroplast genome of *C. sinensis* has also been sequenced (Bausher et al. 2006).

Banana ($2n = 22$) is grown in more than 110 countries throughout the tropics and subtropics and is the most important staple food in the developing world. It has a genome size of about 560-600 Mb (Kamate et al. 2001). The Global *Musa* Genomics Consortium was formed by members from 37 institutions in 24 countries. Four BAC libraries have been developed for *Musa acuminata* (Vilarinhos et al. 2003; Ortiz-Vazquez et al. 2005) and one has been developed for *M. balbisiana* (Safar et al. 2004b). BAC-end sequencing, whole BAC sequencing and reduced representation sequencing has been attempted for sequencing the genome (Roux et al. 2008). Currently, a 1.4 MB sequence of *M. acuminata* has been produced from 13 BAC clones (Lescot et al. 2008). It has been annotated and analyzed along with four previously sequenced BACs (Aert et al. 2004).

6 CONCLUSION

In the past decade, it was a dream and a challenge to have the sequence of a whole genome. But now, with the rapid advancement of molecular techniques and bioinformatics, genome sequencing has become a reality. The time has come to switch the analysis from a single gene to the whole genome. In recent years, development of new tools and innovative technologies for genome research and a subsequent increase in the data has lead to a highly improved and advanced knowledge base. Many integrated national as well as international projects are generating (and will continue to generate) enormous amounts of genome sequence data. The research is more model-organism oriented and genomic resources such as EST database, high-density microarrays, BAC physical maps, and high-density linkage maps, developed by genome sequencing, will help geneticists and plant breeders to overcome the limitations associated with conventional breeding and to manipulate various traits effectively. This chapter has reviewed the current status of whole-genome sequencing research in various crops. With versatile, relevant plant systems, emerging models and a concerted approach, on-going studies in genomics will contribute to many facets of basic plant science while bringing higher-quality products to the consumer with lower environmental impacts (Shulaev et al. 2008). The development of these genomic resources and their applications will give concrete solutions for many intractable problems faced by the world.

It is clearly evident that by combining research resources and by adopting the principle of depositing information in the public domain, freely available to global research partners, the promise of genome research to improve plants, production, and protection from diseases, and enhanced product quality and value, can be realized. The free availability of these tools and materials are truly the key to the success in genomics research (Talon and Gmitter 2008). Future trends in plant genomics should involve the integration of information and resources from different taxa, while research established in model systems will become more applicable to diverse species (Osterlund and Paterson 2002). Bioinformatics and comparative genomics approaches have advanced the discovery and understanding of conserved genes (Lee et al. 2007). It will further enable the application of data from one species to investigations of taxonomically disparate species (Jayashree et al. 2005). However, the challenge remains in truly being able to incorporate the knowledge gained by genome sequencing for developing newer and better cultivars for sustainable agriculture.

References

Aert R, Sagi L, Volckaert G (2004) Gene content and density in banana (*Musa acuminata*) as revealed by genomic sequencing of BAC clones. Theor Appl Genet 109(1): 129-139

Akbari M, Wenzl P, Caig V, Carling J, Xia L, Yang S, Uszynski G, Mohler V, Lehmensiek A, Kuchel H, Hayden MJ, Howes N, Sharp P, Vaughan P, Rathmell B, Huttner E, Kilian A (2006) Diversity arrays technology (DArT) for high-throughput profiling of the hexaploid wheat genome. Theor Appl Genet 113: 1409-1420

Allouis S, Moore G, Bellec A, Sharp R, Faivre RP, et al. (2003) Construction and characterisation of a hexaploid wheat (*Triticum aestivum* L.) BAC library from the reference germplasm "Chinese Spring". Cereals Res Comm 31: 331-337

Ane J, Kiss GB, Riely BK, Penmetsa R, Oldroyd GED, Ayax C, Levy J, Debelle F, Baek J, Kalo P, Rosenberg C, Roe BA, Long SR, Denarie J, Cook DR (2004) *Medicago truncatula* DMI1 required for bacterial and fungal symbiosis in legumes. Science 303: 1364-1367

Ane J, Zhu H, Frugoli J (2008) Recent advances in *Medicago truncatula* genomics. Intl J Plant Genom: DOI:10.1155/2008/256597

Appels R (2003) A consensus molecular genetic map of wheat-a cooperative international effort. In: Pogna NE (ed) Proc 10th Intl Wheat Genet Symp, 1-6 Sept Paestum, Italy, pp 211-214

Appels R, Feuillet C, Gill B, Buell R, Chumley F, Eversole K, Fritz D (2005) Draft white paper on the International Wheat Genome Sequencing Consortium (*http://www.wheatgenome.org/*)

Arumuganathan K, Earle ED (1991) Nuclear DNA content of some important plant species. Plant Mol Biol Rep 9: 211-215

Asamizu E (2007) Tomato genome sequencing: deciphering the euchromatin region of the chromosome 8. Plant Biotechnol 24: 5-9

Asamizu E, Kato T, Sato S, Nakamura Y, Kaneko T, Tabata S (2003) Structural analysis of a *Lotus japonicus* genome. IV. Sequence features and mapping of seventy-three TAC clones which cover the 7.5 Mb regions of the genome. DNA Res 10: 115-122

Bagge M, Xia X, Lubberstedt T (2007) Functional markers in wheat. Curr Opin Plant Biol 10: 1-6

Baird WV, Estager AS, Wells J (1994) Estimating nuclear DNA content in peach and related diploid species using laser flow cytometry and DNA hybridization. J Am Soc Hort Sci 119: 1312-1316

Barnett MJ, Toman CJ, Fisher RF, Long SR (2004) A dual-genome symbiosis chip for coordinate study of signal exchange and development in a prokaryote-host interaction. Proc Natl Acad Sci USA 101: 16636-16641

Barone A, Chiusano ML, Ercolano MR, Giuliano G, Grandillo S, Frusciante L (2008) Structural and functional genomics of tomato. Intl J Plant Genom DOI:10.1155/2008/820274

Bausher MG, Singh ND, Lee SB, Jansen RK, Daniell H (2006) The complete chloroplast genome sequence of *Citrus sinensis* (L.) Osbeck var 'Ridge Pineapple': organization and phylogenetic relationships to other angiosperms. BMC Plant Biol 6: doi:10.1186/1471-2229-6-21

Bedell JA, Budiman MA, Nunberg A, Citek RW, Robbins D, Jones J, Flick E, Rohlfing T, Fries J, Bradford K, McMenamy J, Smith M, Holeman H, Roe BA, Wiley G, Korf IF, Rabinowicz PD, Lakey N, McCombie WR, Jeddeloh JA, Martienssen RA (2005) Sorghum genome sequencing by methylation filtration. PLoS Biol 3(1): 103-115

Bevan MW (2006) Establishing a BAC-based physical map of *Brachypodium distachyon* as an aid to physical mapping in bread wheat. In: Plant Anim Genome XIV Conf, 14-18 Jan, San Diego, CA, USA

Bhat PR, Lukaszewski A, Cui X, Xu J, Svensson JT, Wanamaker S, Waines JG, Close TJ (2007) Mapping translocation breakpoints using a wheat microarray. Nucl Acids Res 35: 2936-2943

Breseghello F, Sorrells ME (2006) Association mapping of kernel size and milling quality in wheat (*Triticum aestivum* L.) cultivars. Genetics 172: 1165-1177

Briggs SP (1998) Plant genomics: More than food for thought. Proc Natl Acad Sci USA 95: 1986-1988

Bronner A, Oliveira J (1990) Creation and study of the Pinot Noir variety lineage. In: Proc 5th Intl Symp, Grape Breed, Sept 1989 St Martin/Pflaz, Germany, Vitis, spl iss, pp 69-80

Cai WW, Reneker J, Chow CW, Vaishnav M, Brabley A (1998) An anchored framework BAC map of mouse chromosome 11 assembled using multiplex oligonucleotide hybridization. Genomics 54: 387-397

Cai X, Wang C, Xu Y, Xu Q, Zheng Z, Zhou X (2007) Efficient gene silencing induction in tomato by a viral satellite DNA vector. Virus Res 125: 169-175

Cannon SB, Crow JA, Heuer ML, Wang X, Cannon EKS, Dwan C, Lamblin A, Vasdewani J, Mudge J, Cook A, Gish J, Cheung F, Kenton S, Kunau TM, Brown D, May GD, Kim D, Cook DR, Roe BA, Town CD, Young ND, Retzel EF (2005) Databases and information integration for the *Medicago truncatula* genome and transcriptome. Plant Physiol 138: 38-46

Choi H-K, Kim D, Uhm T, Limpens E, Lim H, Mun J, Kalo P, Penmetsa RV, Seres A, Kulikova O, Roe BA, Bisseling T, Kiss GB, Cook DR (2004) A sequence-based genetic map of *Medicago truncatula* and comparison of marker colinearity with *M. sativa*. Genetics 166: 1463-1502

Christensen AB, Thordal-Christensen H, Zimmermann G, Gjetting T, Lyngkjær MF, Dudler R, Schweizer P (2004) The Germin like protein GLP4 exhibits superoxide dismutase activity and is an important component of quantitative resistance in wheat and barley. Mol Plant-Micr Interact 17: 109-117

Colbert T, Till BJ, Tompa R, Reynolds S, Steine MN, Yeung AT, McCallum CM, Comai L, Henikoff S (2001) High-throughput screening for induced point mutations. Plant Physiol 126: 480-484

Cook RJ (1998) Toward a successful multinational crop plant genome initiative. Proc Natl Acad Sci USA 95: 1993-1995

Crossa J, Burgueño J, Dreisigacker S, Vargas M, Herrera-Foesse SA, Lillem M, Singh RP, Trethowan R, Warburton M, Franco J, Reynolds M, Crouch JH, Ortiz R (2007) Association analysis of historical bread wheat germplasm using additive genetic covariance of relatives and population structure. Genetics 177: 889-1913

Davuluri GR, van Tuinen A, Fraser PD, Manfredonia A, Newman R, Burgess D, Brummell DA, King SR, Palys J, Uhlig J, Bramley PM, Pennings HMJ, Bowler C (2005) Fruit-specific RNAi-mediated suppression of *DET1* enhances carotenoid and flavonoid content in tomatoes. Nat Biotechnol 23: 890-895

Delalande C, Regad F, Zouine M, Frasse P, Latche A, Pech JC, Bouzayen M (2007) The French contribution to the multinational Solanaceae Genomics Project as integrated part of the European effort. Plant Biotechnol 24: 27-31

Devos KM, Beales J, Nagamura Y, Sasaki T (1999) Arabidopsis-Rice: Will colinearity allow gene prediction across the eudicot-monocot divide? Genome Res 9(9): 825-829

Emmanuel E, Levy AA (2002) Tomato mutants as tools for functional genomics. Curr Opin Plant Biol 5: 112-117

Emberton J, Ma J, Yuan Y, SanMiguel P, Bennetzen JL (2005) Gene enrichment in maize with hypomethylated partial restriction (HMPR) libraries. Genome Res 15: 1441-1446

Endo TR, Gill BS (1996) The deletion stocks of common wheat. J Hered 87: 295-307

Frary A, Xu Y, Liu J, Mitchell S, Tedeschi E, Tanksley SD (2005) Development of a set of PCR-based anchor markers encompassing the tomato genome and evaluation of their usefulness for genetics and breeding experiments. Theor Appl Genet 111: 291-312

Fu D-Q, Zhu B-Z, Zhu H-L, Jiang W-B, Luo Y-B (2005) Virus-induced gene silencing in tomato fruit. Plant J 43: 299-308

Fu D, Uauy C, Blechl A, Dubcovsky J (2007) RNA interference for wheat functional gene analysis. Transgen Res 16: 689-701

Galibert F, Finan TM, Long SR, Puhler A, Abola P, Ampe F, Barloy-Hubler F, Barnett MJ, Becker A, Boistard P, Bothe G, Boutry M, Bowser L, Buhrmester J, Cadieu E, Capela D, Chain P, Cowie A, Davis RW, Dreano S, Federspiel NA, Fisher RF, Gloux S, Godrie T, Goffeau A, Golding B, Gouzy J, Gurjal M, Hernandez-Lucas I, Hong A, Huizar L, Hyman RW, Jones T, Kahn D, Kahn ML, Kalman S, Keating DH, Kiss E, Komp C, Lelaure V, Masuy D, Palm C, Peck MC, Pohl TM, Portetelle D, Purnelle B, Ramsperger U, Surzycki R, Thebault P, Vandenbol M, Vorhölter F, Weidner S, Wells DH, Wong K, Yeh K, Batut J (2001) The composite genome of the legume symbiont *Sinorhizobium meliloti*. Science 293: 668-672

Garvin DF (2007) *Brachypodium distachyon*: A new model system for structural and functional analysis of grass genome. In: Varshney RK, Koebner RMD (eds) Model Plants and Crop Improvement. Taylor and Francis (CRC), Boca Raton, London, New York, pp 109-123

Gidoni D, Fuss E, Burbidge A, Speckmann GJ, James S, Nijkamp D, Mett A, Feiler J, Smoker M, de Vroomen MJ, Leader D, Liharska T, Groenendijk J, Coppoolse E, Smit JJ, Levin I, de Both M, Schuch W, Jones JD, Taylor IB, Theres K, van Haaren MJ (2003) Multifunctional T-DNA/Ds tomato lines designed for gene cloning and molecular and physical dissection of the tomato genome. Plant Mol Biol 51: 83-98

Gill BS, Appels R, Botha-Oberholster A, Buell CR, Bennetzen JL, et al. (2004) A workshop report on wheat genome sequencing: International genome research on wheat consortium. Genetics 168: 1087-1096

Godiard L, Niebel A, Micheli F, Gouzy J, Ott T, Gamas P (2007) Identification of new potential regulators of the *Medicago truncatula-Sinorhizobium meliloti* symbiosis using a large-scale suppression subtractive hybridization approach. Mol Plant-Micr Interact 20: 321-332

Gonsalves D (1998) Control of papaya ringspot virus in papaya: a case study. Annu Rev Phytopathol 36: 415-437

Goyal A, Bandopadhyay R, Sourdille P, Endo TR, Balyan HS, Gupta PK (2005) Physical molecular maps of wheat chromosomes. Funct Integr Genom 5: 260-263

Grain Legume Integrated Project (GLIP) (2004) *http://www.eugrainlegumes.org/documents/documents/Publishable_Final_report_v15.pdf*

Gray J, Picton S, Shabbeer J, Schuch W, Grierson D (1992) Molecular biology of fruit ripening and its manipulation with antisense genes. Plant Mol Biol 19: 69-87

Gupta PK (2008) Ultrafast and low-cost DNA sequencing methods for applied genomics research. Proc Natl Acad Sci India 78: 98-102

Gupta PK, Kulwal PL, Rustgi S (2005) Wheat cytogenetics in the genomics era and its relevance to breeding. Cytogenet Genom Res 109: 315-327

Gupta PK, Mir RR, Mohan A, Kumar J (2008) Wheat genomics: Present status and future prospects. Intl J Plant Genom: DOI:10.1155/2008/896451

Henikoff S, Till BJ, Comai L (2004) TILLING: Traditional mutagenesis meets functional genomics. Plant Physiol 135: 630-636

Heslop-Harrison JS (2000) Comparative genome organization in plants: from sequence and markers to chromatin and chromosomes. Plant Cell 12(5): 617-636

Hill-Ambroz K, Webb CA, Matthews AR, Li W, Gill BS, Fellers JP (2006) Expression analysis and physical mapping of a cDNA library of *Fusarium* head blight infected wheat spikes. Plant Genom (a Suppl to Crop Sci) 46: S14-S26

Hossain KG, Riera-lizarazu O, Kalavacharla V, Vales MI, Rust JL, Maan SS, Kianian SF (2004a) Molecular cytogenetic characterization of an alloplasmic durum wheat line with a portion of chromosome 1D of *Triticum aestivum* carrying the *scsae* gene. Genome 47: 206-214

Hossain KG, Riera-lizarazu O, Kalavacharla V, Vales MI, Maan SS, Kianian SF (2004b) Radiation hybrid mapping of the species cytoplasm-specific (*scsae*) gene in wheat. Genetics 168: 415-423

Huo N, Gu YQ, Lazo GR, Vogel JP, Coleman-Derr D, Luo M, Thilmony R, Garvin DF, Anderson OD (2006) Construction and characterization of two BAC libraries from *Brachypodium distachyon*, a new model for grass genomics. Genome 49: 1099-1108

International Rice Genome Sequencing Project (IRGSP) (2005) The map-based sequence of the rice genome. Nature 436: 793-800

Ito M, Miyamoto J, Mori Y, Fujimoto S, Uchiumi T, Abe M, Suzuki A, Tabata S, Fukui K (2000) Genome and chromosome dimensions of *Lotus japonicus*. J Plant Res 113: 435-442

Jacobs J, Conner A, Bachem C, van Ham R, Visser R (2006) The potato genome sequence consortium. 'Breeding for success: diversity in action'. In: Mercer CF (ed) Proc 13th Aust Plant Breed Conf, 18-21 April, Christchurch, New Zealand, pp 933-936

Jaillon O, Aury J-M, Noel B, Policriti A, Clepet C (2007) The grapevine genome sequence suggests ancestral hexaploidization in major angiosperm phyla. Nature 449: 463-468

Jayashree B, Ferguson M, Ilut D, Doyle J, Crouch JH (2005) Analysis of genomic sequences from peanut (*Arachis hypogaea*). Electronic J Biotechnol: DOI:10.2225/vol8-issue3-fulltext-3 (*http://www.ejbiotechnology.info/content/vol8/issue3/full/3/*)

Jelenkovic G, Harrington E (1972) Morphology of the pachytene chromosomes in *Prunus persica*. Can J Genet Cytol 14: 317-324

Jiang C-Z, Lu F, Imsabai W, Meir S, Reid MS (2008) Silencing polygalacturonase expression inhibits tomato petiole abscission. J Exp Bot 59(4): 973-979

Jordan MC, Somers DJ, Banks TW (2007) Identifying regions of the wheat genome controlling seed development by mapping expression quantitative trait loci. Plant Biotechnol J 5: 1-12

Jung S, Staton M, Lee T, Blenda A, Svancara R, Abbott A, Main D (2008) GDR (Genome Database for Rosaceae): Integrated web-database for Rosaceae genomics and genetics data. Nucl Acids Res 36: D1034-D1040

Kahlau S, Aspinall S, Gray JC, Bock R (2006) Sequence of the tomato chloroplast DNA and evolutionary comparison of solanaceaeous plastid genome. J Mol Evol 63: 194-207

Kalavacharla V, Hossain K, Gu Y, Riera-Lizarazu O, Isabel-Vales M (2006) High resolution radiation hybrid map of wheat chromosome 1D. Genetics 173: 1089-1099

Kalloo G (1991) Genetic Improvement of Tomato. Springer, Berlin, Germany.

Kamate K, Brown S, Durand P, Bureau JM, De Nay D, Trinh TH (2001) Nuclear DNA content and base composition in 28 taxa of *Musa*. Genome 44(4): 622-627

Kaneko T, Asamizu E, Kato T, Sato S, Nakamura Y, Tabata S (2003) Structural analysis of a *Lotus japonicus* genome. III. Sequence features and mapping of sixty-two TAC clones which cover the 6.7 Mb regions of the genome. DNA Res 10: 27-33

Karpechenko GD (1925) On the chromosomes of Phaseolinae. Bull Appl Bot Plant Breed (Leningrad) 14: 143-148

Kato T, Sato S, Nakamura Y, Kaneko T, Asamizu E, Tabata S (2003) Structural analysis of a *Lotus japonicus* genome. V. Sequence features and mapping of sixty-four TAC clones which cover the 6.4 Mb regions of the genome. DNA Res 10: 277-285

Kawakami J (1930) Chromosome numbers in Leguminosae. Bot Mag (Tokyo) 44: 319-328 (in Japanese)

Keller B, Feuillet C (2000) Colinearity and gene density in grass genomes. Trends Plant Sci 5: 246-251

Khan MQ, Narayan RKJ (2007) Phylogenetic diversity and relationships among species of genus *Nicotiana* using RAPDs analysis. Afr J Biotechnol 6(2): 148-162

Knapp S (2002) Tobacco to tomatoes: a phylogenetic perspective on fruit diversity in the Solanaceae. J Exp Bot 53(377): 2001-2022

Kresovich S, Barbazuk B, Bedell JA, Borrell A, Buell CR, Burke J, Clifton S, Cordonnier-Pratt M, Cox S, Dahlberg J, Erpelding J, Fulton TM, Fulton B, Fulton L, Gingle AR, Hash CT, Huang Y, Jordan D, Klein PE, Klein RR, Magalhaes J, McCombie R, Moore P, Mullet JE, Ozias-Akins P, Paterson AH, Porter K, Pratt L, Roe B, Rooney W, Schnable PS, Stelly DM, Tuinstra M, Ware D, Warek U (2005) Toward Sequencing the Sorghum Genome. A

US National Science Foundation-Sponsored Workshop Report. Plant Physiol 138: 1898-1902

Kulikova O, Gualtieri G, Geurts R, Kim D, Cook D, Huguet T, de Jong JH, Fransz PF, Bisseling T (2001) Integration of the FISH pachytene and genetic maps of *Medicago truncatula.* Plant J 27: 49-58

Kulikova O, Geurts R, Lamine M, Kim D, Cook DR, Leunissen J, de Jong H , Roe BA, Bisseling T (2004) Satellite repeats in the functional centromere and pericentromeric heterochromatin of *Medicago truncatula.* Chromosoma 113: 276-283

Labate JA, Baldo AM (2005) Tomato SNP discovery by EST mining and resequencing Mol Breed 16: 343-349

Labate JA et al. 39 authors (2007) Tomato. In: Kole C (ed) Genome Mapping and Molecular Breeding in Plants. Vol 7: Vegetables. Springer, Berlin, Heidelberg, New York, pp 1-125

Lariguet P, Pankhurst CE, Porch T, Silue S, Baudoin J-P, Blair MW, Triplett EW, Broughton WJ (2005) Tilling beans for changes in nodule and seed development. In: Phaseomics IV, Nov 30 - Dec 3, Salta-Argentina, p. 18 (*http:// www.biol.unlp.edu.ar/phaseomicsIV/libroderesumenes.pdf*)

Leader DJ (2005) Transcriptional analysis and functional genomics in wheat. J Cereal Sci 41: 149-163

Lee S, Jo SH, Choi D (2007) Solanaceae genomics: Current status of tomato (*Solanum lycopersicum*) genome sequencing and its application to pepper (*Capsicum* spp.) genome research. Plant Biotechnol 24: 11-16

Lescot M, Piffanelli P, Ciampi AY, Ruiz M, Blanc G, Leebens-Mack J, da Silva FR, Santos CMR, D'Hont A, Garsmeur O, Vilarinhos AD, Kanamori H, Matsumoto T, Ronning CM, Cheung F, Haas BJ, Althoff R, Arbogast T, Hine E, Pappas Jr GJ, Sasaki T, Souza Jr MT, Miller RNG, Glaszmann J, Town CD (2008) Insights into the *Musa* genome: Syntenic relationships to rice and between *Musa* species. BMC Genom 9: 58

Li JR, Zhao W, Li QZ, Ye XG, An BY, Li X, Zhang XS (2005) RNA silencing of waxy gene results in low levels of amylose in the seeds of transgenic wheat (*Triticum aestivum* L.). Acta Genet Sin 32: 846-854

Lijavetzky D, Muzzi G, Wicker T, Keller B, Wing R, Dubcovsky J (1999) Construction and characterization of a bacterial artificial chromosome (BAC) library for the A genome of wheat. Genome 42: 1176-1182

Limpens E, Ramos J, Franken C, Raz V, Compaan B, Franssen H, Bisseling T, Geurts R (2004) RNA interference in *Agrobacterium rhizogenes* roots of *Arabidopsis* and *Medicago truncatula.* J Exp Bot 55: 983-992

Liu J, Blaylock LA, Endre G, Cho J, Town CD, VandenBosch KA, Harrison MJ (2003) Transcript profiling coupled with spatial expression analyses reveals genes involved in distinct developmental stages of an arbuscular mycorrhizal symbiosis. Plant Cell 15: 2106-2123

Liu YG, Shirano Y, Fukaki H, Yanai Y, Tasaka M, Tabata S, Shibata D (1999)

Complementation of plant mutants with large genomic DNA fragments by a transformation-competent artificial chromosome vector accelerates positional cloning. Proc Natl Acad Sci USA 96: 6535-6540

Liu Y, Schiff M, Dinesh Kumar SP (2002) Virus-induced gene silencing in tomato. Plant J 31: 777-786

Loeffler D, Hurst S, McGuire C, Barrios C, Vafeados D (2008) TILLING and Eco-TILLING the tomato expansin *EXP1* gene. In: Plant Biology 2008, 26 June - 1 July, Merida, Mexico

Loukoianov A, Yan L, Blechl A, Sanchez A, Dubcovsky J (2005) Regulation of *VRN-1* vernalization genes in normal and transgenic polyploid wheat. Plant Physiol 138: 2364-2373

Luo M, Ma Y, Huo N (2007) Construction of physical map for *Brachypodium distachyon*. In: Plant Anim Genome XV Conf, 13-17 Jan, San Diego, CA, USA

Martienssen RA, Rabinowicz PD, O'Shaughnessy A, McCombie (2004) Sequencing the maize genome. Curr Opinion Plant Biol 7: 02-107

McCallum CM, Hurst SR, Ann SJ, Colbert TG, Facciotti D, Loeffler D, Steine MN, McGuire C, Jones P, Knauf VC (2008) TILLING for loss of function alleles in the tomato fruit specific polygalacturonase. In: Plant Biology 2008, 26 June -1 July, Merida, Mexico, Abstr 1420

McCartney CA, Somers DJ, Humphreys DG, Lukow O, Ames N, Noll J, Cloutier S, McCallum BD (2005) Mapping quantitative trait loci controlling agronomic traits in the spring wheat cross RL4452 × 'AC Domain'. Genome 48: 870-883

McCartney CA, Somers DJ, Lukow O, Ames N, Noll J, Cloutier S, Humphreys DG, McCallum BD (2006) QTL analysis of quality traits in the spring wheat cross 'RL4452' × 'AC Domain'. Plant Breed 125: 565-575

McIntosh RA, Yamazaki Y, Devos KM, Dubcovsky J, Rogers WJ, Appels R (2003) Catalogue of gene symbols for wheat. In: Pogna NE, Romano M, Pogna E, Galterio G (eds) Proc 10th Intl Wheat Genet Symp. Vol 4. Instituto Sperimentale per la Cerealicotura, Rome, Italy, pp 1-34

McIntosh S, Watson L, Bundock P, Crawford A, White J, Cordeiro G, Barbary D, Rooke L, Henry R, (2007) SAGE of the developing wheat caryopsis. Plant Biotechnol J 5: 69-83

Meissner R, Chague V, Zhu Q, Emmanuel E, Elkind Y, Levy AA (2000) A high throughput system for transposon tagging and promoter trapping in tomato. Plant J 22: 265-274

Menda N, Semel Y, Peled D, Eshed Y, Zamir D (2004) *In silico* screening of a saturated mutation library of tomato. Plant J 38: 861-872

Messing J, Bharti AK, Karlowski WM, Gundlach H, Kim HR, Yu Y, Wei F, Fuks G, Soderlund CA, Mayer KF, Wing RA (2004) Sequence composition and genome organization of maize. Proc Natl Acad Sci USA 101: 14349-14354

Michaelson MJ, Price HJ, Ellison JR, Johnston JS (1991) Comparison of plant

DNA contents determined by feulgen microspectrophotometry and laser flow cytometry. Am J Bot 78(2): 183-188

Ming R, Hou S, Feng Y, Yu Q, Dionne-Laporte A,. Saw JH, Senin P, Wang W, Ly BV, Lewis KLT, Salzberg SL, Feng L, Jones MR, Skelton RL, Murray JE, Chen C, Qian W, Shen J, Du P, Eustice M, Tong E, Tang H, Lyons E, Paull RE, Michael TP, Wall K, Rice DW, Albert H, Wang M, Zhu YJ, Schatz M, Nagarajan N, Acob RA, Guan P, Blas A, Wai CM, Ackerman CM, Ren Y, Liu C, Wang J, Wang J, Na J, Shakirov EV, Haas B, Thimmapuram J, Nelson D, Wang X, Bowers JE, Gschwend AR, Delcher AL, Singh R, Suzuki JY, Tripathi S, Neupane K, Wei H, Irikura B, Paidi M, Jiang N, Zhang W, Presting G, Windsor A, Navajas-Pérez R, Torres MJ, Feltus FA, Porter B, Li Y, Burroughs AM, Luo M, Liu L, Christopher DA, Mount SM, Moore PH, Sugimura T, Jiang J, Schuler MA, Friedman V, Mitchell-Olds T, Shippen DE, dePamphilis CW, Palmer JD, Freeling M, Paterson AH, Gonsalves D, Wang L, Alam M (2008) The draft genome of the transgenic tropical fruit tree papaya (*Carica papaya* Linnaeus). Nature 452: 991-997

Mitra RM, Shaw SL, Long SR (2004) Six non nodulating plant mutants defective for Nod factor-induced transcriptional changes associated with the legume-rhizobia symbiosis. Proc Natl Acad Sci USA 101: 10217-10222

Mohan A, Goyal A, Singh R, Balyan HS, Gupta PK (2007) Physical mapping of wheat and rye EST-SSRs on wheat chromosomes. Plant Genom (a suppl to Crop Sci) 47: S3-S13

Moolhuijzen P, Dunn DS, Bellgard M, Carter M, Jia J, Kong X, Gill BS, Feuillet C, Breen J, Appels R (2007) Wheat genome structure and function: genome sequence data and the international wheat genome sequencing consortium. Aust J Agri Res 58: 470-475

Mueller LA, Tanksley SD, Giovannoni JJ, van Eck J, Stack S, Choi D, Kim BD, Chen M, Cheng Z, Li C, Ling H, Xue Y, Seymour G, Bishop G, Bryan G, Sharma R, Khurana J, Tyagi A, Chattopadhyay D, Singh NK, Stiekema W, Lindhout P, Jesse T, Lankhorst RK, Bouzayen M, Shibata D, Tabata S, Granell A, Botella MA, Giuliano G, Frusciante L, Causse M, Zamir D (2005) The Tomato Sequencing Project, the First Cornerstone of the Intl Solanaceae Project (SOL). Comp Funct Genom 6: 153-158

Mullins MG, Bouquet A, Williams LE (1992) Biology of Grapevine. Cambridge Univ Press, Cambridge, UK

Mun J-H, Kim D-J, Choi H-K, Gish J, Debelle F, Mudge J, Denny R, Endre G, Saurat O, Dudez A, Kiss GB, Roe B, Young ND, Cook DR (2006) Distribution of microsatellites in the genome of *Medicago truncatula*: a resource of genetic markers that integrate genetic and physical maps. Genetics 172: 2541-2555

Nakamura Y, Kaneko T, Asamizu E, Kato T, Sato S, Tabata S (2002) Structural analysis of a *Lotus japonicus* genome. II. Sequence features and mapping of sixty-five TAC clones which cover the 6.5 Mb regions of the genome. DNA Res 9: 63-70

Nam YW, Penmetsa RV, Endre G, Uribe P, Kim D, Cook DR (1999) Construction of a bacterial artificial chromosome library of *Medicago truncula*ta and identification of clones containing ethylene-response gene. Theor Appl Genet 98: 638-646

Opperman CH, Lommel S, Burke M, Carlson J, George C, Gove S, Graham S, Houfek TD, Kalat S, Little PC, Lumpkin A, Redman L, Ross T, Schaffer R, Scholl E, Stephens PJ, Windham E, Zekanis SH, Lakey N, Bidell J, Budiman A (2006) Update on The Tobacco Genome Initiative: A gene discovery platform. In: Plant Anim Genome XIV Conf, 14-18 Jan, San Diego, CA, USA, P 30

Ortiz-Vazquez E, Kaemmer D, Zhang HB, Muth J, Rodriguez-Mendiola M, Arias-Castro C, James A (2005) Construction and characterization of a plant transformation-competent BIBAC library of the black Sigatoka-resistant banana *Musa acuminata* cv. Tuu Gia (AA). Theor Appl Genet 110(4): 706-713

Osterlund MT, Paterson AH (2002) Applied plant genomics: the secret is integration. Curr Opin Plant Biol 5: 141-145

Ozdemir BS, Hernandez P, Filiz E, Budak H (2008) *Brachypodium* Genomics. Intl J Plant Genom: DOI:10.1155/2008/536104

Palmer LE, Rabinowicz PD, O'Shaughnessy A, Balija V, Nascimento L, Dike S, de la Bastide M, Martienssen RA, McCombie WR (2003) Maize genome sequencing by methylation filtration. Science 302: 2115-2117

Parida SK, Kumar KA, Dalal V, Sigh NK, Mohapatra T (2006) Unigene derived microsatellite markers for the cereal genomes. Theor Appl Genet 112: 808-817

Paterson AH (2008) Genomics of sorghum. Intl J Plant Genom: DOI:10.1155/2008/362451

Pedrosa A, Sandal N, Stougaard J, Schweizer D, Bachmair A (2002) Chromosomal map of the model legume *Lotus japonicus*. Genetics 161: 1661-1672

Peng JH, Lapitan NLV (2005) Characterization of EST-derived microsatellites in the wheat genome and development of eSSR markers. Funct Integr Genom 5: 80-96

Peterson DG, Lapitan NLV, Stack SM (1999) Localization of single- and low-copy sequences on tomato synaptonemal complex spreads using fluorescence in situ hybridization (FISH). Genetics 152: 427-439

Poole R, Barker G, Wilson ID, Coghill JA, Edwards KJ (2007) Measuring global gene expression in polyploidy; a cautionary note from allohexaploid wheat. Funct Integr Genom 7: 207-219

Pryer KM, Schneider H, Zimmer EA, Banks JA (2002) Deciding among green plants for whole genome studies. Trends Plant Sci 7(12): 550-554

Qi, LL, Echalier B, Chao S, Lazo GR, Butler GE, Anderson OD, Akhunov ED, Dvorak J, Linkiewicz AM, Ratnasiri A, Dubcovsky J, Bermudez-Kandianis CE, Greene RA, Kantety R, La Rota CM, Munkvold JD, Sorrells SF, Sorrells ME, Dilbirligi M, Sidhu D, Erayman M, Randhawa HS, Sandhu D, Bondareva SN, Gill KS, Mahmoud AA, Ma X.-F, Mift ahudin, Gustafson JP,

Conley EJ, Nduati V, Gonzalez-Hernandez JL, Anderson JA, Peng JH, Lapitan NLV, Hossain KG, Kalavacharla V, Kianian SF, Pathan MS, Zhang DS, Nguyen HT, Choi D-W, Fenton RD, Close TJ, McGuire PE, Qualset CO, and Gill BS (2004) A chromosome bin map of 16,000 expressed sequence tag loci and distribution of genes among the three genomes of polyploidy wheat. Genetics 168: 701-712

Rabinowicz PD, Bennetzen JL (2006) The maize genome as a model for efficient sequence analysis of large plant genomes. Curr Opin Plant Biol 9: 149-156

Ramesh SV, Mishra AK, Praveen S (2007) Hairpin RNA-mediated strategies for silencing t*omato leaf curl virus AC1* and *AC4* genes effective resistance in plants. Oligonucleotides 17: 251-257

Ravel C, Praud S, Murigneux A, Linossier L, Dardevet M, Balfourier F, Dufour P, Brunel D, Charmet G (2006) Identification of *Glu-B1-1* as a candidate gene for the quantity of high-molecular-weight glutenin in bread wheat (*Triticum aestivum* L.) by means of an association study. Theor Appl Genet 112: 738-743

Regina A, Bird A, Topping D, Bowden S, Freeman J, Barsby T, Kosar-Hashemi B, Li Z, Rahman S, Morell M (2006) High-amylose wheat generated by RNA interference improves indices of large-bowel health in rats. Proc Natl Acad Sci USA 103: 3546-3551

Roder MS, Korzun V, Wendehake K, Plaschke J, Tixier MH, Leroy P, Ganal MW (1998) A microsatellite map of wheat. Genetics 149: 2007-2023

Roose ML, Niedz RP, Gmitter FG, Close TJ, Dandekar AM, Cheng JF, Rokhsar DS (2007) Analysis of a 1.2x whole genome sequence of *Citrus sinensis.* In: Plant Anim Genome XV Conf, 13-17 Jan, San Diego, CA, USA

Roux N, Baurens FC, Dolezel J, Hribova E, Heslop-Harrison P, Town C, et al. (2008) Genomics of banana and plantain (*Musa* spp.), major staple crops in the tropics. In: Moore PH, Ming R (eds) Genomics of Tropical Crop Plants. Springer, New York, USA, pp 83-111

Safar J, Bartos J, Janda J, Bellec A, Kubalakova M, Valarik M, Pateyron S, Weiserova J, Tuskova R, Číhalikova J, Vrana J, Šimkova H, Faivre-Rampant P, Sourdille P, Caboche M, Bernard M, Dolezel J, Chalhoub B (2004a) Dissecting large and complex genomes: flow sorting and BAC cloning of individual chromosome types from bread wheat. Plant J 39: 960-968

Safar J, Noa-Carrazana JC, Vrana J, Bartos J, Alkhimova O, et al. 2004b. Creation of a BAC resource to study the structure and evolution of the banana (*Musa balbisiana*) genome. Genome 47(6): 1182-1191

Saisho D, Kawasaki S, Sato K, Takeda K (2002) Construction of a BAC library from Japanese malting barley Haruna Nijo. In: Plant Anim Genome X Conf, 12-16 Jan, San Diego, CA, USA, P 393

Saliba-Colombani V, Causse M, Gervais L, Philouze J (2000) Efficiency of RFLP, RAPD, and AFLP markers for the construction of an intraspecific map of the tomato genome. Genome 43: 29-40

Sato S, Tabata S (2005) *Lotus japonicus* as a platform for legume research. Curr

Opin Plant Biol 9: 128-132

Sato S, Kaneko T, Nakamura Y, Asamizu E, Kato T, Tabata S (2001) Structural analysis of a *Lotus japonicus* genome. I. Sequence features and mapping of fifty-six TAC clones which cover the 5.4 mb regions of the genome. DNA Res 8: 311-318

Schijlen EGWM, Ric de Vos CH, Martens S, Jonker HH, Rosin FM, Molthoff JW, Tikunov YM, Angenent GC, van Tunen AJ, Bovy AG (2007) RNA interference silencing of chalcone synthase, the first step in the flavonoid biosynthesis pathway, leads to parthenocarpic tomato fruits. Plant Physiol 144: 1520-1530

Schweizer P, Pokorny J, Schulze-Lefert P, Dudler R (2000) Double-stranded RNA interferes with gene function at the single-cell level in cereals. Plant J 24: 895-903

Shibata D (2005) Genome sequencing and functional genomics approaches in tomato. J Gen Plant Pathol 71: 1-7

Shulaev V, Korban SS, Sosinski B, Abbott AG, Aldwinckle HS, Folta KM, Iezzoni A, Main D, Arus P, Dandekar AM, Lewers K, Brown SK, Davis TM, Gardiner SE, Potter D, Veilleux RE (2008) Multiple models for Rosaceae genomics. Plant Physiol 147: 985-1003

Skinner DZ, Okubara PA, Baek K, Call DR (2005) Long oligonucleotide microarrays in wheat: evaluation of hybridization signal amplification and an oligonucleotide-design computer script. Funct Integr Genom 5: 70-79

Slade AJ, Fuerstenberg SI, Loeffler D, Steine MN, Facciotti D (2005) A reverse genetic, nontransgenic approach to wheat crop improvement by TILLING. Nat Biotechnol 23: 75-81

Slade AJ, Knauf VC (2005) TILLING moves beyond functional genomics into crop improvement. Transgen Res 14: 109-115

Snape JW, Moore G (2007) Reflections and opportunities: gene discovery in the complex wheat genome. In: Buck HT, Nisi JE, Salomon N (eds) Wheat Production in Stressed Environments. Springer, Netherlands, pp 677-684

Somers D, Edwards KJ, Issac P (2004) A high density microsatellite consensus map for bread wheat (*Triticum aestivum* L.). Theor Appl Genet 109: 1105-1114

Somers DA, Samac DA, Olhoft PM (2003) Recent advances in legume transformation. Plant Physiol 131: 892-899

Somers DJ (2005) Molecular breeding and assembly of complex genotypes in wheat. In: Tsunewaki K (ed) Frontiers of wheat bioscience. Hundredth Memorial Issue of Wheat Information Service. Yokohama: Kihara Memorial Yokohama Foundation for the Advancement of Life Sciences, Yokohoma, Japan, pp 235-246

Sorrells ME, La Rota M, Bermudez-Kandianis CE, Greene RA, Kantety R, Munkvold JD, Miftahudin, Mahmoud A, Ma X, Gustafson PJ, Qi LL, Echalier B, Gill BS, Matthews DE, Lazo GR, Chao S, Anderson OD,

Edwards H, Linkiewicz AM, Dubcovsky J, Akhunov ED, Dvorak J, Zhang D, Nguyen HT, Peng J, Lapitan NLV, Gonzalez-Hernandez JL, Anderson JA, Hossain K, Kalavacharla V, Kianian SF, Choi D, Close TJ, Dilbirligi M, Gill KS, Steber C, Walker-Simmons MK, McGuire PE, Qualset CO (2003) Comparative DNA sequence analysis of wheat and rice genomes. Genome Res 13: 1818-1827

Sourdille P, Singh S, Cadalen T, Brown-Guedira GL, Gay G, Qi L, Gill BS, Dufour P, Murigneux A, Bernard M (2004) Microsatellite-based deletion bin system for the establishment of genetic-physical map relationships in wheat (*Triticum aestivum* L.). Funct Integr Genom 4: 12-25

Starker CG, Parra-Colmenares AL, Smith L, Mitra RM, Long SR (2006) Nitrogen fixation mutants of *Medicago truncatula* fail to support plant and bacterial symbiotic gene expression. Plant Physiol 140: 671-680

Stein N (2007) Triticeae genomics: advances in sequence analysis of large genome cereal crops. Chrom Res 15: 21-31

Stokstad E (2008) Papaya takes on ringspot virus and wins. Science 320: 472

Stougaard J, Sandal N, Jorgensen N, Dam S, Nautrop G, Fredslund J, Hougaard BK, Radutoiu S, Tirichine L, Madsen LH, Madsen EB, Keisuke Y, Romero P, Jurkiewich A, Albrektsen A, Quistgaard EMH, Fukai E (2007) Lotus genetics and genomics: Resources and approaches. Lotus Newsl 37(2): 54-55

Suliman-Pollatschek S, Kashkush K, Shats H, Hillel J, Lavi U (2002) Generation and mapping of AFLP, SSRs and SNPs in *Lycopersicon esculentum*. Cell Mol Biol Lett 7: 583-597

Sun H.-J, Uchii S, Watanabe S, Ezura H (2006) A highly efficient transformation protocol for Micro-Tom, a model cultivar for tomato functional genomics. Plant Cell Physiol 47: 426-431

Swigonova Z, Lai J, Ma J, Ramakrishna W, Llaca V, Bennetzen JL, Messing J (2004) Close split of sorghum and maize genome progenitors. Genom Res 14: 1916-1923

Tadege M, Ratet P, Mysore KS (2005) Insertional mutagenesis: a Swiss army knife for functional genomics of *Medicago truncatula*. Trends Plant Sci 10: 229-235

Tadege M, Wen J, He J, Tu H, Kwak Y, Eschstruth A, Cayrel A, Endre G, Zhao PX, Chabaud M, Ratet P, Mysore KS (2008) Large-scale insertional mutagenesis using the *Tnt1* retrotransposon in the model legume *Medicago truncatula*. Plant J 54: 335-347

Talon M, Gmitter FG Jr (2008) Citrus genomics. Intl J Plant Genom: DOI:10. 1155/2008/528361

Tanksley SD (1993) Linkage map of the tomato (*Lycopersicon esculentum*) (2n = 24). In: O'Brian SJ (ed) Genetic Maps: Locus Maps of Complex Genomes. Cold Spring Harbor Lab Press, Cold Spring Harbor, NY, USA, pp 6.3-6.15

Tanksley SD, Ganal MW, Prince JP, de Vicente MC, Bonierbale MW, Broun P,

Fulton TM, Giovannoni JJ, Grandillo S, Martin GB, Messeguer R, Miller JC, Miller L, Paterson AH, Pineda O, Riider MS, Wing RA, Wu W, Young ND (1992) High density molecular linkage maps of the tomato and potato genomes. Genetics 132(4): 1141-1160

Thoquet P, Gherardi M, Journet E-P, Kereszt A, Ane J, Prosperi J, Huguet T (2002) The molecular genetic linkage map of the model legume *Medicago truncatula*: An essential tool for comparative legume genomics and the isolation of agronomically important genes. BMC Plant Biol 2:1: doi:10. 1186/1471-2229-2-1

Tommasini L, Schnurbusch T, Fossati D, Mascher F, Keller B (2007) Association mapping of *Stagonospora nodorum* blotch resistance in modern European winter wheat varieties. Theor Appl Genet 115: 697-708

Travella S, Klimm TE, Killer B (2006) RNA interference-based gene silencing as an efficient tool for functional genomics in hexaploid bread wheat. Plant Physiol 142: 6-20

Uauy C, Distelfeld A, Fahima T, Blechl A, Dubcovsky J (2006) A NAC gene regulating senescence improves grain protein, zinc, and iron content in wheat. Science 314: 1298-1301

Valentine T, Shaw J, Blok VC, Phillips MS, Oparka KJ, Lacomme C (2004) Efficient virus-induced gene silencing in roots using a modified tobacco rattle virus vector. Plant Physiol 136: 3999-4009

VandenBosch KA, Stacey G (2003) Summaries of legume genomics projects from around the globe. Community resources for crops and models. Plant Physiol 131: 840-865

van der Hoeven R, Ronning C, Giovannoni J, Martin G, Tanksley SD (2002) Deductions about the number, organization, and evolution of genes in the tomato genome based on analysis of a large expressed sequence tag collection and selective genomic sequencing. Plant Cell 14(7): 1441-1456

van Os H (2005) The construction of an ultra-dense genetic linkage map of potato. PhD Thesis, Wageningen Univ, ISBN 90-8504-221-6

Velasco R, Zharkikh A, Troggio M, Cartwright DA, Cestaro A, et al. (2007) A high quality draft consensus sequence of the genome of a heterozygous grapevine variety. PLoS One 2(12): e1326. DOI:10.1371/journal.pone. 0001326

Vilarinhos AD, Piffanelli P, Lagoda P, Thibivilliers S, Sabau X, Carreel F, D'Hont A (2003) Construction and characterization of a bacterial artificial chromosome library of banana (*Musa acuminata* Colla). Theor Appl Genet 106(6): 1102-1106

Villalta I, Reina-Sanchez A, Cuartero J, Carbonell EA, Asins MJ (2005) Comparative microsatellite linkage analysis and genetic structure of two populations of F_6 lines derived from *Lycopersicon pimpinellifolium* and *L. cheesmanii*. Theor Appl Genet 110: 881-894

Way H, Chapman S, McIntyre L, Casu R, Xue GP, Manners J, Shorter R (2005) Identification of differentially expressed genes in wheat undergoing gradual water deficit stress using a subtractive hybridization approach. Plant Sci 168: 661-670

Wei F, Wing RA (2008) A fruitful outcome to the papaya genome project. Genome Biol 9: 227

Werner JE, Endo TR, Gill BS (1992) Towards a cytogenetically based physical map of the wheat genome. Proc Natl Acad Sci USA 89: 11307-11311

Wheat Grain Improvement Network (WGIN) (2008): *http://www.wgin.org.uk/index.php?area=home&page=projectoutline*

Whitelaw CA, Barbazuk WB, Pertea G, Chan AP, Cheung F, Lee Y, Zheng L, van Heeringen S, Karamycheva S, Bennetzen JL, SanMiguel P, Lakey N, Bedell J, Yuan Y, Budiman MA, esnick A, van Aken S, Utterback T, Riedmuller S, Williams M, Feldblyum T, Schubert K, Beachy R, Fraser CM, Quackenbush J (2003) Enrichment of gene-coding sequences in maize by genome filtration. Science 302: 2118-2120

Wicker T, Schlagenhauf E, Graner A, Close TJ, Keller B, Stein N (2006) 454 sequencing put to the test using the complex genome of barley. BMC Genom 7: 275-285

Wilson ID, Barker GLA, Beswick RW, Shepherd SK, Lu C, Coghill JA, Edwards D, Owen P, Lyons R, Parker JS, Lenton JR, Holdsworth MJ, Shewry PR, Edwards KJ (2004) A transcriptomics resource for wheat functional genomics. Plant Biotechnol J 2: 495-506

Wilson ID, Barker GLA, Lu C, Coghill JA, Beswick RW, Lenton J, Edwards KJ (2005) Alteration of the embryo transcriptome of hexaploid winter wheat (*Triticum aestivum* cv. Mercia) during maturation and germination. Funct Integr Genom 5: 144-154

Xie Y, Zhu B, Yang X, Zhang H, Fu D, Zhu H, Shao Y, Li Y, Gao H, Luo Y (2006) Delay of postharvest ripening and senescence of tomato fruit through virus-induced *LeACS2* gene silencing. Postharvest Biol Technol 42: 8-15

Xiong A-S, Yao Q-H, Peng R-H, Li X, Han P-L and Fan H-Q (2005) Different effects on ACC oxidase gene silencing triggered by RNA interference in transgenic tomato. Plant Cell Rep 23: 639-646

Yan L, Loukoianov A, Blechl A, Tranquilli G, Ramakrishna W, SanMiguel P, Bennetzen JL, Echenique V, Dubcovsky J (2004) The wheat *VRN2* gene is a flowering repressor down-regulated by vernalization. Science 303: 1640-1644

Yao YY, Guo GG, Ni ZF, Sunkar R, Du JK, Zhu JK, Sun QX (2007) Cloning and characterization of microRNAs from wheat (*Triticum aestivum* L.). Genome Biol 8: R96

Young ND, Cannon SB, Sato S, Kim D, Cook DR, Town CD, Roe BA, Tabata S (2005) Sequencing the genespaces of *Medicago truncatula* and *Lotus japonicus*. Plant Physiol 137: 1174-1181

Young ND, Roe BA, Town CD (2007) Completing the sequence of *Medicago truncatula*'s gene-rich euchromatin. *plpa.cfans.umn.edu/~neviny/labsite/publicationsf/mt_poster_2007.pdf*

Yu Y, Wing RA (2005) Whole genome sequencing: Methodology and progress in cereals. In: Gupta PK, Varshney RK (eds) Cereal Genomics. Kluwer Academic Publ, Dordrecht, Netherlands, pp 385-423

Yuan Y, SanMiguel PJ, Bennetzen JL (2002) Methylation-spanning linker libraries link gene-rich regions and identify epigenetic boundaries in *Zea mays*. Genome Res 12: 1345-1349

Yue SJ, Li H, Li YW, Zhu YF, Guo JK, Liu YJ, Chen Y, Jia X (2007) Generation of transgenic wheat lines with altered expression levels of 1Dx5 high-molecular-weight glutenin subunit by RNA interference. J Cereal Sci DOI: 10.1016/j.jcs/2007.03.006

Zhang LP, Khan A, Ni˜no-Liu D, Foolad MR (2002) A molecular linkage map of tomato displaying chromosomal locations of resistance gene analogs based on a *Lycopersicon esculentum* × *L. hirsutum* cross. Genome 45: 133-146

Zhu H-L, Zhu B-Z, Shao Y, Wang X-G, Lin X-J, Xie Y-H, Li Y-C, Gao H-Y, Luo Y-B (2006) Tomato fruit development and ripening are altered by the silencing of *LeEIN2* gene. J Integr Plant Biol 48(12): 1478-1485

11 Concepts and Strategies for Reverse Genetics in Field, Forest and Bioenergy Crop Species

Kazuhiro Kikuchi[1], Claire Chesnais[2], Sharon Regan[2] and Thomas P. Brutnell[1*]

[1]Boyce Thompson Institute for Plant Research, Cornell University, Ithaca NY, USA

[2]Department of Biology, Queen's University, Kingston, Ontario, K7L 3N6, Canada

*Corresponding author: tpb8@cornell.edu

1 INTRODUCTION

The goal of this chapter is to introduce several methodologies that have been developed to characterize gene function in crop plants. Increasingly, technologies are exploiting the ease by which genomic DNA sequence can be obtained. This is providing new opportunities to understand genetic and biochemical networks in species that have traditionally been refractory to genetic analysis. Techniques such as association mapping and quantitative trait loci (QTL) analysis (Yu and Buckler 2006) and map-based cloning (Bortiri et al. 2006) are becoming much more attractive due to this increase in sequence data. In addition, techniques such as targeting induced local lesions in genomes (TILLING; Greene et al. 2003; Till et al. 2007) are now being modified to incorporate high-throughput sequencing to reduce the time, labor and cost associated with screens of mutant collections. Sophisticated insertional vectors have also been developed for crop plants, providing new opportunities to introduce transgenes and to perform genetic analysis. The focus of this chapter is on describing some of the emerging technologies that are aiding gene discovery and characterization in a few selected species that represent major world food, forestry and bioenergy crops while understanding that the same or similar technologies could be broadly applied to additional crops. In particular, transposon tagging, activation tagging, expression traps and TILLING resources that are in development will be discussed. We will also compare some of the advantages and disadvantages of these technologies and how they may be applied to developing models for bioenergy crops.

1.1 Challenges of Crop Plants

A number of technical and regulatory challenges need to be considered when developing reverse genetic tools for crop plants. Some of the technical challenges are fairly obvious. Most crop plants have large polyploid genomes, long generation times, are not readily transformable and are physically large in stature. These constraints have led to the adoption of *Arabidopsis thaliana* as the primary model for plant genetics and many seminal discoveries in plant biology have exploited this model system. However, as our understanding of plant development and biochemistry deepens, so too does our need for molecular tools to assay gene function directly in crop plants. New model systems are also required to facilitate the translation of basic discoveries into agronomically valuable traits.

One area that is often neglected by molecular biologists is the consideration of regulatory procedures that have been established for working with genetically modified organisms (GMOs). In the US and Canada, permits from the US Department of Agriculture (USDA) or Canadian Food Inspection Agency are required for the transport of regulated plants and plant products for consumption or propagation. In addition, materials transfer agreements (MTA) must often be signed between parties before seed transfers are possible and this may lead to delays in obtaining materials especially when university or industry intellectual property (IP) is involved. Consideration must also be given to the growth and propagation of regulated materials. For many crop plants, strict regulations are in place in North America to prevent accidental release of transgenic events or pathogens into germplasm collections (see *http://www.aphis.usda.gov/plant_health/permits/index.shtml*). This may require quarantines of materials or growth of plants in greenhouses or growth chambers. These restrictions are often more cumbersome when materials need to be shipped internationally. Although many of these hurdles can be overcome, permits, quarantines and intellectual property (IP) concerns often add significant time delays and significant financial costs to these projects. Thus, collections that exploit endogenous transposons or chemical mutagens to generate materials that are not regulated as "GMO" are likely to be more attractive to users than those that utilize GMOs or require signed material transfer agreements (MTAs). This is especially true, if materials from these screens are to be used in commercial breeding or germplasm development programs.

1.2 Emerging Technologies

Despite many challenges of working with crop plants, several technologies have recently been developed that promise to greatly accelerate progress in developing reverse genetics tools. Perhaps one of the most exciting advances is in DNA sequencing technology. Currently, complete genome sequence is available for rice (Goff et al. 2002; Yu et al. 2002), poplar (Tuskan et al. 2006) and grapevine (Jaillon et al. 2007) and draft sequences of maize (*http://www.maizesequence.org/index.html*) and sorghum (Paterson 2008) genomes have been completed. The pace of plant genome sequence analysis is likely to quicken in the years ahead as new technologies have significantly lowered the cost/base of sequence. Two of the most powerful techniques have been developed by 454 Sequencing (Margulies et al. 2005) and Solexa (Bentley 2006) now owned by Illumina. The new FLX titanium reagents are generating from 400,000 to 600,000 reads of up to 500 bp per instrument run (*http://www.454.com/*). The read length of a Solexa run is considerably shorter—generally 35 to 75 bp, but a single flow cell lane can generate up to thirty million reads. The lower costs associated with Solexa/Illumina have made it the ideal tool for mining small RNA molecules (Berezikov et al. 2006). Although, the short read lengths associated with Solexa can be problematic for the assembly of large genomes, several bioinformatics tools are helping to streamline this process (e.g., Smith et al. 2008) and additional improvements in the Solexa chemistry are likely to result in increased read lengths.

Another area of growth in the past few years has been in the development of sequence-indexed collections of insertions lines. As shown in Tables 1 and 2, extensive collections of insertion lines have been developed for crop plants and many of these collections can be

Table 1 Inventory of T-DNA activation tagged lines in plants other than *Arabidopsis*.

Species (cultivar)	*Type of activation tag*	*Number of lines*	*Institution*	*Reference and database if available*	*Contact*
Rice (Dongjin and Hwayoung)	4× CaMV 35S enhancers	47,932	Pohang University of Science and Technology (POSTECH)	(Jeong et al. 2002, Jeong et al. 2006) http://postech.ac.kr/life/risd	G. An (Email: genean@postech.ac.kr)
Rice (Tainung 67)	8× CaMV 35S enhancers	45,000	Institute of Molecular Biology, Academia Sinica	(Hsing et al. 2007) http://trim.sinica.edu.tw./	Y.C. Hsing (Email: bohsing@gate.sinica.edu.tw)
Rice (Nipponbare)	4× CaMV 35S enhancers + minimal CaMV promoter	13,000	National Institute of Agrobiological Sciences, Japan	(Mori et al. 2007)	M.Mori (E-mail: morimasa@nias.affrc.go.jp)
Poplar (*Populus tremula* × *Populus alba*)	4× 35S CaMV enhancers	1,800	Queen's University, Kingston Ontario	(Harrison et al. 2007)	S. Regan (Email: sharon.regan@queensu.ca)
Poplar (*Populus tremula* × *Populus alba*)	4× 35S CaMV enhancers	627	Department of Forest Science, Oregon State University	(Busov et al. 2003)	S. Strauss (Email: Steve.Strauss@orst.edu)
Poplar (*Populus euramericana*)	2× 35S CaMV enhancers	300	Research Institute of Subtropical Forestry, China	(Wang et al. 2007)	R. Zhuo (Email: zhuory@gmail.com)

Contd.

Table 1 continued

Species (cultivar)	*Type of activation tag*	*Number of lines*	*Institution*	*Reference and database if available*	*Contact*
Potato (*Bintje*)	4× 35S CaMV enhancers	8,613	Canadian Potato	(Duguay et al. 2007) Genome Project	S. Regan (Email: sharon.regan@queensu.ca)
Legume (*Lotus japonicus*)	3× 35S CaMV enhancers + exon trap (CaMV promoter, start codon, splice donor and acceptor sites)	3,500	Department of Applied Biological Sciences, Nihon University, Japan	(Imaizumi et al. 2005) http://www.legumebase.agr.miyazaki-u.ac.jp/index.jsp	T. Aoki (Email: taoki@brs.nihon-u.acjp)
Tomato (Micro-Tom)	4× 35S CaMV enhancers	10,427	Exelixis Plant Sciences, Oregon	(Mathews et al. 2003)	H. Mathews (Email: hmathews@exelixis.com)
Petunia (*Petunia hybrida*)	4× 35S CaMV enhancers	10,000	Leeds Institute for Plant Biotechnology and Agriculture	(Zubko et al. 2002)	P. Meyer (Email: p.meyer@leeds.ac.uk)

Table 2 Genes identified via T-DNA activation tagging in plants other than *Arabidopsis*.

Species	*Gene*	*Product*	*Phenotype*	*Reference*
Blue gem	*cdt-1*	ABA signaling molecule - putative noncoding RNA	Dessication tolerance	(Furini et al. 1997)
Blue gem	*cdt-2*	ABA signaling molecule - putative noncoding RNA	Dessication tolerance	(Smith-Espinoza et al. 2005)
Madagascar periwinkle	*Orca3*	AP2/ERF transcription factor	Resistance to trytophan decarboxylase	(van der Fits et al. 2001)
Petunia	*sho*	Isopentyl transferase-like	Increased cytokinin production	(Zubko et al. 2002)
Poplar	*GA2ox*	GA2-oxidase	Dwarf	(Busov et al. 2003)
Tomato	*ant1*	myb transcription factor	Anthocyanin over-accumulation	(Mathews et al. 2003)
Rice	*GA2ox*	GA2-oxidase	Dwarf	(Hsing et al. 2007)
Rice	*OsAT1*	Acyltransferase	Lesion mimic	(Mori et al. 2007)
Rice	*Sg1*	Not yet determined	Short grain	(Mori et al. 2006)

Table 3 Inventory of T-DNA promoter, enhancer and gene trap lines in plants other than *Arabidiopsis*.

Species (cultivar)	*Type of tag*	*Number of lines*	*Institution*	*Reference and database if available*	*Contact*
Rice (Zhonghua 11, Zhonghua 15, Nipponbare)	Enhancer Trap	128,560	National Center of Plant Gene Research at Huazhong Agricultural University	(Liang et al. 2006, Wu et al. 2003, Zhang et al. 2006) http://rmd.ncpgr.cn/	Q. Zhang (E-mail: qifazh@mail.hzau.edu.cn)
Rice (Nipponbare)	Enhancer Trap	100,000	Biotechnology Research Institute, The Chinese Academy of Agricultural Sciences	(Yang et al. 2004)	L. Tiegang (E-mail: tiegang@caas.net.cn)
Rice (Dongjin and Hwayoung)	Gene Trap En. Gene Trap* Promoter Trap	41,310 13,450 1,590	Pohang University of Science and Technology (POSTECH)	(Jeon et al. 2000, Jeong et al. 2002) http://postech.ac.kr/life/risd	G. An (E-mail: genean@postech.ac.kr)
Rice (Tainung 67)	Promoter Trap En. Pro. Trap*	10,000 45,000	Institute of Molecular Biology, Academia Sinica	(Hsing et al. 2007) http://trim.sinica.edu.tw.	Y.C. Hsing (E-mail bohsing@gate.sinica.edu.tw)
Rice (Nipponbare)	Enhancer Trap Enhancer Trap	29,482 13,881	CIRAD-INRA-SUPAGRO-UMII, Genoplante (France)	(Johnson et al. 2005, Larmande et al. 2008, Sallaud et al. 2004) http://urgi.versailles.inra.fr/OryzaTagLine/	E. Guiderdoni (E-mail:guiderdoni@cirad.fr)

Contd.

Table 3 continued

Species (cultivar)	*Type of tag*	*Number of lines*	*Institution*	*Reference and database if available*	*Contact*
Rice (Zhonghua 11)	Promoter Trap Gene Trap	415 527	Institute of Botany, the Chinese Academy of Sciences	(Chen et al. 2008)	X.L. Cai (E-mail: xlcai@sippe.ac.cn)
Poplar (*Populus tremula* × *Populus alba*)	Gene Trap Enhancer Trap	708 674	Institute of Forest Genetics, Pacific South-west Research Station	(Groover et al. 2004)	A. Groover (Email: agroover@fs.fed.us)
Poplar (*Populus tremula* × *Populus tremuloides*)	Promoter Trap	273	Department of Cell and Molecular Biology, Goteborg University	(Johansson et al. 2003)	O. Olsson (E-mail: olof.olsson@molbio.gu.se)
Legume	Promoter Trap	821	ARC Centre of Excellence for Integrative Legume Research, The University of Queensland, Australia	(Buzas et al. 2005)	D. Buzas (Email: d.buzas@uq.edu.au)
Legume	Promoter Trap	284	Institute of Grassland and Environmental Research, Wales	(Webb et al. 2000)	L. Skot (E-mail: leif.skot@bbsrc.ac.uk)
Legume	Promoter Trap	187	Institut des Sciences du Vegetal, CNRS, France	(Scholte et al. 2002)	P. Ratet (E-mail: Pascal.Ratet@isv.cnrs-gif.fr)

Contd.

Table 3 continued

Species	*Type of tag*	*Number of lines*	*Institution*	*Reference and database if available*	*Contact*
Legume	BiFunc. Pro. Trap*	141	Department of Plant Sciences, University of Hyderabad, India	(Anuradha et al. 2006)	P.B. Kirti (E-mail: pbksl@uohyd.ernet.in)
Tobacco	Promoter Trap	234	Department of Botany, University of Leicester, UK	(Lindsey et al. 1993)	
Potato	Promoter Trap	48	Department of Botany, University of Leicester, UK	(Lindsey et al. 1993)	

* En. Gene Trap = Enhanced Gene Trap; En. Pro. Trap = Enhanced Promoter Trap; BiFunc. Pro. Trap = Bifunctional Promoter Trap.

queried through project web sites. As genome sequence and annotation advance, it will become increasingly important to develop comprehensive, species-specific databases to allow 'one-stop shopping' for genes of interest. It is imperative that websites be developed for crop species to easily search and browse the rapidly increasing collections of genetic materials that are being developed for both forward and reverse genetic screens. Perhaps one of the most comprehensive sites for rice is the OryGenes DB (*http://orygenesdb.cirad.fr/*) that can be used to search for insertions from multiple collections into genes of interest. However, as mentioned above, import restrictions and MTA agreements often severely hinder access to these collections. Streamlined procedures for international transfers of transgenic materials and reduced regulatory restrictions for growing materials for research purposes only could greatly increase access of these collections by the community. A more streamlined process for transfers of germplasm would also likely build international collaboration as seed stocks could more easily be exchanged between research groups.

2 INSERTIONAL MUTAGENESIS

In recent years, there has been an explosion in tool development for functional genomics in crop plants. Hundreds of thousands of insertion lines have been generated for rice alone and many similar resources are under development for other crop plants. There are three general strategies that are currently in use by the plant research community for generating mutants using insertional mutagens. Gene 'knock-out' or insertional mutagenesis generally refers to the process whereby a T-DNA insertion or transposon is used to disrupt the function of a gene. In most recent applications of this technology, the T-DNA or transposon is used as a molecular tag to recover and index the flanking DNA. The second strategy is activation tagging and is used to drive ectopic or increased transcript accumulation of a gene located near the insertional mutagen that carries a strong gene activation sequence. This system has the advantage of being able to help elucidate gene function when gene duplication events mask the phenotype of single gene knockouts. The third general strategy is gene expression trapping. In this method, reporter constructs are introduced into the genome and can be used to detect enhancers, promoters or gene coding sequences. More advanced systems are also under development as discussed below that combine features of all three strategies.

2.1 Transposon Tagging

Although T-DNA mutagenesis has been used widely in *A. thaliana*, transposons have often been the favored tool for mutagenizing crop plants. This is primarily because transformation techniques are still much more cumbersome for crop plants than for *A. thaliana* and thus associated with higher costs and longer recovery times for obtaining transgenic events (e.g., in planta transformation vs. tissue culture). Collections exploiting both endogenous transposons and transgenically introduced elements have been developed for many crop plants (Conrad et al. 2008). In rice, researchers have exploited several active endogenous transposable elements including the *copia*-like retro-transposon *Tos17* (Hirochika et al. 1996), the LINE-type retro-transposon *Karma* (Komatsu et al. 2003), *hAT* family transposons *dTok* (Moon et al. 2006) and *nDart/aDart* (Tsugane et al. 2006; Nishimura et al. 2008); and the MITE transposon *mPing/Ping* (Jiang et al. 2003; Kikuchi et al. 2003; Nakazaki et al. 2003) to generate mutant collections.

In maize, sequenced-indexed collections are still in their infancy but are actively being developed using *Mutator* (McCarty et al. 2005) and *Ac/Ds* (Cowperthwaite et al. 2002; Kolkman et al. 2005; Ahern et al. 2009) transposon family members. These programs will likely rapidly progress with the advent of high-throughput genome sequencing protocols and the completion of the maize genome sequence. Current progress of both the *Mutator* (*http://currant.hos.ufl.edu/mutail/*) and *Ac/Ds* (*http://www.plantgdb.org/prj/AcDsTagging/*) transposon collections can be found at project web sites.

2.1.1 Tos17 in Rice

One of the most extensive reverse genetic resources in rice has exploited the endogenous retro-transposon *Tos17* as an insertional mutagen (Miyao et al. 2007). *Tos17* is a *copia*-like retro-transposon, and is present at low copy number in the cultivar Nipponbare (Hirochika et al. 1996). *Tos17* is inactive in most lines, but it can be activated through tissue culture and inactivated again following regeneration of plants. Interestingly, the copy number of *Tos17* increases with the time in tissue culture, providing a mechanism to control the proliferation of the element throughout the genome (Hirochika 1997). As *Tos17* moves through an RNA intermediate, insertions are stably inherited in subsequent generations. Like all transposons, *Tos17* displays some insertion site preferences and this can limit the range of target sites recovered in large-scale mutagenesis programs. It prefers a weak target site consensus of ANGTT-TSD-

AACNT and displays some insertion site bias (Miyao et al. 2007). In a study of two independently derived *Tos17* collections, several insertion hot spots were identified that mapped to subtelomeric regions of the chromosomes (Piffanelli et al. 2007). In addition, germinal reversion events do not occur, necessitating the recovery of additional mutant alleles. Another important consideration when using *Tos17* lines or any materials that have been derived from tissue culture is that the tissue culture process is highly mutagenic, inducing a number of genomic changes including deletions, insertions and base exchanges generally referred to as somaclonal variation. Indeed, it has been estimated that fewer than 10% of the phenotypes identified in *Tos17* collections are caused by the *Tos17* insertion (Miyao et al. 2007).

Despite these limitations, *Tos17* offers a powerful tool for reverse genetics in rice. To date, nearly 50,000 *Tos17* insertion lines have been generated at the National Institute for Agrobiological Science (NIAS, Japan). With an average copy number of 10 new *Tos17* insertions/line, the population likely carries approximately 500,000 insertions (Miyao et al. 2007). These populations have been screened extensively using PCR-based methods (e.g., Takano et al. 2001, 2005). To date, over 50,000 lines have been phenotyped and 20,000 *Tos17* flanking sequences have been assigned to the rice genome sequence (Miyao et al. 2003, 2007). Additional *Tos17* insertions are also likely to be found in existing T-DNA collections that were mobilized during the tissue culture process that was used to introduce the T-DNA construct. Yue-Ie (Hsing et al. 2007) has generated 55,000 T-DNA insertion lines in the cultivar Tainung 67 and screened these populations for new *Tos17* insertions. Although *Tos17* was active, the average copy number of *Tos17* in these lines was very low, approximately 3.12 insertions/line (the original Tainung 67 cultivar contained 3 *Tos17* insertions). In a similar study, Piffanelli and colleagues characterized several thousand flanking sequence tags (FSTs) of *Tos17* from the *Oryza* Tag Line (OTL) T-DNA mutant library in the cultivar Nipponbare (Piffanelli et al. 2007). Again, the goal was to mine new *Tos17* insertions that may have been activated during the tissue culture process. Mapping of the FSTs revealed a strikingly similar distribution of insertions in the NAIS collection; namely, that most were to gene-rich regions of the genome (78%) with a distribution that suggested an insertion site bias. Of the approximately 9,000 insertions that mapped to non-transposable element genes, only 2,773 genes were disrupted (3.2 inserts/gene on average). Thus, *Tos17* may be an ideal mutagen for generating multiple alleles of particular genes, but is unlikely to be sufficient for saturation mutagenesis of the rice genome.

2.1.2 mPing in Rice

mPing was the first active miniature inverted-repeat transposable element (MITE) discovered in plants. In most rice accessions, the element does not appear to actively transpose, but can be activated using cell culture (Jiang et al. 2003), anther culture (Kikuchi et al. 2003) or γ-irradiation (Nakazaki et al. 2003) treatments. The non-autonomous element is 430 bp with 15 bp terminal inverted repeats that generates a target site duplication of TAA or TTA. Recently, Yang et al. (2007) showed that a cDNA derived from the autonomous *Ping* element and a full-length copy of the related *Pong* element are both capable of mobilizing *mPing* in *Arabidopsis*. Thus, it appears that both *Ping* and *Pong* may encode the transposase needed for *mPing* transposition in rice.

The copy number of *mPing* varies dramatically between the two subspecies of rice, *Japonica* and *Indica* (Jiang et al. 2003). For instance, the *Japonica* cultivar Nipponbare contains more than 50 *mPing* elements, whereas the *Indica* cultivar 93-11 carries less than 10 elements. Interestingly, the Aikoku landraces, that includes the Gimbozu cultivar of *Japonica* rice, carries a markedly higher copy number of *mPing* (>1,000 copies) than some other temperate varieties (Naito et al. 2006). Importantly, many of the *mPing* insertions identified in these lines are within coding regions of the rice genome and recently, Ohmori and colleagues reported that an autonomous *Ping* element integrated into the *YABBY* gene *DROOPING LEAF* (Ohmori et al. 2008). These data suggest that *Ping*/*mPing* may serve as useful tools for selectively mutagenizing coding regions of the genome.

As mentioned above, one of the advantages of using endogenous elements as insertional mutagens is in the ease of distribution due to fewer regulatory restrictions. To develop a series of *mPing* insertions lines for rice, Brutnell and colleagues have introduced an active copy of the *Ping* transposon from Nipponbare into one of the most widely grown US cultivars of rice, Kaybonnet. The low copy number of *mPing* insertions in this Kaybonnet (less than five copies) suggests that the *mPing* elements are not active in this variety. Thus, by introducing an active *Ping* element from Nipponbare into Kaybonnet, it may be possible to mobilize *mPing* in a US variety. To define *mPing* insertion sites, PCR-based protocols were developed to selectively amplify and sequence *mPing* insertion sites using 454 sequencing technology (Kikuchi and Brutnell, unpublished). In a pilot study approx. 789 *mPing* insertions were mapped to the rice genome from 169 different *Japonica* cultivars. Thus, 454 sequencing technologies offer a powerful tool in developing

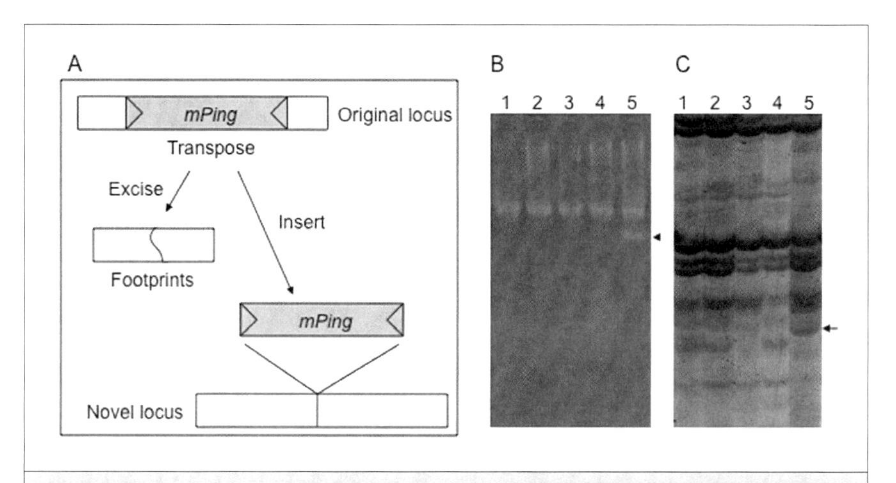

Fig. 1 *mPing* transposition in F3 progeny of a cross between Nipponbare x Kaybonnet. (A) Molecular signatures of *mPing* insertion and excision. (B) PCR products using primers that flank an *mPing* insertion site. Arrowhead indicates a product amplified following *mPing* excision. (C) Transposon display. A novel *mPing* insertion was identified using inverse PCR (arrow).

sequence-indexed collections of transposon insertions.

Currently, the F_3 generation of the Nipponbare x Kaybonnet crosses is being examined for evidence of transposition (Fig. 1). Several putative *mPing* footprint alleles were detected in some lines and novel insertions also were detected using transposon display, a method that was developed to selectively amplify sequences flanking MITEs (Casa et al. 2000). Through continued backcrosses to the Kaybonnet parent, accompanied by selection for *Ping* activity (e.g., *mPing* excision and transposition) materials are being generated for a large-scale *mPing* mutagenesis using a US cultivar of rice. These non-transgenic lines carrying *mPing* insertions will be sequence-indexed and provided to plant scientists and breeders in the US and abroad without the need for MTA agreements or APHIS permits.

2.2 Activation Tagging

Activation tagging refers to a form of mutagenesis whereby a strong promoter is used to drive ectopic gene expression. This technique has been widely used in several plants as a tool for functional genomics. One of the first large-scale programs was developed in *Arabidopsis* using four copies of an enhancer element from the cauliflower mosaic virus 35S

(CaMV 35S) promoter which then interacts with the transcription machinery of neighboring genes to enhance their expression (Weigel et al. 2000). Most gene activation events occur when the activation tag inserts within 5 kb of either side of a gene (Walden 2002). However, there have been cases where the ectopically expressed gene is up to 11 kb away from the enhancers (Hsing et al. 2007) and, occasionally, the enhancers can cause over-expression of more than one gene (Van der Graaf et al. 2002).

In contrast to insertion mutagenesis that generally results in the recovery of loss-of-function phenotypes, activation tagging has the advantage of producing dominant or gain-of-function mutant phenotypes in the first generation. This means that it is possible to identify a phenotype even if the gene has redundant functions with other members of a gene family or if alternative metabolic pathways exist, both of which are especially common in plants. For example, it is estimated that 27-66% of the rice genome exists as duplicated segments (Jung et al. 2008), suggesting that many single gene knock-outs would not lead to observable phenotypes due to redundancy. Redundancy is problematic even in the model species *A. thaliana* where it has been estimated that up to 17% of all genes are arranged in tandem arrays and at least 65% of the genes have at least one paralog (Arabidopsis 2000). Indeed, in a survey of 200 knock-out lines of *Arabidopsis* less than 2% of gene disruptions led to an observable mutant phenotype (Bouche and Bouchez 2001). Thus, the ability to uncover the function of a gene through mis- or over-expression suggests that activation tagging could be a particularly powerful tool in complex polyploids such as potato and wheat. Table 1 lists a number of T-DNA lines that have been developed for crop plants to date.

Despite the potential of activation tagging, relatively few activation tagged lines have been characterized to date (Table 2). In *Arabidopsis*, for instance, over 40 activation tagging mutants have been reported, but these mutants represent a relatively small number of the thousands of characterized T-DNA and transposon-induced mutants (Ayliffe and Pryor 2007). This relatively small number of lines likely reflects the inefficiencies of the first generation of activation tagging constructs. Initial estimates of activation-tagged phenotypes in *Arabidopsis* varied from 0.08% to 1.1% of individuals screened (Wilson et al. 1996; Weigel et al. 2000; Marsch-Martinez et al. 2002). A similar frequency of 1-2% was obtained in screens of activation-tagged populations of poplar (Busov et al. 2005; Harrison et al. 2007). Activation tagging populations have also been generated in rice (Jeong et al. 2002, 2006; Hsing et al. 2007; Mori et al. 2007; Wan et al. 2008) and barley (Ayliffe et al. 2007). In rice the frequency of dominant phenotypes recovered has been estimated to be approximately 1.2% (Wan et al. 2008). However, the frequency of altered phenotypes is continually increasing as screens are conducted for specific traits.

There are several reasons why activation tag inserts might not produce a detectable phenotype. If the insert is near a promoter that is not responsive to enhancers or is adjacent to intervening sequences that insulate the promoter from the enhancers, the construct may fail to drive ectopic expression (Weigel et al. 2000). For example, An and colleagues found over-expression of nearby genes in rice in only 52% of 112 randomly selected activation tagged lines upon analysis of flanking sequences (Jeong et al. 2006). There also appears to be some tissue specificity in the 35S enhancer since they have been reported to have poor activity in roots (Weigel et al. 2000). In other cases, it is likely that transcript levels may not be limiting for the expression of the gene product or are not ectopically expressed in the tissues or developmental times where the gene normally functions. Another concern is that the enhancers may activate a gene in a specific tissue or at a specific point in development that results in a very subtle phenotype. In field screens of rice activation tagging populations, a higher number of mutant phenotypes were observed than in greenhouse screens (Wan et al. 2008), suggesting that abiotic or biotic stresses may reveal phenotypes that are masked in greenhouse screens.

2.2.1 Alternatives to 35S Enhancers

To overcome the possible limitations of the 35S enhancers, other strategies can be used such as alternative enhancers or promoters. Different enhancers may interact differently with promoters, and the use of endogenous enhancers may even reduce silencing induced by the presence of foreign sequences (Chalfun et al. 2003). The 35S promoter has been used as well, in combination with tetramerized 35S enhancers, to eliminate the problems associated with position effect (insulator sequences, promoter selectivity) and detection problems due to tissue-specific expression (Mori et al. 2007). Using this system a rice mutant with spotted leaf blades was identified due to mis-expression of the acyltransferase *OsAT1* and it was suggested that a single 35S enhancer would not have revealed a phenotype for this gene because of its low expression and tissue specificity.

When a tag containing the 35S promoter inserts in a gene instead of upstream of a gene, the flanking sequence is still forcibly expressed. In this situation, the lack of a start codon means there is usually no translational activation and presumably no mutant phenotype. To create translational activation, a start codon and splice donor and acceptor sites can be included downstream of the promoter. This type of tag has been dubbed an 'exon trap' (Jingu et al. 2004). The enhancer and exon trap strategies can also be combined, such that if the tag does not land in a gene, the 35S enhancers can still activate nearby genes in the endogenous pattern (Imaizumi et al. 2005).

It is also possible to use tissue-specific enhancers and promoters to activate genes in the tissue of interest. For example, to identify genes involved in desiccation tolerance, Furini and colleagues used a promoter that is expressed in actively dividing cells to screen callus rather than mature plant tissues (Furini et al. 1997). Although *Craterostigma plantagineum* (resurrection plant) is extremely desiccation tolerant, the tissues derived from callus are not. Thus screens for viable callus lines were performed following transformation with the activation cassette and a dehydration treatment. From this screen a single line was identified that was resistant to desiccation, leading to a cloning of a gene that functions in ABA signal transduction.

The use of inducible promoters can also be helpful if constitutive expression is lethal or affects viability. Two types of inducible promoters include synthetic promoters, which can be turned on and off by the addition or removal of exogenously applied chemicals (for a review, see Wang et al. 2003) and heat-activated promoters. An estrogen-inducible promoter was used to recover *Arabidopsis* gain-of-function mutants in the cytokinin signal transduction pathway (Zuo et al. 2002; Hirochika et al. 2004). To drive gene expression following heat-shock treatment, a region of the *Arabidopsis* HSP18.2 promoter was used to drive flanking genes expression. Following an incubation at 37°C, plants can be screened for phenotypes or evidence of ectopic gene expression (Matsuhara et al. 2000).

2.2.2 *Strategies that Allow Screening for Gain-of-function Mutants in Specific Processes*

Elegant strategies have been developed that allow for the identification of genes involved in a specific process of interest. These strategies take advantage of special starter lines. For example, to identify genes involved in disease resistance, starter lines were used that contained a luciferase reporter gene fused to the *PR-1* promoter (Grant et al. 2003). The PR genes code for pathogenesis-related proteins, which are involved in systematic acquired resistance (SAR) and help provide resistance to a broad range of pathogens. The starter lines were then transformed with an activation-tagging vector. In the absence of a SAR response, the starter construct was not expressed and luciferase did not accumulate. However, if a gene involved in the positive regulation of the *PR-1* gene was ectopically expressed by the 35S enhancers, then luciferase would accumulate, potentially defining a gene involved in pathogen defense. In another example, a sugar-inducible promoter from the *sporamin* gene of sweet potato was fused to luciferase and a GUS reporter. Transgenic plants carrying this reporter were used to identify a transcriptional activator that regulates expression of a subset of sugar-inducible genes (Masaki et al. 2005).

2.3 Expression Traps

As discussed above, mis-expression will only result in mutant phenotypes for a subset of genes. However, all genes can be categorized by their expression patterns, including redundant and essential genes. There are three basic types of transformation vectors used to detect elements that direct expression: the promoter trap, the gene trap, and the enhancer trap (Fig. 2). Similar to knockouts and activation tags, these traps are based on the random integration of a T-DNA or transposon vector, but contain reporter genes, which will reflect the expression of the endogenous gene under the control of that promoter or enhancer. The promoter trap is the simplest type of trap, consisting of a reporter gene only. Gene traps consist of a reporter gene with splice acceptor and donor sites to form gene-reporter fusions. Enhancer traps consist of a reporter gene and a minimal promoter. These expression traps can accelerate analysis of gene function by using a relatively simple phenotypic assay (e.g., GUS expression) to monitor tissue- or cell-specific gene expression at different developmental stages, under a variety of stresses or after the application of hormones and other elicitors.

Since promoter and gene traps do not have promoter sequences, their reporter genes must become part of a transcriptional fusion with an endogenous gene in order to be expressed. The reporter gene of a promoter trap must insert within an exon for this to occur. Gene traps are engineered with splice acceptor sites upstream of the reporter, such that a translational fusion occurs if the reporter lands within an intron. Some gene traps are engineered to create a translational fusion if the

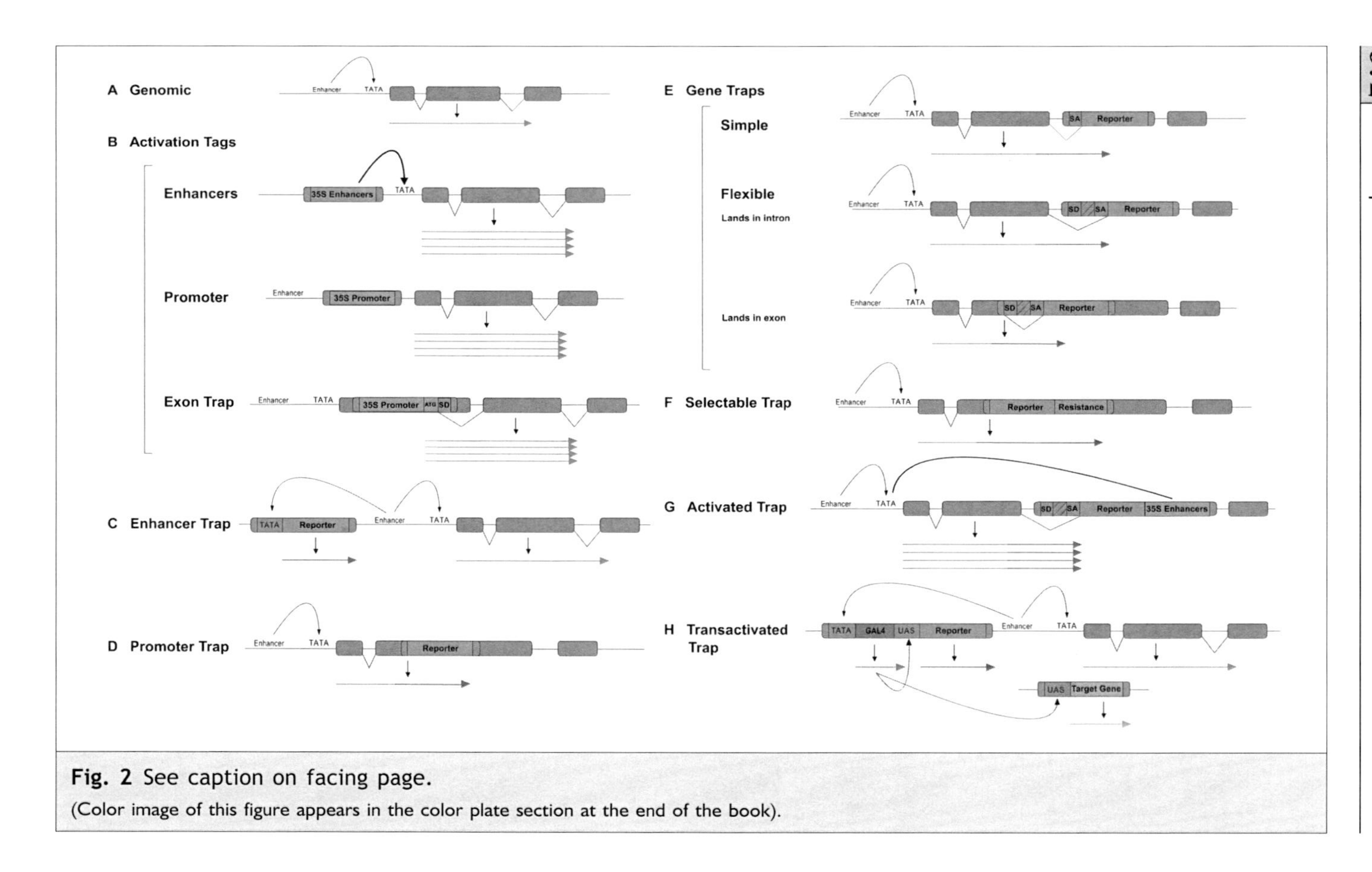

Fig. 2 See caption on facing page.
(Color image of this figure appears in the color plate section at the end of the book).

Fig. 2 Diversity of T-DNA constructs used for insertional mutagenesis in plants (Adapted from Springer et al. 2000). Green boxes are exons; lines are introns; grey boxes delineateT-DNA borders; thatched grey boxes are introns within the T-DNA; "SA" is splice acceptor site; "SD" is splice donor site; "UAS" is upstream activation sequence; arrows represent transcripts that are produced.

(A) Typical gene structure.

(B) Activation tagging vectors. 35S enhancers are most commonly used to promote transcription, but the 35S promoter can also be used. An endogenous gene is overexpressed when the 35S enhancers interact with the endogenous promoter elements, or when the 35S promoter drives transcription. In an exon trap, an ATG and splice donor (SD) within the T-DNA may allow for an in-frame gene fusion.

(C) Enhancer trap. The reporter gene is expressed when an endogenous enhancer element (downstream or upstream) activates the minimal promoter (TATA).

(D) Promoter trap. The reporter gene is expressed when the T-DNA lands within the exon of an endogenous gene. The transcript and translated product are fusions between the endogenous gene and the reporter gene.

(E) Gene Traps. In a simple gene trap, the reporter gene is expressed when the T-DNA inserts within the intron of an endogenous gene, providing a splice acceptor site (SA) that functions with an endogenous SD site. In a flexible gene trap, the reporter gene is expressed when the T-DNA lands in either an exon or an intron, because the construct contains both SA and SD sites.

(F) Selectable Trap. In this version of a promoter trap, the resistance gene is included downstream of the reporter gene and creates a fusion product, such that antibiotic selection can be applied to identify successful promoter trap insertions.

(G) Activated Trap. In this version of a gene trap, the reporter gene is expressed to a greater extent due to the activity of the 35S enhancers.

(H) Transactivated Trap. In this version of an enhancer trap, the GAL4/UAS system is used to ectopically express target genes in the same pattern as any given trap line. GAL4 is overexpessed when an endogenous enhancer element (downstream or upstream) activates its minimal promoter (TATA). GAL4 binds to upstream activator sequences (UAS), which activate the reporter gene in the same transcriptional pattern as GAL4. A target gene of interest can be introduced under the control of a UAS element, which will also be expressed in the same pattern as GAL4 and the reporter gene.

reporter lands in either an intron or an exon, and can increase the frequency of reporter expression by 10- to 100-fold (Springer 2000). These gene traps contain splice donor sites in all three frames and an intron upstream of the splice acceptor sites, such that if the reporter lands in an exon, a chimeric gene-reporter fusion is generated (Groover et al. 2004).

Since an enhancer trap has a minimal promoter, its reporter gene does not have to create a transcriptional fusion with an endogenous gene to be expressed. This means the reporter can land anywhere, in any orientation, so long as it is close enough to an enhancer to drive expression. This has both limitations and benefits. Enhancer traps lead to higher frequencies of reporter gene expression than promoter and gene traps; screens of enhancer lines revealed 25-29% GUS positive lines (Wu et al. 2003) compared to 1.6-2% GUS expression in gene trap lines (Jeon et al. 2000). One complicating factor in interpreting the results of enhancer trapping is the long distances over which enhancers can act, making it difficult to define the sequence that is acting as the enhancer responsible for the expression pattern. This problem is exacerbated in organisms with large genomes and low gene density (Busov et al. 2005).

One modification to the promoter trap design allows direct selection of successful promoter traps in the tissue culture process. These "bifunctional" or selectable traps consist of a fusion between the reporter gene and a resistance gene to an antibiotic used in selection during tissue culture. Only those lines in which the T-DNA has landed downstream of an endogenous promoter will have a resistance gene product and survive selection. The reporter gene can be used later to screen for expression patterns as usual. This system was used to produce 20 promoter tagged lines of oilseed rape (*Brassica napus*) using a GUS reporter and the *npt*II (kanamycin resistance) gene (Bade et al. 2003).

An improved gene trap design (activation trap) was accidentally discovered when dual-purpose vectors were designed for gene trapping and activation tagging (with a gene trap at one end of the T-DNA and the 35S enhancers at the other end of the T-DNA) (Jeong et al. 2002). In screens that have used these constructs, GUS expression was found to be two to three times higher than in similar lines lacking the enhancers (Jeong et al. 2002; Hsing et al. 2007).

Promoters that have been identified from gene traps and that drive expression in specific patterns (e.g., temporally or spatially), can be used to drive selectable marker genes or to ectopically express candidate genes. A special enhancer trap system (transactivated trap), first deve-

Table 4 Genes identified via enhancer, promoter or gene traps in plants other than *Arabidopsis*

Species	*Expression screen*	*Gene*	*Product*	*Proposed Role*	*Reference*
Rice	anther	Udt1	transcription factor	early anther development	(Jung et al. 2005)
Rice	anther	RIP1	nuclear protein	pollen maturation and germination	(Han et al. 2006)
Rice	anther	OsCP1	cysteine protease	pollen development	(Lee et al. 2004a)
Rice	anther	Wda1		wax production in anther tapetum and epidermis	(Jung et al. 2006)
Rice	cold stress	OsLRK1	protein kinase	cold stress tolerance	(Lee et al. 2004b)
Rice	cold stress	OsKMKT1	methyltranserase	cold stress tolerance	(Lee et al. 2004b)
Rice	cold stress	OsPTR1	putative peptide transporter	stress signaling pathway	(Kim et al. 2004)
Rice	salt stress	OsGSK1	glycogen synthase	stress signal transduction and floral development	(Koh et al. 2007)
Rice	salt stress	OsP5CS2		salt and cold stress tolerance	(Hur et al. 2004)
Rice	Leaf, in chlorophyll-deficient knockout mutants	OsCHLH	Mg-chelatase	chlorophyll production	(Jung et al. 2003)
Poplar	root	ET304	putative transcription factor	adventitious root development	(Filichkin et al. 2006)
Poplar	cambial region	PttRPS18	ribosomal protein	cambial growth	(Johansson et al. 2003)
Lotus japonicus	root	LjCbp1	calcium-binding protein	root nodule formation	(Webb et al. 2000)

Table 5

	Advantages	*Disadvantages*
T-DNA	• Wide genome coverage targets larger range of genes. • Stable upon integration • Insertion pattern better for activation tags	• Efficient transformation not yet feasible in all species • Transformation procedures which require tissue culture can produce unlinked phenotypes due to somaclonal variation • Complex insertions (multiples, tandem repeats) can complicate analysis
Transposon	• Transformation not required or required for only a few starter lines • Mutant reversion accomplished easily by excision	• Insertion pattern better for knockouts and gene traps • Insertion preference for particular regions makes genome saturation difficult • Unstable; achieving stability requires extra measures

loped in *Drosophila,* facilitates this process and has been adapted for use in *Arabidopsis* and rice (Wu et al. 2003; Johnson et al. 2005). The concept is depicted in Fig. 2. GAL4, a yeast transcriptional activator, is fused to the minimal promoter and randomly inserted into the genome. The expression pattern of GAL4 is thus determined by endogenous enhancer elements. The GAL4 protein binds to upstream activator sequences 'UAS', and drives downstream reporter gene expression. Thus, the reporter gene is expressed in the same temporal and spatial specificity as GAL4. Once a mutant line has been isolated with a reporter gene expression pattern of interest (called a 'pattern line'), the line can be crossed to a line containing a target gene fused to the UAS sequences. Through the action of GAL4, the target gene will then be expressed in the same pattern as GAL4 and of the reporter gene. This system was used for functional analysis of the *WUSCHEL* gene, a key player in the formation of shoot apical meristems (Gallois et al. 2004). When *WUSHCEL* was ectopically expressed in *Arabidopsis* roots, leaves formed on the root without the addition of external stimulus, which led the researchers to conclude that *WUSCHEL* establishes stem cells with intrinsic shoot identity. This method bypasses the need to create a new transgenic line for each experiment involving a different promoter and/ or experimental gene.

Genes expressed in distinct patterns can also be used as 'markers'.

Their expression during normal development can be characterized and used to relate expression patterns of mutant phenotypes. Marker genes are especially useful in studying cell lineages during development, epidermal patterning, leaf and flower development, and embryogenesis (Springer 2000). An inventory of promoter, gene and enhancer traps developed for crop plants is shown in Table 3 and a list of genes identified using these systems is shown in Table 4.

2.4 Choice of Reporter Gene

There are three main reporter genes used currently: β-glucoronidase (GUS), green fluorescent protein (GFP) and luciferase (LUC) (reviewed in de Ruijter et al. 2003). GUS expression is revealed using a simple histochemical staining assay and provides resolution at the cellular level (Koo et al. 2007). GUS also offers high sensitivity; its low rate of protein turnover results in cumulative activity, which allows detection of genes with weak expression (An et al. 2005). However, the staining and de-staining procedures destroy tissue. Because of this, stained tissues are unavailable for downstream analysis and GUS reporter systems can't be used to image plants in real time. In addition, the substrate required for GUS staining is expensive and is not equally available to all cell types (An et al. 2005). GFP and LUC, on the other hand, are non-invasive and thus useful for live, real-time imaging. GFP offers some advantages over LUC in that it does not requires a substrate and like GUS provides high spatial resolution (Koo et al. 2007). However, some plant structures, especially secondary plant cell walls, autofluoresce at a wavelength similar to the optimal for GFP, making it difficult to detect (Koo et al. 2007), and illumination may be impractical for large plants (Springer 2000). LUC offers the best temporal and conditional resolution due to its rapid protein turnover (de Ruijter et al. 2003). To reap the benefits of the different types of reporters, dual constructs such as GUS-GFP, LUC-GFP and GUS-LUC are now being created. For example, researchers using a GUS-LUC fusion found it was efficient to use LUC in the initial screen to identify successful traps and to use GUS for more specialized screening in the next generation (Koo et al. 2007).

2.5 Considerations for Insertional Mutagenesis

As discussed above, there are two general types of biological tools, which can be used for mutagenesis: T-DNA and transposons. *Agrobacterium tumefaciens* is a bacteria, which has evolved a mechanism to transfer a segment of DNA (called Transfer DNA, hence T-DNA) from its Ti plasmid to the genomes of plant hosts. Vectors used for mutagenesis exploit this mechanism of insertion by including elements from the T-DNA, the right and left border inverted repeats. Of the major crop species, the most extensive collections of T-DNA insertion lines have been developed for rice (Jeon et al. 2000; Eamens et al. 2004; Sallaud et al. 2004; Zhang et al. 2006; Hsing et al. 2007) and many of these have been sequence-indexed (Tanaka et al. 2008). As discussed above, transposons such as *Tos17* have been widely exploited in rice, as have non-endogenous transposons such as *Ac/Ds*. However, there are several factors that should be taken into consideration when choosing a vector for insertional mutagenesis. These include the mechanism of insertion, insertion site complexity, insertion stability, and genomic distribution. Table 5 summarizes the advantages and disadvantages of both types of vectors.

2.5.1 Insertion Mechanisms

There are difficulties associated with both T-DNA and transposon insertion mechanisms. First, in certain plants such as barley, wheat and many other grasses, transformation procedures are inefficient and labor intensive (Qu et al. 2008) and the generation of large populations of independent transformants is costly. This may change as transformation efficiencies improve; for example, over-expression of histone genes shows some promise to increase efficiency (Walden 2002). Since each plant line in a population of T-DNA mutants must be generated by an independent transformation, low transformation efficiency can be a limiting factor. Transposon-based lines can be generated by transposition, thus avoiding the problem. However, transposons are sometimes introduced on a T-DNA backbone, especially when heterologous systems are used.

A second problem is the presence of unlinked mutations, arising either from the tissue culture process (somaclonal variation) or from the transformation or transposition event itself. Genetic mutations that have been observed due to tissue culture include translocations, insertions and deletions, ploidy changes, point mutations, and activation of dormant transposons. Epigenetic changes in methylation patterns have also been observed following tissue culture procedures. It has been hypothesized that these changes occur because meiosis and fertilization are bypassed in the tissue culture process; since it is during these developmental processes that epigenetic patterns are thought to be 'reset', the normal epigenetic patterns may not be established in regenerated plants (Kaeppler et al. 2000). Somaclonal variation may thus arise for the same reasons that mammals cloned from somatic cells have developmental problems. The fact that mutant phenotypes caused by tissue culture are usually developmental (for example, changes in leaf or flower morphology, time of flowering, and sterility) supports this notion (Bhat and Srinivasan 2002). When creating T-DNA or transposon lines via transformation, a tissue culture step to induce callus formation is often required. Fortunately, T-DNA insertion populations of *Arabidopsis* have been created without tissue culture, using seed transformation, vacuum infiltration, or floral dip methods, and are associated with a much lower proportion of untagged mutants (Pereira 2000). Somaclonal variation can also be a problem during the activation of transposition in transposon-based systems. For example, tissue culture is the process by which *Tos17* elements are mobilized in rice (Piffanelli et al. 2007).

Both T-DNA and transposons can generate mutations that are unlinked to the insertional vector. Transposons can excise and leave a mutant 'footprint' (Topping and Lindsey 1995) or induce more complex rearrangements of genome sequences (Zhang and Peterson 1999, 2005). In maize, it has been estimated that only 10% of the mutants recovered in *Ac*-mutagenesis programs were linked to the autonomous element (Kolkman et al. 2005). This may reflect the ability of transposons to increase the mutagenic capacity of the genome or the mobilization of non-autonomous elements. Regardless of the mechanisms, this phenomenon is likely to apply broadly across many transposable element families. Aborted T-DNA insertions can also result in additions or deletions at the site of an aborted insertion or induce chromosomal rearrangements (Walden 2002). Chromosomal rearrangements occur when T-DNAs interact as a result of multiple insertions, but can also accompany single insertions (Tax and Vernon 2001). Besides creating unlinked mutations, chromosomal rearrangements such as these can complicate subsequent genetic analysis (Tax and Vernon 2001).

Unlinked mutations should theoretically be less of a problem in populations where phenotypes are screened in the first generation and must therefore be dominant, as is the case with activation tagging and gene traps. However, dominant phenotypes have been observed that were unlinked to the T-DNA tag in rice (Jeong et al. 2002) or as unstable mutants in *Arabidopsis* (Weigel et al. 2000). Thus, secondary screens are often necessary to confirm dominant mutant phenotypes through co-segregation analysis or where possible by recapitulation of the mutant phenotype.

2.5.2 Insertion Site Complexity

Multiple or complex insertions can complicate the analysis of mutant lines. Both T-DNAs and transposons can insert at multiple loci within the genome; for example, some high copy transposon families such as *Mutator* in maize are often associated with up to 10 new insertions per generation (May et al. 2003). In gene trap lines, multiple insertions can confound interpretations of reporter gene expression (Springer 2000). Somatic insertions of active transposons can also lead to false positives in reverse genetic screens (May et al. 2003). T-DNAs can also insert multiple times at any given loci, forming either direct or inverted repeats. The insertions sometimes carry vector backbone sequence, due to the fact that left border integration is not that precise or the T-DNA sequence can be incomplete, possibly from nuclease digestion prior to insertion (Walden 2002). Inverted repeats or complex insertions at a single loci are problematic in that reporter gene expression can occur from promoters introduced along with the T-DNA instead of endogenous promoters (Springer 2000). Complex insertion sites including vector backbone sequence can also make analysis of flanking sequences difficult, especially for large scale strategies (Ramachandran and Sundaresan 2001).

2.5.3 Stability and Silencing

One of the advantages of working with T-DNA is that the insertions are physically and chemically stable, while transposons can often excise out of an integration site. However, numerous two-component systems have been developed for transposon systems, where a non-autonomous element is under the control of the autonomous element (e.g., Bancroft et al. 1992). In these cases, segregation of the autonomous element away from the non-autonomous element stabilizes the insertion. The instability of transposons can also be advantageous. Remobilization of the transposon can be used to generate alleles encoding proteins with altered activities (Giroux et al. 1996), to create genetic chimeras (Jenik and Irish 2001) or produce an allelic series following imprecise excision events (Bai et al. 2007). In transposon activation tagged lines, it may also be possible to quickly create knockouts in genes that are over-expressed in activation tagging populations by exploiting the preferential transposition of transposons to closely linked sites (Weigel et al. 2000).

Neither T-DNAs nor transposons are immune to the unpredictable effects of silencing. Silencing has been shown to be associated with multiple insertions, DNA rearrangements, and over-expression (Bhat and Srinivasan 2002). The underlying cause of silencing may be enhancer methylation. One group investigated this possibility by analyzing a subset of Weigel's activation-tagged mutants (Chalfun et al. 2003). They found that the 35S enhancers were often methylated, especially when the T-DNA had inserted as an inverted repeat. Posited mechanisms that would attract the plants methylation machinery include DNA-DNA pairing of the repeated enhancers as well as the perception of the T-DNA insertion components as foreign.

2.5.4 Genomic Distribution

Both T-DNA and transposons show insertion site preferences within the genome. Comparative analyses of T-DNA and transposon insertion sites have found that transposons have higher insertion site densities in exons than T-DNAs, which favor regions upstream of the start codon, 3'UTRs, and intergenic regions (Pan et al. 2005; Hsing et al. 2007). The site preferences are thought to be related to GC content, where transposons insert into high GC content regions and T-DNAs insert into low GC content regions (Pan et al. 2005). The general conclusion is that since transposons often insert in exons, they are well suited for gene disruptions and promoter or gene traps, whereas T-DNAs are better suited for activation tagging.

3 TARGETING INDUCED LOCAL LESIONS IN GENOMES (TILLING)

TILLING is a high-throughput reverse genetic tool that detects single-nucleotide polymorphisms (SNPs) in a gene of interest using chemically induced mutations. McCallum and colleagues (McCallum et al. 2000) were the first to describe the TILLING method, which combines the efficiency of ethyl methanesulfonate (EMS) chemical mutagenesis with denaturing high-performance liquid chromatography (DHPLC) to detect base pair changes by heteroduplex analysis. Colbert and colleagues later established a high-throughput screen by using an endonuclease *Cel*I that recognizes and cleaves mismatched DNA on heteroduplex formation that is detectable following fractionation on acrylamide gels (Colbert et al. 2001). TILLING is a simple and rapid technique for reverse genetics and has been utilized to identify single gene mutations in a number of crop species including maize (Till et al. 2004), rice (Wu et al. 2005; Till et al. 2007; Suzuki et al. 2008) soybean (Cooper et al. 2008), barley (Caldwell et al. 2004; Talame et al. 2008) and wheat (Slade et al. 2005).

The most common mutagens for TILLING programs are the alkylating agents ethyl methanesulfonate (EMS) and N-methyl-N-nitrosourea (MNU). Both cause predominantly G to A transition mutations and can result in a range of allelic variation from severe mutant alleles if a splice acceptor or donor site is affected to relatively weak alleles generated through single amino acid substitutions. In *Arabidopsis*, it was estimated that at least 700 G:A transition mutations are created in each M_1 line following saturation mutagenesis with EMS (Jander et al. 2003). In polyploid genomes such as wheat, much higher mutational loads are tolerated due to genetic redundancy. For comparison, it has been estimated that frequency of EMS induced mutations generated in TILLING projects for *A. thaliana* was approximately one mutation per 170 kb (Greene et al. 2003) compared to one mutation per 40 kb in tetraploid wheat and one mutation in 24 kb in hexaploid wheat (Slade et al. 2005). These findings suggest that genome size is not a limitation in developing TILLING screens.

3.1 Methodologies

A critical step in establishing a TILLING population is efficient mutagenesis. In rice and many plants with perfect flowers, seed treatments are preferred due to the difficulty of harvesting pollen. However, in maize, abundant pollen and the ease of outcrossing make pollen treatment the preferred method (Till et al. 2004). The M_1 plants are self-fertilized, and the thousands of individuals in the M_2 generation are used to prepare DNA samples while their M_3 seeds are stored. DNA samples from 4 to 8 M_2 individuals are often pooled to increase throughput and reduce costs.

For the TILLING assay, dye-labeled PCR primers are designed to specifically amplify a single gene target fragment of up to ~1.5 kb. The codons optimized to detect deleterious lesion (CODDLE; *http://www.proweb.org/input*) program allows requestors to design PCR primers that are optimized for the TILLING process. After PCR amplification, these fragments are denatured and annealed to form heteroduplexes between mutant and wild type DNA strands. Heteroduplex formation is identified through recognition and cleavage of mismatched DNA by the *Cel*I endonuclease (Oleykowski et al. 1998). Cleavage products can be visualized by fractionation on an acyrlamide gel using primers that are fluorescently-labeled. A secondary screen is then conducted from the positive pool (4 to 8 M_2 individuals), and M_3 seed is picked from stock for phenotypic analysis.

The original TILLING detection system by using fluorescent PCR primers is able to detect one mutant line in a mixed DNA pool of eight lines (Till et al. 2003). More recently, Suzuki and colleagues developed a system that exploits ethidium bromide intercalation to detect SNPs without the use of labeled primers and a capillary gel electrophoresis system that can separate DNA from 150 to 1,500 bp within 10 min (Suzuki et al. 2008). As mutant populations are relatively easy to generate and high-throughput detection methods are becoming more cost-effective, it is likely that TILLING will gain increasing popularity as a tool for functional genomics in crop plants. In practical terms, this makes it possible to generate a working reverse genetics population in approximately one and a half years from starting the mutagenesis to beginning the mutant screen. Importantly these collections are generated without transformation, regeneration, or selection of transformants. As the EMS mutants are non-transgenic, subsequent generations can be grown under field conditions, without restrictions.

3.2 Platforms for Crop Species

3.2.1 Maize TILLING

For maize, with relatively few insertion lines available, TILLING offers one of the best tools for generating an allelic series in genes of interest (Till et al. 2004). In a pilot experiment, pollen from the maize inbred B73 was EMS mutagenized and applied to the silks of B73 ears. Pools of DNA were screened for mutations in 1-kb segments from 11 different genes, and 17 independent mutations were obtained from a population of 750 pollen-mutagenized maize plants. In recent years, mutagenized populations have been expanded to include both B73 and W22 inbreds to capture additional genetic diversity. In addition, the 26 founder lines for the maize diversity project have been included for Ecotilling as described below. The Maize TILLING public service is open with a cost of $2,500 per target as of January 1, 2008 (*http://genome.purdue.edu/maizetilling/*).

3.2.2 Rice TILLING

Rice is one of the major plant models with a sequenced genome (Goff et al. 2002; Yu et al. 2002) and two groups have developed mutagenized populations and TILLING DNA libraries. Till et al. (2003) has developed two mutagenized rice populations of *Oryza sativa* ssp. *Japonica* cv. Nipponbare using different chemical mutagens (Till et al. 2007). As a result, of the 10 target genes screened, 27 nucleotide changes in the EMS-treated population and 30 nucleotide changes in the Azacytidine (AZ)-MNU population were identified from each of the 768 unique individual samples in pools of eight individuals. This mutation rate is equivalent to one nucleotide change in every 530 kb by EMS treatment and 497 kb by AZ-MNU treatment. This Rice TILLING service for the community will be accessible at the URL: *http://tilling.ucdavis.edu*. Suzuki et al. (2008) developed another population of *Oryza sativa* ssp. *Japonica* cv. Taichung 65, by treating single zygotic cells with MNU (Suzuki et al. 2008). The mutation rate was one nucleotide change every 135 kb. This TILLING population will be gradually increased and available to users in the rice research community. The International Rice Research Institute (IRRI) also produced a large mutagenized population of *Indica* rice (IR64) as a public resource for gene discovery (Wu et al. 2005). Together, these populations will provide a powerful resource for functional genomics in rice.

3.2.3 Soybean TILLING

The soybean (*Glycine max* L. Merr.) genome is large (1.1-1.15 Gb) and is a partially diploidized tetraploid (Shultz et al. 2006). In addition, transformation of soybean remains difficult and few insertion lines have been generated in the public sector. However, chemical mutagenesis has been successfully used for many phenotypic screens in soybean (Zhu et al. 1995). Thus, TILLING promises to be a very effective reverse genetic tool in this species.

Cooper et al. (2008) developed four soybean TILLING populations using both EMS and MNU mutagens (Cooper et al. 2008). They screened seven target genes in each population and discovered a total of 116 induced mutations. The mutation rates of MNU and one EMS population were as high as approximately one nucleotide change every 140 kb. The mutation frequencies in these soybean populations were higher than in the other TILLING populations of barley and maize and may reflect the increased genetic redundancy associated with the paleoploid genome of soybean. Although, individual mutations may not result in phenotypic changes in initial screens of TILLING populations, crosses between plants carrying mutations in duplicated genes could produce mutant phenotypes. Thus, TILLING can be applied to soybeans and other polyploids even with extensively duplicated genomes. A soybean TILLING service is accessible at the URL: *http://www.soybeantilling.org/index.jsp*.

3.2.4 Barley TILLING

Barley is one of the few truly diploid member of the Triticeae tribe, though it has a large genome (5,300 Mbp), 80% of which is composed of repetitive DNA (Sreenivasulu et al. 2008). The Scottish Crop Research Institute generated a large TILLING population in the barley cultivar Optic with EMS induced mutations (Caldwell et al. 2004). This library is composed of 9,216 plants and is assembled into 1,152 eight-plant pools. The mutation rate of this population is approximately one nucleotide change in every one million base pairs or ~5,000 mutations per genome. This significantly lower frequency relative to *Arabidopsis* may be due to the poor transmission of many EMS mutations owing to the diploid genome. However, Talame et al. (2008) developed a sodium azide mutagenized population of the cultivar Morex with a mutation rate of one nucleotide change in every 374 kb (Talame et al. 2008), suggesting that the barley genome may be refractory or extremely sensitive to mutagensis by alkylating agents. The barley TILLING and phenotype database in the Scottish Crop Research Institute is accessible at the URL: *http://www.scri.ac.uk/research/genetics/BarleyTILLING*.

3.2.5 Wheat TILLING

Wheat (*Triticum aestivum* L.) is an allohexaploid (AABBDD) with a very large genome (16,000 Mb) containing ~80% repetitive sequence. It is one of the most challenging crop plants for functional genomics (Francki and Appels 2002; Gupta et al. 2008). Slade et al. (2005) developed a TILLING population of both hexaploid wheat, cultivar Express and tetraploid wheat, cultivar Kronos by treatment with EMS to induce mutations (Slade et al. 2005). They identified 196 alleles of the waxy genes (2,114 kb of *Wx-A1* and 1,345 kb of *Wx-B1*) in screens of 1,152 M_2 hexaploid plants, and 50 alleles of the waxy genes (1,232 kb of *Wx-A1* and 487 kb of *Wx-B1*) in 768 M_2 tetraploid wheat plants. The majority of these mutations (244/246) were transitions (G to A or C to T), as expected from G-residue alkylation by EMS. The mutation frequency is approximately one mutation per 24 kb in hexaploid wheat and one in 40 kb in tetraploid wheat.

3.3 Ecotilling

Ecotilling refers to the use of heteroduplex analysis to study genetic variation in natural populations, usually in the form of single-nucleotide polymorphisms (SNPs) but also including insertions and deletions (indels). The proof of concept study was carried out in *Arabidopsis*, where 55 haplotypes were identified in screens of approximately 1 kb segments from five genes in an assay of 192 individuals (Comai et al. 2004). More recently, ecotilling was successfully applied to survey natural diversity in heterozygous poplar trees. In this study, nine loci were examined in 41 black cottonwood trees from 41 different geographical populations, and 63 SNPs were discovered (Gilchrist et al. 2006). Ecotilling with heterozygous individuals is a special case, because it requires two duplexing reactions. The first is conducted without the 'wild type' or reference individual, where dupluxes that occur indicate heterozygous sites within an individual. The second run includes the reference individual to uncover haplotype variation. Ecotilling has since been used to mine for natural variants in resistance alleles in barley, melon and mungbean. In barley, variants in the *mlo* and *mla* genes were identified, which confer resistance to the fungal pathogen of barley powdery mildew (Mejlhede et al. 2006). In melon, accessions were mined for variants in the *eIF4E* gene, which has been linked to the resistance response to melon necrotic spot virus (Nieto et al. 2007). Ecotilling has also been used to survey genetic diversity in mungbean, an important crop and food staple in developing countries (Barkley et al. 2008).

In summary, TILLING represents a promising approach for reverse genetics in a number of crop species. TILLING is generally applicable to plants with both small and large genomes, diploid or polyploids, and thus has great potential as a reverse genetic tool in any species. Furthermore, recent developments in DNA sequence technology (e.g., 454, Solexa and Solid) are now being exploited to detect mutations in TILLING or ecotilling populations using strategies that are likely to increase throughput and decrease costs—still one of the greatest limitations to this technology.

4 PROSPECTS FOR TOOL DEVELOPMENT IN EMERGING BIOENERGY CROP MODELS

Recent research into lignocellulosic ethanol production has led to renewed interest in several grasses as biofuel feedstocks and to a need for new model systems. Unfortunately, two of the most promising grass feedstocks, *Panicum virgatum* (switchgrass) and *Miscanthus* x *gigantea* (Miscanthus) are not readily tractable for genetic analysis. Most widely-grown switchgrass varieties are polyploid and self-incompatible. The Miscanthus variety that is currently grown for biofuel production is also problematic as it is an infertile triploid with no genetic variation. Transformation of both species is difficult and the research community working on these grasses is small. Nevertheless, rapid improvements in sequencing technologies are greatly facilitating marker development for switchgrass (*http://www.jgi.doe.gov/sequencing/why/50008.html*), which can be applied to breeding programs. Programs are also underway to develop new varieties of Miscanthus that have increased genetic diversity. However, it is still not clear if TILLING or insertional mutagenesis will be attractive tools for mutagenesis in these species in the near term as the downstream analysis of any mutant phenotypes will be exceedingly difficult in the absence of the ability to self-pollinate. One possible exception would be in the development of activation tagging lines. Here, the greatest limitation is in the transformation process, but once constructs have been introduced, screens can be performed in the first generation. In addition, generation of dominant negative mutants may also prove fruitful for investigating numerous developmental and biochemical processes based on known pathways in other plant systems.

In contrast to these biofuel feedstocks, the monocot grass *Brachypodium distachyon* has several characteristics that make it an attractive model system (Opanowicz et al. 2008). Unlike *Arabidopsis*, *Brachypodium* is likely to share cell wall characteristics, disease susceptibilities and several developmental strategies with the bioenergy grasses that have evolved since the divergence of monocot and dicot lineages nearly 200 million years ago. It is a member of the Pooideae and is sister to several agronomically important temperate grasses including wheat, barley, rye and oats. It has a small (~300 Mbp) and sequenced (8x coverage) genome, diploid accessions, a small size, self-fertility, a short lifecycle, and simple growth requirements (see *http://www.brachypodium.org/*).

With transformation efficiencies approaching 75% (John Vogel, pers comm) and a generation time of approximately eight weeks, several possibilities are open for developing functional genomics tools in this species. It is easy to envision insertional mutagenesis programs that exploit either T-DNA or transposon vectors as described above. Although no endogenous transposons have been described for *Brachypodium*, it should be possible to exploit transposons from other systems as insertional mutagens. Perhaps one of the most readily available technologies for development in *Brachypodium* is TILLING. Protocols have been developed for EMS mutagenesis and the small size and short generation time of *Brachypodium* make it ideal for conducting large population screens.

As the world population continues to grow and petroleum reserves begin to decline, plant science will play an increasingly important role in both food and energy security. Limitations in land, water and fertilizer usage will demand a second green revolution that will rely on a much deeper understanding of plant biochemicals, physiological and genetic regulatory networks than we currently possess. Thus, we can look forward to a challenging and exciting era of plant biology, where functional genomics will help guide the development of the next generation of field, forest and bioenergy crops.

References

Ahern KR, Deewatthanawong P, Schares J, Muszynski M, Weeks R, Vollbrecht E, Duvick J, Brendel VP, Brutnell TP (2009) Regional mutagenesis using *Dissociation* in maize. Methods 49: 248-254

An G, Jeong DH, Jung KH, Lee S (2005) Reverse genetic approaches for functional genomics of rice. Plant Mol Biol 59: 111-123

Anuradha TS, Jami SK, Datla RS, Kirti PB (2006) Genetic transformation of peanut (*Arachis hypogaea* L.) using cotyledonary node as explant and a promoterless gus :: nptII fusion gene based vector. J Biosci 31: 235-246

Arabidopsis GI (2000) Analysis of the genome sequence of the flowering plant *Arabidopsis thaliana*. Nature 408: 796-815

Ayliffe MA, Pryor AJ (2007) Activation tagging in plants-generation of novel, gain-of-function mutations. Aust J Agri Res 58: 490-497

Ayliffe MA, Pallotta M, Langridge P, Pryor AJ (2007) A barley activation tagging system. Plant Mol Biol 64: 329-347

Bade J, van Grinsven E, Custers J, Hoekstra S, Ponstein A (2003) T-DNA tagging in Brassica napus as an efficient tool for the isolation of new promoters for selectable marker genes. Plant Mol Biol 52: 53-68

Bai L, Singh M, Pitt L, Sweeney M, Brutnell TP (2007) Generating novel allelic variation through Activator insertional mutagenesis in maize. Genetics 175: 981-992

Bancroft I, Bhatt AM, Sjodin C, Scofield S, Jones JD, Dean C (1992) Development of an efficient two-element transposon tagging system in *Arabidopsis thaliana*. Mol Gen Genet 233: 449-461

Barkley NA, Wang ML, Gillaspie AG, Dean RE, Pederson GA, Jenkins TM (2008) Discovering and verifying DNA polymorphisms in a mung bean [V. radiata (L.) R. Wilczek] collection by EcoTILLING and sequencing. BMC Res Notes 1:28

Bentley DR (2006) Whole-genome re-sequencing. Curr opin Genet Dev 16: 545-552

Berezikov E, Cuppen E, Plasterk RH (2006) Approaches to microRNA discovery. Nat Genet 38 (Suppl):S2-S7

Bhat SR, Srinivasan S (2002) Molecular and genetic analyses of transgenic plants: Considerations and approaches. Plant Sci 163: 673-681

Bortiri E, Chuck G, Vollbrecht E, Rocheford T, Martienssen R, Hake S (2006) *ramosa2* encodes a LATERAL ORGAN BOUNDARY domain protein that determines the fate of stem cells in branch meristems of maize. Plant Cell 18: 574-585

Bouche N, Bouchez D (2001) Arabidopsis gene knockout: phenotypes wanted. Curr Opin Plant Biol 4: 111-117

Busov VB, Meilan R, Pearce DW, Ma C, Rood SB, Strauss SH (2003) Activation tagging of a dominant gibberellin catabolism gene (GA 2-oxidase) from poplar that regulates tree stature. Plant Physiol 132:1283-1291

Busov V, Fladung M, Groover A, Strauss S (2005) Insertional mutagenesis in Populus: relevance and feasibility. Tree Genet Genom 1: 135-142

Buzas DM, Lohar D, Sato S, Nakamura Y, Tabata S, Vickers CE, Stiller J, Gresshoff PM (2005) Promoter trapping in *Lotus japonicus* reveals novel root and nodule GUS expression domains. Plant Cell Physiol 46: 1202-1212

Caldwell DG, McCallum N, Shaw P, Muehlbauer GJ, Marshall DF, Waugh R (2004) A structured mutant population for forward and reverse genetics in Barley (*Hordeum vulgare* L.). Plant J 40: 143-150

Casa AM, Brouwer C, Nagel A, Wang L, Zhang Q, Kresovich S, Wessler SR (2000) Inaugural article: the MITE family heartbreaker (Hbr): Molecular markers in maize. Proc Natl Acad Sci USA 97: 10083-10089

Chalfun A, Mes JJ, Mlynarova L, Aarts MGM, Angenent GC (2003) Low frequency of T-DNA based activation tagging in Arabidopsis is correlated with methylation of CaMV 35S enhancer sequences. FEBS lett 555: 459-463

Chen SY, Wang AM, Li W, Wang ZY, Cai XL (2008) Establishing a gene trap system mediated by T-DNA(GUS) in rice. J Int Plant Biol 50: 742-751

Colbert T, Till BJ, Tompa R, Reynolds S, Steine MN, Yeung AT, McCallum CM, Comai L, Henikoff S (2001) High-throughput screening for induced point mutations. Plant Physiol 126: 480-484

Comai L, Young K, Till BJ, Reynolds SH, Greene EA, Codomo CA, Enns LC, Johnson JE, Burtner C, Odden AR, et al. (2004) Efficient discovery of DNA polymorphisms in natural populations by Ecotilling. Plant J 37: 778-786

Conrad LJ, Kikuchi K, Brutnell TP (2008) Transposon tagging in cereal crops. In: Kahl, G, Meksem K (eds) The Handbook of Plant Functional Genomics. Wiley-Blackwell, Weinheim, UK

Cooper JL, Till BJ, Laport RG, Darlow MC, Kleffner JM, Jamai A, El-Mellouki T, Liu S, Ritchie R, Nielsen N, et al. (2008) TILLING to detect induced mutations in soybean. BMC Plant Biol 8:9

Cowperthwaite M, Park W, Xu Z, Yan X, Maurais SC, Dooner HK (2002) Use of the transposon Ac as a gene-searching engine in the maize genome. Plant Cell 14: 713-726

de Ruijter NCA, Verhees J, van Leeuwen W, van der Krol AR (2003) Evaluation and comparison of the GUS, LUC and GFP reporter system for gene expression studies in plants. Plant Biol 5: 103-115

Duguay JL, XiuQing L, Regan S (2007) The potential for genomics in potato improvement. In: CAB Rev: Perspectives in Agriculture, Veterinary Science, Nutrition and Natural Resources. Vol 2, p 13

Eamens AL, Blanchard CL, Dennis ES, Upadhyaya NM (2004) A bidirectional gene trap construct suitable for T-DNA and Ds-mediated insertional mutagenesis in rice (*Oryza sativa* L.). Plant Biotechnol J 2: 367-380

Filichkin SA, Wu Q, Busov V, Meilan R, Lanz-Garcia C, Groover A, Goldfarb B, Ma CP, Dharmawardhana P, Brunner A, et al. (2006) Enhancer trapping in woody plants: Isolation of the ET304 gene encoding a putative AT-hook motif transcription factor and characterization of the expression patterns conferred by its promoter in transgenic *Populus* and *Arabidopsis*. Plant Sci 171: 206-216

Francki M, Appels R (2002) Wheat functional genomics and engineering crop improvement. Genome Biol 3:reviews1013.1-1013.5

Furini A, Koncz C, Salamini F, Bartels D (1997) High level transcription of a member of a repeated gene family confers dehydration tolerance to callus tissue of *Craterostigma plantagineum*. EMBO J 16: 3599-3608

Gallois JL, Nora FR, Mizukami Y, Sablowski R (2004) WUSCHEL induces shoot stem cell activity and developmental plasticity in the root meristem. Genes Dev 18: 375-380

Gilchrist EJ, Haughn GW, Ying CC, Otto SP, Zhuang J, Cheung D, Hamberger B, Aboutorabi F, Kalynyak T, Johnson L, et al. (2006) Use of Ecotilling as an efficient SNP discovery tool to survey genetic variation in wild populations of *Populus trichocarpa*. Mol Ecol 15: 1367-1378

Giroux MJ, Shaw J, Barry G, Cobb BG, Greene T, Okita T, Hannah LC (1996) A single mutation that increases maize seed weight. Proc Natl Acad Sci USA 93: 5824-5829

Goff SA, Ricke D, Lan TH, Presting G, Wang R, Dunn M, Glazebrook J, Sessions A, Oeller P, Varma H, et al. (2002) A draft sequence of the rice genome (*Oryza sativa* L. ssp. *japonica*). Science 296: 92-100

Grant JJ, Chini A, Basu D, Loake GJ (2003) Targeted activation tagging of the Arabidopsis NBS-LRR gene, *ADR1*, conveys resistance to virulent pathogens. Mol Plant-Micr Interact 16: 669-680

Greene EA, Codomo CA, Taylor NE, Henikoff JG, Till BJ, Reynolds SH, Enns LC, Burtner C, Johnson JE, Odden AR, et al. (2003) Spectrum of chemically induced mutations from a large-scale reverse-genetic screen in Arabidopsis. Genetics 164: 731-740

Groover A, Fontana JR, Dupper G, Ma CP, Martienssen R, Strauss S, Meilan R (2004) Gene and enhancer trap tagging of vascular-expressed genes in poplar trees. Plant Physiol 134: 1742-1751

Gupta PK, Mir RR, Mohan A, Kumar J (2008) Wheat genomics: Present status and future prospects. Int J Plant Genom 2008: 896451

Han MJ, Jung KH, Yi G, Lee DY, An G (2006) Rice Immature Pollen 1 (RIP1) is a regulator of late pollen development. Plant Cell Physiol 47: 1457-1472

Harrison EJ, Bush M, Plett JM, McPhee DP, Vitez R, O'Malley B, Sharma V, Bosnich W, Seguin A, MacKay J, et al. (2007) Diverse developmental mutants revealed in an activation-tagged population of poplar. Can J Bot 85: 1071-1081

Hirochika H (1997) Retrotransposons of rice: Their regulation and use for genome analysis. Plant Mol Biol 35: 231-240

Hirochika H, Sugimoto K, Otsuki Y, Tsugawa H, Kanda M (1996) Retrotransposons of rice involved in mutations induced by tissue culture. Proc Natl Acad Sci USA 93: 7783-7788

Hirochika H, Guiderdoni E, An G, Hsing YI, Eun MY, Han CD, Upadhyaya N, Ramachandran S, Zhang Q, Pereira A, et al. (2004) Rice mutant resources for gene discovery. Plant Mol Biol 54: 325-334

Hsing YI, Chern CG, Fan MJ, Lu PC, Chen KT, Lo SF, Sun PK, Ho SL, Lee KW, Wang YC, et al. (2007) A rice gene activation/knockout mutant resource for high throughput functional genomics. Plant Mol Biol 63: 351-364

Hur J, Jung KH, Lee CH, An GH (2004) Stress-inducible *OsP5CS2* gene is essential for salt and cold tolerance in rice. Plant Sci 167: 417-426

Imaizumi R, Sato S, Kameya N, Nakamura I, Nakamura Y, Tabata S, Ayabe SI, Aoki T (2005) Activation tagging approach in a model legume, *Lotus japonicus*. J Plant Res 118: 391-399

Jaillon O, Aury JM, Noel B, Policriti A, Clepet C, Casagrande A, Choisne N, Aubourg S, Vitulo N, Jubin C, et al. (2007) The grapevine genome sequence suggests ancestral hexaploidization in major angiosperm phyla. Nature 449: 463-467

Jander G, Baerson SR, Hudak JA, Gonzalez KA, Gruys KJ, Last RL (2003) Ethylmethanesulfonate saturation mutagenesis in Arabidopsis to determine frequency of herbicide resistance. Plant Physiol 131: 139-146

Jenik PD, Irish VF (2001) The Arabidopsis floral homeotic gene *APETALA3* differentially regulates intercellular signaling required for petal and stamen development. Development 128: 13-23

Jeon JS, Lee S, Jung KH, Jun SH, Jeong DH, Lee J, Kim C, Jang S, Yang K, Nam J, et al. (2000) T-DNA insertional mutagenesis for functional genomics in rice. Plant J 22: 561-570

Jeong DH, An S, Kang HG, Moon S, Han JJ, Park S, Lee HS, An K, An G (2002) T-DNA insertional mutagenesis for activation tagging in rice. Plant Physiol 130: 1636-1644

Jeong DH, An S, Park S, Kang HG, Park GG, Kim SR, Sim J, Kim YO, Kim MK, Kim J, et al. (2006) Generation of a flanking sequence-tag database for activation-tagging lines in japonica rice. Plant J 45: 123-132

Jiang N, Bao Z, Zhang X, Hirochika H, Eddy SR, McCouch SR, Wessler SR

(2003) An active DNA transposon family in rice. Nature 421: 163-167

Jingu F, Shirase T, Ohtomo I, Imai A, Komeda Y, Takahashi T (2004) The plant exon finder: a tool for precise detection of exons using a T-DNA-based tagging approach. Gene 338: 267-273

Johansson AM, Wang C, Stenberg A, Hertzberg M, Little CH, Olsson O (2003) Characterization of a PttRPS18 promoter active in the vascular cambium region of hybrid aspen. Plant Mol Biol 52: 317-329

Johnson AA, Hibberd JM, Gay C, Essah PA, Haseloff J, Tester M, Guiderdoni E (2005) Spatial control of transgene expression in rice (*Oryza sativa* L.) using the GAL4 enhancer trapping system. Plant J 41: 779-789

Jung KH, Hur J, Ryu CH, Choi Y, Chung YY, Miyao A, Hirochika H, An G (2003) Characterization of a rice chlorophyll-deficient mutant using the T-DNA gene-trap system. Plant Cell Physiol 44: 463-472

Jung KH, Han MJ, Lee YS, Kim YW, Hwang I, Kim MJ, Kim YK, Nahm BH, An G (2005) Rice Undeveloped Tapetum1 is a major regulator of early tapetum development. Plant Cell 17: 2705-2722

Jung KH, Han MJ, Lee DY, Lee YS, Schreiber L, Franke R, Faust A, Yephremov A, Saedler H, Kim YW, et al. (2006) Wax-deficient anther1 is involved in cuticle and wax production in rice anther walls and is required for pollen development. Plant Cell 18: 3015-3032

Jung KH, An GH, Ronald PC (2008) Towards a better bowl of rice: assigning function to tens of thousands of rice genes. Nat Rev Genet 9: 91-101

Kaeppler SM, Kaeppler HF, Rhee Y (2000) Epigenetic aspects of somaclonal variation in plants. Plant Mol Biol 43: 179-188

Kikuchi K, Terauchi K, Wada M, Hirano HY (2003) The plant MITE *mPing* is mobilized in anther culture. Nature 421: 167-170

Kim JY, Lee SC, Jung KH, An GH, Kim SR (2004) Characterization of a cold-responsive gene, *OsPTR1*, isolated from the screening of beta-glucuronidase (GUS) gene-trapped rice. J Plant Biol 47: 133-141

Koh S, Lee SC, Kim MK, Koh JH, Lee S, An G, Choe S, Kim SR (2007) T-DNA tagged knockout mutation of rice OsGSK1, an orthologue of Arabidopsis BIN2, with enhanced tolerance to various abiotic stresses. Plant Mol Biol 65: 453-466

Kolkman JM, Conrad LJ, Farmer PR, Hardeman K, Ahern KR, Lewis PE, Sawers RJ, Lebejko S, Chomet P, Brutnell TP (2005) Distribution of *Activator* (*Ac*) throughout the maize genome for use in regional mutagenesis. Genetics 169: 981-995

Komatsu M, Shimamoto K, Kyozuka J (2003) Two-step regulation and continuous retrotransposition of the rice LINE-type retrotransposon Karma. Plant Cell 15: 1934-1944

Koo J, Kim Y, Kim J, Yeom M, Lee IC, Nam HG (2007) A GUS/luciferase fusion reporter for plant gene trapping and for assay of promoter activity with luciferin-dependent control of the reporter protein stability. Plant Cell Physiol 48: 1121-1131

Larmande P, Gay C, Lorieux M, Perin C, Bouniol M, Droc G, Sallaud C, Perez P, Barnola I, Biderre-Petit C, et al. (2008) *Oryza* Tag Line, a phenotypic mutant database for the Genoplante rice insertion line library. Nucl Acids Res 36: D1022-1027

Lee S, Jung KH, An G, Chung YY (2004a) Isolation and characterization of a rice cysteine protease gene, *OsCP1*, using T-DNA gene-trap system. Plant Mol Biol 54: 755-765

Lee SC, Kim JY, Kim SH, Kim SJ, Lee K, Han SK, Choi HS, Jeong DH, An GH, Kim SR (2004b) Trapping and characterization of cold-responsive genes from T-DNA tagging lines in rice. Plant Sci 166: 69-79

Liang D, Wu C, Li C, Xu C, Zhang J, Kilian A, Li X, Zhang Q, Xiong L (2006) Establishment of a patterned GAL4-VP16 transactivation system for discovering gene function in rice. Plant J 46: 1059-1072

Lindsey K, Wei WB, Clarke MC, McArdle HF, Rooke LM, Topping JF (1993) Tagging genomic sequences that direct transgene expression by activation of a promoter trap in plants. Transgen Res 2: 33-47

Margulies M, Egholm M, Altman WE, Attiya S, Bader JS, Bemben LA, Berka J, Braverman MS, Chen YJ, Chen Z, et al. (2005) Genome sequencing in micro-fabricated high-density picolitre reactors. Nature 437: 376-380

Marsch-Martinez N, Greco R, Van Arkel G, Herrera-Estrella L, Pereira A (2002) Activation tagging using the *En-I* maize transposon system in Arabidopsis. Plant Physiol 129: 1544-1556

Masaki T, Tsukagoshi H, Mitsui N, Nishii T, Hattori T, Morikami A, Nakamura K (2005) Activation tagging of a gene for a protein with novel class of CCT-domain activates expression of a subset of sugar-inducible genes in *Arabidopsis thaliana*. Plant J 43: 142-152

Mathews H, Clendennen SK, Caldwell CG, Liu XL, Connors K, Matheis N, Schuster DK, Menasco DJ, Wagoner W, Lightner J, et al. (2003) Activation tagging in tomato identifies a transcriptional regulator of anthocyanin biosynthesis, modification, and transport. Plant Cell 15: 1689-1703

Matsuhara S, Jingu F, Takahashi T, Komeda Y (2000) Heat-shock tagging: A simple method for expression and isolation of plant genome DNA flanked by T-DNA insertions. Plant J 22: 79-86

May BP, Liu H, Vollbrecht E, Senior L, Rabinowicz PD, Roh D, Pan X, Stein L, Freeling M, Alexander D et al. (2003) Maize-targeted mutagenesis: A knockout resource for maize. Proc Natl Acad Sci USA 100: 11541-11546

McCallum CM, Comai L, Greene EA, Henikoff S (2000) Targeted screening for induced mutations. Nat Biotechnol 18: 455-457

McCarty DR, Settles AM, Suzuki M, Tan BC, Latshaw S, Porch T, Robin K, Baier J, Avigne W, Lai J et al. (2005) Steady-state transposon mutagenesis in inbred maize. Plant J 44: 52-61

Mejlhede N, Kyjovska Z, Backes G, Burhenne K, Rasmussen SK, Jahoor A (2006) EcoTILLING for the identification of allelic variation in the powdery mildew resistance genes *mlo* and *Mla* of barley. Plant Breed 125: 461-467

Miyao A, Tanaka K, Murata K, Sawaki H, Takeda S, Abe K, Shinozuka Y, Onosato K, Hirochika H (2003) Target site specificity of the *Tos17* retrotransposon shows a preference for insertion within genes and against insertion in retrotransposon-rich regions of the genome. Plant Cell 15: 1771-1780

Miyao A, Iwasaki Y, Kitano H, Itoh J, Maekawa M, Murata K, Yatou O, Nagato Y, Hirochika H (2007) A large-scale collection of phenotypic data describing an insertional mutant population to facilitate functional analysis of rice genes. Plant Mol Biol 63: 625-635

Moon S, Jung KH, Lee DE, Jiang WZ, Koh HJ, Heu MH, Lee DS, Suh HS, An G (2006) Identification of active transposon *dTok*, a member of the hAT family, in rice. Plant Cell Physiol 47: 1473-1483

Mori M, Nakamura H, Ichikawa H (2006) Characterization of the short grain mutant (*Sg1*) isolated by rice activation tagging. Plant Cell Physiol 47: S177-S177

Mori M, Tomita C, Sugimoto K, Hasegawa M, Hayashi N, Dubouzet JG, Ochiai H, Sekimoto H, Hirochika H, Kikuchi S (2007) Isolation and molecular characterization of a *Spotted leaf 18* mutant by modified activation-tagging in rice. Plant Mol Biol 63: 847-860

Naito K, Cho E, Yang G, Campbell MA, Yano K, Okumoto Y, Tanisaka T, Wessler SR (2006) Dramatic amplification of a rice transposable element during recent domestication. Proc Natl Acad Sci USA 103: 17620-17625

Nakazaki T, Okumoto Y, Horibata A, Yamahira S, Teraishi M, Nishida H, Inoue H, Tanisaka T (2003) Mobilization of a transposon in the rice genome. Nature 421: 170-172

Nieto C, Piron F, Dalmais M, Marco CF, Moriones E, Gomez-Guillamon ML, Truniger V, Gomez P, Garcia-Mas J, Aranda MA, et al. (2007) EcoTILLING for the identification of allelic variants of melon *eIF4E*, a factor that controls virus susceptibility. BMC Plant Biol 7:34

Nishimura H, Ahmed N, Tsugane K, Iida S, Maekawa M (2008) Distribution and mapping of an active autonomous aDart element responsible for mobilizing nonautonomous *nDart1* transposons in cultivated rice varieties. Theor Appl Genet 116: 395-405

Ohmori Y, Abiko M, Horibata A, Hirano HY (2008) A transposon, *Ping*, is integrated into intron 4 of the *DROOPING LEAF* gene of rice, weakly reducing its expression and causing a mild drooping leaf phenotype. Plant Cell Physiol 49: 1176-1184

Oleykowski CA, Bronson Mullins CR, Godwin AK, Yeung AT (1998) Mutation detection using a novel plant endonuclease. Nucl Acids Res 26: 4597-4602

Opanowicz M, Vain P, Draper J, Parker D, Doonan JH (2008) *Brachypodium distachyon*: making hay with a wild grass. Trends Plant Sci 13: 172-177

Pan X, Li Y, Stein L (2005) Site preferences of insertional mutagenesis agents in Arabidopsis. Plant Physiol 137: 168-175

Paterson AH (2008) Genomics of sorghum. Int J Plant Genom 2008:362451

Pereira A (2000) A transgenic perspective on plant functional genomics. Transgen Res 9: 245-260

Piffanelli P, Droc G, Mieulet D, Lanau N, Bes M, Bourgeois E, Rouviere C, Gavory F, Cruaud C, Ghesquiere A, et al. (2007) Large-scale characterization of *Tos17* insertion sites in a rice T-DNA mutant library. Plant Mol Biol 65: 587-601

Qu S, Desai A, Wing R, Sundaresan V (2008) A versatile transposon-based activation tag vector system for functional genomics in cereals and other monocot plants. Plant Physiol 146: 189-199

Ramachandran S, Sundaresan V (2001) Transposons as tools for functional genomics. Plant Physiol Biochem 39: 243-252

Sallaud C, Gay C, Larmande P, Bes M, Piffanelli P, Piegu B, Droc G, Regad F, Bourgeois E, Meynard D, et al. (2004) High throughput T-DNA insertion mutagenesis in rice: A first step towards *in silico* reverse genetics. Plant J 39: 450-464

Scholte M, d'Erfurth I, Rippa S, Mondy S, Cosson V, Durand P, Breda C, Trinh H, Rodriguez-Llorente I, Kondorosi E, et al. (2002) T-DNA tagging in the model legume Medicago truncatula allows efficient gene discovery. Mol Breed 10: 203-215

Shultz JL, Kurunam D, Shopinski K, Iqbal MJ, Kazi S, Zobrist K, Bashir R, Yaegashi S, Lavu N, Afzal AJ, et al. (2006) The Soybean Genome Database (SoyGD): A browser for display of duplicated, polyploid, regions and sequence tagged sites on the integrated physical and genetic maps of *Glycine max*. Nucl Acids Res 34: D758-765

Slade AJ, Fuerstenberg SI, Loeffler D, Steine MN, Facciotti D (2005) A reverse genetic, nontransgenic approach to wheat crop improvement by TILLING. Nat Biotechnol 23: 75-81

Smith AD, Xuan Z, Zhang MQ (2008) Using quality scores and longer reads improves accuracy of Solexa read mapping. BMC Bioinform 9:128

Smith-Espinoza CJ, Phillips JR, Salamini F, Bartels D (2005) Identification of further *Craterostigma plantagineum cdt* mutants affected in abscisic acid mediated desiccation tolerance. Mol Genet Genom 274: 364-372

Springer PS (2000) Gene traps: Tools for plant development and genomics. Plant Cell 12: 1007-1020

Sreenivasulu N, Graner A, Wobus U (2008) Barley genomics: An overview. Int J Plant Genom 2008:486258

Suzuki T, Eiguchi M, Kumamaru T, Satoh H, Matsusaka H, Moriguchi K, Nagato Y, Kurata N (2008) MNU-induced mutant pools and high performance TILLING enable finding of any gene mutation in rice. Mol Genet Genom 279: 213-223

Takano M, Kanegae H, Shinomura T, Miyao A, Hirochika H, Furuya M (2001) Isolation and characterization of rice phytochrome A mutants. Plant Cell 13: 521-534

Takano M, Inagaki N, Xie X, Yuzurihara N, Hihara F, Ishizuka T, Yano M, Nishimura M, Miyao A, Hirochika H, et al. (2005) Distinct and cooperative functions of phytochromes A, B, and C in the control of deetiolation and flowering in rice. Plant Cell 17: 3311-3325

Talame V, Bovina R, Sanguineti MC, Tuberosa R, Lundqvist U, Salvi S (2008) TILLMore, a resource for the discovery of chemically induced mutants in barley. Plant Biotechnol J 6: 477-485

Tanaka T, Antonio BA, Kikuchi S, Matsumoto T, Nagamura Y, Numa H, Sakai H, Wu J, Itoh T, Sasaki T, et al. (2008) The Rice Annotation Project Database (RAP-DB): 2008 update. Nucl Acids Res 36:D1028-1033

Tax FE, Vernon DM (2001) T-DNA-associated duplication/translocations in Arabidopsis. Implications for mutant analysis and functional genomics. Plant Physiol 126: 1527-1538

Till BJ, Colbert T, Tompa R, Enns LC, Codomo CA, Johnson JE, Reynolds SH, Henikoff JG, Greene EA, Steine MN, et al. (2003) High-throughput TILLING for functional genomics. Meth Mol Biol 236: 205-220

Till BJ, Reynolds SH, Weil C, Springer N, Burtner C, Young K, Bowers E, Codomo CA, Enns LC, Odden AR, et al. (2004) Discovery of induced point mutations in maize genes by TILLING. BMC Plant Biol 4:1

Till BJ, Cooper J, Tai TH, Colowit P, Greene EA, Henikoff S, Comai L (2007) Discovery of chemically induced mutations in rice by TILLING. BMC Plant Biol 7:19

Topping JF, Lindsey K (1995) Insertional mutagenesis and promoter trapping in plants for the isolation of genes and the study of development. Transgen Res 4: 291-305

Tsugane K, Maekawa M, Takagi K, Takahara H, Qian Q, Eun CH, Iida S (2006) An active DNA transposon *nDart* causing leaf variegation and mutable dwarfism and its related elements in rice. Plant J 45: 46-57

Tuskan GA, Difazio S, Jansson S, Bohlmann J, Grigoriev I, Hellsten U, Putnam N, Ralph S, Rombauts S, Salamov A, et al. (2006) The genome of black cottonwood, Populus trichocarpa (Torr. & Gray). Science 313: 1596-1604

van der Fits L, Hilliou F, Memelink J (2001) T-DNA activation tagging as a tool to isolate regulators of a metabolic pathway from a genetically non-tractable plant species. Transgen Res 10: 513-521

Van der Graaf E, Hooykass PJJ, Keller B (2002) Activation tagging of the two closely linked genes LEP and VAS independantly affects vascular cell number. Plant J 32: 819-830

Walden R (2002) T-DNA tagging in a genomics era. Crit Rev Plant Sci 21: 143-165

Wan S, Wu J, Zhang Z, Sun X, Lv Y, Gao C, Ning Y, Ma J, Guo Y, Zhang Q, et al. (2008) Activation tagging, an efficient tool for functional analysis of the rice genome. Plant Mol Biol 69: 69-80

Wang RH, Zhou XF, Wang XZ (2003) Chemically regulated expression systems and their applications in transgenic plants. Transgen Res 12: 529-540

Wang S-G, Luan W-J, Qiao G-R, Zhuo R-Y, Sun Z-X (2007) Construction of T-DNA activation tagging from Populus functional genomics. For Res 20: 586-590

Webb KJ, Skot L, Nicholson MN, Jorgensen B, Mizen S (2000) *Mesorhizobium loti* increases root-specific expression of a calcium-binding protein homologue identified by promoter tagging in *Lotus japonicus*. Mol Plant-Micr Interact 13: 606-616

Weigel D, Ahn JH, Blazquez MA, Borevitz JO, Christensen SK, Fankhauser C, Ferrandiz C, Kardailsky I, Malancharuvil EJ, Neff MM, et al. (2000) Activation tagging in Arabidopsis. Plant Physiol 122: 1003-1013

Wilson K, Long D, Swinburne J, Coupland G (1996) A Dissociation insertion causes a semidominant mutation that increases expression of *TINY*, an Arabidopsis gene related to *APETALA2*. Plant Cell 8: 659-671

Wu C, Li X, Yuan W, Chen G, Kilian A, Li J, Xu C, Zhou DX, Wang S, Zhang Q (2003) Development of enhancer trap lines for functional analysis of the rice genome. Plant J 35: 418-427

Wu JL, Wu C, Lei C, Baraoidan M, Bordeos A, Madamba MR, Ramos-Pamplona M, Mauleon R, Portugal A, Ulat VJ, et al. (2005) Chemical- and irradiation-induced mutants of *indica* rice IR64 for forward and reverse genetics. Plant Mol Biol 59: 85-97

Yang G, Zhang F, Hancock CN, Wessler SR (2007) Transposition of the rice miniature inverted repeat transposable element *mPing* in *Arabidopsis thaliana*. Proc Natl Acad Sci USA 104: 10962-10967

Yang YZ, Peng H, Huang HM, Wu JX, Ha SR, Huang DF, Lu TG (2004) Large-scale production of enhancer trapping lines for rice functional genomics. Plant Sci 167: 281-288

Yu J, Buckler ES (2006) Genetic association mapping and genome organization of maize. Curr Opin Biotechnol 17: 155-160

Yu J, Hu S, Wang J, Wong GK, Li S, Liu B, Deng Y, Dai L, Zhou Y, Zhang X, et al. (2002) A draft sequence of the rice genome (*Oryza sativa* L. ssp. *indica*). Science 296: 79-92

Zhang J, Peterson T (1999) Genome rearrangements by nonlinear transposons in maize. Genetics 153: 1403-1410

Zhang J, Peterson T (2005) A segmental deletion series generated by sister-chromatid transposition of *Ac* transposable elements in maize. Genetics 171: 333-344

Zhang J, Li C, Wu C, Xiong L, Chen G, Zhang Q, Wang S (2006) RMD: A rice mutant database for functional analysis of the rice genome. Nucl Acids Res 34: D745-748

Zhu B, Gu A, Deng X, Geng Y, Lu Z (1995) Effects of caffeine or EDTA post-treatment on EMS mutagenesis in soybean. Mutat Res 334: 157-159

Zubko E, Adams CJ, Machaekova I, Malbeck J, Scollan C, Meyer P (2002) Activation tagging identifies a gene from Petunia hybrida responsible for the production of active cytokinins in plants. Plant J 29: 797-808

Zuo J, Niu QW, Frugis G, Chua NH (2002) The *WUSCHEL* gene promotes vegetative-to-embryonic transition in Arabidopsis. Plant J 30: 349-359

12 Genetic Variation Identified through Gene and Genome Sequencing

Erica G. Bakker
Center for Genome Research and Biocomputing
Oregon State University, Corvallis, OR, USA
Current address: Dow AgroSciences, Portland, OR, USA
E-mail: Bakker2@dow.com

1 INTRODUCTION

Genetic variation at the gene and genome level can be observed as patterns of single nucleotide polymoprhisms (SNPs) and indels (insertions and deletions). Whereas genome sequencing projects focus on obtaining the full genome sequence of a single organism as a representative of a species, polymorphism studies focus on genome-wide variation present within a species. Although the full genome sequence of a species provides valuable information regarding homology and distribution of genes across a genome, it does not provide information regarding the dynamic processes that affect a genome and the genes therein.

This chapter will focus on SNPs and their patterns in the genomes of the model plant *Arabidopsis thaliana*, fruitfly (*Drosophila* species) and human (The Human HapMap Project). Whereas the main focus of this book is on plants, a section on genome-wide SNP studies in other organisms (fruitfly and human) is included in this chapter to provide perspective, since genome-wide SNP studies in plants are limited to the model plant *Arabidopsis thaliana*. SNPs can be detected by most technologies and have widespread applications ranging from association mapping to genome-wide searches for natural selection. Applications of genome-wide SNP studies in crops are discussed at the end of this chapter.

2 SINGLE NUCLEOTIDE POLYMORPHISMS

2.1 SNPs, Alleles and Haplotypes

A single nucleotide polymorphism (SNP) is a mutation that differentiates DNA sequences of two or more individuals at a single nucleotide position. SNPs are interchangeably used with the term alleles when the individuals that are being compared are part of the same species. An SNP generally represents a single mutation event at a genomic position resulting in the presence of two different nucleotides in a population of related individuals. Multiple mutation events at the same genomic position can lead to either more than two nucleotides or the reversal of the mutation to its original state (reverse or back mutations). Both these events are rare for within-species comparisons but are more likely for comparisons between species, especially when common ancestry between these two species dates back for significant periods of time.

SNPs can occur in any genomic region, however, they are differentiated based on their occurrence in coding and non-coding regions of the genome. SNPs in non-coding DNA are also referred to as silent mutations, since these mutations have no effect on amino acid composition. SNPs in coding regions can either be silent (synonymous) or result in amino acid replacement (non-synonymous) dependent on the position in the codon. In general, mutations at third codon positions are synonymous, since the majority of amino acids are coded by four codons that vary for their third position as specified by the universal genetic code (Table 1). Mutations at first and second codon positions result, in general, in non-synonymous changes.

A genomic region containing a number of SNPs that have been co-inherited over time is referred to as a haplotype. Linkage disequilibrium (LD) between these SNPs is a result of an absence of recombination events, which can be explained by recent co-ancestry of individuals and/or a recent selective sweep. In addition, life history characteristics like breeding mechanism can affect haplotype structure: inbreeding organisms, such as *Arabidopsis thaliana,* have a more pronounced haplotype structure as compared to outbreeding organisms, such as *Drosophila melanogaster*.

Table 1 The universal genetic code.

Codon	*Amino acid*	*Codon*	*Amino acid*	*Codon*	*Amino acid*	*Codon*	*Amino acid*
UUU	Phe	UCU	Ser	UAU	Tyr	UGU	Cys
UUC	Phe	UCC	Ser	UAC	Tyr	UGC	Cys
UUA	Leu	UCA	Ser	UAA	Stop	UGA	Stop
UUG	Leu	UCG	Ser	UAG	Stop	UGG	Trp
CUU	Leu	CCU	Pro	CAU	His	CGU	Arg
CUC	Leu	CCC	Pro	CAC	His	CGC	Arg
CUA	Leu	CCA	Pro	CAA	Gln	CGA	Arg
CUG	Leu	CCG	Pro	CAG	Gln	CGG	Arg
AUU	Ile	ACU	Thr	AAU	Asn	AGU	Ser
AUC	Ile	ACC	Thr	AAC	Asn	AGC	Ser
AUA	Ile	ACA	Thr	AAA	Lys	AGA	Arg
AUG	Met	ACG	Thr	AAG	Lys	AGG	Arg
GUU	Val	GCU	Ala	GAU	Asp	GGU	Gly
GUC	Val	GCC	Ala	GAC	Asp	GGC	Gly
GUA	Val	GCA	Ala	GAA	Glu	GGA	Gly
GUG	Val	GCG	Ala	GAG	Glu	GGG	Gly

2.2 Detection Methods of Known SNPs

Population screening for known SNPs requires fast and inexpensive single-SNP detection methods, which can be subdivided into two main categories. The first category involves the capture, cleavage or mobility change during electrophoresis or liquid chromatography of DNA molecules. These methods are, in general, of low cost and have been extensively used in polymorphism and mutation studies. The disadvantage of these methods is that the exact nucleotide change cannot be (accurately) determined. The second category of SNP detection methods is sequence-specific (Kwok 2001). Within this category of sequence-specific SNP detection methods four general mechanisms for allelic discrimination can be distinguished: allele-specific hybridization, allele-specific nucleotide incorporation, allele-specific oligonucleotide ligation, and allele-specific invasive cleavage (Fig. 1A; Kwok 2000).

Allele-specific hybridization involves hybridization of two probes that differ by one nucleotide. The probes are immobilized on solid

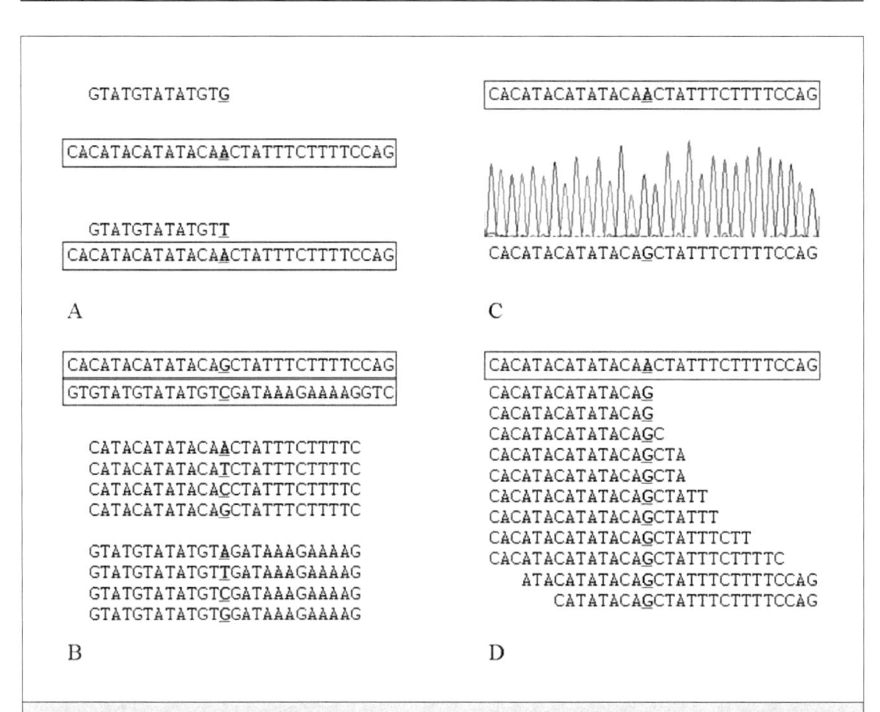

Fig. 1 Different SNP detection methods. (A) SNP detection through hybridization with a probe. A match with the reference genome results in a hybridization signal, whereas a SNP in the other genome results in an absence of a hybridization signal. (B) Whole-genome re-sequencing arrays use eight 25 bp probes varying in the center nucleotide. A match to the forward/reverse reference genome results in a hybridization signal, identifying a potential SNP at that position. (C) Comparison of dideoxy sequencing results with the reference genome identifies potential SNPs. (D) Massive parallel sequencing generates multiple short fragments that can be aligned to the reference genome which allows for the identification of potential SNPs. Reference genome sequences are boxed, SNPs are bold face underlined.

support and a hybridization reaction is carried out with fluorescently labeled target DNA. Probes are designed in such a way that only a perfect match gives a hybridization signal while a one base mismatch does not provide any signal. In this way, a large number of individuals can be screened for known SNPs segregating for two alleles. There are a number of genotyping assays that are based on this principle but that differ in the way they report the hybridization event. The 5′ nuclease assay involves the binding of a probe to amplified target DNA. Cleavage of the probe during the polymerase chain reaction (PCR) identifies whether hybridization has occurred (Livak 1999). The molecular beacon technology involves the opening and closing of the stem-loop structure dependent on whether hybridization has occurred (Kostrikis et al. 1998; Tyagi et al. 1998). Another method involves peptide nucleic acid (PNA) oligomers, which are labeled with a thiazole orange derivative that emits a signal when they specifically hybridize to complementary nucleic acids (Svanvik et al. 2000).

Allele-specific nucleotide incorporation, also called primer extension, is a sequencing approach where after hybridization of a primer to the target DNA the primer is extended with a nucleotide that can later be examined for its identity. A variant of this approach is an allele-specific PCR approach that is based on perfect complimentarity of different primers with the target DNA resulting only in a PCR product when a specific SNP is present in the target DNA (Kwok 2001).

Allele-specific ligation is based on the ability of DNA ligase to be highly specific in repairing nicks in the DNA molecule. This method involves annealing of two probes adjacent to each other to target DNA. Only when hybridization of both probes is perfect a ligation reaction takes place, which can be identified through the ligation chain reaction (LCR) (Barany 1991) or circularized ligation probes (Baner et al. 1998; Lizardi et al. 1998).

Allele-specific invasive cleavage is based on availability of a cleavage site in a complex of two overlapping oligonucleotide probes. The probes are designed in such a way that they overlap at the polymorphic site. The correct cleavage site will only be available when the allele-specific probe has formed the overlapping structure.

2.3 Whole-Genome Arrays

Genome-wide detection of novel SNPs requires genome-wide re-sequencing methods different from the methods mentioned in the previous section. One approach involves whole-genome arrays (WGA) (Mockler and Ecker 2005). The basic principle behind WGAs is similar to allele-specific hybridization and is based on probes attached to a solid surface that are hybridized with target DNA. Three main methods are used which represent low, medium and high resolution detection of genome-wide SNPs and differ based on the array design. Low levels of resolution are obtained through comparative genome hybridization (CGH), which is suitable for the detection of large insertions and deletions (Pinkel et al. 1998). Medium resolution is obtained through single-feature polymorphism (SFP) detection, which provides information regarding identity of non-overlapping 25-mers to a reference genome (Borevitz et al. 2003; Winzeler et al. 2003). The highest level of resolution is obtained through tiling array-based genome re-sequencing that allows for interrogation of >99.99% of bases in a genome (Clark et al. 2007). This technique involves query of each position in the genome with forward and reverse strand probe quartets with lengths of 25 bases, where the central nucleotide varies (Fig. 1B). This implies that for tightly linked SNPs all eight probes will have at least one mismatch with the target DNA resulting in reduced hybridization signals and confounded SNP detection. Another disadvantage of this technique is fluctuation in hybridization signals which might result in high false discovery rates (Clark et al. 2007).

2.4 Dideoxy Sequencing

The most informative SNP detection method is dideoxy re-sequencing (Fig. 1C). As opposed to tiling array-based genome re-sequencing, standard dideoxy re-sequencing can detect any SNP both in high and low variable regions in a genome. Although this method generates valuable data on all positions in a given sequenced region, the method is too laborious and costly to re-sequence an entire genome in multiple individuals (in order to obtain 1x coverage of the *Arabidopsis* genome of 125 Mb with 800 bp fragments >155 thousand fragments need to be sequenced in both the forward and reverse direction). For this reason, SNP detection projects using dideoxy re-sequencing focus on re-sequencing of a set of ~500 bp fragments, which are randomly distributed over the genome (Consortium 2003; Nordborg et al. 2005; Schmid et al. 2005). Fragments can be chosen based on existing data (sequence-tagged sites) (Schmid et al. 2005), or to be equally spaced over the entire genome (Nordborg et al. 2005) or to represent areas of interest as identified through patterns of association between nearby markers (Consortium 2003).

2.5 Novel Genome Sequencing Technologies

Several novel genome sequencing technologies have emerged recently as an alternative to the dideoxy sequencing method. All these technologies share the same principle of massive parallel sequencing which involves parallel sequencing of millions of short DNA fragments. Bioinformatics software takes care of the assembly of these short fragments, in general guided by a reference genome (Fig. 1D).

The first company to announce this new type of sequencing was 454

Life Sciences, which uses molecular beads to perform parallel sequencing reactions of DNA fragments of ~100 bp (Margulies et al. 2005). At the moment various massive parallel sequencing technologies have been developed either based on the molecular bead principle (Applied Biosystems, Inc.) or based on solid-phase amplification and sequencing of short DNA fragments (60 bp; bridge amplification technology; Illumina).

These new sequencing technologies have resulted in a shift in focus from generating sequencing data to the assembly of short reads. De novo assembly of short fragments is only possible for simple genomes without much repetitive DNA (e.g., bacteria). Available software like SSAKE (Warren et al. 2006) and SHARCGS (Dohm et al. 2007) assist in de novo data assembly. However, for more complex genomes containing vast amounts of repetitive DNA, as is the case for plants, de novo assembly is impossible for short DNA fragments. A reference genome is necessary in order to assemble the sequencing data and to extract information for downstream applications like SNP detection. Additional features of massive parallel sequencing technologies involve paired-end read modules, which will improve the potential of de novo and reference guided assembly for more complex genomes.

For SNP detection these new genome sequencing technologies are promising at various levels. First of all, they allow for the detection of all or most of the SNPs that differentiate a genome from the reference genome. This was so far impossible based on available technologies and could only be accomplished by doing traditional shotgun sequencing of an entire genome. Second, for large genomes sequencing can be focused on cDNA, reducing the total sequencing cost while allowing detection of important SNPs in coding regions. Third, with future improvements of the technology longer fragments can be sequenced possibly allowing for de novo assembly of coding regions in plant genomes. This will open up possibilities of SNP discovery in plant species for which no reference genome is available.

3 POLYMORPHISM PATTERNS IN THE GENOMES OF FRUITFLY AND HUMAN

3.1 *Drosophila* Population Genomics Projects

The fruitfly, *Drosophila melanogaster,* has been one of the most important genetic models in modern biology. This model organism continues taking the lead in the era of genomics with 12 published genome sequences and large-scale population genomic projects. New SNP detection technologies and methodological resources developed for *Drosophila* will soon find their applications for future genome-wide SNP studies in other organisms, including wild plants and agricultural crops.

After the publication of the *Drosophila melanogaster* annotated genome sequence (Adams et al. 2000) and the *D. pseudoobscura* genome (Richards et al. 2005) focus has shifted towards comparative genomics and population genomics. The *Drosophila* Species Genomes project (*http://insects.eugenes.org/species/*; Consortium 2007a) involved the sequencing of genomes of 10 *Drosophila* species in addition to the published genomes of *D. melanogaster* (Adams et al. 2000) and *D. pseudoobscura* (Richards et al. 2005). Comparative genomics of these 12 *Drosophila* species has provided evidence for dynamic expansion and contraction of gene families among species with an average of 17 gene duplications fixed in a genome every million years (Hahn et al. 2007).

Population genomic analysis of whole-genome polymorphism in seven *D. simulans* lines compared with genome sequences of *D. melanogaster* and *D. yacuba* revealed large-scale fluctuations of polymorphism and divergence along chromosome arms. In addition, genes experiencing adaptive evolution were identified and genomic patterns of base composition for coding and non-coding sequence were characterized (Begun et al. 2007). The relatively small sample size used in this study limited the number of population genetic analyses that could be performed. The *Drosophila* Population Genomic project (*http://www.dpgp.org/*) proposes to identify common polymorphisms in a much larger set of individuals (50 *D. melanogaster* individuals), but will only focus on part of the *D. melanogaster* genome: two large well-annotated genomic segments of 7.3 Mb in total length. This project is mainly focused on the development of new methods and concepts in the field of population genomics. Various re-sequencing technologies will be developed and evaluated. In addition, methodological resources will be developed for the analysis of genomic variation in natural populations and its functional role in the inheritance of complex traits. Proposed methodological resources include tools for genome-wide SNP data quality analysis, tools for easy access of genome-wide re-sequencing data, tools for genomic scale extensions of commonly used population genetic tools, incorporation of functional annotation into population genomic analysis and the application of genome-wide SNP data for the study of complex traits.

3.2 The International HapMap Project

The objective of the International HapMap Project (IHP) is to study haplotype patterns in the human genome. The project aims to detect a majority of common SNPs as identified through haplotype structure in a large set of individuals. In order to cover most variation, 270 individuals were sampled from three different continents, Africa (90 Yoruba individuals from Nigeria), Asia (45 Japanese and 45 Han Chinese individuals) and Europe (90 individuals of a US Utah population with Northern and Western European ancestry) (Consortium 2003).

SNP detection is targeted towards the detection of common variants, which constitute ~90% of all sequence variation in the human genome (Kruglyak and Nickerson 2001). Common SNPs are expected to be responsible for disease risk, however, some diseases are caused by rare SNPs. This should not be a problem since comparison between groups of healthy individuals and groups of individuals affected by the disease has revealed extended haplotype patterns in the disease group which implies that even rare SNPs can be identified due to considerable linkage disequilibrium in the region containing the gene(s) of interest (Kerem et al. 1989; Hastbacka et al. 1992).

With every new mutation a new haplotype is formed since each new SNP is initially associated with SNPs in flanking locations on the chromosome region where it emerged. Additional mutations and recombination events result in the breakdown of the initial haplotype into a mosaic-like structure of smaller haplotypes representing the parental haplotypes (Paabo 2003). SNP alleles that are located within a certain distance on the genome have a high chance to be passed on to the next generation as the same haplotype due to strong associations between these alleles in the population (strong LD). When the distance between two SNPs increases, the likelihood of a recombination event occurring between those two SNPs increases, which is the reason that with distance associations between SNP alleles decrease. For the human genome, it has been observed that LD varies significantly across the genome (Gabriel et al. 2002; Reich et al. 2003; Castaneda et al. 2005). For this reason a genome-wide SNP study of the human genome cannot be based on randomly selected or evenly spaced SNPs, but rather SNPs need to be chosen based on empirical studies of LD. Empirical evidence has shown that most genetic variation is represented by 10 million common SNPs, which can be detected by genotyping 200,000 to 1,000,000 tag SNPs across the genome (Gabriel et al. 2002).

When the project started the public database dbSNP already

contained 2.8 million SNPs. Additional SNPs were detected through random shotgun sequencing from whole-genome and whole-chromosome libraries, by making use of high-density oligonucleotide arrays (Perlegen) and sequence traces (Applied Biosystems, Inc.) (Venter et al. 2001). Further SNP genotyping in regions with low SNP densities was done using five high-throughput genotyping technologies (Third Wave Invader; Illumina BeadArray; Sequenom MassExtend; ParAllele MIP; PerkinElmer AcycloPrime-FP) (Consortium 2005).

Since common SNPs tend to be older than rare SNPs their associations on the genome reflect historical recombination and demographic events, which has resulted in chromosomes appearing as mosaics of ancestral chromosome regions (Paabo 2003). Because of these historical relationships among parts of the human genome, these genomic regions are shared among different individuals within and among populations (Tishkoff et al. 1996). This provides the foundation for the human haplotype map (HapMap), which captures common SNPs and their associations that can be used in applications for detection of SNPs involved in human diseases.

SNP patterns across the human genome have identified recombination hotspots, a block-like structure of linkage disequilibrium and low haplotype diversity (Consortium 2005). The second generation HapMap, which includes more than 3.1 million SNPs has an SNP density of around one per kb and contains around 25-35% of all 9-10 million common SNPs in the human genome (Consortium 2007b). The increased SNP density has resulted in improved coverage of common variation, whereas coverage of rare variation is relatively low. Based on this information a minimum number of 0.5-1 million tag SNPs will be required to capture all phase II SNPs with $r^2 > 0.8$ dependent on the population (Consortium 2007b). Arrays have been developed for tag SNP identification in disease studies. The second generation HapMap has provided information regarding varying recombination rates around genes and between genes of different function. Also, an increased differentiation at non-synonymous as compared to synonymous SNPs was observed indicating the effects of natural selection between populations (Consortium 2007b).

HapMap data have been used in large-scale studies that have identified novel loci that are involved in multiple complex diseases (Altshuler and Daly 2007; Bowcock 2007). HapMap data have provided further insight in the distribution and causes of recombination hotspots (Myers et al. 2005) and detailed information regarding structural variation

(Conrad et al. 2006) including copy-number variation (Redon et al. 2006) and the identity of genes under adaptive selection (Voight et al. 2006). HapMap data have also been informative for studies regarding the relationship between SNP variation and human leukocyte antigen (HLA) types (de Bakker et al. 2006) and heritable influences on gene expression (Cheung et al. 2005).

4 POLYMORPHISM PATTERNS IN *ARABIDOPSIS THALIANA*

4.1 Polymorphisms Studies in *Arabidopsis*

Arabidopsis thaliana was the first plant to have its genome sequenced (Initiative 2000). Because of the availability of this reference genome sequence in combination with its small genome size and its status as a model species, this plant species has been extensively studied for polymorphism patterns. Based on polymorphism studies in *Arabidopsis* deductions can be made regarding patterns of polymorphism in other (related) plant species, including crops.

Initial polymorphism studies in *Arabidopsis* used dideoxy sequencing in order to re-sequence a set of fragments randomly distributed over the *Arabidopsis* genome (Nordborg et al. 2005; Schmid et al. 2005). The study of Schmid et al. (2005) was based on 334 (average sequence length 414 bp) sequence-tagged sites (short unique genomic regions used as anchors in mapping projects) sequenced in 12 *A. thaliana* accessions, whereas the study of Nordborg et al. (2005) was based on 876 (1,213 sequenced fragments in total at the end of the project) randomly chosen short fragments (average sequence length of 583 bp) sequenced in 96 *A. thaliana* accessions. A third study used whole-genome re-sequencing arrays and was able to investigate SNP variation at >99.99% of the *Arabidopsis* genome (Clark et al. 2007). Whereas the study of Clark et al. (2007) was able to detect SNPs over the total genome sequence of 20 *A. thaliana* accessions, the studies of Schmid et al. (2005) and Nordborg et al. (2005) were limited to a relatively small set of fragments randomly distributed over the genome. The genome-wide SNP information of the study of Clark et al. (2007) allowed for detailed information regarding patterns of variation in groups of genes. However, because the tiling-array approach does not allow for the (accurate) detection of more than one SNP within each interrogated 25-mer, this approach does not provide accurate information regarding polymorphism patterns in highly variable regions. Another disadvantage of the tiling-array technology is that deleted and duplicated regions are not easily detected. Since low hybridization signal can be associated both with deleted and hypervariable regions, advanced analysis is required to distinguish both categories (Clark et al. 2007). Duplicated regions cannot be detected since the array only contains probes representing the *Arabidopsis* Col-0 genome. Although the studies of Schmid et al. and Nordborg et al. were not able to investigate SNPs over the entire *Arabidopsis* genome, their dideoxy sequencing approach allowed them to obtain full information regarding SNPs even in highly variable regions. The other advantage is that the false discovery rate is low because of sequencing of both forward and reverse strands. Tiling-array studies face problems with hybridization signal detection and advanced statistical tests need to be applied in order to identify SNPs with high probability levels (Clark et al. 2007). Also, regional differences in repeat content cause bias in array-based re-sequencing (Clark et al. 2007). With these limitations these three studies were able to describe different features regarding the pattern of polymorphism in *Arabidopsis*; for many features these three studies were confirmatory.

4.2 Patterns of Polymorphism Across the *Arabidopsis* Genome

In general, polymorphism levels as detected by tiling arrays were lower as compared to dideoxy re-sequencing, which is due to the inability to detect SNPs in highly polymorphic regions. Neutral variation at synonymous sites and intergenic regions shows a non-random pattern across the *Arabidopsis* genome with high levels of polymorphism near the centromeres and clusters of NB-LRR genes (Nordborg et al. 2005; Clark et al. 2007). This non-random pattern of variation could be attributed to either selection on linked sites or differences in mutation rates between chromosomal regions because of differences in base composition. NB-LRR gene clusters have a strong effect on polymorphism levels, even when they are excluded from the analysis. This effect can be explained by (transient) balancing selection acting on the majority of NB-LRR genes over intermediate periods of time (Bakker et al. 2006a). Three of these NB-LRR genes (*RPM1, RPS2, RPS5*) have been studied in detail and all share a signature of selection characterized by a sharp increase in polymorphism at the locus (Caicedo et al. 1999; Stahl et al. 1999; Tian et al. 2002; Mauricio et al. 2003).

Regions with high levels of polymorphism contain in addition to NB-LRR genes, receptor-like kinase (RLK) genes and F-box genes (Clark et al. 2007). RLK genes have diverse functions (Shiu and Bleecker 2001), but also appear to be involved in race-specific pathogen defense (Song et al. 1995), like NB-LRR genes, which indicates that the high levels of polymorphism observed in this group of genes is the result of plant-pathogen co-evolution (Stahl et al. 1999). It is known that F-box genes have undergone rapid birth and death in the *A. thaliana* genome (Thomas 2006; Tuskan et al. 2006). Their high polymorphism levels provide additional support for presumed involvement in response to pathogen pressure during their evolutionary history (Thomas 2006).

Regions with low levels of variation are presumably the result of selective sweeps as has been identified for the region surrounding the flowering time gene *FRI* (Toomajian et al. 2006). Tiling-array data identified a near-complete species-wide sweep for a 500 kb region on chromosome 1, which contains 50 annotated genes. Additional candidates for selective sweeps have affected smaller numbers of accessions (Clark et al. 2007).

Comparison of θ_S (number of polymorphic sites) with θ_P (average number of pairwise differences) shows that θ_S is consistently higher (Nordborg et al. 2005; Schmid et al. 2005). This pattern is the result of an unusually high ratio of rare to common alleles, which could have been caused by recent population growth (Innan et al. 1997) and effects of natural selection. Due to these effects *Arabidopsis* appears to be no different than other organisms in the observation that genome-wide polymorphism data do not fit the standard neutral model (Andolfatto and Przeworski 2000; Stephens et al. 2001; Akey et al. 2004). These observations might affect tests for selection (see Section 5.1).

4.3 Population Structure as Inferred from Genome-wide Polymorphism Data

Genome-wide polymorphism data based on re-sequenced fragments of the *Arabidopsis* genome allowed for detailed investigation of population structure. Analysis of population structure of *A. thaliana* based on micro-satellites and SNPs in housekeeping genes was unable to distinguish a clear population structure in the set of 96 *A. thaliana* accessions that were used for the genome-wide polymorphism study (Bakker et al. 2006a). Analysis based on 876 re-sequenced DNA fragments in the same set of 96 *A. thaliana* accessions, however, could detect eight regional groups using the program Structure (Pritchard et al. 2000; Nordborg et al. 2005). This indicates that although spread of *A. thaliana* over the world is relatively recent, genome-wide patterns of polymorphism reflect geographic regions, however, genetic differences between regions are in general small.

4.4 Genome-wide Patterns of Recombination and Gene Conversion

Whole-genome tiling array data show that linkage disequilibrium decays within 10 kb (at 3-4 kb it reaches 50% of its initial value (Kim et al. 2007). Genome-wide polymorphism data also show that the extent of LD is highly variable as a result of an unequal distribution of recombination rates over the *Arabidopsis* genome, the action of natural selection, gene conversion and multiple mutations (Nordborg et al. 2005; Kim et al. 2007).

Genome-wide observations of too little short-range LD relative to long-range LD could be the result of gene conversion (Kim et al. 2007). It has been estimated that gene conversion is about five times more common than crossing over based on genome-wide polymorphism data for *Arabidopsis*, which confirms previous results (Copenhaver et al. 2002; Haubold et al. 2002; Nordborg et al. 2005).

The genome-wide average estimate for the population recombination rate ρ (Hudson 2001) is 0.8 per kb, which is about twice the estimate for humans (Padhukasahasram et al. 2006), but an order of magnitude lower as compared to *Drosophila melanogaster*, which is an outcrosser with a large effective population size (Haddrill et al. 2005). The population recombination rate ρ varies extensively across the *Arabidopsis* genome and appears to correlate with polymorphism level, GC content, and gene density (Kim et al. 2007). In total 260 recombination hotspots have been identified, which is much higher than expected by chance. Hotspots are 1-2 kb long and 200-times stronger than the background genome. Their location is preferentially in intergenic regions (Kim et al. 2007).

5 APPLICATIONS OF GENOME-WIDE POLYMORPHISM DATA

5.1 Tests for Natural Selection

Until recently, tests for natural selection involved small numbers of genes using the standard neutral model as a null model. However, since genome-wide polymorphism data do not conform to the neutral model—mostly due to demographic effects—tests for selection using the neutral model will, in general, falsely identify increased numbers of genes under natural selection (Nordborg et al. 2005). An alternative would be to include demographic effects in the null model as has been done by Akey et al. (2004) for human polymorphism data on 132 genes primarily involved in inflammation, blood clotting and blood pressure regulation and Schmid et al. (2005) for 334 STS loci in *Arabidopsis*. Although Akey et al. (2004) were able to fit part of the polymorphism data to a model involving a bottleneck occurring 40,000 years ago, the confidence intervals for the observed summary statistics were broad and various aspects of the data were consistent with other models. Demographically robust estimates were obtained by comparing all the candidate genes with all five tested demographic models. In this way, only eight demographically robust genes were identified to be under selection as compared to the 20 genes that were previously identified using the standard neutral model (Akey et al. 2004).

Arabidopsis genome-wide polymorphism data have not been consistent with parameter estimates of various demographic models (Schmid et al. 2005). These observations infer that use of the standard neutral model as a null model in tests for selection will result in the incorrect identification of genes under selection. It is expected that this is the main cause for the much higher fraction of genes that have been reported to be under selection in *A. thaliana* as compared to other species (Wright and Gaut 2005). Future population genomics studies are expected to focus on finding models that explain the bulk of the data (Nordborg et al. 2005).

Instead of using a demographic model as a reference to test for natural selection, it is possible to use genome-wide polymorphism data to generate empirical distributions for various population genetic summary statistics. This approach would involve comparison of population genetic summary statistics obtained for a gene or group of genes with corresponding empirical distributions of these summary statistics calculated based on genome-wide polymorphism patterns (Kreitman 2000). This approach was used in a genome-wide study of F_{ST} values (a measure of population differentiation) of individual SNPs in the human genome, which were compared with the empirical genome-wide distribution of F_{ST} (Akey et al. 2002). In this way, 174 candidate genes were identified as targets for natural selection (Akey et al. 2002). In *Arabidopsis*, this approach was used to test for natural selection on 27 NBS-LRR genes (Bakker et al. 2006b) and 27 defense response genes

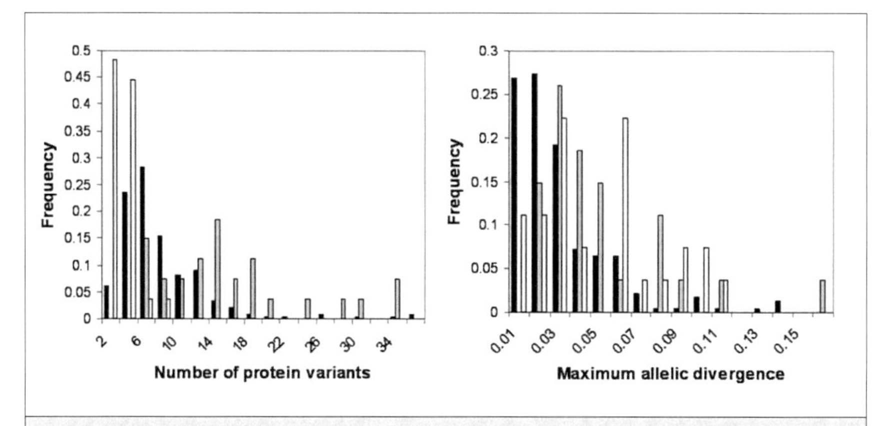

Fig. 2 Genome-wide comparisons of number of protein variants and maximum allelic divergence between empirical distributions of these population genetic summary statistics for 876 random genomic fragments (black bars), 27 *R* genes (shaded bars) and 27 defense response genes (white bars).

(Bakker et al. 2008). In both studies, empirical distributions generated based on 876 random fragments from the Nordborg et al. (2005) study were used to infer extent and type of natural selection in a set of 27 *R* genes and 27 defense response genes, respectively, that were sequenced in the same set of 96 *A. thaliana* accessions. This approach identified NBS-LRR genes as highly polymorphic, whereas defense response genes maintain low levels of variation mainly as a result of purifying selection (Bakker et al. 2006b, 2008) (Fig. 2).

Several tests for selection require an outgroup sequence. *Arabidopsis lyrata* is commonly used as outgroup for *A. thaliana* evolutionary research. *A. thaliana* and *A. lyrata* differ from each other by 0.087-0.111 substitutions per base (as identified for the *Adh* and *Chs* genes corresponding with a divergence time of 5 million years ago) (Koch et al. 2000). This implies that on average in a 20-bp primer binding site 1-2 nucleotides in the *A. lyrata* genome are expected to differ from the *A. thaliana* genome, which makes amplification and sequencing of corresponding *A. lyrata* sequence challenging, especially for highly variable regions. An additional problem is that *A. lyrata* is an outcrosser which means that most loci are heterozygous as opposed to the inbreeder *A. thaliana*. Whereas *A. thaliana* fragments can be directly sequenced from PCR products, *A. lyrata* fragments first need to be cloned in order to obtain single stranded template for sequencing. Genome-wide searches for natural selection would be possible if the full genome sequence of this outgroup species is available as is the case of fruitfly (Consortium 2007a). With the expected release of the *A. lyrata* genome in 2009, tests for natural selection can be performed at a genome-wide scale. However, since most tests use the neutral model, care needs to be taken to either use a demographic model that fits the data or to use empirical distributions as suggested by Schmid et al. (2005).

5.2 Association Mapping

In contrast to conventional mapping, which makes use of members of a pedigree or experimental cross, association mapping makes use of a general population of 'unrelated' individuals (Nordborg and Tavare 2002). The basic principle behind association mapping is that unrelated individuals are always related at some distance because they share alleles identical by descent that are surrounded by shared ancestral haplotypes that can be identified in genome-wide polymorphism studies. Association mapping can be performed based on a genome-wide approach when there is no prior knowledge regarding genomic associations with the trait of interest or a candidate gene approach, when there is biological knowledge that allows for a focus on parts of the genome (Collins et al. 1997).

Association mapping has the advantage that no pedigree or cross is needed. In some cases crosses produce only limited levels of variation segregating in the progeny or choice of crossing parents is limited. In cases where crosses are impossible as for human studies, pedigrees are limiting. A second advantage of association mapping is the high level of resolution reached because of high numbers of recombinations that have accumulated over time separating individuals since their common ancestry. A disadvantage of association mapping is that it does not involve a controlled experiment. Since the decay of linkage disequilibrium is noisy and the genetic architecture of a trait is unknown, it is impossible to predict power, which is an additional disadvantage (Weiss and Terwilliger 2000; Nordborg and Tavare 2002; Zondervan and Cardon 2004). Also, population structure has negative effects on the false discovery rate since it can affect association studies through strong spurious correlations (Weiss and Terwilliger 2000; Lander and Schork 1994; Weiss and Terwilliger 2000).

Dependent on the extent of linkage disequilibrium decay with distance, hundreds—thousands of individuals need to be genotyped with a large number of SNPs. For *Arabidopsis,* it has been determined that in order to detect all 341,602 SNPs as identified in a whole-genome tiling array study (Clark et al. 2007) at least 140,351 SNPs need to be identified. It is expected that this number will be sufficient to detect all SNPs in additional samples since they are assumed to be more related to each other (i.e. longer haplotype blocks) than the group of 20 accessions used in the tiling array. Although the tiling-array SNPs represent only a fraction of all SNPs in the *Arabidopsis* genome (which is predicted to be ~1.5 million SNPs based on dideoxy data) various analyses have shown that the tiling array data sufficiently represent all SNPs in the *Arabidopsis* genome (Kim et al. 2007).

Association mapping is used for mapping of complex disease on the human genome (Johnson and Todd 2000; Risch 2000). An initial genome-wide association mapping study by Aranzana et al. (2005) used polymorphism data generated for 876 random fragments in 95 *A. thaliana* accessions (Nordborg et al. 2005) to test whether association mapping was feasible with this dataset, which was known to have a regional demographic structure. The presence of a regional population structure caused an extremely high rate of false positives. However, association mapping resulted in the identification of major genes for all tested phenotypes, showing the potential for association studies with an appropriate genomic control (Aranzana et al. 2005). Recently, improved statistical methods (Yu et al. 2006) have been developed and application of association mapping for gene discovery in crops is promising (Yu and Buckler 2006; Yu et al. 2008).

6 SNP STUDIES IN CROPS AND APPLICATIONS FOR PLANT BREEDING

6.1 SNP Studies in Crops

Genome-wide SNP studies have so far been done in fruitfly, human and *Arabidopsis*. Genome-wide SNP studies in crops are planned for the near future and will mainly focus on the generation of SNPs for use in association mapping (*http://www.panzea.org;* Yu et al. 2008). Future genome-wide SNP studies are expected to be limited to species for which a (partially completed) reference genome and/or extensive genomics resources are available. This means that future SNP studies are not only possible for crops that have their genomes completed such as rice (Goff et al. 2002), grape (Characterization, 2007), black cottonwood (Tuskan et al. 2006) and sorghum (*http://www.jgi.doe.gov/sequencing/cspseqplans 2006.html*), but also for crops with partially completed or nearly completed genomes such as cotton and cassava (*http://www.jgi.doe.gov/*

sequencing/cspseqplans2007.html), foxtail millet and eucalyptus tree (*http://www.jgi.doe.gov/sequencing/cspseqplans2008.html*), tomato (*http://www.sgn.cornell.edu/about/tomato_sequencing.pl)*, maize (*http://www.maizegenome.org/*), cabbage and turnip (*http://www.brassica.info/*). In addition, there are 17 additional crops for which genomics resources (mainly expressed sequence tags, ESTs) are available as part of the Gene Index Project (*http://compbio.dfci.harvard.edu/tgi/plant.html*).

6.2 Applications in Plant Breeding

Because of the high cost of dideoxy sequencing and re-sequencing array technology, genome-wide SNP studies have so far only been possible for model organisms. However, with the availability of novel genome sequencing technologies which produce millions of bases of sequence data in a matter of days for a fraction of the cost of the conventional technologies, genome-wide SNP studies in crops are becoming more feasible.

A potential direct application of genome-wide SNP studies in plant breeding is the generation of markers for the construction of (dense) genetic linkage maps. Massive parallel sequencing of the genomes of a set of two crossing parents followed by comparative analysis of their consensus sequences will reveal genomic regions of interest for which the two crossing parents differ, which could be used to identify informative molecular markers. In a similar fashion, germplasm collections can be screened for SNPs in coding regions, which will facilitate future choice of crossing parents.

Another application is association mapping. Association mapping could be an alternative to conventional mapping when denser maps are desired or when available crossing parents are too related, generating insufficient segregating variation in the crossing population. At the moment, association mapping is not commonly used in crops because of several disadvantages of the method (see Section 5.2). Current efforts that combine linkage mapping and association mapping (nested association mapping) appear to be a cost-effective alternative, which is promising for the dissection of complex traits, while avoiding the confounding effects of population structure (Yu et al. 2008).

Other applications of genome-wide SNP studies involve extrapolation of genome-wide SNP information available for model species to related crops. *Brassica* crops are likely to benefit from genome-wide SNP studies in *A. thaliana*, whereas cereal crops are expected to benefit from genome-wide SNP studies in rice and the upcoming new model species for the grasses, *Brachypodium distachyon* (*www.brachypodium.org*). In addition, genome-wide tests for selection through genome-wide comparisons between phenotypically divergent accessions could be considered, thus identifying genes of interest for future breeding programs.

7 CONCLUSION

Genome-wide SNP studies in fruitfly, human, and *Arabidopsis* have provided us with a deeper understanding about the processes that affect genome-wide patterns of polymorphism. It is now clear that levels of nucleotide diversity and linkage disequilibrium are not constant across a genome, but fluctuate as a result of various processes. Applications of genome-wide SNP data are genome-wide tests for selection and association mapping. Both applications are promising for the detection of resistance genes in crops because of their typical pattern of polymorphism. With the availability of genome sequences and other genomic resources for a large set of crops together with cost-effective novel sequencing methods it is expected that genome-wide SNP studies will find many applications in plant breeding.

References

Adams MD, Celniker SE, Holt RA, Evans CA, Cocayne JD, Amanatides PG, Scherer SE, et al. (2000) The genome sequence of *Drosophila melanogaster*. Science 287: 2185-2195

Akey JM, Zhang G, Zhang K, Jin L, Shriver MD (2002) Interrogating a high-density SNP map for signatures of natural selection. Genome Res 12: 805-1814

Akey JM, Eberle MA, Rieder MJ, Carlson CS, Shriver MD, Nickerson DA, Kruglyak L (2004) Population history and natural selection shape patterns of genetic variation in 132 genes. PLoS Biol 2: 1591-1599

Altshuler D, Daly M (2007) Guilt beyond reasonable doubt. Nat Genet 39: 13-815

Andolfatto P, Przeworski M. 2000. A genome-wide departure from the standard neutral model in natural populations of *Drosophila*. Genetics 156: 257-268

Aranzana MJ, Kim S, Zhao K, Bakker E, Horton M, Jakob K, Lister C, Molitor J, Shindo C, Tang C, Toomajian C, Traw B, Zheng H, Bergerlson J, Dean C, Marjoram P, Nordborg M (2005) Genome-wide association mapping in Arabidopsis identifies previously known flowering time and pathogen resistance genes. PLos Genet 1: 531-539

Bakker EG, Stahl E, Toomajian C, Nordborg M, Kreitman M, Bergelson J (2006a) Distribution of genetic variation within and among local populations of

Arabidopsis thaliana over its species range. Mol Ecol 15: 405-1418

Bakker EG, Toomajian C, Kreitman M, Bergelson J (2006b) A genome-wide survey of R gene polymorphisms in Arabidopsis. Plant Cell 18: 1803-1818

Bakker EG, Traw MB, Toomajian C, Kreitman M, Bergelson J (2008) Low levels of polymorphism in genes that control the activation of defense response in *Arabidopsis thaliana*. Genetics 178: 2031-2043

Baner J, Nilsson M, Mendel-Hartvig M, Landegren U (1998) Signal amplification of padlock probes by rolling circle replication. Nucl Acids Res 26: 5073-5078

Barany F (1991) Genetic disease detection and DNA amplification using cloned thermostable ligase. Proc Natl Acad Sci USA 88: 189-193

Begun DJ, Holloway AK, Stevens K, Hillier LD, Poh Y-P, Hahn MW, Nista PM (2007) Population genomics: Whole-genome analysis of polymorphism and divergence in Drosophila simulans. PLoS Biol 5

Borevitz JO, Liang D, Plouffe D, Chang HS, Zhu T, Weigel D, Berry CC, Winzeler E, Chory J (2003) Large-scale indenification of single-feature polymorphisms in complex genomes. Genom Res 13: 513-523

Bowcock AM (2007) Genomics: guild by association. Nature 447: 645-646

Caicedo AL, Schaal BA, Kunkel BN (1999) Diversity and molecular evolution of the *RPS2* resistance gene in *Arabidopsis thaliana*. Proc Natl Acad Sci USA 96: 302-306

Castaneda A, Reddy JD, El-Yacoubi B, Gabriel DW (2005) Mutagenesis of all eight *avr* genes in *Xanthomonas campestris* pv. *campestris* had no detected effect on pathogenicity, but one *avr* gene affected race specificity. Mol Plant-Micr Interact 18: 1306-1317

Characterization TF-IPCfGG (2007) The grapevine genome sequence suggests ancestral hexaploidization in major angiosperm phyla. Nature 449: 463-468

Cheung VG, Spielman RS, Ewens KG, Weber TM, Morley M, Burdick JT (2005) Mapping determinants of human gene expression by regional and genome-wide association. Nature 437: 1365-1369

Clark RM, Schweikert G, Toomajian C, Ossowski S, Zeller G, Shinn P, Warthmann N, Hu TT, Fu G, et al. (2007) Common sequence polymorphisms shaping genetic diversity in *Arabidopsis thaliana*. Science 317:338-342

Collins FS, Guyer MS, Charkarvarti A (1997) Variations on a theme: cataloging human DNA sequence variation. Science 278: 1580-1581

Conrad DF, Andrews TD, Carter NP, Hurles ME, Pritchard JK (2006) A high-resolution survey of deletion polymorphism in the human genome. Nat Genet 38: 75-81

Consortium DGSaA (2007a) Evolution of genes and genomes in the context of the *Drosophila* phylogeny. Nature 450: 203-218

Consortium TIH (2003) The international HapMap project. Nature 426: 789-796

Consortium TIH (2005) A haplotype map of the human genome. Nature 437: 1299-1320

Consortium TIH (2007b) A second generation human haplotype map of over 3.1 million SNPs. Nature 449: 851-861

Copenhaver GP, Housworth EA, Stahl FW (2002) Crossover interference in *Arabidopsis*. Genetics 160: 1631-1639

de Bakker PIW, McVean G, Sabeti PC, Miretti MM, Green T, Marchini J et al. (2006) A high-resolution HLA and SNP haplotype map for disease association studies in the extended human MHC. Nat Genet 38: 1166-1172

Dohm JC, Lottaz C, Borodina T, Himmelbauer H (2007) SHARCGS, a fast and highly accurate short-read assembly algorithm for de novo genomic sequencing. Genome Res 17: 1697-1706

Gabriel SB, Schaffner SF, Nguyen H, Moore JM, Roy J, Blumenstiel B, Higgins J et al. (2002) The structure of haplotype blocks in the human genome. Science 296: 2225-2229

Goff SA, Ricke D, Lan TH, Presting G, Wang RL, Dunn M, Glazebrook J, Sessions A, Oeller P et al. (2002) A draft sequence of the rice genome (*Oryza sativa* L. ssp. *japonica*). Science 296: 92-100

Haddrill PR, Thornton KR, Charlesworth B, Andolfatto P (2005) Multilocus patterns of nucleotide variability and the demographic and selection history of *Drosophila melanogaster* populations. Genome Res 15: 790-799

Hahn MW, Han MV, Han S-G (2007) Gene family evolution across 12 *Drosophila* genomes. PLos Genet 3

Hastbacka J, Chapelle A, Kaitila I, Sistonen P, Weaver A, Lander E (1992) Linkage disequilibrium mapping in isolated founder populations: diastrophic dysplasia in Finland. Nat Genet 2: 204-211

Haubold B, Kroymann J, Ratzka A, Mitchell-Olds T, Wiehe T (2002) Recombination and gene conversion in a 170-kb genomic region of *Arabidopsis thaliana*. Genetics 161: 1269-1278

Hudson RR (2001) Two-locus sample distributions and their applications. Genetics 159: 1805-1817

Initiative TAG (2000) Analysis of the genome sequence of the flowering plant *Arabidopsis thaliana*. Nature 408: 796-815

Innan H, Terauchi R, Miyashita NT (1997) Microsatellite polymorphism in natural populations of the wild plant *Arabidopsis thaliana*. Genetics 146: 1441-1452

Johnson GC, Todd JA (2000) Strategies in complex disease mapping. Curr Opin Genet Dev 10: 330-334

Kerem B, Rommens JM, Buchanan JA, Markiewicz D, Cox TK, Chakravarti A, Buchwald M, Tsui LC (1989) Identification of the cystic fibrosis gene: genetic analysis. Science 245: 1073-1080

Kim S, Plagnol V, Hu TT, Toomajian C, Clark RM, Ossowski S, Ecker JR, Weigel D, Nordborg M (2007) Recombination and linkage disequilibrium in *Arabidopsis thaliana*. Nat Genet 39: 1151-1155

Koch MA, Haubold B, Mitchell-Olds T (2000) Comparative evolutionary analysis of chalcone synthase and alcohol dehydrogenase loci in Arabidopsis, Arabis, and related genera (Brassicaceae). Mol Biol Evol 17: 1483-1498

Kostrikis LG, Tyagi S, Mhlanga MM, Ho DD, Kramer FR (1998) Spectral genotyping of human alleles. Science 279: 1228-1229

Kreitman M (2000) Methods to detect selection in populations with applications to the human. Annu Rev Genom Hum Genet 1: 539-559

Kruglyak L, Nickerson DA (2001) Variation is the spice of life. Nat Genet 27: 234-236

Kwok P-Y (2000) High-throughput genotyping assay approaches. Pharmacogenomics 1: 95-100

Kwok P-Y (2001) Methods for genotyping single nucleotide polymorphisms. Annu Rev Genom Hum Genet 2: 235-258

Lander ES, Schork NJ (1994) Genetic dissection of complex traits. Science 265: 2037-2048

Livak KJ (1999) Allelic discrimination using fluorogenic probes and the 5' nuclease assay. Genet Anal 14: 143-149

Lizardi PM, Huang X, Zhu Z, Bray-Ward P, Thomas DC, Ward DC (1998) Mutation detection and single-molecule counting using isothermal rolling-circle amplification. Nat Genet 19: 225-232

Margulies M, Egholm M, Altman WE, Attiya S, Bader JS, Bemben LA, Berka J et al. (2005) Genome sequencing in microfabricated high-density picolitre reactors. Nature 437: 376-380

Mauricio R, Stahl EA, Korves T, Tian D, Kreitman M, Bergelson J (2003) Natural selection for polymorphism in the disease resistance gene *Rps2* of *Arabidopsis thaliana*. Genetics 163: 735-746

Mockler TC, Ecker JR (2005) Applications of DNA tiling arrays for whole-genome analysis. Genomics 85: 1-15

Myers S, Bottolo L, Freeman C, McVean G, Donnelly P (2005) A fine-scale map of recombination rates and hotspots across the human genome. Science 310: 321-324

Nordborg M, Tavare S (2002) Linkage disequilibrium: What history has to tell us. Trends Genet 18: 83-90

Nordborg M, Hu TT, Ishino Y, Jhaveri J, Toomajian C, Zheng H, Bakker E, Calabrese P, Gladstone J, Goyal R, Jakobsson M, Kim S, Morozov Y, Padhukasahasram B, Plagnol V, Rosenberg NA, Shah C, Wall JD, Wang J, Zhao K, Kalbfleisch T, Schulz V, Kreitman M, Bergelson J (2005) The pattern of polymorphism in *Arabidopsis thaliana*. PLoS Biol 3: 1289-1299

Paabo S (2003) The mosaic that is our genome. Nature 421: 409-412

Padhukasahasram B, Wall JD, Marjoram P, Nordborg M (2006) Estimating recombination rates from single-nucleotide polymorphisms using summary statistics. Genetics 174: 1517-1528

Pinkel D, Segraves R, Dudar D, Clark S, Poole I, Kowbel D, Collins C, Kuo W-L, et al. (1998) High resolution analysis of DNA copy number variation using comparative genomic hybridization to microarrays. Nat Genet 20: 207-211

Pritchard JK, Stephens M, Donnelly P (2000) Inference of population structure using multilocus genotype data. Genetics 155: 945-959

Redon R, Ishikawa S, Fitch KR, Feuk L, Perry GH, Andrews TD et al. (2006) Global variation in copy number in the human genome. Nature 444: 444-454

Reich DE, Cargill M, Bolk S, Ireland J, Sabeti PC (2003) Linkage disequilibrium in the human genome. Nature 411: 199-204

Richards S, Liu Y, Bettencourt BR, Hradecky P, Letovsky S, Nielsen R, Thornton K, Hubisz MJ, et al. (2005) Comparative genome sequencing of *Drosophila pseudoobscura*: chromosomal, gene, and *cis*-element evolution. Genome Res 15: 1-18

Risch NJ (2000) Searching for genetic determinants in the new millennium. Nature 405: 847-856

Schmid KJ, Ramos-Onsins S, Ringys-Beckstein H, Weisshaar B, Mitchell-Olds T (2005) A multilocus sequence survey in *Arabidopsis thaliana* reveals a genome-wide departure from a neutral model of DNA sequence polymorphism. Genetics 169: 1601-1615

Shiu SH, Bleecker AB (2001) Plant receptor-like kinase gene family: Diversity, function, and signaling. Science STKE 2001: re22

Song WY, Wang GL, Chen LL, Kim HS, Pi LY, Holsten T, Gardner J, Wang B, Zhai WX, Zhu LH, Fauquet C, Ronald P (1995) A receptor kinase-like protein encoded by the rice disease resistance gene, *Xa21*. Science 270: 1804-1806

Stahl EA, Dwyer G, Mauricio R, Kreitman M, Bergelson J (1999) Dynamics of disease resistance polymorphism at *Rpm1* locus of Arabidopsis. Nature 400: 667-671

Stephens JC, Schneider JA, Tanguay DA, Choi J, Acharya T, et al. (2001) Haplotype variation and linkage disequilibrium in 313 human genes. Science 293: 489-493

Svanvik N, Stahlberg A, Sehlstedt U, Sjoback R, Kubista M (2000) Detection of PCR products in real time using light-up probes. Anal Biochem 287: 179-182

Thomas JH (2006) Adaptive evolution in two large families of ubiquitin-ligase adapters in nematodes and plants. Genome Res 16: 1017-1030

Tian D, Araki H, Stahl E, Bergelson J, Kreitman M (2002) Signature of balancing selection in Arabidopsis. Proc Natl Acad Sci USA 99: 11525-11530

Tishkoff SA, Dietzsch E, Speed W, Pakstis AJ, Kidd JR, Cheung K, et al. (1996) Global patterns of linkage disequilibrium at the *CD4* locus and modern human origins. Science 271: 1380-1387

Toomajian C, Hu TT, Aranzana MJ, Lister C, Tang C, Zheng H, Zhao K, Calabrese P, Dean C, Nordborg M (2006) A nonparametric test reveals selection for rapid flowering in the Arabidopsis genome. PLoS Bi

Tuskan GA, DiFazio S, Jansson S, Bohlmann J, Grigoriev I, Hellsten U, P N, et al. (2006) The genome of black cottonwood, *Populus trichocarpa* (& Gray). Science 313: 1596-1604

Tyagi S, Bratu DP, Kramer FR (1998) Multicolor molecular beacons for all discrimination. Nat Biotechnol 16: 49-53

Venter JC, Adams MD, Myers EW, Li PW, Mural RJ, Sutton GG, Holt RA (2001) The sequence of the human genome. Science 291: 1304-1351

Voight BF, Kudarvalli S, Wen X, Pritchard JK (2006) A map of recent positive selection in the human genome. PLoS Biol 4: e72.

Warren RL, Sutton GG, Jones SJM, Holt RA (2006) Assembling millions of short DNA sequences using SSAKE. Bioinformatics 23: 500-501

Weiss KM, Terwilliger JD (2000) How many diseases does it take to map a gene with SNPs? Nat Genet 26: 151-157

Winzeler EA, Castillo-Davis CI, Oshiro G, Liang D, Richards DR, Zhou Y, Hartl DL (2003) Genetic diversity in yeast assessed with whole-genome oligonucleotide arrays. Genetics 163: 79-89

Wright SI, Gaut BS (2005) Molecular population genetics and the search for adaptive evolution in plants. Mol Biol Evol 22: 506-519

Yu J, Buckler ES (2006) Genetic association mapping and genome organization of maize. Curr Opin Biotechnol 17: 155-160

Yu J, Pressoir G, Briggs WH, Vroh Bi I, Yamasaki M, Doebley J, McMullen MD, Gaut BS, Nielsen DM, Holland JB, Kresovich S, Buckler ES (2006) A unified mixed-model method for association mapping that accounts for multiple levels of relatedness. Nat Genet 38: 203-208

Yu J, Holland JB, McMullen MD, Buckler ES (2008) Genetic design and statistical power of nested association mapping in maize. Genetics 178: 539-551

Zondervan KT, Cardon LR (2004) The complex interplay among factors that influence allelic association. Nat Rev Genet 5: 89-100

13 Transcriptional Profiling

Sam R. Zwenger and Chhandak Basu*
School of Biological Sciences, University of Northern Colorado, Greeley, Colorado 80639, USA
*Corresponding author: chhandak.basu@unco.edu

1 INTRODUCTION

Within the last decade, there has been rapid advancement in expression profiling of all organisms, including plants. The study of transcripts, primarily those of mRNA origin, has become one of many 'omics' characters to appear on the stage of molecular biology. Transcriptomics, the specific discipline, which aims to characterize organisms' transcripts, is heavily involved in understanding the patterns of gene expression. These patterns are variable under certain environmental conditions, for example, during different life stages, or within a particular tissue. Research has come to show that although housekeeping genes may be constitutively expressed, other genes are not necessarily expressed all of the time (e.g., stress-response genes). While one might only focus on an organism's mRNA transcripts, other factors (promoters, cis- and trans-acting elements, transcription factors, enhancers, siRNA etc.) are necessarily included in the broader definition of transcriptomics.

Although plant transcriptomics research has begun to reveal basic environmental stress-responses, evolutionary trends and growth and development, fertile ground remains for exploratory and unique experiments designed to better understand unique transcriptional states and biochemical pathways. Often, libraries representing an entire transcriptional state can be obtained, which has led to the emergence of databases, such as the expressed sequence tag database. In addition, advancements in the computational power and *in silico* methods, necessarily accompanied by wet lab data, have played a major role in understanding plant transcription. These can often be separated into two categories: single gene expression profiling and global expression profiling. Global expression profiling can be broken down further into sequence-based profiling and hybridization-based profiling. Incorporating more than one profiling technique bolsters a research experiment. The steps of transcriptional profiling include searching for or identifying a set of genes in which the researcher is interested, choosing a profiling technique, and searching for significant patterns. This chapter explores necessary technology and methods, which have helped to facilitate a better understanding of plant transcriptomics. In addition, future techniques and technologies are covered, although as many can attest to, the pace at which technology has been progressing means that the tools used in molecular biology can rapidly change.

2 SINGLE GENE-BASED TRANSCRIPTIONAL PROFILING METHODS

2.1 Northern Blot

Northern blot, first described by Alwine et al. (1977), is a single gene expression profiling method. In this technique, RNA is extracted from the plant and separated via gel electrophoresis. To prohibit the formation of RNA secondary structures a formaldehyde gel is often used. A positively charged membrane is then applied to the gel, transferring the RNA to the membrane. Labeled probes are created through reverse transcribing target RNA that incorporates the proper amount of either radioactive dNTPs (e.g., ^{32}P) or dNTPs conjugated to a non-radioactive, fluorescent molecule (e.g., digoxigenin). Additionally, labeled probes can be ordered pre-made through a variety of companies. Following a series of wash steps, the labeled probes are added to the membrane. Adequate time, typically overnight, is allowed for the labeled probes to hybridize with the gene of interest, if it is present on the membrane. The membrane can then be viewed and will depend on the method of labeling used (radioactive or non-radioactive). The presence of bands can indicate the gene of interest was expressed in the experiment. Subsequently, the membrane can be stripped and re-probed, if necessary (Mi et al. 2008).

Although first described in 1980, plant research laboratories continue to find useful applications for northern blot; often it is in order to confirm gene expression. Frizzi et al. (2008) confirmed expression of an endosperm-specific promoter in transgenic corn through the use of northern blot analysis. They isolated RNA from 25-day old kernels and used polymerase chain reaction (PCR) products from cDNA as probes (radioactively labeled). Mizuno and Yamashino (2008) investigated the relationship of hormone-associated gene expression and photoperiodism. They used northern blot to help confirm abscisic acid and methyl jasmonate related genes were indeed related to circadian rhythm in plants. Bu et al. (2008) also investigated the role of jasmonate in plant physiology. However, they studied jasmonate's relationship to plant defense signaling. They used northern blot to verify over-expression of an important transcription factor (ANACO19).

2.2 RNA Dot Blot

Many variations of northern blot exist, such as RNA dot blot. This technique is similar to the northern blot but avoids separating the RNA based on size and instead relies on total RNA samples being spotted directly onto the positively charged membrane. Nasmith et al. (2008) used RNA dot blot to determine the prevalence of Dutch elm disease (*Ophiostoma novo-ulmi*), an important fungal pathogen that has decimated American elm tree (*Ulmus americana*) populations. Upon obtaining RNA samples from leaf midribs, they used northern blot to assay the presence or absence of fungal marker genes.

2.3 Reverse Northern Blot

In another variation of northern blot termed reverse northern blot, genomic DNA fragments or target cDNAs are fixed onto a membrane. An RNA probe is then incorporated for determining presence of a gene. Li et al. (2008) used reverse northern to help determine the regulation mechanisms of an important fungal pathogen (*Sclerotinia sclerotiorum*) of plants.

Hsu et al. (2008) used an RNA dot blot and a reverse dot blot analysis in their study of gibberellin-related genes in developing lily (*Lilium longiflorum*) anthers. A reverse dot blot uses cDNA as a probe. It has been demonstrated that reverse transcribing RNA to make cDNA helps overcome some of the difficulties associated with working with RNA (e.g., degradation, sensitivity and stability) (Jaakola et al. 2001).

2.4 Differential Display

In a technique known as differential display, which is similar to northern analysis, mRNA is isolated and random oligo dT primers are used to amplify the various mRNA transcripts (Liang and Pardee 1992). Advantages of using differential display include being able to analyze the expression levels of two or more genes and doing so without having prior sequence knowledge. Kavitha and Thomas (2008) used differential display to better understand the resistance mechanisms of ginger (*Zingiber zerumbet*) to soft rot (*Pythium aphanidermatum*). They used 68 combinations of primers to help find potential genes involved in ginger defense mechanisms. Hongbo et al. (2008) used differential display to characterize gene expression of *Chrysanthemum* species subjugated to sound stress (a form of mechanical stress). They used 100-decibel sound waves before freezing plant tissue in liquid nitrogen and subsequent mRNA extraction. The methods used, namely differential display accompanied with northern dot blot, allowed for determination of the turning on or off of stress-induced genes. Their results also helped confirm the standing hypothesis that sound waves can influence cell division rates, and thus, plant growth in general.

2.5 PCR Methods

Expression levels of a particular plant gene can be analyzed and quantified through the use of quantitative real-time polymerase chain reaction (qRT-PCR). This technique amplifies an unknown amount of DNA starting material using a properly designed primer, as in classical PCR. Built into the PCR machine is a fluorescence detector, which measures intensity of fluorescence as the PCR product accumulates. The first fluorescent dye to be used was ethidium bromide. However, this has been used to a lesser degree and more commonly a cyanine dye such as SYBR green is used, which often gives a better signal to noise ratio. Both the dyes report the accumulation of product by binding to DNA. Ethidium bromide intercalates between the bases of DNA, while SYBR green I binds to DNA's small groove. By measuring the fluorescence through different cycle times of the PCR, a graph can be constructed, which conveys the amount of double-stranded DNA present and the hybridization efficiency. Unfortunately, this method may have interpretation errors. For example, ethidium bromide and SYBR green I might bind to primer dimers, contaminant DNA, or other foreign materials in the reaction tube.

In a study by Stepanova et al. (2008), auxin catabolic pathway was better understood after creating a mutant line of *Arabidopsis* and then analyzing expression of the gene of interest through qRT-PCR methods. Giraud et al. (2008) investigated light and drought stress in the *alternative oxidase1a* (*aox1a*) mutant of *Arabidopsis* and used qRT-PCR to confirm expression of stress-related genes. Others have used qRT-PCR to study genes involved with leaf morphology (Kimura et al. 2008) and finding associations between cold-induced genes and circadian clock in *Arabidopsis* (Bieniawska 2008). Readers should also refer to Fleige and Pfaffl (2006) for a useful review on how RNA quality influences qRT-PCR.

A common probe used in qRT-PCR is TaqMan, where a flourophore reporter at the 5′ end and a quencher dye at the 3′ end of the probe are incorporated. Due to the proximity of the two ends, the fluorescence of the 5′ end is quenched by the 3′ end. After annealing and extension cycles, the 5′ exonuclease activity of *Taq* polymerase cuts the probe, which allows fluorescence of the 5′ end by distancing the quencher. The reporter dye signal is recorded through each cycle given in real time results. Another approach similar to TaqMan is to use molecular beacon probes. In this method, a hairpin probe (shaped like a small beacon) has a fluorophore at one end and a quencher at another end. As stated previously, because of the proximity of the two ends, the fluorescence of the 5′ end is quenched by the 3′ end. However, upon annealing to a target sequence, the distance increases and fluorescence is permitted. Hybridization probes can also be used and these consist of two oligonucleotides, each with a flourophore as described above. Each oligo must bind within 1-5 nucleotides of the other oligo. Fluorescence is only given off, if the two probes are within the necessary distance.

Several researcher groups have used these techniques. For example, Tambong et al. (2008) used ligation-mediated PCR with TaqMan and designed a protocol to detect the presence and/or levels of genetically modified content in maize. In a more comprehensive study that discusses detection of genetically modified organisms, Gasparic et al. (2008) compared different real-time PCR techniques (i.e., SYBR Green, Light Upon eXtension, Plexor, Locked Nucleic Acids, and Cycling Probe Technology). It should be noted here that pathogen attack can also be studied using TaqMan. For example, Hongyun et al. (2008) investigated cucumber green mottle mosaic virus and determined that as few as 50 RNA transcript molecules provided adequate detection. They also outlined a relatively quick method for future cucumber green mottle mosaic virus outbreaks, or perhaps other harmful pathogens.

qRT-PCR has been applied in global expression profiling. The method behind analyzing multiple transcripts rests on the fact that fluorescent probes can be constructed using different fluorophores, thereby allowing multiple amplifications simultaneously. However, it is important to remember that each target may not have the same efficiency of hybridization.

2.6 Ribonuclease Protection Assay

Ribonuclease protection assay (RPA) works on the principle that certain enzymes like RNAseA will not degrade double stranded RNA, whereas single stranded RNA will be degraded. After adding a labeled RNA probe to a pool of RNA, hybridization might occur, upon which RNAseA is added and is unable to degrade the target-probe hybrid. RPA is not limited to nuclear gene expressions; it has been used to confirm presence of sequence-specific ribosomal RNA (Matyasek et al. 2007). Additionally, Hacham et al. (2006) used RPA as well as qRT-PCR to investigate the N-terminus region of cystathionine γ-synthase, an enzyme involved in the first stage of methionine biosynthesis in higher plants.

Asakura and Barkan (2007) discussed how the CRM (chloroplast RNA splicing and ribosome maturation) domain, which has been identified as an RNA binding domain, participates in three different classes of RNA molecules (group I and II introns and the 23S rRNA). They used RPA and an RNA dot blot to help monitor the splicing of an intron (*rpoC1*) in *Arabidopsis* plants. Understanding these splicing pathways is sure to help understand transcriptional fates.

3 GLOBAL EXPRESSION PROFILING

High-throughput methods provide the ability to simultaneously analyze the expression of thousands of genes. Global expression profiling techniques include sequence-based profiling and hybridization-based profiling and each are distinct yet interrelated methods of multiplexing (i.e. studying more than one gene). Multiplexing has pushed technology and innovation to provide new methods, each with distinct advantages. Coughlan et al. (2004) has provided a review of these techniques and their applications in *Arabidopsis* and an even more recent review has been provided by Vega-Sanchez et al. (2007). Global expression profiling can yield overwhelming amounts of data, which in turn has led to development of computer-based analysis of data (bioinformatics). High-throughput technologies have decreased the time required to run a reaction and the number of reactions running simultaneously. It is important to realize these methods can be either open or closed systems. Open systems allow for discovery of previously uncharacterized genes while closed systems are limited to analysis of known genes or gene sets.

3.1 Expressed Sequence Tags

Sequencing-based profiling of gene expression involves determining the sequence of a transcript and relating that to previously determined functions. Although there are multiple methods for sequencing, all give essentially the same result. However, rather than sequencing an entire genome, an expressed sequence tag (EST) library can be constructed. The EST library can be taken from a pool of cDNA that originally came from reverse transcribing extracted mRNA. A single EST sequence represents a partial mRNA transcript and only needs to be a few hundred base pairs in length. That is to say, the EST sequence can be a partial sequence read of the mRNA. The library, therefore, represents only genes, or parts of genes that are expressed at the time of mRNA extraction.

Often, an EST library is referred to as a 'poor man's genome', referring to the relative inexpensiveness compared to sequencing an entire genome (Rudd 2003). Upon extraction of total RNA, the mRNA must be separated from other RNA molecules such as the tRNA, rRNA, mtRNA and other RNA molecules. This can be accomplished because of the presence of the 3′ polyA tail attached during post-translational modifications to yield the mature mRNA. Polythymidine-beads (oligo dTs), which have been synthesized and attached to small magnetic beads or a column, hybridize to the 3′ polyA tail of the mRNA. The hybridized molecules can be retained and a series of salt-specific washes removes all molecules except the mRNA. The final elution, therefore, will contain the desired transcripts.

The mRNA is reverse transcribed making a double stranded cDNA molecule, which is much more stable than its RNA counterpart. This occurs through a process of first strand and second strand synthesis, respectfully. Next the cDNA can be ligated (cloned) into a vector for further manipulation such as sequencing. The cDNA sequence is used as a reference as to which genes were expressed at that given time period. The expression then, can be tissue-specific or specific for that developmental stage. According to NCBI's (National Center for Biotechnology Information, Bethesda, Maryland, USA) EST database (*dbEST* at *http://www.ncbi.nlm.nih.gov/dbEST/)* 62,344, 458 ESTs have been deposited (as of

Table 1 Examples of some plant ESTs deposited in the EST database. (Source: *http://www.ncbi.nlm.nih.gov/dbEST/dbEST_summary.html*)

Organism	*Number of ESTs*
Zea mays (maize)	2,018,634
Arabidopsis thaliana (thale cress)	1,527,298
Glycine max (soybean)	1,386,618
Oryza sativa (rice)	1,248,955
Triticum aestivum (wheat)	1,066,998
Brassica napus (oilseed rape)	632,344
Hordeum vulgare subsp. *vulgare* (barley)	501,366
Panicum virgatum (switchgrass)	436,535
Phaseolus coccineus (runner bean)	391,138
Vitis vinifera (wine grape)	353,941

July 10, 2009) from an array of organisms including many plant species. Table 1 shows the top ten plants (by rank) with current EST library construction obtained from dbEST release 071009.

3.2 Serial Analysis of Gene Expression (SAGE)

Serial analysis of gene expression (SAGE) is based on a similar process to creating an EST library but has the advantage of using much shorter tags from the transcripts, decreasing the expense for sequencing (Velculescu et al. 1995). RNA transcripts are reverse transcribed using a biotin labeled oligo-dT primer. The 3′ polyA end of the transcript is useful because it helps to distinguish it from similar transcripts, as that part of the transcript is less conserved, and thus more diverse. The total group of transcripts is divided into two separate groups. These are ligated with special linkers, which have recognition cut sites that allows later cutting with a restriction enzyme to generate the SAGE tag attached to the linker. After ligating two tags by their ends, ditags result, which can be PCR amplified and then digested with the restriction enzyme *Nla*III, which removes the linkers, leaving the ditags. Ditags can then be concatemerized to form joined ditags. The ditags are inserted into a plasmid vector and sequenced. Because the tags become incorporated into a cloning vector as a long sequence composed of multiple tags, a continuous flow of information is produced; 25-50 tags can be read simultaneously on a DNA sequencer lane.

SAGE has not only reduced costs in sequencing, but has shown to be more effective at producing a full cDNA sequence. This is because SAGE can sequence multiple transcripts in a single pass. In addition to determining expression of a gene, SAGE tags can also measure transcript abundance by correlating number of times the tag is present in a sequencing run.

Variations of SAGE have emerged and can be categorized based on the length of the tag created. One of these is LongSAGE, developed by Saha et al. (2002). This uses the restriction enzyme *Mme*I, which creates tags up to 21 bp and seems to be more efficient than the original SAGE. Robust LongSAGE (RL-SAGE) was an adaptation of LongSAGE, where fewer PCR reactions are needed to produce nearly twice as many tags than LongSAGE (Gowda et al. 2006).

An even more recent technique, SuperSAGE, uses a distinct restriction enzyme, *EcoP*15I, to create 26 bp fragments, which have been shown to have greater reliability when pairing up with ESTs in public databases (Matsumura et al. 2005).

3.3 Massively Parallel Sequencing Signatures

An alternative to using the SAGE method is using massively parallel sequence signatures (MPSS) (Nakano et al. 2006). This is a combination of sequencing and hybridization processes. Developed by Nobel Laureate Sydney Brenner while at Lynx therapeutics in Colorado, this technology is able to count each individual transcript. It relies on amplifying, cloning, and sequencing millions of transcripts simultaneously, hence the descriptor, 'in parallel'. High-throughput sequencing methods (next generation sequencing technologies) have excelled in obtaining plant sequence information. The most well known methods are briefly discussed below.

3.4 Next Generation Sequencing

Many high-throughput, massively parallel, 'non-Sanger based' (Schuster 2008) sequencing technologies have emerged known as next generation sequencing. The next generation sequencing technology is pioneered by three technologies, namely, 454 sequencing technology by Roche Applied Sciences, Solexa genome analysis system by Illumina, and SOLiD system by Applied Biosystems. Each system approaches the sequencing differently; 454 uses pyrosequencing, Solexa uses bridge amplification, and ABI uses a ligation approach. For a more comprehensive discussion on these technologies see Chapter 5. However, a mention is given here because these technologies can be used in transcriptional profiling, such as sequencing cDNA libraries.

3.4.1 454 Sequencing

This sequencing method relies on pyrosequencing, a method that has been described extensively in many sequencing review papers (Diggle and Clarke 2004; Ahmadian 2006; Hudson 2008). In this method, the polonies or polymerase colonies are attached to ~30 μm beads. A base-specific chemiluminescence is released when DNA polymerase (three additional enzymes are used as well) incorporates a nucleotide. 454 sequencing has been used heavily in re-sequencing projects, where one genome of the species has previously been sequenced. Importantly, 454 sequencing can generate hundreds of thousands of ESTs

It has been used somewhat in plant transcriptomics studies by Wicker et al. (2006). Weber et al. (2007) used it to sample the transcriptome of two *Arabidopsis* seedlings. Upon analysis of sequence reads the authors found only a few of the sequences were of mitochondrial or chloroplast origin, indicating the importance of purifying mRNA prior to sequencing. In addition to obtaining longer sequence reads, 454 sequencing also yielded more reads than traditional Sanger sequencing. In regard to the quality and reliability, Barbazuk et al. (2007) reported similar benefits in searching for SNPs. They sequenced two lines (B73 and Mo17) of maize shoot apical meristems and created an expression library with 260,887 and 287,917 ESTs, respectively.

3.4.2 Solexa

This company's sequencing technology uses an 8-channel flow cell for a sequencing surface, where clusters of DNA are randomly located. Sequence detection relies on a 4-channel fluorescence reaction. It also relies on the sequencing by synthesis approach, but uses a reversible terminator.

This has not been as extensively used as 454 sequencing, primarily because it has not been available nearly as long. However, the technology has been implicated in finding unique RNAs in *Arabidopsis* (Lu et al. 2005). Sequencing technologies can indeed be used for a single research project. For example, Addo-Quaye et al. (2008) used both 454 and Solexa sequencing in their study of the *Arabidopsis* degradome.

Solexa was one of many companies present at the European Science Foundation-Wellcome Trust Conference on Crop Genomics, Trait

Analysis and Breeding (Hinxton, UK), which focused on using plant genomics to enhance crop plants (Bevan and Waugh 2007). Weber et al. (2007) used shotgun bisulphate sequencing to determine the methylome of *Arabidopsis*. Methylation patterns can be one of the factors of epigenomics, controlling gene expression.

3.4.3 ABI SOLiD

Applied Biosystems uses sequencing by oligonucleotide ligation and detection (SOLiD). The support used for this closed system technology is a ~1 µm bead which are attached to a sequencing surface consisting of a single slide. It relies on 4-channel florescence detection and ligation of sequence-specific labeled oligos for its sequencing chemistry.

3.4.4 Increasing Throughput in the Future

The main goal of many of the above mentioned sequencing technologies is to increase throughput and decrease cost. Indeed, the general trend of the last decade has seen a drop in sequencing price. A recent breakthrough in sequencing throughput has been from Helicos. This company's system (called HeliScope) works by fragmenting nucleotide molecules into 100-200 base strands. After addition of a polyA primer and fluorescing tag, the strands are hybridized to immobilized oligo dTs. Due to its high density arrangement, 100 million templates per centimeter are attached. The sample is loaded into the Helicos sequencing machine allowing billions of bases to be sequenced in a single run. This technology claims it has also achieved a never before seen 'single molecule sequencing'.

Perhaps the most interesting prospect is sequencing in real time. In a recent report by Eid et al. (2009), they observed the real-time activity of individual DNA polymerase molecules. The sequencing technology only seems to be getting more interesting and appears to be promising for transcriptional profiling.

3.5 Rapidly Amplified cDNA Ends

Rapidly amplified cDNA ends (RACE) is the process of reverse transcribing RNA molecules to produce cDNA. Those are then amplified through PCR and the sequence is determined. RACE allows for amplification of a transcript for which only part of the sequence is known, with specific interest in the 3′ or 5′ end. Therefore, depending on the primers used, either the 5′ or 3′ end of the transcript can be sequenced.

This also means that unknown transcript sequences might be able to be determined using RACE, if a homologous transcript is thought to be present in another organism. For instance, Yu et al. (2008) determined a presumably novel gene involved in wheat (*Triticum aestivum*) hypersensitivity response to stripe rust. After a BLAST search to find a sequence with high similarity in barley (*Hordeum vulgare*), they developed primers specific for the homologous gene in wheat. They subsequently used 3′ and 5′ RACE to develop full-length cDNAs.

RACE has also been used in plant stress studies. Zdunek-Zastocka (2008) studied pea (*Pisum sativum*) and used RACE to help characterize three aldehyde oxidase genes. Xie et al. (2005) characterized *MIRNA* genes and used 5′ RACE as part of their study to help identify miRNA transcripts in *Arabidopsis*. The results provided a broader understanding of miRNA and gene silencing in plants.

3.6 MicroRNA Expression Profiling

Some species of RNA molecules have become recognized as a player in transcription, and has led to a new understanding of gene silencing. This has prompted further investigation into sequencing the genes or the microRNAs (miRNAs). The miRNA molecules function by hybridizing with mRNA, allowing for degradation by a degradation pathway, which destroys the mRNA. The existence of these molecules has stimulated research in the area of miRNA functioning, yielding interesting results. For example, some miRNA molecules have been shown to have the same sequence in plants as in animals (Hikosaka et al. 2007). Fahlgren et al. (2007) sequenced a pool of RNA molecules from *Arabidopsis* and used a bioinformatics approach to identify miRNA molecules in order to find new miRNA loci. Based on their sequence results, they hypothesized that there might be a frequent fluctuation of miRNA genes disappearing and reappearing. The miRNA platforms for performing arrays, similar to microarrays (described below), have been developed. Although miRNA is found in small amounts making it difficult to study them, there have been methods recently developed to analyze less than a femtomole of small RNA in tissues from less than 1 μg of RNA sample. Shingara et al. (2005) have provided such description of a platform in humans and it will be interesting to see if can be applied to plants for an even more in depth understanding of transcript silencing.

Similar to miRNA molecules, yet differing in the mode of activity, siRNA (short interfering RNA) can cause gene silencing. Kasschau et al. (2007) have analyzed wild type and mutant *Arabidopsis* lines to reveal genome-wide influences of siRNA. Their results provide a somewhat better understanding of how the location of the siRNA genes might influence gene expression.

3.7 nCounter

Future technology will be expected to sequence the total nucleotides present in transcripts. This desire, to more fully understand all sequences in a cell, has led to deep sequencing and ultra-deep sequencing (identifies more than 99% of the nucleotides). Undoubtedly, this will lead to a more thorough understanding of previously undetected workings of a cell.

More recently, a technique called nCounter, developed by NanoString Technologies, has emerged (Fortina and Surrey 2008). This has shown to be as sensitive or, perhaps, even more sensitive than current technologies, including microarrays and qPCR, for transcript profiling. What makes this technology unique is that it allows for detection of a fractional fold changes in transcript levels. In addition, an increase in the multiplexing capability lowering the cost even further than previously described next generation technologies. The nCounter Analysis System is currently being marketed by NanoString Technologies.

3.8 Microarray

Microarray (Schena et al. 1995) is a technology to measure and analyze the expression of thousands of genes expressed at any given time. This makes microarray a whole transcriptome analysis technology as it captures a great majority of the cell's mRNA molecules.

In microarray production, short nucleotide sequences are immobilized on a glass surface (e.g., on a glass microscope slide) by robots. These types of microarray chips can be purchased from many sources (Box 1). The nucleotide sequences can be either oligonucleotide sequences representing genes of an organism or cDNA sequences of an organism. The cDNA sequence-based microarrays are relatively inexpensive. Affymetrix produces oligonucleotide based microarrays under the GeneChip trademark. Fluorescently labeled cDNAs from 'control' and 'test' tissues are synthesized and hybridized on the glass slide with 'spotted genes' on it (Fig. 1). cDNAs can be labeled by many types of dyes including cyanidine dyes (Cy-3 and Cy-5). The slide is then scanned with laser, which excites the dyes on the spots and the fluorescence emitted from each spot is captured using a computer for

Box 1 Sources of microarray chips for various plant species.

***Arabidopsis*:**
Affymetrix: *http://www.affymetrix.com/products/arrays/specific/arab.affx*
ArrayIt: *http://arrayit.com/Products/Microarrays/*
Harvard University: *http://chip.dfci.harvard.edu/pricing.php*
University of Arizona: *http://www.ag.arizona.edu/microarray/*

Barley:
Affymetrix: *http://www.affymetrix.com/products/arrays/specific/barley.affx*
Harvard University: *http://chip.dfci.harvard.edu/pricing.php*

Cotton:
Affymetrix: *http://www.affymetrix.com/products/arrays/specific/cotton.affx*

Grape:
who: *http://www.affymetrix.com/products/arrays/specific/vitis.affx*
Harvard University: *http://chip.dfci.harvard.edu/pricing.php*

Maize:
Affymetrix: *http://www.affymetrix.com/products/arrays/specific/maize.affx*
Iowa State University: *http://www.plantgenomics.iastate.edu/maizechip/*
Harvard University: *http://chip.dfci.harvard.edu/pricing.php*

Medicago:
Affymetrix: *http://www.affymetrix.com/products/arrays/specific/medicago.affx*
Harvard University: *http://chip.dfci.harvard.edu/pricing.php*

Poplar:
Affymetrix: *http://www.affymetrix.com/products/arrays/specific/poplar.affx*
Picme: *http://www.picme.at*

Pine:
Picme: *http://www.picme.at*

Rice:
Affymetrix: *http://www.affymetrix.com/products/arrays/specific/rice.affx*
GreenGene Bio Tech: *http://www.ggbio.com/*
Harvard University: *http://chip.dfci.harvard.edu/pricing.php*

Soybean:
Affymetrix: *http://www.affymetrix.com/products/arrays/specific/soybean.affx*
Harvard University: *http://chip.dfci.harvard.edu/pricing.php*

Sugarcane:
Affymetrix: *http://www.affymetrix.com/products/arrays/specific/sugarcane.affx*
Harvard University: *http://chip.dfci.harvard.edu/pricing.php*

Tomato:
Affymetrix: *http://www.affymetrix.com/products/arrays/specific/tomato.affx*
Harvard University: *http://chip.dfci.harvard.edu/pricing.php*

Wheat:
Affymetrix: *http://www.affymetrix.com/products/arrays/specific/wheat.affx*
Harvard University: *http://chip.dfci.harvard.edu/pricing.php*

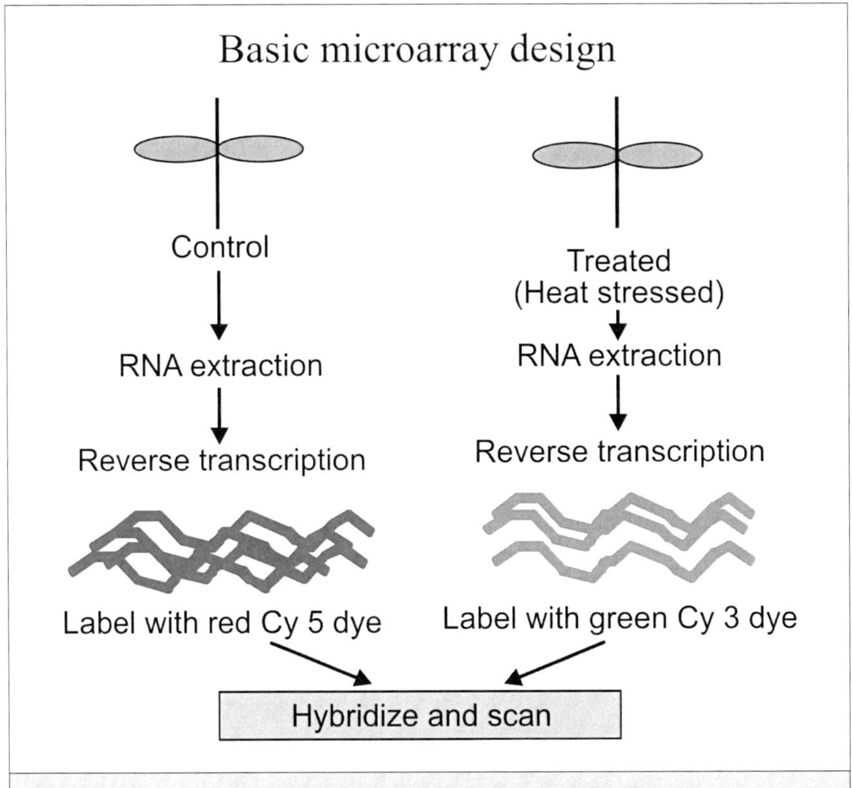

Fig. 1 Schematic representation of design of a basic microarray experiment.

data analysis. The fluorescence intensities of the spots on the microarray correspond to gene activities of a cell.

Microarray publications should follow some standardized procedures known as MIAME (Minimum information about a microarray experiment; Brazma et al. 2001). Although microarrays provide a lot of information about gene expression, this technology does have some limitations. One of the disadvantages of microarray is that this technology is not suitable for detection of novel genes; the gene sequence information must already be known. All microarray experiments must be validated by real-time reverse transcriptase PCR, northern blot, ribonuclease protection assay, etc. A poor hybridization on the slide may also generate false positive results. Inspite of these limitations microarray remains a popular choice for global transcriptional profiling in plants.

4 PLANT BIOINFORMATICS AND TRANSCRIPTION ANALYSIS

Bioinformatics is a new dynamic filed of science, which involves applications of tools of omics, software programs, molecular genetics, etc. to understand gene structure and function. Bioinformatics is extensively used in various genome sequencing projects.

National Institute of Health (*http://www.bisti.nih.gov/)* defines bioinformatics as: "Research, development, or application of computational tools and approaches for expanding the use of biological, medical, behavioral or health data, including those to acquire, store, organize, archive, analyze, or visualize such data". Tools of bioinformatics can be successfully used to understand the mechanisms of transcription, gene structure and function, analysis of *cis* and *trans* acting elements in DNA, etc. A comprehensive information about plant bioinformatics database is available (Edwards 2008). Some of the freely available online tools and databases, which could be used for plant transcriptional analysis, are mentioned here.

4.1 TAIR Database: Resource for *Arabidopsis*

The TAIR (The *Arabidopsis* Information Resource) database at *www.arabidopsis.org* is a comprehensive database of *Arabidopsis* genomic information. The database includes the complete *Arabidopsis* genome sequence along with various tools for gene annotations. One can browse various stocks of mutant seeds germplasms, cDNAs, etc. through this website. Some of the useful tools available in this website are mentioned below.

GBrowse: This is a genome browser. The user can search for promoter elements, orthologs, transposable elements associated with a gene.

AraCyc: This is a useful tool, which gives a graphical view of a biochemical pathway.

Chromosome map tool: This feature allows the user to pinpoint gene/s on a chromosome and generate an image of chromosomes with the gene/s of interest.

The TAIR website also contains microarray data obtained from innumerable experiments and published results. Another very valuable genome annotation tool available is 'Plant Promoter and Regulatory Element Resources' at *http://www.arabidopsis.org/portals/genAnnotation/genome_annotation_tools/cis_element.jsp*. This webpage is a comprehensive resource for many valuable genome annotation tools including transcription factor binding sites, *cis* acting regulatory elements including promoters etc.

4.2 Plant Genome Central: Comprehensive Resource for Many Plants

The Plant Genome Central website (*http://www.ncbi.nlm.nih.gov/genomes/PLANTS/PlantList.html*) hosted by NCBI contains information about partially and fully sequenced plant genome, plant EST databases, genetic maps etc.

4.3 Gramene: The Grass Gene Database

The Gramene (*http://www.gramene.org/*) database is an open source resource for grass genomics including wheat, rice, barley, oat, foxtail millet, wild rice, sorghum and *Brachypodium*. One can search for genes and proteins in this database. The 'Plant Metabolic Pathways' resource allows the user to study various metabolic pathways of many plants including grasses.

4.4 HarvEST: EST Database for Many Plants

HarvEST (*http://harvest.ucr.edu*) is an EST database for many plant species including barley, *Brachypodium*, rice, coffee, soybean, wheat, and cowpea. HarvEST can be downloaded onto a personal computer, avoiding the need for an Internet connection.

Besides the above examples, Table 2 contains information about additional plant gene expression databases available over the Internet.

5 CONCLUSIONS

Future plant transcriptional analysis might combine different types of transcript profiling techniques for the most reliable results (e.g., comparative expression profiling in a plant species derived from frequency analysis of ESTs and MPSS). Undoubtedly, it is increasingly likely that, given the pace of sequencing and transcript profiling capabilities, newer technologies will continue to increase the throughput capacities. This will have several far-reaching effects on molecular biology and transcriptomics in particular. First, a more complete understanding of the transcriptional states of plants will emerge.

From what we have described in this chapter, and is evident throughout this book, our knowledge on the inner workings of plants

Table 2 Examples of some plant gene expression databases

Name of the database and website address	*Plants covered*	*Comment about the database*
PED (Plant Gene Expression Database) *http://bioinfo.ucr.edu/projects/Unknowns/external/express.html*	*Arabidopsis*	Microarray expression data
PLEXdb (Plant Expression Database) *http://www.plexdb.org/*	Barley, Rice, Maize, Citrus, Grape, Cotton, Medicago, Poplar, Soybean, Tomato, Wheat	Gene expression data for plants and plant pathogens
RMD (Rice Mutant Database) *http://rmd.ncpgr.cn/*	Rice	Information of about 129,000 T-DNA insertion rice lines
SGED (Soybean Expression Database) *http://aragon.ncsa.uiuc.edu/soybean-microbe*	Soybean	Microarray expression data
SoyXpress http://soyxpress.agrenv.mcgill.ca	Soybean	377,095 soybean EST collection
ACT (*Arabidopsis* Co-expression Tool) *http://www.arabidopsis.leeds.ac.uk/act/coexpanalyser.php*	*Arabidopsis*	Database for analysis for co-expression of genes from microarray data
Genevestigator *http://www.genevestigator.ethz.ch/*	*Aarabidopsis* and rice	Microarray expression data
Plant Gene Indices *http://compbio.dfci.harvard.edu/tgi/plant.html*	Many plants	EST collection and gene information

Contd.

Table 2 continued

Name of the database and website address	*Plants covered*	*Comment about the database*
AthaMap *http://www.athamap.de/*	*Arabidopsis*	Transcription factor binding sites
WhETS: Wheat Estimated Transcript Server *http://www4.rothamsted.bbsrc.ac.uk/whets/cgi-bin/whets1.2/whets_home.pl*	Wheat	Estimation of a wheat transcript sequence
YOGY: eukarYotic OrtholoGY *http://www.sanger.ac.uk/PostGenomics/S_pombe/YOGY/*	*Arabidopsis*	Detection of orthologous proteins from *Arabidopsis* and other nine eukaryotes and one prokaryote
GDR (Genome Database for Rosaceae) *http://www.bioinfo.wsu.edu/gdr/*	Rosaceae family plants including almond, peach and strawberry	EST database and physical maps

has steadily increased. Transcriptional profiling will play a large role, perhaps in the entire human community, as we seek to understand plants and their responses to environmental stress, for example climate change. Additionally, the methods described in this chapter will give insight into plant metabolic pathways and might allow tapping into plant secondary metabolite pathways to produce desired plant products. The limits to what can be achieved through plant biotechnology and genetic engineering are also expanding, with much credit owed to our understanding of transcriptional profiling.

Disclaimer

Many software, tools, technologies mentioned in this chapter may be proprietary. The authors declare no competing interests and are not affiliated with any of the developers of these products. This chapter was written for academic and educational purposes only.

References

Addo-Quaye C, Eshoo TW, Bartel DP, Axtell MJ (2008) Endogenous siRNA and miRNA targets identified by sequencing of the *Arabidopsis* degradome. Curr Biol 18(10): 758-762

Ahmadian A, Ehn M, Hober S (2006) Pyrosequencing: history, biochemistry and future. Clin Chim Acta 363(1-2): 83-94

Alwine JC, Kemp DJ, Stark GR (1977) Method for detection of specific RNAs in agarose gels by transfer to diazobenzyloxymethyl-paper and hybridization with DNA probes. Proc Natl Acad Sci USA 74(12): 5350-5354

Asakura Y, Barkan A (2007) Arabidopsis orthologs of maize chloroplast splicing factors promote splicing of orthologous and species-specific group II introns. Plant Physiol 142(4): 1656-1663

Barbazuk WB, Emrich SJ, Chen HD, Li L, Schnable PS (2007) SNP discovery via 454 transcriptome sequencing. Plant J 51(5): 910-918

Bevan M, Waugh R (2007) Applying plant genomics to crop improvement. Genome Biol 8(2): 302

Bieniawska Z, Espinoza C, Schlereth A, Sulpice R, Hincha DK, Hannah MA (2008) Disruption of the *Arabidopsis* circadian clock is responsible for extensive variation in the cold-responsive transcriptome. Plant Physiol 147(1): 263-279

Brazma A, Hingamp P, Quackenbush J, Sherlock G, Spellman P, Stoeckert C, Aach J, Ansorge W, Ball CA, Causton HC, et al. (2001) Minimum information about a microarray experiment (MIAME)-toward standards for microarray data. Nat Genet 29(4): 365-371

Bu Q, Jiang H, Li C, Zhai Q, Zhang J, Wu X, Sun J, Xie Q, Li C (2008) Role of the *Arabidopsis thaliana* NAC transcription factors ANAC019 and ANAC055 in regulating jasmonic acid-signaled defense responses. Cell Res 18(7): 756-767

Coughlan SJ, Agrawal V, Meyers B (2004) A comparison of global gene expression measurement technologies in *Arabidopsis thaliana*. Comp Funct Genom 5(3): 245-252

Diggle MA, Clarke SC (2004) Pyrosequencing: sequence typing at the speed of light. Mol Biotechnol 28(2): 129-137

Edwards D (ed) (2008) Plant Bioinformatics: Methods and Protocols. Humana Press, Totowa, New Jersey, USA

Eid J, Fehr A, Gray J, Luong K, Lyle J, Otto G, Peluso P, Rank D, Baybayan P, Bettman B (2009) Real-time DNA sequencing from single polymerase molecules. Science 323: 133-138

Fahlgren N, Howell MD, Kasschau KD, Chapman EJ, Sullivan CM, Cumbie JS, Givan SA, Law TF, Grant SR, Dangl JL, Carrington JC (2007) High-throughput sequencing of *Arabidopsis* microRNAs: evidence for frequent birth and death of MIRNA genes. PLoS ONE 2(2): e219

Fleige S, Pfaffl MW (2006) RNA integrity and the effect on the real-time qRT-PCR performance. Mol Aspects Med 27(2-3): 126-139

Frizzi A, Huang S, Gilbertson LA, Armstrong TA, Luethy MH, Malvar TM (2008) Modifying lysine biosynthesis and catabolism in corn with a single bifunctional expression/silencing transgene cassette. Plant Biotechnol J 6(1): 13-21

Fortina, P and Surrey, S (2008) Digital mRNA profiling. Nat Biotechnol 26: 293-294

Gasparic MB, Cankar K, Zel J, Gruden K (2008) Comparison of different real-time PCR chemistries and their suitability for detection and quantification of genetically modified organisms. BMC Biotechnol 8: 26

Giraud E, Ho LHM, Clifton R, Carroll A, Estavillo G, Tan Y, Howell KA, Ivanova A, Pogson BJ, Millar AH, Whelan J (2008) The absence of ALTERNATIVE OXIDASE1a in *Arabidopsis* results in acute sensitivity to combined light and drought stress. Plant Physiol 147(2): 595-610

Gowda M, Li H, Alessi J, Chen F, Pratt R, Wang G (2006) Robust analysis of 5'-transcript ends (5'-RATE): a novel technique for transcriptome analysis and genome annotation. Nucl Acids Res 34(19): e126

Hacham Y, Schuster G, Amir R (2006) An in vivo internal deletion in the N-terminus region of *Arabidopsis* cystathionine gamma-synthase results in CGS expression that is insensitive to methionine. Plant J 45(6): 955-967

Hikosaka A, Takaya, K, Jinno M, Kawahara A (2007) Identification and expression-profiling of *Xenopus tropicalis* miRNAs including plant miRNA-like RNAs at metamorphosis. FEBS Lett 581(16): 3013-3018

Hongbo S, Biao L, Bochu W, Kun T, Yilong L (2008) A study on differentially expressed gene screening of *Chrysanthemum* plants under sound stress. Crit Rev Biol 331(5): 329-333

Hongyun C, Wenjun Z, Qinsheng G, Qing C, Shiming L, Shuifang Z (2008) Real time TaqMan RT-PCR assay for the detection of Cucumber green mottle mosaic virus. J Virol Meth 149(2): 326-329

Hudson ME (2008) Sequencing breakthroughs for genomic ecology and evolutionary biology. Mol Ecol Resour 8: 3-17

Hsu Y, Tzeng J, Liu M, Yei F, Chung M, Wang C (2008) Identification of anther-specific/predominant genes regulated by gibberellin during development of lily anthers. J Plant Physiol 165(5): 553-563

Jaakola L, Pirttilä AM, Hohtola A (2001) cDNA blotting offers an alternative method for gene expression studies. Plant Mol Biol Rep 19: 125-128

Kasschau KD, Fahlgren N, Chapman EJ, Sullivan CM, Cumbie JS, Givan SA, Carrington JC (2007) Genome-wide profiling and analysis of *Arabidopsis* siRNAs. PLoS Biol 5(3): e57

Kavitha PG, Thomas G (2008) Defence transcriptome profiling of *Zingiber zerumbet* (L.) Smith by mRNA differential display. J Biosci 33(1): 81-90

Kimura S, Koenig D, Kang J, Yoong FY, Sinha N (2008) Natural variation in leaf morphology results from mutation of a novel KNOX gene. Curr Biol 18(9): 672-677

Li H, Fu Y, Jiang D, Li G, Ghabrial SA, Yi X (2008) Down-regulation of *Sclerotinia sclerotiorum* gene expression in response to infection with *Sclerotinia sclerotiorum* debilitation-associated RNA virus. Virus Res 135(1): 95-106

Liang P, Pardee AB (1992) Differential display of eukaryotic messenger RNA by means of the polymerase chain reaction. Science 257(5072): 967-971

Lu C, Tej SS, Luo S, Haudenschild CD, Meyers BC, Green PJ (2005) Elucidation of the small RNA component of the transcriptome. Science 309: 1567-1569

Matsumura H, Ito A, Saitoh H, Winter P, Kahl G, Reuter M, Krüger DH, Terauchi R (2005) SuperSAGE. Cell Microbiol 7(1): 11-18

Matyásek R, Tate JA, Lim YK, Srubarová H, Koh J, Leitch AR, Soltis DE, Soltis PS, Kovarík A (2007) Concerted evolution of rDNA in recently formed Tragopogon allotetraploids is typically associated with an inverse correlation between gene copy number and expression. Genetics 176(4): 2509-2519

Mi S, Cai T, Hu Y, Chen Y, Hodges E, Ni F, Wu L, Li S, Zhou H, Long C, Chen S, Hannon GJ, Qi Y (2008) Sorting of small RNAs into *Arabidopsis* argonaute complexes is directed by the 5' terminal nucleotide. Cell 133(1): 116-127

Mizuno T, Yamashino T (2008) Comparative transcriptome of diurnally oscillating genes and hormone-responsive genes in *Arabidopsis thaliana:* Insight into circadian clock-controlled daily responses to common ambient stresses in plants. Plant Cell Physiol 49(3): 481-487

Nakano M, Nobuta K, Vemaraju K, Tej SS, Skogen JW, Meyers BC (2006) Plant MPSS databases: signature-based transcriptional resources for analyses of mRNA and small RNA. Nucl Acids Res 34: D731-735

Nasmith C, Jeng R, Hubbes M (2008) A comparison of *in vivo* targeted gene expression during fungal colonization of DED-susceptible *Ulmus Americana*. For Pathol 38: 104-112

Rudd S (2003) Expressed sequence tags: alternative or complement to whole genome sequences? Trends Plant Sci 8(7): 321-329

Saha S, Sparks AB, Rago C, Akmaev V, Wang CJ, Vogalstein B, Kinzler KW, Velculescu VE (2002) Using the transcriptome to analyze the genome. Nat Biotechnol 19: 508-512

Schena M, Shalon D, Davis RW, Brown PO (1995) Quantitative monitoring of gene expression patterns with a complementary DNA microarray. Science 270(5235): 467-70

Schuster SC (2008) Next-generation sequencing transforms today's biology. Nat Meth 5: 16-18

Shingara J, Keiger K, Shelton J, Laosinchai-Wolf W, Powers P, Conrad R, Brown D, Labourier E (2005) An optimized isolation and labeling platform for accurate microRNA expression profiling. RNA 11(9): 1461-1470

Stepanova AN, Robertson-Hoyt J, Yun J, Benavente LM, Xie D, Dolezal K, Schlereth A, Jürgens G, Alonso JM (2008) TAA1-mediated auxin biosynthesis is essential for hormone crosstalk and plant development. Cell 133(1): 177-191

Tambong, JT, Mwange, KN, Bergeron, M, Ding, T, Mandy, F, Reid, LM, Zhu X. (2008) Rapid detection and identification of the bacterium Pantoea stewartii in maize by TaqMan real-time PCR assay targeting the cpsD gene. J Appl Microbiol 104(5): 1525-1537

Vega-Sanchez ME, Gowda M, Want G (2007) Tag-based approaches for deep transcriptome analysis in plants. Plant Sci 173: 371-380

Velculescu VE, Zhang L, Vogelstein B, Kinzler KW (1995) Serial analysis of gene expression. Science 270(5235): 484-487

Weber APM, Weber KL, Carr K, Wilkerson C, Ohlrogge JB (2007) Sampling the Arabidopsis transcriptome with massively parallel pyrosequencing. Plant Physiol 144(1): 32-42

Wicker T, Schlagenhauf E, Graner A, Close TJ, Keller B, Stein N (2006) 454 sequencing put to the test using the complex genome of barley. BMC Genom 7: 275

Xie Z, Allen E, Fahlgren N, Calamar A, Givan SA, Carrington JC (2005) Expression of *Arabidopsis* MIRNA genes. Plant Physiol 138(4): 2145-2154

Yu X, Yu X, Qu Z, Huang X, Guo J, Han Q, Zhao J, Huang L, Kang Z (2008) Cloning of a putative hypersensitive induced reaction gene from wheat infected by stripe rust fungus. Gene 407(1-2): 193-198

Zdunek-Zastockax E (2008) Molecular cloning, characterization and expression analysis of three aldehyde oxidase genes from *Pisum sativum* L. Plant Physiol Biochem 46(1): 19-28

14 Advanced Bioinformatics Tools and Strategies

Matthew A. Hibbs
The Jackson Laboratory, 600 Main Street, Bar Harbor, ME 04609, USA
E-mail: matt.hibbs@jax.org

1 INTRODUCTION

Computational approaches are becoming increasingly important for many aspects of biology as we generate exponentially growing amounts of data at many levels, from genome sequences to protein structures to whole-genome expression studies to population dynamics. The sheer volume and size of the datasets being generated requires analysis approaches very different from those applied in traditional biology settings. In many respects, computer science is a field devoted to studying scalability, and as such, it is well suited to bring many insights and advances to the problems faced in modern biology.

In this chapter, we will describe and explore many aspects of computational biology and bioinformatics, but we will do so at a level targeted for biologists interested in understanding and applying these techniques in their own works. As such, we will not focus on specific (though important) technical or algorithmic details, but we will provide references and links where interested readers can gain a deeper understanding of the topics that are covered in this chapter. First, we will look at what bioinformatics and computational biology are at a high level (i.e., What problems can computational approaches help address or not help address? What areas are ripe for further computational study?). Second, we will discuss supervised machine learning methods for classification and prediction, which have been successfully applied to biological problems ranging from clinical diagnosis to functional genomics. Third, we will examine unsupervised approaches for data exploration and organization, including clustering techniques. And lastly, we will discuss future directions for this field and how computationalists and biologists can work together to address these problems.

2 A PRIMER ON COMPUTATIONAL BIOLOGY

Computational biology and bioinformatics are very broad fields covering a variety of problems and solutions. The main thread binding all of these works together is the marriage of a biological motivation with the use of computational tools. While there is no universally agreed upon distinction, here we will consider bioinformatics to be primarily concerned with the organization of biological information (e.g., databases, ontologies, etc.) whereas computational biology is focused on investigating specific biological issues using computational approaches. In this chapter, we will focus on the later of these two categories. (However, a discussion of bioinformatics resources can be found in Chapter 9, Volume I of this book series.)

Defined in this way, computational biology can largely be thought of as techniques that utilize data in order to draw inferences about an area of interest. Abstractly, this process is very similar to an experimental biologist making observations, forming new hypotheses, and devising new experiments. As such, some of the most useful computational tools for biology stem from the field of *artificial intelligence*, such as machine learning and data mining techniques. These tools are especially valuable in situations where it is difficult or impossible for a single biologist to understand or interpret the available data due to its scale and scope. This issue has become progressively more important over the past decade, as high-throughput whole-genome techniques have grown in popularity, cost effectiveness, and availability.

While this explosion in the volume of data production is a relatively recent phenomenon for biology, computer science has been addressing issues of scalability for decades. In fact, much of theoretical computer science research is focused on developing approaches and algorithms that can scale to increasingly large amounts of data (Cormen et al. 2001). One active area of computational research relevant to biological problems is in *computability*, i.e. what problems can and cannot be calculated in a tractable amount of time. The major intuition here is that for many general problems (e.g., sorting a list of objects, finding the shortest path through a network) there are provable bounds on the amount of time required to find the solution. These bounds are typically expressed as functions of the number of objects being processed (i.e., to sort n objects requires at least $n \log n$ units of time). However, for some classes of problems, the fastest known algorithms require too much time to handle even moderately large numbers of objects. In Table 1, we show some common lower bounds on processing time and how those bounds scale as the number of objects increases.

Table 1 Examples of computational bounds and their scalability.

Bound	$n = 2$	$n = 10$	$n = 100$	$n = 1000$	$n = 10000$
$n \log_2 n$	2	33	664	9966	132877
n^2	4	100	10000	10^6	10^8
n^3	8	1000	10^6	10^9	10^{12}
2^n	4	1024	$\sim 10^{30}$	$\sim 10^{300}$	∞
$n!$	2	$\sim 10^6$	$\sim 10^{158}$	∞	∞

From this small table, we can see that these bounds can make an incredible difference in the amount of time required to perform a calculation as the number of objects increases. Note that there is a vast difference between the bounds in the first three rows of this table (referred to as polynomial bounds) and the last two rows (non-polynomial bounds). In general, problems with bounds in the polynomial class are considered possible to compute, even for large numbers of objects; while problems with bounds that are non-polynomial are largely intractable, since for even modestly high numbers of objects the amount of time required becomes prohibitively large. As an example, one classic problem with a surprisingly large time bound is known as the 'traveling salesman problem,' where our goal is to find the shortest route traveling through n cities where each city is visited at least once (Schrijver 2005). At first, it may seem that this should be easy, but as the number of cities becomes larger, the number of possible paths that visit each city becomes very large—in fact, there are on the order of $n!$ possible paths. The fastest known solutions to this problem are still non-polynomial, hence it belongs to a class known as 'NP-hard' problems. To give some intuition for this scale, if every grain of sand on Earth were as powerful as the world's fastest supercomputer, their combined power would need more than a thousand years to find the exact solution for the traveling salesman problem with 1,000 cities.

There are a large number of important problems that fall into this class, but for even moderately large values of n, the exact solutions to these problems are impractical to compute. When dealing with biological problems, the number of objects often corresponds to the number of nucleotides or genes in a genome; for numbers this large, non-polynomial algorithms are clearly intractable, and even fast growing polynomial bounds (e.g., n^4, or even n^3) can prove impractical. As such, it is important to realize the limits of what problems can be exactly solved for biological applications.

Despite these disturbingly long bounds for exact solutions, if we are willing to have solutions that are 'almost' right and/or 'probably' correct, then we can use a number of heuristics and tricks to practically find solutions. For example, in the case of the traveling salesman problem, polynomial time algorithms have been developed that can provide solutions whose distances are within a reasonable factor of the optimal solutions with high probability (Rosenkrantz et al. 1977). Many important biological problems have solutions that employ heuristics in order to actually achieve solutions, but it is important to consider the impact of these heuristics in a particular area of interest. For example, some heuristics convert continuously valued data (e.g., gene expression levels) into binary data (e.g., on/off) in order to calculate approximate solutions. In some cases, this procedure will produce good results, but in other cases this assumption throws away too much information to be practically useful.

There are many biologically useful algorithms that are either polynomial or have solid heuristics that allow for their successful application to biology (some of which are discussed below). However, an important area of current research is in developing heuristics and approximations that are motivated from the biology in order to achieve tractable running times as well as biologically relevant solutions. Developing these types of algorithms typically requires collaborations between computationalists and biologists in order to meet all of the goals required. Despite these challenges, many computational approaches have proven their utility for biology over the past decade, including the areas of machine learning and data mining, which we discuss in the next sections.

3 SUPERVISED MACHINE LEARNING FOR CLASSIFICATION AND PREDICTION

The basic premise of *machine learning* techniques is that given known examples we can form rules about those examples that are generalizable to new examples. For instance, given a collection of microarray data and a list of genes known to be involved in cell cycle progression, we can form a picture of what the expression pattern of a cell cycle gene looks like. Then, given the expression profile of a gene of unknown function, we can determine if that gene resembles the cell cycle pattern or not.

In this section we will discuss *supervised* methods for *classification* and prediction. The classification problem is defined as predicting membership of an example in two or more classes (e.g., deciding if a gene is cell cycle regulated, or not cell cycle regulated). In general, a supervised machine learning method requires example vectors of data whose classes are known. Each data point of the example vectors is often called a *feature* (e.g., gene expression measurement). *Labels* are associated with each data vector to represent which known examples fall into which class (e.g., cell cycle, or not cell cycle). Supervised classification methods are given labeled *training data* from which they attempt to learn rules, and then those rules are applied to new examples to predict their label. There are a variety of methods for supervised classification, but here we will examine three of the most popular: Decision Trees, Support Vector Machines (SVMs) and Bayesian Networks.

These classification methods can be applied to a broad range of biological data and questions. For example, functional genomics researchers often make predictions of the role or function of individual genes given the available high-throughput data for those genes. In this case we can treat functional categories, such as those defined by the Gene Ontology (Ashburner et al. 2000) as labels for training. We can also construct a vector of data points for each gene from sources such as gene expression data, protein-protein interaction data, genetic interaction data, etc. Several laboratories have developed systems based on classification approaches to generate high quality predictions of protein function for a wide variety of organisms. The same general approach can be applied to any situation where suitable data and labels are available, such as predicting physical protein interactions, drug targets, or clinical diagnoses.

3.1 Decision Trees

Decision trees are one of the simplest types of supervised learning algorithm, but they are still very powerful and useful. Through a process called *induction*, decision trees are able to learn a collection of nested rules from example data that can be used to classify new examples (Breiman et al. 1984). Decision trees can be used for classification with an arbitrary number of categories, but they are most often applied in situations where there are exactly two possible data labels (e.g., yes/no, 0/1, etc.). The vectors of data used by decision trees can be continuous, but for simplicity we will consider the common case where the input data is *discrete*, meaning that each feature has a fixed number of possible values that it can take. The rules learned by decision tree induction are organized in a nested fashion much like a flowchart.

For illustration, we will examine a sample collection of training data and the decision tree that can be learned from that data. Suppose we wish to classify whether or not proteins localize to the mitochondria (i.e., our labels will be yes or no), given a small collection of gene expression data under a few conditions. We will also assume that we have discretized these expression values into three classes: low, medium, and high. So our labeled training data would have the form shown in Table 2.

Table 2 Example labeled training data for a decision tree.

Gene ID	*Label*	*Data 1*	*Data 2*	*Data 3*	*Data 4*
A	Yes	Low	Low	Medium	High
B	Yes	Medium	High	Medium	High
C	Yes	Medium	Medium	High	Low
D	No	Medium	High	High	High
E	No	Low	High	Low	High
F	No	Medium	Low	Low	Low
G	No	High	Medium	Medium	Medium
H	No	Medium	Low	Medium	Low

We will see shortly that a decision tree learned from this training set could look like Fig. 1. Based on this tree, we can see that each of the examples used for training are correctly classified. For example, looking at gene C, we first examine its value for data point 3, since this value is the root of the decision tree. Since gene C has a 'High' value for this data point, we traverse the tree and then examine data point 2, for which

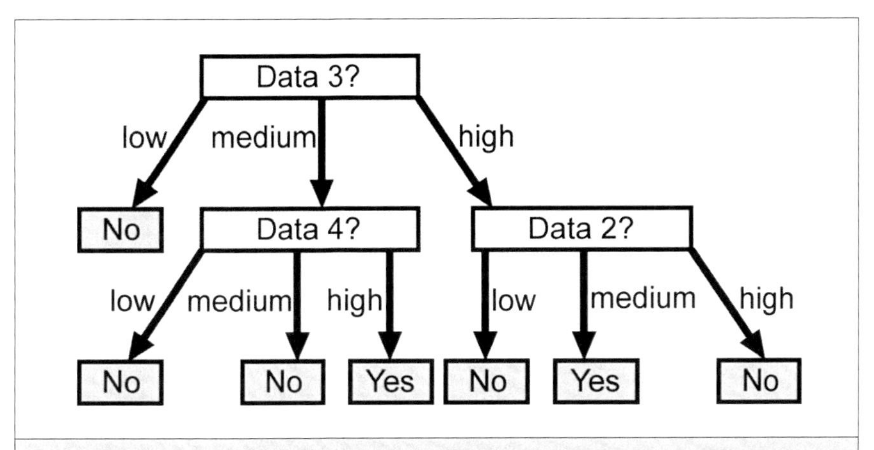

Fig. 1 An example decision tree based on the data in Table 2. Starting with the root node ('Data 3?'), paths are followed until a classification node (Yes/No) is reached.

gene C has a 'Medium' value, thus we would classify gene C as a 'Yes', which is the correct classification given our known labels. Following the tree for all of our training examples produces the correct classification result, so we would say that this tree has perfect *training accuracy*. There are cases, particularly when the training data is ambiguous or contradictory, where a learned decision tree will have less than perfect training accuracy. However, a much better way of assessing how well a decision tree (or any classifier) performs is to learn the classifier from a subset of the available training examples, and evaluate performance on examples that were not used in training. (These types of evaluation are discussed in Section 3.4). Once a decision tree has been created from a training set, it can then be used to classify new, unlabeled examples by following the tree structure in the same fashion as shown in Table 3.

Table 3 Classification results on new data. By following the branches of the tree in Fig. 1, we can classify each of these three examples as shown in the rightmost column.

Gene ID	*Data 1*	*Data 2*	*Data 3*	*Data 4*	*Predicted class*
X	Low	Medium	High	High	Yes
Y	Medium	Medium	Medium	Medium	No
Z	High	High	High	Low	No

The basic algorithm used for learning a decision tree from training data is a recursive process that follows these basic steps:

1. Select the most informative data feature not previously appearing in this branch
2. For each possible value of the selected feature:
   a) If all examples with this value share the same label, place that label in the tree
   b) Otherwise, place a new decision point in the tree by repeating this process from step 1

To perform step 1, the amount of *information* gained by selecting a data feature is typically used to determine which feature is the most informative at each iteration. Here, information refers more specifically to Shannon information, one of the founding concepts of the field of information theory (MacKay 2003). Intuitively, we can think of the information gain as how much closer to an optimal solution a choice will take you. For example, from our training examples, choosing 'Data 3' as the first decision point allows us to immediately classify two training examples as 'No' because all of our examples that were 'Low' at this feature have the 'No' label. If we selected any other feature for the initial decision point, we could not immediately classify two examples (for 'Data 1' and 'Data 4' we could immediately classify one example; for 'Data 2' we could not immediately classify any). In this way, 'Data 3' provides the most information about our labels at this first step.

Following this procedure will result in a decision tree with perfect training accuracy, so long as the training data is not contradictory (i.e. two identical data vectors have different labels). However, following this basic procedure can produce very complicated trees that are tailored so specifically to the training data that they may lose their ability to generalize and accurately classify additional examples. This problem is called *overfitting*, and can occur in almost all classification methods (Russell et al. 2002). We will discuss general methods to assess and combat overfitting in Section 3.4. In the specific case of decision trees, there are several approaches designed to learn a tree that may not have perfect training accuracy, but performs better when presented with new examples for classification. The general premise of these approaches is that simpler trees tend to generalize better. Many of these techniques remove branches that only slightly decrease training accuracy after the tree is constructed; these are called *pruning* approaches (Helmbold and Schapire 1997; Mansour 1997).

Another popular variant of the decision tree method, call... *stumps*, uses just a single decision feature for classification, ...ponding to just the root node of a full decision tree. While a ... decision stump will generally perform poorly, they can be effectiv... used in an *ensemble*, or collection of classifiers. One ensemble approach, called *boosting*, can efficiently use a large collection of decision stumps to form highly accurate classifiers (Freund and Schapire 1999). The basic idea of boosting is to assign a weight to every training example that represents how important it is to correctly classify that training example. Initially, all examples start with the same weight. A decision stump is learned that best classifies the examples, then the weight of all incorrectly classified examples is increased, and the weight of all correctly classified examples is decreased. Then another stump is learned, but this time due to the changes in weights the new stump 'cares more' about correctly classifying the examples that were previously incorrectly classified. By repeating this process of giving higher priority to incorrectly classified examples and creating another decision stump, boosting creates a large ensemble of different stumps. To classify a new example, each stump 'votes' on the new example and the consensus of all the stumps is used to assign a label. It has been shown that boosting can greatly decrease the problems of overfitting.

Decision trees, with and without boosting, have been successfully applied to many biological problems. In general, decision trees are well suited for learning classification rules that are easily interpretable and translatable to meaningful features. For example, several groups have used the trees to create intuitive descriptions and guidelines for clinical decisions (Dettling and Buhlmann 2003; Povalej et al. 2005; Chan et al. 2006) and others have used decision trees to form hypotheses of molecular-level activity, such as gene regulatory programs (Middendorf et al. 2004; Kundaje et al. 2006).

3.2 Support Vector Machines

Support Vector Machines (SVMs) are a powerful and commonly utilized method for classification. The general premise of SVMs is to consider the features or data vectors of an example as a point in very high dimensional space. When learning an SVM, each training example is a labeled point in space, and the goal of learning is to define the best linear boundary between the examples with different labels. Then when presented with a new example, an SVM classifies it based on which side of the boundary the example falls. This basic idea can be combined with

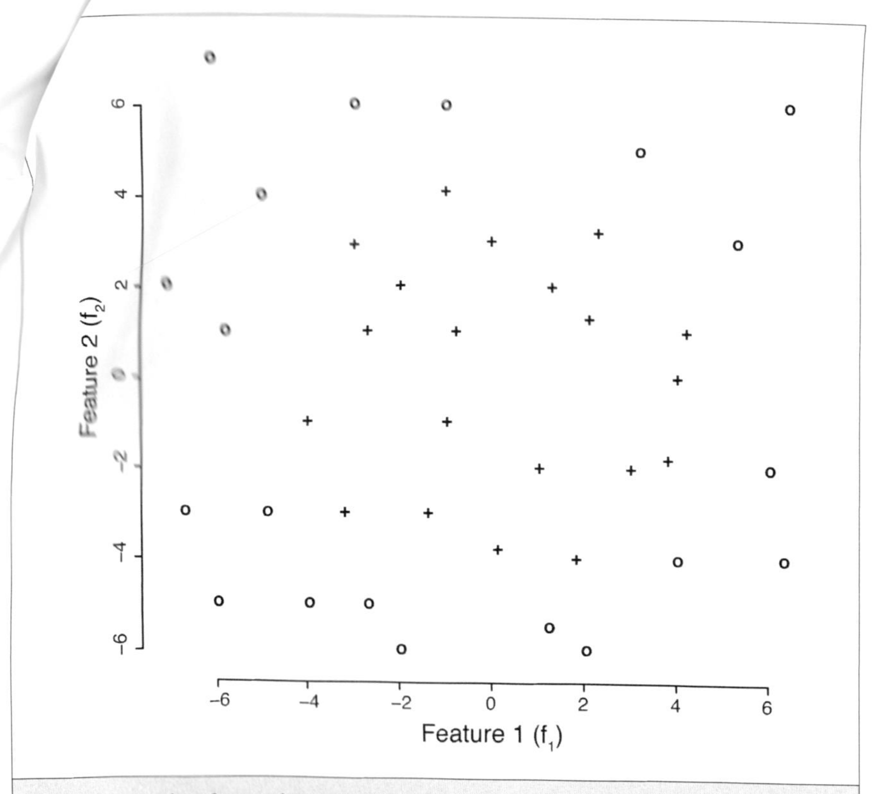

Fig. 2 Example of two-dimensional data vectors that are not linearly separable. Here data vectors consist of two features, one plotted on the *x*-axis and the other on the *y*-axis. Positive examples are shown as '+'s and negative examples are shown as 'o's. There is no single line that can perfectly separate the positive from the negative examples.

a mathematical data transformation (the *kernel trick*) to create highly accurate and extensible classifiers (Vapnik 1995).

To illustrate the idea of SVMs, we will consider a simple example where our data consists of just two features or two dimensions as shown in Fig. 2. In this case, positive examples are shown as '+'s and negative examples are shown as 'o's. An SVM will find the best linear separator between the training examples in the different classes. However, we can see that in this case, there is no straight line that can be drawn to separate the positives from the negatives. This task is not hopeless though. SVMs can perform transformations of the data to project it into a higher dimensional space where the examples are linearly separable.

For example, we can project this data into a three dimensional space where our new dimensions are defined by the formulas:

$$\begin{cases} d_1 = f_1^2 \\ d_2 = f_2^2 \\ d_3 = \sqrt{2} f_1 f_2 \end{cases}$$

Although we will not discuss the mathematical details of SVM kernels here, this transformation corresponds to a two-degree polynomial kernel, which is relatively simple but also powerful and commonly used (Russell et al. 2002). However, for many biological applications, designing a kernel appropriate for a specific problem often produces better results (Noble 2004, 2006).

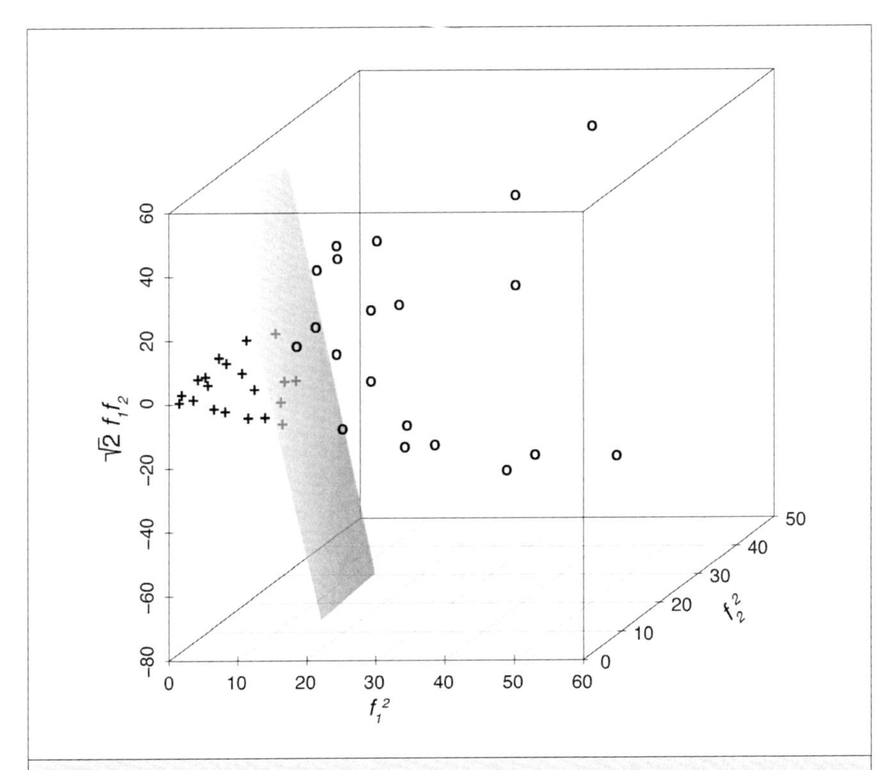

Fig. 3 Example of two-dimensional data vectors projected into a kernel space. These are the same data points as in Fig. 2, but the features have been projected using the kernel formulas shown labeling the three axes. In this space, the positive examples are linearly separable from the negative.

By applying this transformation to our original data, we obtain the three dimensional projection shown in Fig. 3. We can see now that by performing this transformation a linear plane exists that separates the positive examples from the negative examples. In general, for an arbitrary number of dimensions a linear object is called a *hyperplane*. So for cases with more than three features or where those features are projected to a space with more than three dimensions, an SVM finds the optimal separating hyperplane.

This brings us to how SVMs determine which possible separating hyperplane is optimal. In many cases there are a variety of possible separating hyperplanes with equal accuracy applied to the training examples (see Fig. 4A). However, since we plan to use the learned hyperplane to classify new examples, we want to make the separation as generalizable as possible. SVMs do this by choosing the hyperplane that is furthest away from the examples that are closest to the hyperplane (see Fig. 4B). This distance between the closest positive and closest negative examples to the hyperplane is called the *margin* and is the quantity that SVMs maximize to choose the optimal separator. In fact, by finding these examples closest to the boundary, SVMs can use a small number of parameters to define the separating hyperplane. These closest examples are called the *support vectors*, and by attaching a parameter to each of these examples an SVM can fully define the optimal hyperplane without regard to the remaining examples.

The SVM learning algorithm is performed by maximizing a formula related to the size of the margin through a process called *quadratic programming*. The details of this process are beyond the scope of this chapter, but several implementations of this algorithm are freely available and can be applied to many types of data to predict classifications including SVMlight (Joachims 1999) and the Matlab SVM Toolbox (Cawley 2000). It is also common when presented with new examples to calculate their distance to the hyperplane as a proxy for confidence in the classification. For instance, the classification of a new data point that lies very near the hyperplane or within the margin might be interpreted as dubious, while the classification of a data point very far from the hyperplane or well outside of the margin can be thought of as very confident.

SVMs have been applied to a wide variety of problems in computational biology with great success (Noble 2004). This includes work in identifying protein homology (Jaakkola et al. 1999; Ding and Dubchak 2001; Liao and Noble 2003), locating important DNA features (Zien et al.

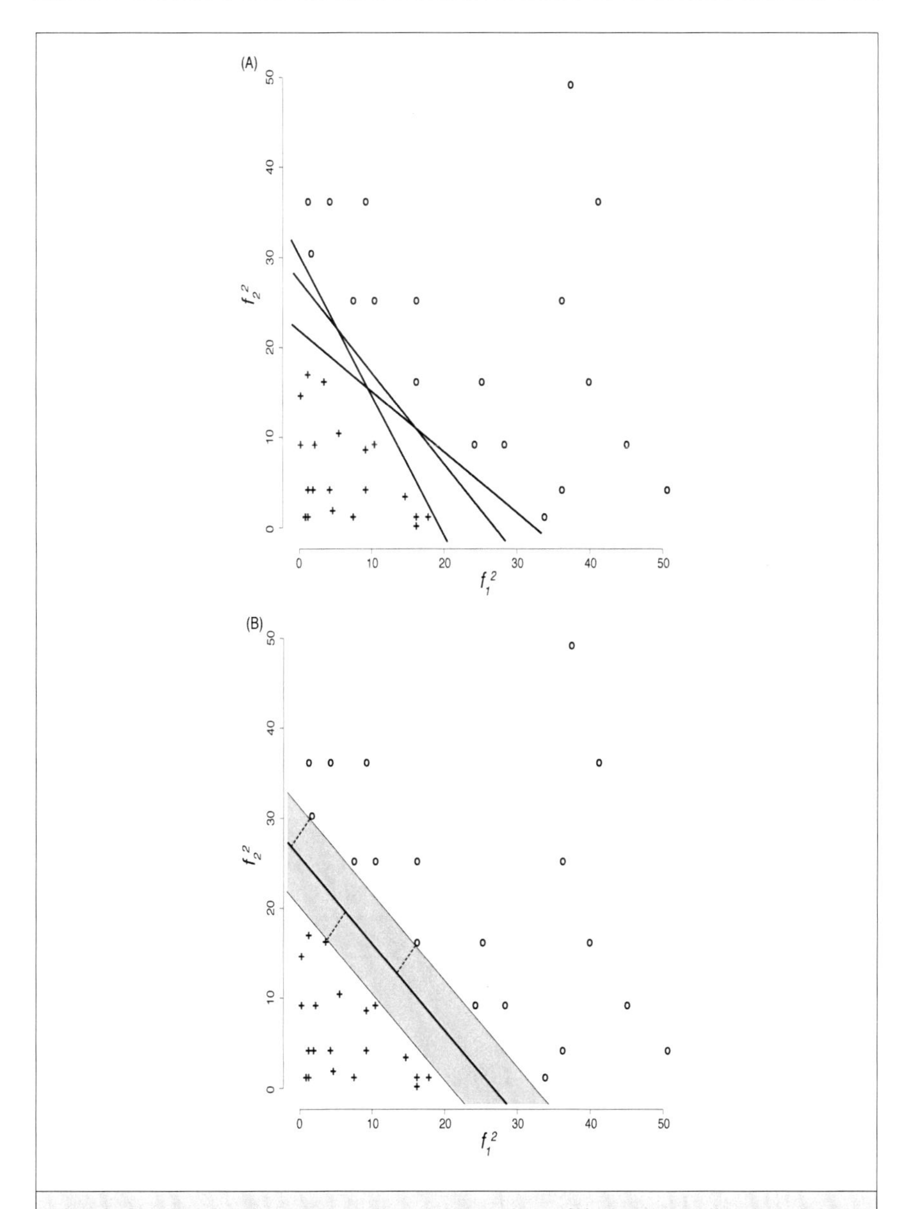

Fig. 4 Choice of the optimal separating hyperplane. (A) shows a collection of three possible classifying lines, each of which perfectly separates positive from negative examples. (B) shows the optimal separating line, chosen by maximizing the distance to the nearest positive and negative examples on each side. The dotted lines indicate which examples are closest to the classifying hyperplane and are used as the support vectors.

2000; Degroeve et al. 2002), predicting functional properties of proteins (Brown et al. 2000; Vert 2002; Lanckriet et al. 2004), and disease diagnosis (Moler et al. 2000; Model et al. 2001; Segal et al. 2003). Much of this work tailors the choice of kernel to the particular problem being investigated, but each of these efforts center on the basic methodology presented here.

3.3 Bayesian Networks

Another technique to perform many types of prediction is *Bayesian networks*, or Bayes nets. These networks are a formal, probabilistic way to organize pieces of knowledge and encode statistical dependence relationships between these pieces of knowledge (Russell et al. 2002). For example, knowing if a protein is localized to the mitochondria or to the nucleus changes our expectation whether or not a protein is involved in the TCA cycle. More formally, a Bayes net is constructed from *nodes* that represent random variables (e.g., protein localization, gene expression, etc.) and *directed edges* between those nodes that represent dependence relationships. Each node has a probability distribution attached to it that quantitatively defines the dependences of the edges. However, edges cannot form cycles in a Bayes net (e.g., A→B, B→C, and C→A cannot all be in the same network), and as such a Bayesian network is a *directed acyclic graph*, or DAG. Once a Bayesian network has been constructed, then given values for some of the nodes in the network (called *evidence* variables or *observed* variables), statistical inference can be performed to predict the remaining nodes (called *hidden* variables or *unobserved* variables).

As an illustrative example, we will examine one biological area where Bayesian networks have been successfully applied, which is in data integration for gene function prediction (Troyanskaya et al. 2003). The goal in this case is to infer the probability that two genes play similar roles, participate in the same pathway, or are otherwise functionally related. As such, we will have a hidden variable node (that we hope to infer) representing whether or not a pair of genes are functionally related. In order to perform this inference, we will also have a number of evidence variable nodes that correspond to specific data that can inform whether a pair of genes are functionally related (e.g., a specific physical binding assay). As some of these specific data sources are assaying the same phenomenon, we can also have a level of intermediate hidden variables that represent slightly more abstract concepts (e.g., whether two proteins physically interact). A simplified example of this type of Bayesian network is shown in Fig. 5.

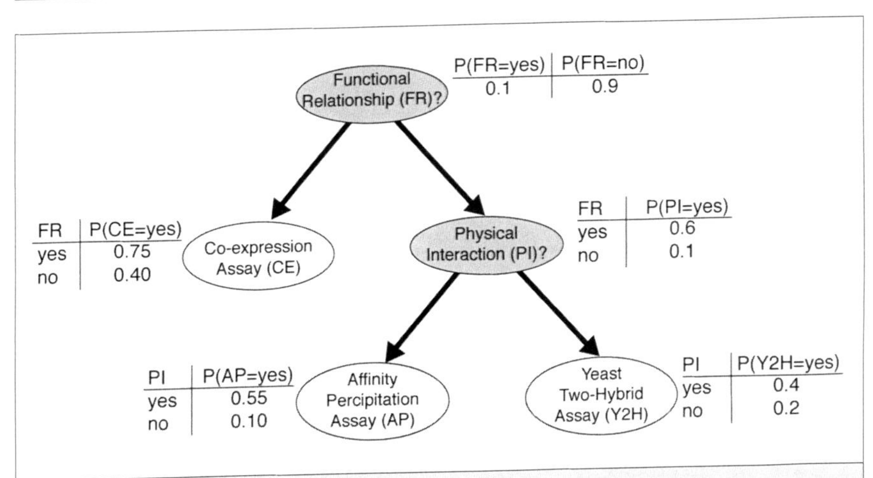

Fig. 5 Example of a simple Bayesian network. Each circle represents a node, or random variable, in the network and each arrow represents an edge, or dependence relationship, in the network. The shaded nodes represent hidden variables, and the white nodes represent evidence variables. Sample conditional probability tables are shown next to each node.

In this example, we have restricted all nodes to take only binary (yes/no) values, though in general Bayesian networks can have multiple discrete values or even continuous values per node. In the discrete case, such as here, each node has a *conditional probability table*, or CPT, attached to it that quantifies the dependency relationships of the edges. For example, for the functional relationship (FR) node here, we have a *prior* probability that there is a 10% chance that a pair of genes has a functional relationship and a 90% chance that a pair of genes is not functionally related. This is called a prior probability because it represents our belief that two genes are functionally related before we consider any additional evidence of that relationship. For the co-expression assay (CE) node, the CPT indicates that for a pair of genes that functionally interact, we expect a 75% chance that those genes will be co-expressed in this assay (and an implied 25% chance they will not co-express). But for a pair of genes that is not functionally related, then we expect just a 40% chance that the genes are co-expressed (and a 60% chance they are not). The CPTs of the remaining nodes in this network encode analogous relationships.

Many of the initial uses of Bayesian networks in computational biology used panels of experts to define the CPTs and edge structure of the network. Much of the initial expert-driven work was performed in clinical diagnosis, but more recently this has expanded to areas such as gene function prediction. While these expert systems have proven valuable, the experts are not always accurate in their decisions about the probabilistic relationships between variables. More recent work has used labeled training examples to empirically learn these CPTs and/or edge structures; we will discuss these learning algorithms shortly. Regardless of how the structure and probabilities are determined, once we have a complete Bayesian network, we can infer the values of hidden nodes given the values of the evidence nodes. For example, if we know that a pair of genes is co-expressed (CE = yes), was interacting in the affinity precipitation assay (AP = yes), and was interacting in the yeast two-hybrid assay (Y2H = yes), then we can infer the probability that the genes are functionally related (FR). Formally, this question can be written as the formula:

$$P(FR \mid CE = yes, AP = yes, Y2H = yes)$$

In order to perform this calculation, we can apply *Bayes' rule*, which is the statistical theorem that relates conditional probabilities and is the namesake of Bayesian networks. As a reminder, Bayes' rule states that $P(A \mid B)$ is proportional to $P(B \mid A)P(A)$. In order to apply this theorem to a general Bayesian network, we must sum over all possible values of the other hidden nodes, so in this case we would obtain the formula:

$$\alpha\ P(FR)\,P(CE = yes \mid FR) \sum_{PI} P(PI \mid FR)\,P(AP = yes \mid PI)\,P(Y2H = yes \mid PI)$$

Where α is a proportionality constant, and the summation over *PI* is the sum over *PI* = yes and *PI* = no. Based on the CPTs in our network, evaluating this formula for the presence of a functional relationship (*FR* = yes) produces 0.0105α, and for *FR* = no we obtain 0.0216α. Since the probabilities of an event occurring and not occurring must sum to 1, we can normalize out the constant α, and obtain that there is a 32.7% chance that our proteins are functionally related, and a 67.3% chance they are not. These are called the *posterior* probabilities since they are determined after the evidence is observed. So for this simple case the chance of a pair of genes being functionally related rises from 10% to 32.7% when these three assays are observed.

We were able to exactly calculate the posterior probabilities in this simple example, but in general for much more complex Bayesian networks, performing this calculation can be impractical. In such situations, we can apply an approximate inference approach, such as a Monte Carlo algorithm (Neal 1993). These algorithms sample from the possible values of all of the hidden nodes based on the CPTs and the observed variables. By sampling the network a very large number of times, probabilities are determined based on the fraction of samples where a node took on a value of interest. For large enough samples and properly formed networks, it can be shown that these types of approaches converge on the result obtained from exactly evaluating the probability equation.

Now that we have seen what a Bayesian network is, and how they can be used, we will consider how they can be automatically learned from data. First, we will consider an even simpler Bayesian network structure as shown in Fig. 6. This is called a *naïve Bayesian network* as it consists of a single hidden variable that all evidence variables are dependent on. These networks are considered 'naïve' because the structure implies that all of the evidence variables are independent of one another, which can be a dangerous (though greatly simplifying) assumption. In such a network, the learning algorithm boils down to simple counting. Given a set of unbiased, labeled training examples, a CPT can be constructed for a hidden node (e.g., FR) by calculating the fraction of training examples with each label. For the evidence nodes (e.g., D1, D2) CPTs can be constructed by counting the fraction of examples with each label that take on each possible data value. For example, assuming a binary network, if out of 1000 training examples, 800 had no functional relationship, then $P(FR = \text{no})$ would be 0.8 and $P(FR = \text{yes})$ would be 0.2. Then if 600 of these 800 training examples had a yes value for evidence data 3, we would assign $P(D3 = \text{yes} \mid FR = \text{no})$ to be 0.75. A similar process could be applied to completely determine all values in the CPTs of a naïve network.

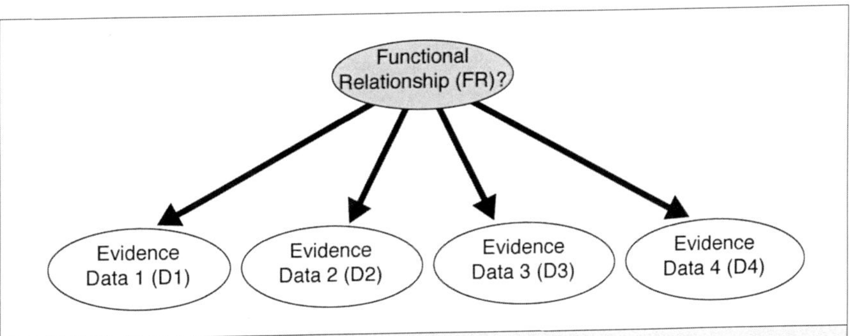

Fig. 6 Example of a naïve Bayesian network structure. In a naïve network there is one root node, and all other nodes are children of that node. Learning and inference are much simpler in a network with this structure.

For more complicated Bayesian networks with interior hidden nodes, learning the probability distributions associated with each hidden variable is more complex. One of the most commonly used approaches to this problem is the *expectation-maximization*, or EM, algorithm. The basic premise of EM is to alternate between an expectation step, which calculates the expected value of the hidden parameters, and a maximization step, which maximizes the likelihood of the data assuming the values from the expectation step are accurate. This alternating process generally continues until the estimated parameters converge. It is more difficult to learn the structure of a Bayesian network, often simply called *structure learning*, but this can also be performed. However, since the space of possible network structures is so large (for a network with n nodes, there are even more than $n!$ possible structures), accurately learning network structures requires a potentially very large number of training examples, which can make it an impractical proposition for many situations. Despite these concerns, there has been some success applying structure learning to relatively small biological networks.

There are several available implementations of various Bayesian learning and inference approaches, including the University of Pittsburgh's SMILE library, the Bayes Net Toolbox (Murphy 2001), and the Sleipnir library (Huttenhower et al. 2008). As previously mentioned, Bayesian networks have been applied to perform clinical diagnoses (Waxman and Worley 1990) and for data integration and function prediction (Troyanskaya et al. 2003; Myers et al. 2005; Lee et al. 2007; Guan et al. 2008). Additional applications include determining gene regulatory relationships (Beal et al. 2005; Sachs et al. 2005) and inferring physical protein interactions (Jansen et al. 2003; Bradford et al. 2006).

3.4 Evaluation of Classification and Prediction Approaches

In the previous sections, we have presented several algorithms that can learn rules based on training data that can be used to predict classifications or probabilities for new examples. For each of these approaches a potential problem is *overfitting*, or matching the rule so closely to the training data that it does not properly generalize and correctly classify new examples (Russell et al. 2002). While there are techniques specific to each learning approach to combat overfitting, there are general approaches to detect and attempt to rectify this problem.

One of these approaches is *cross validation*. The basic premise of cross validation is to partition all of the known, labeled examples into separate *training* and *test* sets (Kohavi 1995). The learning algorithm is applied using just the training set of examples, and once learned the effectiveness of the model is determined by applying it to the held-out test set of examples. If the classification performance is significantly worse for the test set than the training set, this is a strong indication that the algorithm is overfitting.

There are several approaches to cross validation, but one of the most commonly used is *k-fold cross validation*. In this case, the labeled examples are divided into k groups, or *folds*, each containing $1/k^{th}$ of the examples. Then learning and evaluation is performed k times, each time holding out a different fold for use as the test set and using the other k-1 folds for training (Fig. 7 shows this for $k = 5$). The choice of k varies greatly, though values between 3 and 10 are often used. In the extreme, k can be set to the number of available examples, such that iterations are performed leaving out just one labeled example for testing (this is

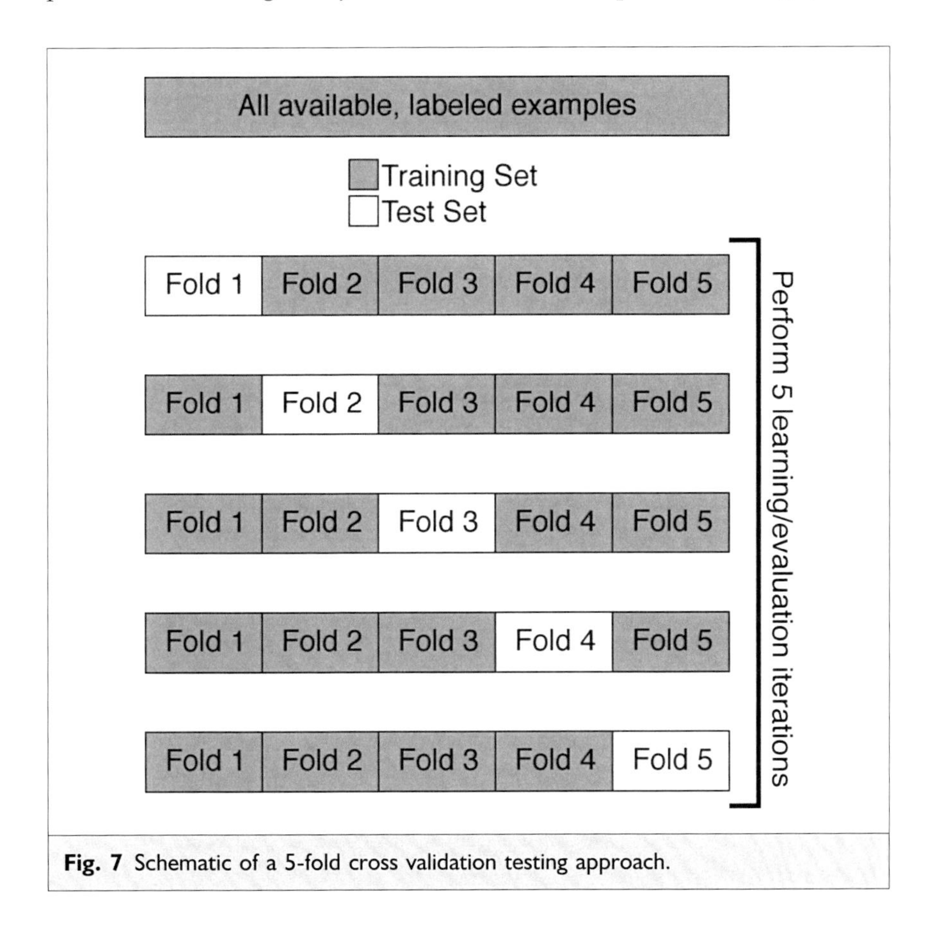

Fig. 7 Schematic of a 5-fold cross validation testing approach.

sometimes called *jackknifing*). After these learning/evaluation iterations are performed, we can select the model produced by the iteration with the best performance on the held out training data for later use. Another option is to create an *ensemble* model where each of the learned models is given a 'vote' and used for prediction and/or classification. In some cases, these approaches can combat overfitting by preventing the model from very tightly constraining to the available training data.

It is straightforward to measure the performance of a model performing classification (Jansen and Gerstein 2004; Myers et al. 2006). In the case of binary classification, we can simply count four quantities: the number of examples correctly classified as positive/yes (often called *true positives*, or TP), the number correctly classified as negative/no (called *true negatives*, or TN), the number incorrectly classified as positive/yes (*false positives*, or FP), and the number incorrectly classified as negative/no (*false negatives*, or FN). False positives are also referred to as *type I errors*, and false negatives are also referred to as *type II errors*. Given these quantities, there are a few commonly used metrics to arrive at a single number encapsulating the classification performance:

Accuracy measures the overall performance and is simply the fraction of examples correctly classified.

$$Accuracy = \frac{TP + TN}{TP + FP + TN + FN}$$

Precision, or *positive predictive value* (PPV), only penalizes false positives and is defined as the fraction of examples correctly classified as positive out of all examples called positive.

$$Precision = \frac{TP}{TP + FP}$$

Specificity penalizes false negatives, and is defined as the fraction of examples correctly classified as negative out of all negative examples.

$$Specificity = \frac{TN}{TN + FP}$$

Recall, or *sensitivity*, is a measure of what fraction of all possible true positives were correctly classified as positive.

$$Recall = \frac{TP}{TP + FN}$$

For many simple classification approaches, evaluating performance using accuracy is sufficient. For more specialized tasks, it may be useful to place more emphasis on correctly classifying either positives or negatives. For example, if we were predicting whether a patient has a disease, then we might be willing to accept the occasional false positive, but we would not want to misdiagnose a sick patient as healthy, and would want to minimize false negatives. In such a case, specificity may be a better a choice of evaluation metric. However, in a situation where false positives are more costly, such as selecting candidate genes for expensive or time consuming follow-up experiments, then precision may be a better metric choice.

When we have confidences in the classification results, or probabilities of belonging to each class, then evaluation is a bit more complex. In such cases in order to calculate one of the above evaluation metrics we must define a cutoff of confidence or probability to decide which examples are predicted positive and which are predicted negative. However, the proper choice of cutoff is not always apparent, and can often change depending on the situation. As such, there are techniques that examine all possible choices of cutoff to characterize the performance of predictions. These approaches take a sorted list of confidences and slide the cutoff from the extreme position of classifying all examples as negatives though the sorted examples to the other extreme of classifying all examples as positives. Throughout this sweep, evaluation metrics are calculated and plotted, typically focusing on either precision or specificity. *Precision-recall curves* (PR-curves) are generated by plotting recall (typically on the x-axis) versus precision (typically on the y-axis) for each possible cutoff. *Receiver operating characteristic* (ROC) curves are created by plotting 1 - specificity (on the x-axis) versus sensitivity (on the y-axis). Examples of these two types of plots are shown in Fig. 8.

These curves allow us to observe more detailed characteristics of the trade offs inherent in choosing a cutoff than the single number based metrics. For example, we could claim that only our most confident prediction as a positive prediction, in which case we are likely to have perfect precision but very low recall. On the other hand, we could predict that all examples are positive, in which case we would have perfect recall, but very low precision. In general, these curves can show the dynamics of the tradeoffs between these extreme cases. The choice between precision-recall and ROC plots is generally the same as between precision and sensitivity (Davis and Goadrich 2006). For a situation

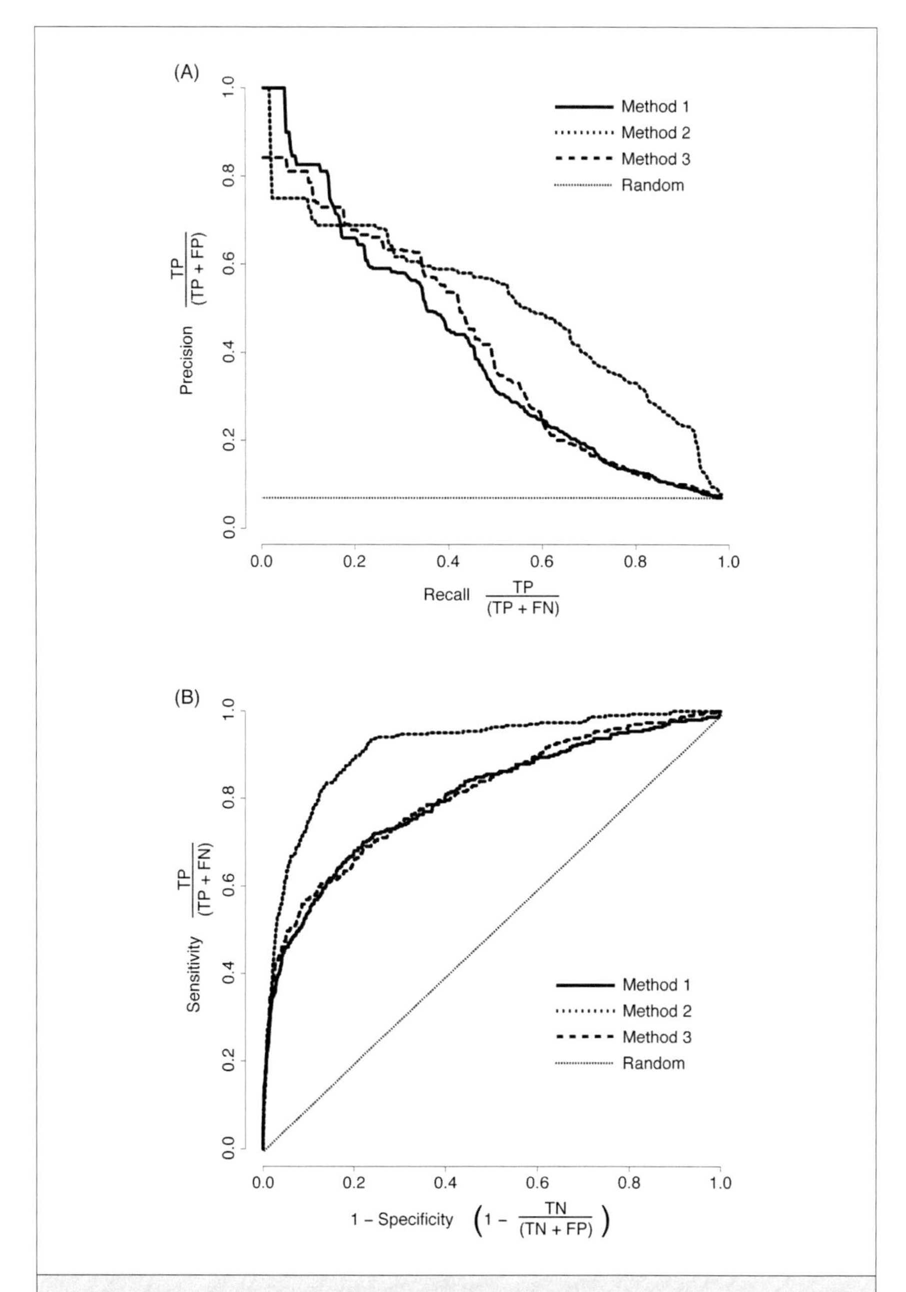

Fig. 8 Examples of Precision-Recall and ROC curves. Here prediction data is shown from a recent paper predicting protein function (Hibbs et al. 2009) using both (A) a precision-recall curve, and (B) an ROC curve.

where false positives are most costly, such as gene function prediction, then precision-recall plots are preferable. While for situations where false negatives are the most costly, then ROC plots are more appropriate.

4 UNSUPERVISED METHODS FOR DATA EXPLORATION

The supervised methods discussed in the previous section can very accurately and usefully perform classification and prediction tasks. However, supervised approaches require labeled training data in order to learn a model that can then be generalized and applied to new examples. There are many situations where these labels are unknown or unreliable, and in these cases we are unable to apply a supervised algorithm. Fortunately, there are alternatives, called *unsupervised* methods, which do not require labeled training examples (Russell et al. 2002; MacKay 2003). The basis of unsupervised analysis is to let the data 'organize itself' and then observe and interpret patterns or structure within that data. There are many unsupervised approaches useful in biological applications, but here we will discuss two of the most popular: clustering and singular value decomposition. Then we will also discuss some commonly used procedures that can be used to evaluate the results of an unsupervised approach.

4.1 Clustering Methods to Organize Data

The basic concept of *clustering* approaches is to rearrange data such that similar data points are next to each other (Jardine and Sibson 1968; Sherlock 2000; MacKay 2003). For our purposes, we will consider gene expression microarray data in our examples and discussion, but the general concepts can be applied to any data where a number of measurements are made for many different objects. It is useful to think of the data we will be clustering as a matrix. In the case of expression data, we will consider rows of our matrix to correspond to genes, and the columns of our matrix will correspond to experimental conditions. Thus, the high level goal of clustering is to reorder the genes (or conditions) of a dataset to place similar patterns of expression next to each other. This can be very useful to do in order to find higher level structure in the data that may be biologically meaningful. For example, with expression data, if a group of genes all behave in a similar manner across many conditions, they may be functionally related to each other (this is sometimes called the 'guilt by association' hypothesis). There are nearly as many clustering methods as there are computational biology researchers. However, most of these methods are some variation on two of the most popular techniques, hierarchical clustering and *k*-means clustering. We will examine these two methods in some detail, and then more briefly discuss some of the variations on these approaches.

4.1.1 Distance Metrics

Before we describe these clustering algorithms, we must decide how we will determine whether two vectors of data are similar to each other (i.e., in order to place similar data next to each other, we must define what 'similar' means). This is typically achieved by choosing a *distance metric* that can provide a measure of similarity or difference between two vectors of data. There are many possible choices of metrics, and different distance metrics can capture different types of relationships between data vectors (Wilson and Martinez 1997). Three of the most popular distance metrics are *Euclidean distance, Pearson correlation,* and *Spearman correlation.* Each of these three metrics identifies different types of similarity between data vectors. As an example, we will consider some

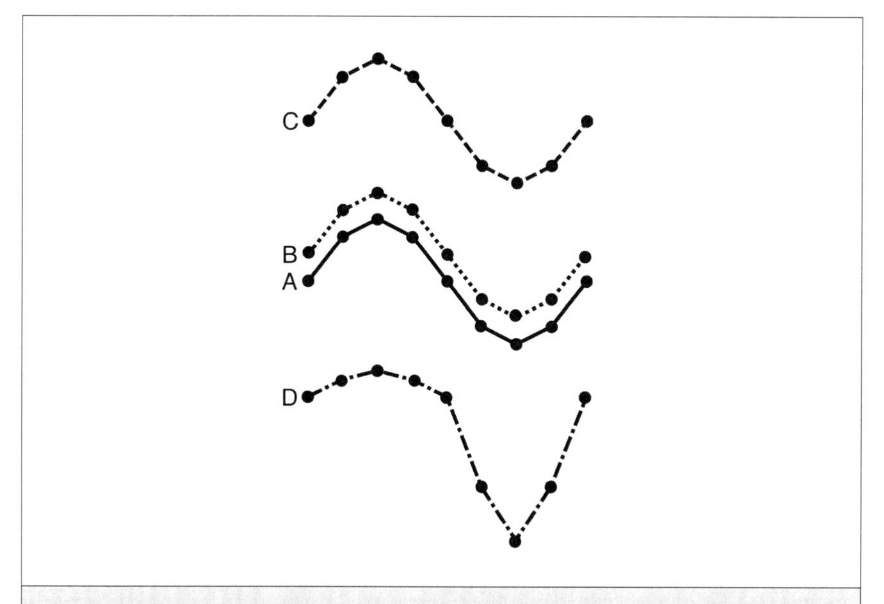

Fig. 9 Examples of data vectors whose similarity is greatly dependent on the choice of distance metric. The pair A-B is similar using Euclidean distance; the pairs A-B and A-C are similar by Pearson correlation; the pairs A-B, A-C, and A-D are similar by Spearman correlation.

sample vectors of nine data points as shown in Fig. 9. These sets of vectors could, for example, represent the expression levels of different genes over a time course, or some other phenotypic measure.

Euclidean distance most closely corresponds to what we normally think of as a 'distance' in a geometric sense. It measures the absolute separation in space between points or sets of points using the formula:

$$Euclidean(x, y) = \sqrt{\sum_{i=1}^{n}(x_i - y_i)^2}$$

Where n is the length of the data vectors, x and y are the vectors being compared, and x_i and y_i are the ith element of each respective data vector. Among our example vectors, the pair A-B would have a very small Euclidean distance between them, while the pairs A-C and A-D would have a much higher distance.

Pearson correlation is a measure of how linearly correlated two vectors are, which corresponds to determining how similarly the values change over the course of the vector (Rodgers and Nicewander 1988). The calculation of the Pearson correlation involves subtracting the average of each vector and normalizing by the variance of the vectors using the formula:

$$Pearson(x, y) = \frac{1}{n-1}\sum_{i=1}^{n}\frac{(x_i - \mu_x)(y_i - \mu_y)}{\sigma_x \sigma_y}$$

Where n, x, y, x_i, and y_i are as above, μ_x and μ_y are the means of each data vector, and σ_x and σ_y are the standard deviations of each data vector. Correlation values vary between -1 and 1, where 1 represents perfectly correlated, 0 completely uncorrelated, and -1 anti-correlated (e.g. when one vector is high the other is low, and where one vector is low the other is high). As such, correlations near 1 represent greater similarity, so to treat this measure more like a true distance, it is sometimes subtracted from 1. In our example vectors, the pair A-B is still very similar by this metric, but the pair A-C is now also very similar, because these two patterns show the same up and down shifts across all nine data points. The pair A-D, however, is still considered not very similar because while these track up and down similarly, the degree of those shifts are inconsistent.

Spearman correlation is similar in spirit to the Pearson correlation, but it considers correlation in *rank transformed* data (Spearman 1904). Rank transforming data means that each value is replaced by its overall rank when the values were sorted, so the smallest value would be replaced with a rank of 1, the next lowest with 2, and so on until the highest value is replaced with n, the length of the data vector. Once the data has been rank transformed, the Spearman correlation is calculated by the formula:

$$Spearman(x,y) = 1 - \frac{6\sum_{i=1}^{n}(rank[x_i] - rank[y_i])^2}{n(n^2-1)}$$

Where $rank[x_i]$ and $rank[y_i]$ represent the results of the rank transformation for the ith element of each data vector. Like Pearson correlation, Spearman correlation produces values from -1 to 1, with 1 representing the highest correlation, so for use as a distance it is often subtracted from 1. Using this metric, the pairs A-B and A-C are still closely related, but now the pair A-D is also considered closely related. By rank transforming the data, the skewed or shifted pattern of the D vector is removed as all 4 patterns share the same rank order.

Other distance metrics have been used in computational biology (notably Manhattan distance, Kendall's τ, and mutual information), but the majority of clustering is based on one of the three metrics presented here. Clearly the choice of distance metric can greatly affect clustering analysis, since different patterns may or may not be considered to be similar depending on which metric is used. The proper choice of metric depends on which biological questions we are interested in exploring. In the case of gene expression data, it may be appropriate to use Euclidean distance if we are concerned with identifying genes that are all over-expressed or under-expressed under the same conditions. However, if we want to identify genes that are changing in similar ways over the course of conditions, then one of the correlation metrics is more appropriate. Regardless of our choice of distance metric, the clustering procedures are the same (though the outcomes will be different).

4.1.2 Hierarchical Clustering

The basic premise of *hierarchical clustering* is to create a tree-like structure defined by the similarity of data points to each other. This structure enforces an amount of order on the underlying data that is both flexible and potentially informative. While there are many variations on hierarchical clustering, one of the most commonly used approaches is called *agglomerative* hierarchical clustering (Sokal and Michener 1958). The basic algorithm of this approach is as follows:

1. Initialize a set of nodes with one node for every data vector to cluster
2. Calculate all pair-wise distances between every node
3. Choose the two nodes with the smallest distance between them
   a) Connect these two nodes together
   b) Remove these two from the set of all nodes
   c) Place a new node in the set representing the union of these two joined nodes
4. Repeat steps 2-3 until only one node remains

In order to better understand this approach, we will first consider a simple example of this process as shown in Fig. 10. In this example, we are hierarchically clustering six data vectors shown as line patterns using a distance metric that compares overall patterns. Initially, all six vectors await clustering in the node set. Then, after comparing all pairs of

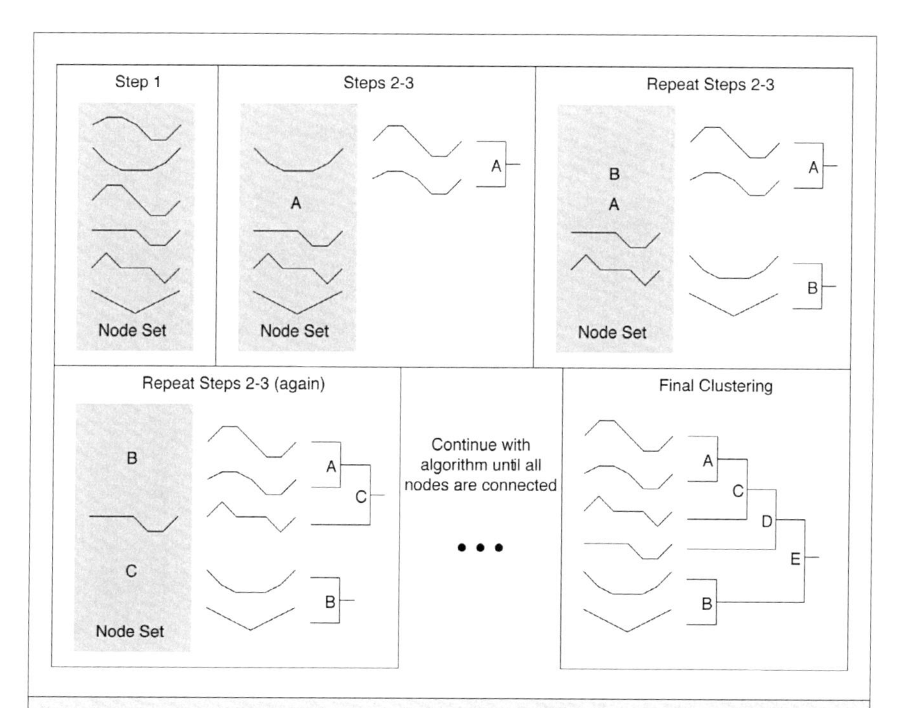

Fig. 10 Example of the hierarchical clustering procedure. In the upper left, we show six example data vectors to cluster. In the upper center, we have selected the two closest vectors to connect, removed them from the node set, and added a new node (A) that represents the union of the joined vectors. This process continues until the final clustering shown in the lower right is reached.

vectors, we choose the two most similar and connect them. Those two vectors are removed from the node set, and in their place we put a special node (marked A) that represents both of the vectors linked together in this step. This process is repeated until the node set is empty and all data vectors have been linked together into a full tree structure.

One important remaining detail of hierarchical clustering is concerned with how to determine similarity scores between nodes representing multiple data vectors (such as the distance between nodes A and B in our example). There are three strategies commonly used to calculate these distances as shown in Fig. 11. *Average-linkage clustering* calculates this similarity as the mean of all possible pair-wise connections spanning the two sets of data vectors. *Single-linkage clustering* uses the minimum distance of possible pair-wise connections as the similarity (i.e. the single

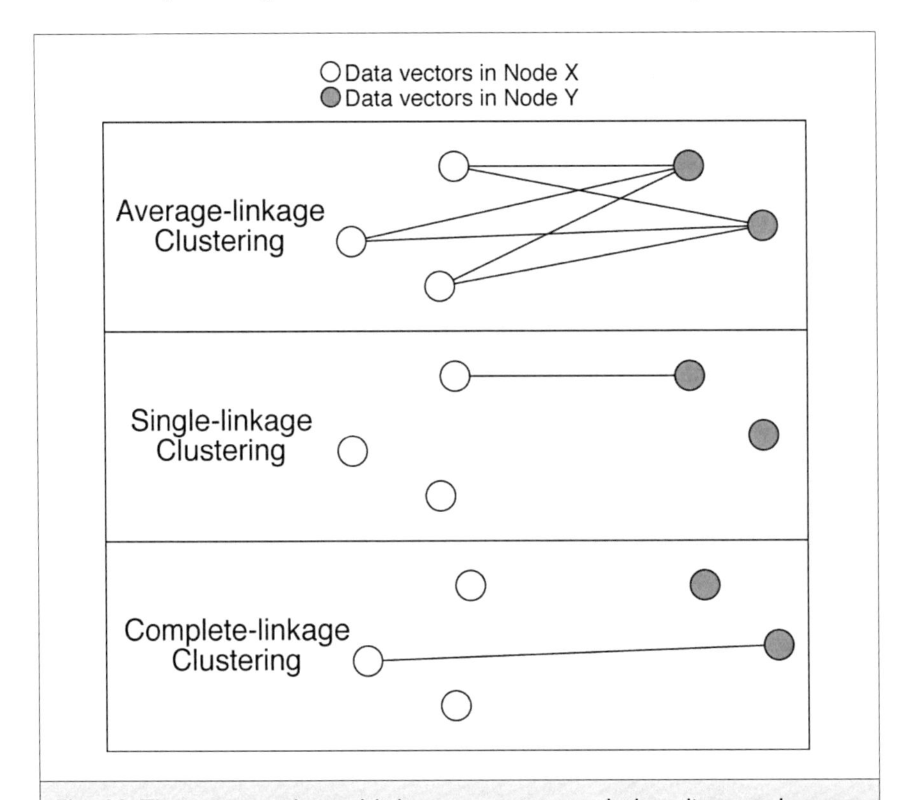

Fig. 11 Three commonly used linkage strategies to calculate distances between interior nodes in a hierarchical clustering. The open circles correspond to data vectors in one node, and the shaded circles correspond to data vectors in a different node.

best distance). Lastly, *complete-linkage clustering* uses the maximum distance of all pair-wise connections. In practice, this choice rarely affects the overall clustering patterns using common distance metrics, but there are specific instances where one method may be more appropriate than another.

Hierarchical clustering is often used to visualize data for publication and for initially exploring a dataset to gain a better understanding of what signals are present in the dataset as a whole. To aid with both of these tasks, the connecting trees of the clustering, called *dendrograms*, are typically drawn in a way that encodes the distances between the clustered objects (Seo and Shneiderman 2002). For example, in Fig. 12 we show a portion of a gene expression dataset whose genes have been hierarchically clustered using a Pearson correlation distance metric. We can see in this example that data vectors with the most similar patterns have shorter 'branches' connecting them, while for less similar pairs a greater horizontal distance must be traveled before the genes are linked through the tree.

There are several limitations and drawbacks to hierarchical clustering, as well as several approaches that attempt to rectify these problems. One of these potential issues is that there are a large number of valid ways to draw the results of hierarchical clustering by 'pivoting' the tree around an interior node, as shown in Fig. 13. Since there are as many interior nodes as original data vectors, it is intractable to examine all possible combinations of pivots to produce an optimally oriented tree. However, there are several heuristic approaches to produce useful, approximately-optimal orderings (Bar-Joseph et al. 2001; Saeed et al. 2003). Another potential issue with hierarchical clustering is that the initial joins have a great impact on the rest of the tree. For example, if the distance calculations are noisy, that noise is compounded many times over as the rest of the tree is constructed. To address this issue, several methods for *divisive* (rather than agglomerative) hierarchical clustering have been developed. These approaches begin with all of the data together and make partitioning decisions to form the tree in a 'top-down' manner, rather than building from the 'bottom-up' as we have shown here (Sherlock 2000; Varshavsky et al. 2008).

Despite these limitations, hierarchical clustering is a widely used technique for gene expression data analysis. Applications of hierarchical clustering have been successful in areas ranging from indicating protein functions (Eisen et al. 1998) to diagnozing cancer types (Garber et al. 2001). The general hierarchical clustering technique is applicable far

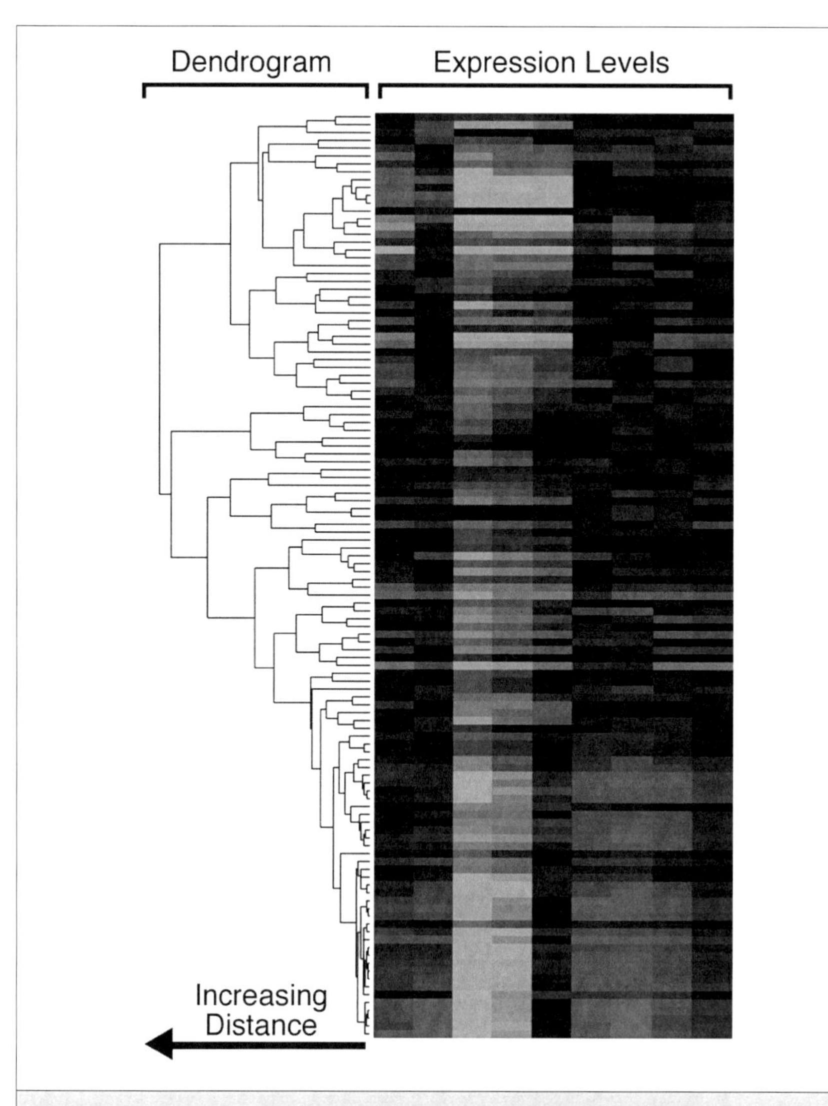

Fig. 12 Example of a hierarchical dendrogram. Data vectors correspond to rows, and these vectors were hierarchically clustered. The tree structure shown on the left, called a dendrogram, encodes the distances between vectors clustered together. The greater the height of the branches in the tree structure that link vectors together, the greater the distance between them. For example, the short branches near the bottom show that these vectors are very similar, while the vectors at the top and bottom are more dissimilar since they are connected only at the top of the dendrogram.

(Color image of this figure appears in the color plate section at the end of the book).

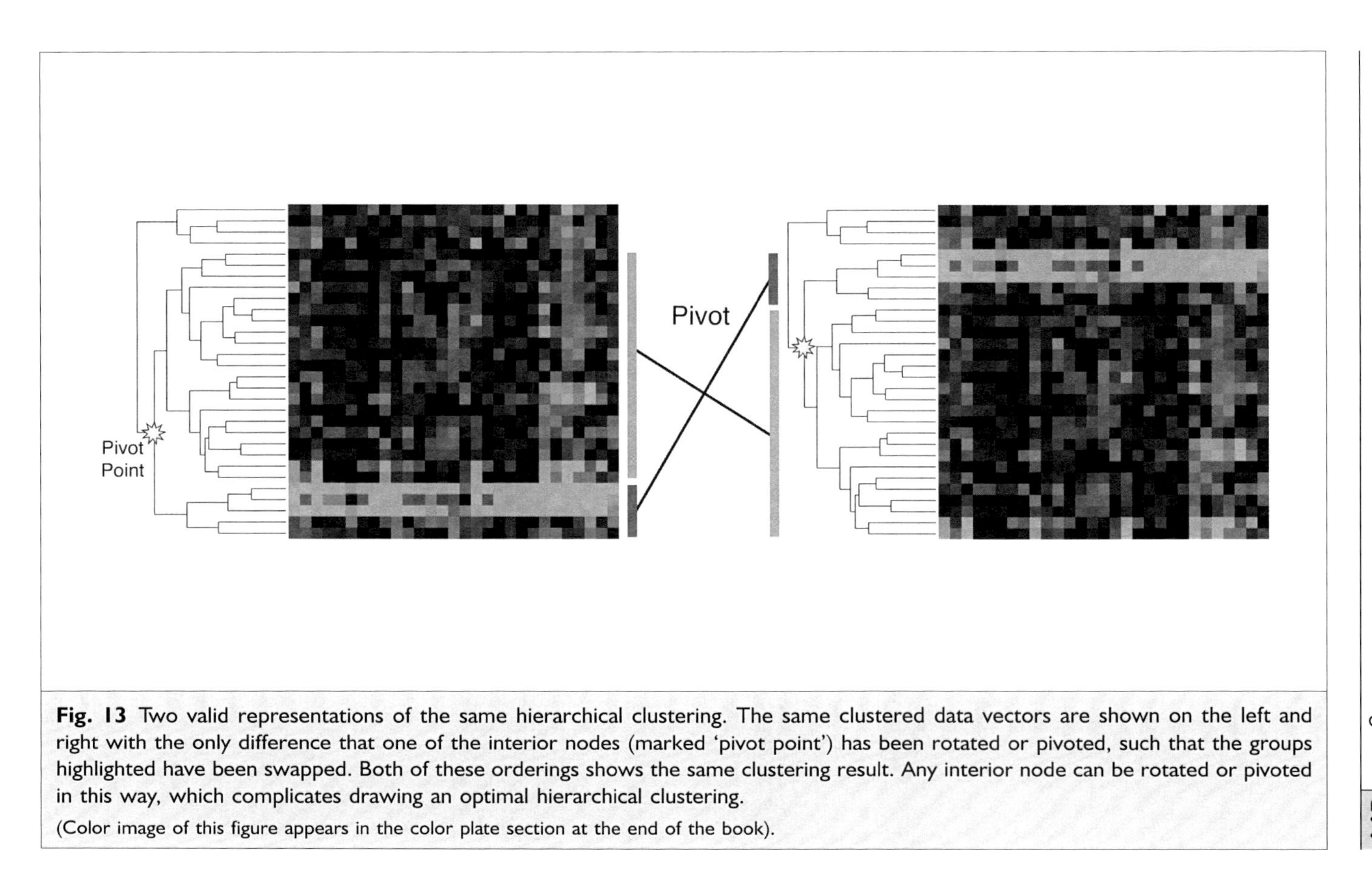

Fig. 13 Two valid representations of the same hierarchical clustering. The same clustered data vectors are shown on the left and right with the only difference that one of the interior nodes (marked 'pivot point') has been rotated or pivoted, such that the groups highlighted have been swapped. Both of these orderings shows the same clustering result. Any interior node can be rotated or pivoted in this way, which complicates drawing an optimal hierarchical clustering.

(Color image of this figure appears in the color plate section at the end of the book).

beyond expression data, and has been successfully used in studying protein-protein interactions (Arnau and Marin 2003), synthetic genetic interactions (St Onge et al. 2007), and a variety of other types of data. There are several implementations of hierarchical clustering and visualization approaches available, including the original Cluster/TreeView program (Eisen et al. 1998), JavaTreeView (Saldanha et al. 2004), HIDRA (Hibbs et al. 2007), HCE (Seo and Shneiderman 2002), and MeV (Saeed et al. 2003).

4.1.3 K-means Clustering

While hierarchical clustering and its variants globally organize data into a structure, other clustering approaches separate data more explicitly into distinct groups, or *clusters*. One of the simplest methods for this type of clustering is called *k-means clustering*. The goal of *k*-means is to divide a set of data vectors into exactly *k* groups, each of which is coherent as possible (Sherlock 2000; Russell 2002). The general algorithm for performing this approach is as follows:

1. Choose a value for *k*
2. Initialize *k* clusters with a random data vector
3. Assign an average vector to each cluster that is the mean of all data vectors currently assigned to the cluster
4. Remove all genes from all clusters
5. For all data vectors, place the vector in the cluster whose average is most similar
6. Repeat steps 3-5 until a stop condition is met

A simple example of performing this procedure where $k = 2$ is shown in Fig. 14. There are several possibilities for determining when to stop iterating the *k*-means algorithm, but the most commonly used methods attempt to identify when the results of the algorithm converge. This is usually defined as when the data vectors partition into the same clusters two iterations in a row (as in the example in Fig. 14), or when the average data vectors for each cluster change very little. There are cases where neither of these conditions occurs and in these situations alternate stop conditions are used, such as defining a maximum number of iterations to perform.

As with most clustering techniques, *k*-means clustering has its own advantages and disadvantages. In particular, the strengths of *k*-means include that the algorithm is easy to perform using a variety of distance metrics and the results are easily interpretable as the best way to separate the data vectors into *k* separate groups. One of the largest

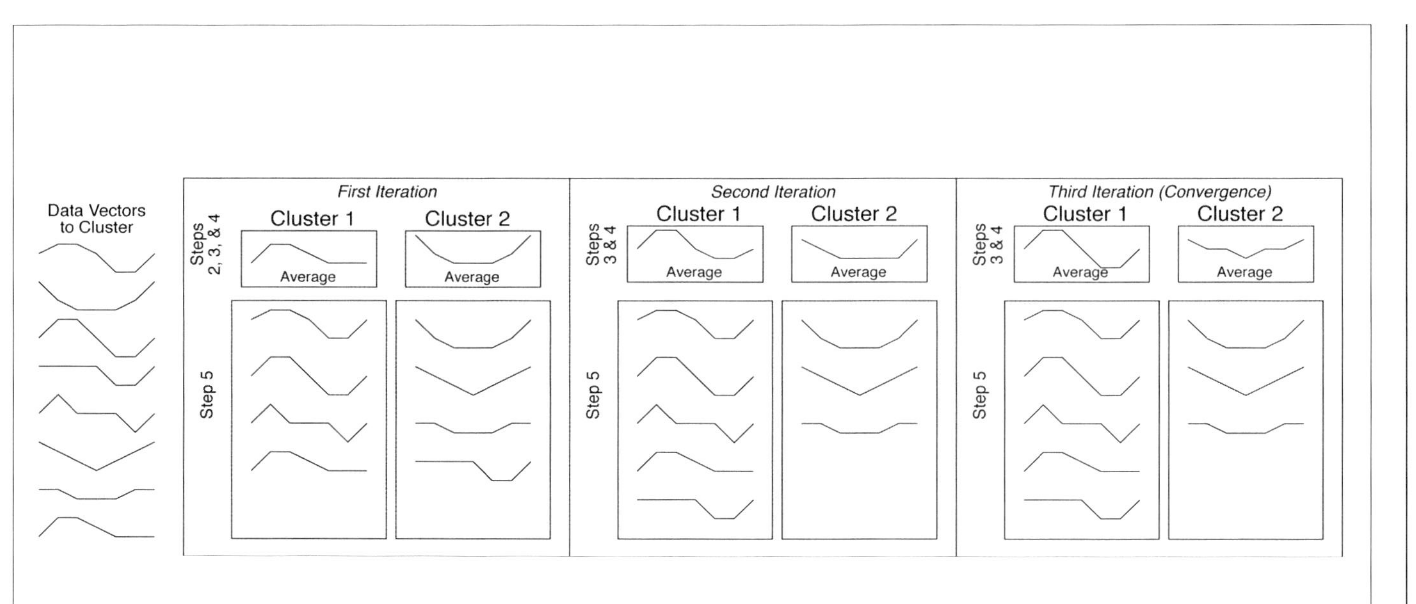

Fig. 14 Example of *k*-means clustering procedure. Beginning on the left, we have 8 data vectors to cluster. In this case we have chosen $k = 2$, and picked two vectors at random to represent the initial cluster averages. As shown in the first iteration panel, each data vector was assigned to the cluster to whose average it is most similar. Given these assignments, the cluster averages were updated as shown at the top of the second iteration panel. The process of assigning vectors and updating averages is repeated until a stop condition is met. In this case the same assignment occurs during the second and third iterations, so the algorithm has converged.

potential issues with k-means clustering is that it is often not obvious what the proper choice is for the parameter k. In practice, k can vary within a large range of values from as small as three to many dozens of clusters. Several modifications and adaptations of k-means clustering have been proposed to address this issue, including approaches that use many different values of k and then choose the best result by a variety of metrics and others that group genes incrementally (Ben-Dor et al. 1999; Heyer et al. 1999). A further issue with k-means clustering (and many others) is that clusters are defined very rigidly, meaning that each data vector must belong to exactly one cluster. In many biological situations it may be more meaningful for a data vector to not belong to any clusters, or to belong to multiple clusters. One approach to dealing with this issue is the notion of *fuzzy clustering* (Gasch and Elisen 2002), where data vectors probabilistically belong to individual clusters (e.g., a vector can be 70% in one cluster and 30% in a different cluster).

4.1.4 Additional Clustering Techniques and Implementations

Many other clustering techniques can be thought of as extensions or modifications to either hierarchical clustering or k-means clustering. For example, *self organizing maps* (SOM) uses concepts from both k-means clustering and hierarchical clustering. Rather than creating k entirely separate clusters, SOMs arrange k clusters in space, such that nearby clusters have influence over each other (Tamayo et al. 1999). The result is a set of clusters with membership properties similar to k-means, but with a global organization of the data vectors in a spirit similar to hierarchical methods.

One important class of clustering algorithms fundamentally different from those discussed so far is *bi-clustering* techniques. Bi-clustering slightly re-frames the goal of clustering by looking for similar patterns in just a subset of the features in a data vector. For example, if two data vectors agree very well for half of their feature points but diverge greatly in the other half, then normal clustering approaches may not group these vectors together. But a bi-clustering approach would attempt to identify the subset of conditions where these vectors agree and base the clustering procedure on that subset, regardless of the remaining features. This can be particularly useful for many biological problems, especially in cases where similar data patterns are only expected in a subset of all measurements. For example, in gene expression data it is often the case that two genes may be strongly co-expressed under a particular environmental condition, but their expression may be wholly unrelated under other conditions.

Unfortunately, the general bi-clustering problem is too computationally intensive to practically compute (the problem is NP-hard). As such, most bi-clustering methods use a variety of heuristics, simplifications, and assumptions in order to perform the calculations (Madeira and Oliveira 2004). In some cases these simplifications can divorce the results from the original data in damaging ways, but in other cases these approaches can produce excellent results. Several recent approaches have begun to use data mining techniques to effectively re-frame the bi-clustering problem as a search procedure (Owen et al. 2003; Hibbs et al. 2007; Mostafavi et al. 2008). These approaches generally start with a small set of data vectors that are known to be related, and based on that information additional vectors can be identified that relate to the starting group in a significant subset of the available features.

There are a large number of software packages available to perform a variety of clustering and bi-clustering algorithms. These include Cluster (Eisen et al. 1998) (hierarchical, *k*-means, SOM), MultiExperimentViewer (Saeed et al. 2003) (hierarchical and variants, *k*-means and variants, SOM, QT clustering), Expander (Sharan et al. 2003) (variants of *k*-means, bi-clustering, expression specific approaches), and programing languages such as R and Matlab. Any of these tools can be used to generate clusterings from input data that can be useful for exploring data and locating potentially interesting patterns.

4.2 Singular Value Decomposition and Principal Component Analysis

Another commonly used unsupervised analysis technique is *singular value decomposition* (SVD). At an intuitive level, SVD separates out the dominant signals in a data matrix into their base patterns, and associates each data vector with these patterns (Golub and Kahan 1965; Wall et al. 2003). Stated differently, SVD performs a change of basis operation where the new basis corresponds to common signals in the original data. This procedure can be useful both in quantifying what underlying patterns are important and/or over-represented in the data, and in quantitatively associating groups of data vectors with each of these underlying patterns. The basic calculation of SVD produces the matrix equation:

$$X_{m\times n} = U_{m\times n}\Sigma_{n\times n}V^{t}_{n\times n}$$

Where X is an m-by-n matrix constructed with m data vector rows each consisting of n columns representing each of the features in the data vectors. This matrix is decomposed into three matrices: U is an m-by-n matrix containing the *left eigenvectors*, Σ is an n-by-n matrix that is 0 everywhere except for its diagonal which contains the *singular values*, and V^t is an n-by-n matrix containing the *right eigenvectors*.

More intuitively, V^t contains the most prominent patterns present in the original data in order from the most dominant to the least. These patterns are all orthonormal and form the basis for a new projection of the original data. The singular values in Σ each correspond to one of the vectors in V^t and are related to the amount of variation in the original data that each V^t vector captures. Lastly, U contains the coefficients of each original data vector in the new space defined by V^t. A simple example of SVD in just two dimensions is shown in Fig. 15.

In this example, we have a collection of 2D data vectors plotted as usual in part A. The process of SVD essentially identifies the dimension of greatest variance through the space of all data points and uses that direction as the first eigenvector in V^t as shown in part B of the example. SVD then finds the dimension orthogonal to the first that captures the most remaining variance, and so on until the original data space is

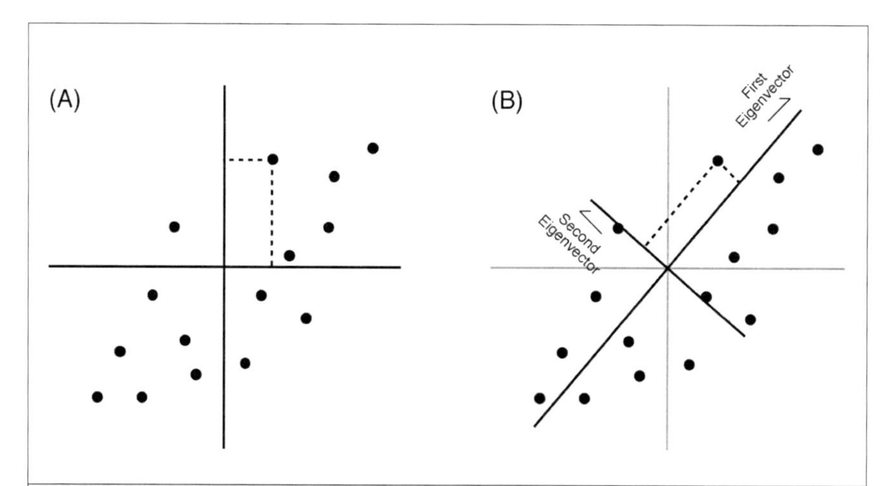

Fig. 15 Example of singular value decomposition. (A) shows the original two-dimensional data vectors laid out in space. SVD projects these data points into a new basis, shown in (B), where the first vector captures the most variation in the original space, the second vector the next most variation, etc. The coefficients for one data point in both the original and new basis are indicated by dotted lines.

entirely covered by the new eigenvectors. Each of the eigenvectors in V^t are orthonormal (i.e., 1 unit long and orthogonal), and the diagonal of Σ captures the amount of variation covered by each eigenvector. The matrices are constructed so that the singular values are in decreasing order, which means that the first eigenvector captures the most variation, the second eigenvector the next most variation, and so on. Finally, the U matrix contains the coordinates of the original data points in the new space defined by Σ and V^t, as shown with dotted lines for one data point in the example.

When SVD is applied to feature-centered data matrices, the procedure is called *principal component analysis* (PCA). Here, a feature-centered matrix is one where the average value of each column is subtracted from each value in that column, which places the mean of all data vectors at zero. When PCA is performed, the right eigenvectors are then referred to as *principal components*. A schematic of the matrices produce by SVD or PCA is shown in Fig. 16.

Both SVD and PCA have been successfully applied in many biological situations involving very high dimensional data, such as gene

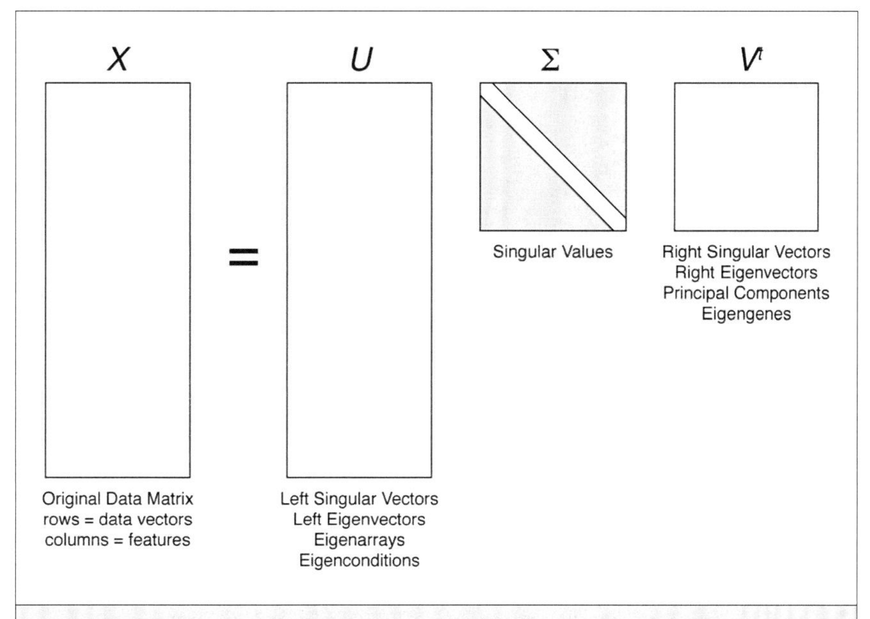

Fig. 16 Schematic of the matrices produced by singular value decomposition and principal component analysis. Matrices are labeled by some of the terms used to describe them in different disciplines and applications.

expression microarray data. In the case of expression data, the original data matrix typically consists of rows corresponding to genes and columns corresponding to experimental samples or conditions. As such, the right eigenvectors or principal components are often called the *eigengenes*, while the left eigenvectors are referred to as the *eigenconditions*. Since SVD can be interpreted as separating out data into its constituent parts, there are several ways that it is often used to process and visualize data (Sturn et al. 2002; Saeed et al. 2003; Hibbs et al. 2005).

SVD can be used for both dimensionality reduction and noise reduction by disregarding the eigenvectors that capture either little variation or undesired variation in the original data. For dimensionality reduction, the desired columns of the U matrix or of $U\Sigma$ can simply be plotted against one another. By plotting the first two columns of U against each other the data is displayed such that the most possible variation is shown in two dimensions. For noise reduction, the right eigenvectors thought to correspond to noise are removed (i.e. replaced with zeros), and the matrices are multiplied back together to produce a reconstruction of the original matrix.

It can frequently be informative to study the right eigenvectors of the decomposition for biological significance. For example, the eigengenes of an expression experiment are often highly correlated with one or more experimental factors of the study. These factors can include intentional experimental effects (Raychaudhuri et al. 2000), but have also been shown to sometimes include unintended effects (Alter et al. 2000). When a right eigenvector corresponds to an interesting signal, examining the corresponding coefficients in the left eigenvectors can identify groups of data points where that signal is present. In this way, SVD and PCA can be used in a manner similar to clustering to identify groups of related genes.

There are also a number of approaches related to SVD with useful applications in biological problems. One of these approaches, called iterative signature analysis, performs a modified form of the biclustering problem by identifying and removing orthogonal sources of information from datasets (Bergmann et al. 2003; Kloster et al. 2005). A microarray search method, called SPELL, in part utilizes the results of SVD to re-weight the signals present in a dataset in order to provide an equal contribution from all of the available orthogonal signals (Hibbs et al. 2007). Lastly, the surrogate variable analysis method systematically applies SVD in order to remove un-modeled or unintended signals from datasets (Leek and Storey 2007). These approaches are just a sampling of the potential applications of SVD and PCA to biological data analysis. From this work and others, it is clear that due to the complex nature of many biological data sources, methods to break apart and characterize signals are particularly valuable.

4.3 Assessing Groups of Genes

Regardless of the technique employed, the effective outputs of many unsupervised approaches are lists or groups of related data points. However, the interpretation of these groups can be problematic. As such, there are several techniques to assess the quality and meaning of these groupings. One class of these approaches assesses cluster quality largely independent of the types of data vectors that were grouped together. These approaches tend to focus on either statistical measures of how cohesive the signals are within each group and/or how different the signals are between groups (Yeung et al. 2001; Mendez et al. 2002; Datta and Datta 2003). For example, if we separated data vectors into two distinct groups, we could measure the similarity of the vectors in each cluster and compare that to the similarities between the two clusters. In the simplest case, we could use some distance metric to calculate all possible pair-wise distances and then compare the within-cluster distances to the between-cluster distances. We would consider our groups to be of higher quality if the within-cluster differences were significantly less than the between-cluster distances. While performing this particular procedure is fairly simple, several more sophisticated approaches based on this concept have been used in practice.

Another common approach to assessing clustering results is based on identifying over- or under-representation of known qualities of the data within clusters. For example, if we were clustering a gene expression dataset we could produce clusters that correspond to lists of genes. We can then determine if the genes in a cluster share known properties, such as biochemical functions, involvement in cellular processes, localization, upstream regulatory elements, etc. There are a variety of approaches of this type (Edelman et al. 2006; Subramanian et al. 2005), but one simple, effective method is to perform analysis using the *hypergeometric test* (Boyle et al. 2004; Sealfon et al. 2006). For example, we could assess over-representation of mitochondrially-localized genes in a cluster using this approach. Given the number of all genes (N), the number of genes known to be mitochondrially-localized (M), the number of genes in our cluster (n), and the number of genes in our cluster known to be mitochondrially-localized (x), we can calculate a p-value for how surprising this result is using the formula:

$$p - value = \sum_{j=x}^{n} \frac{\binom{M}{j}\binom{N-M}{n-j}}{\binom{N}{n}}$$

The intuition of this approach is to look at the number of possible ways to create a cluster containing n genes, and of those ways determine what fraction have as many or more mitochondrially-localized genes as the cluster that we observed.

This approach is not limited to just assessing localization; any property based on known memberships to a set can be analyzed using this style of test. In the case of gene clusters, for example, the annotations maintained by the Gene Ontology (GO) Consortium (Ashburner et al. 2000), the KEGG pathway database (Kanehisa 2002), or many other sources could be used to define the known properties. One important consideration when performing this type of analysis is to apply *multiple hypothesis correction* if more than one p-value is calculated. For example, if we tested a cluster for enrichment of 300 different GO categories, then at a raw 0.05 p-value (95%) confidence interval, we would expect to observe 15 significant enrichments based purely on chance. As such, these p-values are typically increased in order to reduce this chance. Two of the most common techniques to perform this correction are the very conservative Bonferroni correction (Boyle et al. 2004) and the less conservative false discovery rate (Benjamini and Yekutieli 2001) procedure.

Assessing clusters for their general quality and for over-represented properties can lead to important hypotheses based on a 'guilt by association' assumption. For example, given a set of well-defined, coherent, and distinct clusters of genes, if we observe highly significant enrichment for a particular property in one of these groups, then we can suppose that the entire group captures some real information about biological relationships. As such, this type of analysis is excellent for generating new hypotheses about the entities that have been clustered.

5 FUTURE DIRECTIONS OF COMPUTATIONAL BIOLOGY

All of the techniques and approaches presented and discussed in this chapter use the strengths of computer science to analyze biological data and form computational hypotheses. However, it must be emphasized that the results of these machine learning techniques are *only hypotheses*. Additional follow-up laboratory work is required in order to definitively confirm any of these computational predictions. Further, the results produced from any machine learning approach are only as reliable as the assumptions and data used to generate those results. In computer science this is often called the GIGO principle—'garbage in, garbage out'. This is not to imply that computational approaches are flawed, or that biological data is unreliable. Rather, it is important for computationalists to understand the biological nature of the data they consume and the problems they address in order to tailor their methods appropriately. And it is important for biologists to understand the assumptions, heuristics, and simplifications employed by computational approaches in order to properly interpret the results.

Much of the best work in computational biology occurs when biologists and computer scientists work together from beginning to end on a project. For example, there are vast benefits of involving a computational expert in experimental design as well as involving a biological expert in computational analysis. Computational machine learning techniques can provide great benefits to biology, including identifying more reliable candidate genes, placing new data in a broader context, and quickly analyzing experimental results. Moving forward, the beneficial integration of computation and biology is likely to increase even more, so long as researchers in these two fields continue to work together towards common goals.

References

Alter O, Brown PO, Botstein D (2000) Singular value decomposition for genome-wide expression data processing and modeling. Proc Natl Acad Sci USA 97: 10101-10106

Arnau V, Marin I (2003) A hierarchical clustering strategy and its application to proteomic interaction data. Proc 1st Conf Pattern Recog Img Anal, pp 62-69

Ashburner M, Ball CA, Blake JA, Botstein D, Butler H, et al. (2000) Gene ontology: Tool for the unification of biology. The Gene Ontology Consortium. Nat Genet 25: 25-29

Bar-Joseph Z, Gifford DK, Jaakkola TS (2001) Fast optimal leaf ordering for hierarchical clustering. Bioinformatics 17 (Suppl 1): S22-S29

Beal MJ, Falciani F, Ghahramani Z, Rangel C, Wild DL (2005) A Bayesian approach to reconstructing genetic regulatory networks with hidden factors. Bioinformatics 21: 349-356

Ben-Dor A, Shamir R, Yakhini Z (1999) Clustering gene expression patterns. J Comput Biol 6: 281-297

Benjamini Y, Yekutieli D (2001) The control of the false discovery rate in multiple testing under dependency. Ann Stat 29: 1165-1188

Bergmann S, Ihmels J, Barkai N (2003) Iterative signature algorithm for the analysis of large-scale gene expression data. Phys Rev E Stat Nonlin Soft Matter Phys 67: 031902

Boyle EI, Weng S, Gollub J, Jin H, Botstein D, et al. (2004) GO::TermFinder - open source software for accessing Gene Ontology information and finding significantly enriched Gene Ontology terms associated with a list of genes. Bioinformatics 20: 3710-3715

Bradford JR, Needham CJ, Bulpitt AJ, Westhead DR (2006) Insights into protein-protein interfaces using a Bayesian network prediction method. J Mol Biol 362: 365-386

Breiman L, Friedman JH, Olshen RA, Stone CJ (1984) Classification and Regression Trees. Tech Rep, Wadsworth Intl, Monterey, CA, USA

Brown MP, Grundy WN, Lin D, Cristianini N, Sugnet CW et al. (2000) Knowledge-based analysis of microarray gene expression data by using support vector machines. Proc Natl Acad Sci USA 97: 262-267

Cawley GC (2000) Matlab Support Vector Machine Toolbox: *http://theoval.sys.uea.ac.uk/svm/toolbox/*

Chan ALF, Chen J-X, Wang H-Y (2006) Application of data mining to predict the dosage of vancomycin as an outcome variable in a teaching hospital population. Int J Clin Pharmacol Ther 44: 533-538

Cormen TH, Leiserson CE, Rivest RL, Stein C (2001) Introduction to Algorithms. MIT Press, Cambridge, MA, USA

Datta S, Datta S (2003) Comparisons and validation of statistical clustering techniques for microarray gene expression data. Bioinformatics 19: 459-466

Davis J, Goadrich M (2006) The relationship between Precision-Recall and ROC curves. Proc 23rd Int Conf on Mach Learn, pp 233-240

Degroeve S, Baets BD, de Peer YV, Rouzé P (2002) Feature subset selection for splice site prediction. Bioinformatics 18 (Suppl 2): S75-S83

Dettling M, Bühlmann P (2003) Boosting for tumor classification with gene expression data. Bioinformatics 19: 1061-1069

Ding CH, Dubchak I (2001) Multi-class protein fold recognition using support vector machines and neural networks. Bioinformatics 17: 349-358

Edelman E, Porrello A, Guinney J, Balakumaran B, Bild A, et al. (2006) Analysis of sample set enrichment scores: assaying the enrichment of sets of genes for individual samples in genome-wide expression profiles. Bioinformatics 22: e108-e116

Eisen MB, Spellman PT, Brown PO, Botstein D (1998) Cluster analysis and display of genome-wide expression patterns. Proc Natl Acad Sci USA 95: 14863-14868

Freund Y, Schapire R (1999) A Short Introduction to Boosting. Jap Soc Artif Intel 14: 771-780

Garber ME, Troyanskaya OG, Schluens K, Petersen S, Thaesler Z, et al. (2001) Diversity of gene expression in adenocarcinoma of the lung. Proc Natl Acad Sci USA 98: 13784-13789

Gasch AP, Eisen MB (2002) Exploring the conditional coregulation of yeast gene expression through fuzzy k-means clustering. Genome Biol 3(11): RESEARCH0059

Golub G, Kahan W (1965) Calculating the Singular Values and Pseudo-Inverse of a Matrix. J Soc Indust Appl Math 2: 205-224

Guan Y, Myers CL, Lu R, Lemischka IR, Bult CJ, Troyanskaya OG (2008) A genomewide functional network for the laboratory mouse. PLoS Comput Biol 4: e1000165

Helmbold D, Schapire R (1997) Predicting nearly as well as the best pruning of a decision tree. Mach Learn 27: 61-68

Heyer LJ, Kruglyak S, Yooseph S (1999) Exploring expression data: Identification and analysis of coexpressed genes. Genome Res 9: 1106-1115

Hibbs MA, Dirksen NC, Li K, Troyanskaya OG (2005) Visualization methods for statistical analysis of microarray clusters. BMC Bioinformat 6: 115

Hibbs MA, Hess DC, Myers CL, Huttenhower C, Li K, Troyanskaya OG (2007) Exploring the functional landscape of gene expression: directed search of large microarray compendia. Bioinformatics 23: 2692-2699

Hibbs MA, Myers CL, Huttenhower C, Hess DC, Li Kai, et al. (2009) Directing experimental biology: a case study in mitochondrial biogenesis. PLoS Comput Biol 5(3): e1000322

Huttenhower C, Schroeder M, Chikina MD, Troyanskaya OG (2008) The Sleipnir library for computational functional genomics. Bioinformatics 24: 1559-1561

Jaakkola T, Diekhans M, Haussler D (1999) Using the Fisher kernel method to detect remote protein homologies. Proc Int Conf Intell Syst Mol Biol, pp 149-158

Jansen R, Gerstein M (2004) Analyzing protein function on a genomic scale: the importance of gold-standard positives and negatives for network prediction. Curr Opin Microbiol 7: 535-545

Jansen R, Yu H, Greenbaum D, Kluger Y, Krogan NJ, et al. (2003) A Bayesian networks approach for predicting protein-protein interactions from genomic data. Science 302: 449-453

Jardine N, Sibson R (1968) The construction of hierarchic and non-hierarchic classifications. Comput J 11: 177-184

Joachims T (1999) Making large-scale SVM learning practical. In: Scholkopf B, Burges C, Smola A (eds) Advances in Kernel Methods - Support Vector Learning. MIT Press, Boston, MA, USA

Kanehisa M (2002) The KEGG database. Novartis Found Symp 247: 91-101;

discussion 101-3, 119-28, 244-52

Kloster M, Tang C, Wingreen NS (2005) Finding regulatory modules through large-scale gene-expression data analysis. Bioinformatics 21: 1172-1179

Kohavi R (1995) A Study of Cross-Validation and Bootstrap for Accuracy Estimation and Model Selection. Proc Int Joint Conf Artif Intell, pp 1137-1145

Kundaje A, Middendorf M, Shah M, Wiggins CH, Freund Y, Leslie C (2006) A classification-based framework for predicting and analyzing gene regulatory response. BMC Bioinformat 7 (Suppl 1): S5

Lanckriet GR, Deng M, Cristianini N, Jordan MI, Noble WS (2004) Kernel-based data fusion and its application to protein function prediction in yeast. Pac Symp Biocomput, pp 300-311

Lee I, Li Z, Marcotte EM (2007) An improved, bias-reduced probabilistic functional gene network of baker's yeast, Saccharomyces cerevisiae. PLoS ONE 2(10): e988

Leek J, Storey J (2007) Capturing heterogeneity in gene expression studies by surrogate variable analysis. PLoS Genet 3(9): e161

Liao L, Noble WS (2003) Combining pairwise sequence similarity and support vector machines for detecting remote protein evolutionary and structural relationships. J Comput Biol 10: 857-868

MacKay DJC (2003) Information Theory, Inference, and Learning Algorithms. Cambridge Univ Press, Cambridge, UK

Madeira SC, Oliveira AL (2004) Biclustering algorithms for biological data analysis: A survey. IEEE/ACM Trans Comput Biol Bioinform 1: 24-45

Mansour Y (1997) Pessimistic decision tree pruning based on tree size. Proc 14th Intl Workshop on Mach Learn, pp 195-201

Mendez MA, Hödar C, Vulpe C, González M, Cambiazo V (2002) Discriminant analysis to evaluate clustering of gene expression data. FEBS Lett 522: 24-28

Middendorf M, Kundaje A, Wiggins C, Freund Y, Leslie C (2004) Predicting genetic regulatory response using classification. Bioinformatics 20 (Suppl 1): i232-i240

Model F, Adorján P, Olek A, Piepenbrock C (2001) Feature selection for DNA methylation based cancer classification. Bioinformatics 17 (Suppl 1): S157-S164

Moler EJ, Chow ML, Mian IS (2000) Analysis of molecular profile data using generative and discriminative methods. Physiol Genom 4: 109-126

Mostafavi S, Ray D, Warde-Farley D, Grouios C, Morris Q (2008) GeneMANIA: a real-time multiple association network integration algorithm for predicting gene function. Genome Biol 9 (Suppl 1): S4

Murphy K (2001) The Bayes Net Toolbox for Matlab. Comput Sci Stat 33: 200-201

Myers CL, Robson D, Wible A, Hibbs MA, Chiriac C, et al. (2005) Discovery of biological networks from diverse functional genomic data. Genome Biol 6: R114

Myers CL, Barrett DR, Hibbs MA, Huttenhower C, Troyanskaya OG (2006) Finding function: Evaluation methods for functional genomic data. BMC Genom 7: 187

Neal RM (1993) Probabilistic Inference using Markov Chain Monte Carlo Methods. Tech Rep CRG-TR-93-1, Univ of Toronto, Canada

Noble WS (2004) Support vector machine applications in computational biology. In: Schoelkopf B, Tsuda K, Vert JP (eds) Kernel Methods in Computational Biology. MIT Press, Boston, MA, USA, pp 71-92

Noble WS (2006) What is a support vector machine? Nat Biotechnol 24: 1565-1567

Owen AB, Stuart J, Mach K, Villeneuve AM, Kim S (2003) A gene recommender algorithm to identify coexpressed genes in *C. elegans*. Genome Res 13: 1828-1837

Povalej P, Lenic M, Zorman M, Kokol P, Dinevski D (2005) Accuracy of intelligent medical systems. Comput Meth Prog Biomed 80 (Suppl 1): S95-S105

Raychaudhuri S, Stuart JM, Altman RB (2000) Principal components analysis to summarize microarray experiments: Application to sporulation time series. Pac Symp Biocomput, pp 455-466

Rodgers J, Nicewander W (1988) Thirteen ways to look at the correlation coefficient. Am Stat 42: 59-66

Rosenkrantz DJ, Stearns RE, Lewis PM, II (1977) An analysis of several heuristics for the Traveling Salesman Problem. SIAM J Comput 6: 563-581

Russell N (2002) Artificial Intelligence: A Modern Approach. Prentice Hall, Englewood Cliffs, NJ, USA

Sachs K, Perez O, Pe'er D, Lauffenburger DA, Nolan GP (2005) Causal protein-signaling networks derived from multiparameter single-cell data. Science 308: 523-529

Saeed AI, Sharov V, White J, Li J, Liang W et al. (2003) TM4: a free, open-source system for microarray data management and analysis. Biotechniques 34: 374-378

Saldanha AJ, Brauer MJ, Botstein D (2004) Nutritional homeostasis in batch and steady-state culture of yeast. Mol Biol Cell 15: 4089-4104

Schrijver A (2005) On the history of combinatorial optimization (till 1960). In: Aardal K, Nemhauser GL, Weismantel R (eds) Handbook of Discrete Optimization. Elsevier, Amsterdam, Netherlands, pp 1-68

Sealfon RS, Hibbs MA, Huttenhower C, Myers CL, Troyanskaya OG (2006) GOLEM: An interactive graph-based gene-ontology navigation and analysis tool. BMC Bioinformat 7: 443

Segal NH, Pavlidis P, Antonescu CR, Maki RG, Noble WS, et al. (2003) Classification and subtype prediction of adult soft tissue sarcoma by functional genomics. Am J Pathol 163: 691-700

Seo J, Shneiderman B (2002) Interactively exploring hierarchical clustering results [geneidentification]. Computer 35: 80-86

Sharan R, Maron-Katz A, Shamir R (2003) CLICK and EXPANDER: A system for clustering and visualizing gene expression data. Bioinformatics 19: 1787-1799

Sherlock G (2000) Analysis of large-scale gene expression data. Curr Opin Immunol 12: 201-205

Sokal RR, Michener CD (1958) A statistical method for evaluating systematic relationships. Univ Kansas Sci Bull 28: 1409-1438

Spearman C (1904) The proof and measurement of association between two things. Am J Psychol 15: 72-101

St Onge RP, Mani R, Oh J, Proctor M, Fung E et al. (2007) Systematic pathway analysis using high-resolution fitness profiling of combinatorial gene deletions. Nat Genet 39: 199-206

Sturn A, Quackenbush J, Trajanoski Z (2002) Genesis: Cluster analysis of microarray data. Bioinformatics 18: 207-208

Subramanian A, Tamayo P, Mootha VK, Mukherjee S, Ebert BL, et al. (2005) Gene set enrichment analysis: A knowledge-based approach for interpreting genome-wide expression profiles. Proc Natl Acad Sci USA 102: 15545-15550

Tamayo P, Slonim D, Mesirov J, Zhu Q, Kitareewan S, et al. (1999) Interpreting patterns of gene expression with self-organizing maps: methods and application to hematopoietic differentiation. Proc Natl Acad Sci USA 96: 2907-2912

Troyanskaya OG, Dolinski K, Owen AB, Altman RB, Botstein D (2003) A Bayesian framework for combining heterogeneous data sources for gene function prediction (in *Saccharomyces cerevisiae*). Proc Natl Acad Sci USA 100: 8348-8353

Vapnik VN (1995) The Nature of Statistical Learning Theory. Springer, New York, NY, USA

Varshavsky R, Horn D, Linial M (2008) Global considerations in hierarchical clustering reveal meaningful patterns in data. PLoS ONE 3: e2247

Vert JP (2002) Support vector machine prediction of signal peptide cleavage site using a new class of kernels for strings. Pac Sym Biocomput: 649-660

Wall E, Rechtsteiner, Rocha M (2003) Singular value decomposition and principal component analysis. In: Berrar P, Dubitzky, Granzow (eds) A Practical Approach to Microarray Data Analysis. Kluwer Academic Publ, Boston, MA, USA, pp 91-109

Waxman HS, Worley WE (1990) Computer-assisted adult medical diagnosis: subject review and evaluation of a new microcomputer-based system. Medicine (Baltimore) 69: 125-136

Wilson D, Martinez T (1997) Improved Heterogeneous Distance Functions. J Artif Intel Res 6: 1-34

Yeung KY, Haynor DR, Ruzzo WL (2001) Validating clustering for gene expression data. Bioinformatics 17: 309-318

Zien A, Rätsch G, Mika S, Schölkopf B, Lengauer T, Müller KR (2000) Engineering support vector machine kernels that recognize translation initiation sites. Bioinformatics 16: 799-807

15 Application of Genomics to Plant Breeding

Thomas Lübberstedt* and Madan K. Bhattacharyya
Iowa State University, Department of Agronomy,
Agronomy Hall, 50011-1010 Ames, IA, USA
*Corresponding author: thomasl@iastate.edu

1 INTRODUCTION

Lederberg and McGray (2001) traced the origin of the term "Genomics" to the foundation of the journal with same name in 1977. Whereas the initial focus was on mapping and sequencing of genomes, todays definitions are broader. Besides exploring the structure of complete genomes (structural genomics), genomics strives to study the function of all genes in an organism (functional genomics) in a multiparallel manner (Lander 1996). An essential component of current genomics projects is the use of tools allowing massively parallel studies of several or all genes of an organism such as microarrays for transcriptome expression profiling experiments. Moreover, the close phylogenetic relationships of species and their genomes within systematic families led to the development of both comparative genomic approaches and the model species concept. The underlying idea is that species with small genomes such as *Arabidopsis thaliana* and rice in plants can be more easily addressed by expensive structural and functional genomics approaches. Information gathered in model species can then be transferred to related species by use of syntenic relationships of their genomes (Devos 2005), both for hypothesis-driven research or application in crop species. However, with the advent of new sequencing technologies (e.g., FLX and Solexa) and progress in genomic tool development, genomic tools and information (such as completely sequenced genomes) have become increasingly available in all major crop species (e.g, *http://www.jgi. doe.gov/genom projects/pages/projects.jsf?taxonomy=Eukaryote%2C+Large*). Depending on the definition, genomics includes epigenomics, computational genomics, proteomics, structural genomics for protein structures and metabolomics, which will not be specifically discussed below.

Although genomics has undoubtedly helped to understand the structure of genomes and the function of genes, its usefulness for crop sciences and in particular plant breeding has been debated (Cooper et al. 2004). However, meanwhile all major plant breeding established directly or by collaboration high-throughput pipelines for molecular markers, various genomic tools, and transgene technology (e.g., Monsanto R&D pipeline 2008). Thus, the question of applying genomic tools and approaches in plant breeding has evolved to: which of the various genomic options is most efficient to address a particular step during the breeding process, and is it more efficient than traditional conventional options (such as phenotypic selection). According to Hallauer and Miranda (1988), plant breeding can be divided into (base) population

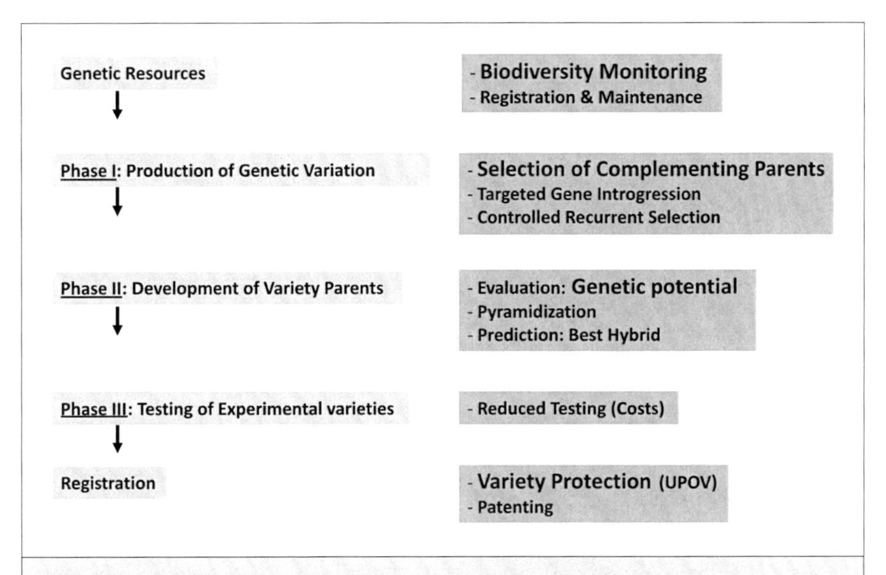

Fig. 1 Application of tools and approaches derived from genomics to plant breeding. On the left side, the major steps in crop breeding are displayed, valid for any breeding category (i.e., line, clone, population, and hybrid breeding). On the right side, various approaches are assigned to these basic steps in plant breeding based on or benefiting from tools or resources provided by plant genomics.

improvement and variety development, the latter subdivided into generation of genetic diversity, development of variety parents (e.g., inbred lines in maize breeding), and testing of experimental varieties (Fig. 1). The objectives of this chapter are to (1) present the major genomic tools and approaches with relevance to plant breeding, (2) discuss their application in plant breeding, (3) present examples for application of genomics to improve input and output traits, and (4) discuss prospects and limitations of genomic versus conventional alternatives in plant breeding.

2 PLANT GENOMICS AND GENOMICS TOOLS FOR PLANT BREEDING

One of the major contributions of genomics research to plant breeding has been in the area of structure and organization of plant genomes. Through whole-genome and comparative mapping studies, much is known on how genes in euchromatic and heterochromatic areas are organized in chromosomes. With the advent of molecular marker in 1980 (Botstein et al. 1980), it became feasible to develop genetic maps of the whole genome of plant species from single crosses. The first whole-genome molecular map was developed for tomato. Subsequently, molecular maps were constructed for other crop species including potato, capsicum, eggplant, wheat, rice, maize, sorghum, soybean, etc. Meanwhile, molecular maps are available for most plant species of agricultural importance. Molecular mapping studies subsequently revealed many important aspects of plant genomes. The first high-density maps of over 1,000 molecular markers constructed for tomato and potato genomes led to the understanding of both genome structures and evolutionary relationship between species within Solanaceae (Tanksley et al. 1992). Crossing-over rates of different chromosomes were positively correlated to the length of chromosomes. However, crossing-over events were unevenly distributed within a chromosome. Integration of the cytogenic map with molecular and classical maps in tomato revealed 10-fold reduced recombination rates for centromeres and centromeric heterochromatin, and in some telomeric regions as compared to the rest of the tomato genome. Surprisingly, a significant proportion (28%) of molecular markers including gene transcripts mapped to these recombination-suppressed regions (Tanksley et al. 1992). High-resolution physical maps of recombination events based on frequency and distribution of recombination nodules on tomato synaptonemal complexes (SCs) revealed similar results (Sherman and Stack

1995). In these physical maps, that were developed based on location of recombination nodules on pachytene chromosomes, exact locations of recombination events were depicted. Furthermore, how a crossing-over affects the subsequent crossing-overs within a chromosome (interference), and how both positive and negative interferences were distributed within the long arm of chromosomes, was determined and discussed (Sherman and Stack 1995). These early studies in the model plant tomato have clearly demonstrated the distribution of recombination events in the plant genome.

Whole genome studies with molecular markers also revealed homoelogous regions within plant genomes suggesting polyploid nature of crop species. In an elaborate study with nine segregating soybean populations developed from both intra- (*Glycine max* x *G. max*) and interspecific (*G. max* x *G. sojae*) crosses, homoelogous regions ranging from 1.5-106.4 cM, with an average size of 45.3 cM were identified (Shoemaker et al. 1996). Some of the segments were found to occur six times with an average of 2.55 duplications per segment. Earlier, pairing of chromosomes during meiosis suggested presence of extensive homoeology and thus polyploid nature of the soybean genome (Crane et al. 1982). Mapping of expressed sequence tags (ESTs) suggested that the soybean genome has undergone two rounds of polyploidization (Shoemaker et al. 1996; Schlueter et al. 2004). Comparably, maize is a segmental allotetraploid developed most likely from three rounds of duplication (Gaut and Doebley 1997; Schlueter et al. 2004), whereas rice, sorghum, barley, tomato, potato, and *Medicago* have likely undergone two rounds of polyploidization (Gaut and Doebley 1997; Schlueter et al. 2004).

In hexaploid bread wheat (*Triticum aestivum*), investigation of deletion lines allowed discrimination of gene-rich and gene-poor chromosomal segments (Sandhu and Gill 2002). For example, mapping of 1,110 ESTs to 2,600 loci revealed two gene-rich islands that were surrounded by gene-poor regions on both short and long arms of wheat chromosome 2 (Conley et al. 2004). Comparison of gene-rich and gene-poor regions with the extent of crossing over revealed that the rate of recombination was positively correlated with the proportion of genes in the gene-rich regions, with suppression of recombination in gene-poor and centromeric regions (Sandhu and Gill 2002). To date, near-complete genome sequences have been obtained for only two plants: the eudicot *Arabidopsis thaliana* and the moncot rice (*Oryza sativa* and *O. japonica*). A draft sequence of black cottonwood (*Populus trichocarpa*) is also available

(Tuskan et al. 2006). Genome sequencing of soybean (*Glycine max*) and maize (*Zea mays*) is in an advanced stage. Once available, sequence of these two crop species will provide a complete picture of the genome structures in these crop species.

Genome mapping studies led to the discovery of colinearity of gene orders in chromosomes between related crop species or genera within a family. The first study to demonstrate colinearity of gene orders was conducted in solanaceious species, tomato (*Lycopersicon esculentum*) and potato (*Solanum tuberosum*) using cDNA markers (Bonierbale et al. 1988). Tomato (*Lycopersicon esculentum*) is more closely related to potato (*Solanum tuberosum*) and eggplant (*Solanum melongena*) than to pepper (*Capsicum* spp.) (Bonierbale et al. 1988; Tanksley et al. 1988, 1992; Doganlar et al. 2002). Conservation of gene order was also observed among the members of the gramineae family. A large degree of colinearity was recorded within 5-10 cM regions of grass species. This led to development of a circular genome map to illustrate colinearity of chromosome segments of wheat, maize, rice, oat, sorghum, and foxtail millet (Devos and Gale 1997). As compared to the members of solanaceae and gramineae families, members of leguminosae and brassicaceae families are more diverse and show colinearity to a lesser degree (Lagercrantz 1998; Cannon 2008). For example, colinearity between the model legume *Medicago truncatula* and soybean is limited to small genetic intervals (Choi et al. 2004).

Conservation of gene orders among members of a family, described by *comparative genomics* studies, bears important implications in the field of both genetics and plant breeding. We can address the issue of plant evolution through comparative genomics studies. Gene identification from a complex species can be facilitated through comparative genomics studies with species carrying less complex genomes (e.g., Sandhu and Gill 2002). Development of molecular markers for a new species can be expedited through its possible colinearity with extensively characterized species (Hougaard et al. 2008).

Tightly-linked genes are inherited in clusters from generation to generation. As result, tightly-linked molecular marker loci cosegregate in a population, and particular alleles occur together more (or less) often than expected by chance, which is known as linkage disequilibrium (LD). However, due to random mating alleles of loosely-linked genes recombine and reach linkage equilibrium. Thus, LD decays with time and only very tightly-linked loci are able to stay together and show LD. In the open-pollinated crop maize, LD decays at a much faster rate than the self-pollinated crop soybean or the allopolyploid species sugarcane (Flint-Garcia et al. 2003; Hyten et al. 2007). LD between tightly-linked molecular markers and quantitative trait loci (QTL) allow identification of useful markers for selection of QTL through association mapping studies on a large set of diverse germplasm or advance generations of crosses. In maize, this approach has been applied in mapping starch composition and oleic acid contents (Wilson et al. 2004; Belo et al. 2008).

Finding function of individual genes is the final frontier of genomics research and will have immense implication in plant biology, genetics and plant breeding. *Functional genomics* approaches allow us to investigate the function of potentially the whole gene set of an organism rather than single genes. As a result, gene networks carrying out various physiological functions are being discovered. As a first step towards conducting functional genomics studies, transcripts of genes (EST) were sequenced. For some crop species, such as maize, rice and wheat over one million ESTs are available (*http://www.ncbi.nlm.nih.gov/dbEST/dbEST_summary.html*). Availability of ESTs not only allowed uncovering the genic sequences of the genome, but also contributed towards developing experimental tools for studying expression patterns of thousands of genes in single biological samples, such as leaves, roots, stems, flowers, petals, infected and uninfected plant tissues, etc. These experiments are now applicable to a large number of economically important crop species (*http://www.plexdb.org/plex.php?database=Barley*).

This high-throughput transcript profiling functional genomics approach has important implication in plant breeding. One can apply this approach in identifying candidate genes that underlie QTLs (e.g., Druka et al. 2008). This approach requires transcript profiling of a set of recombinant inbred lines (RILs) that are characterized for QTL. This method was coined *genetical genomics* (Jansen and Nap 2001). The loci that differ quantitatively for expression of transcripts among RILs are known as expression QTL (eQTL). Thus, tight association between eQTL and QTL in a set of recombinant inbred lines (RILs) suggests eQTL to carry genes that are essential for the QTL under study.

Reverse genetics and genomics will play a pivotal role in finding the function of candidate genes for agronomic attributes. Transposon and T-DNA insertion mutatgenesis as well as point mutations induced by chemical mutagens, *targeting induced local lesions in genomes* (TILLING), RNA interference (RNAi) and virus-induced gene silencing (VIGS) are being applied in defining the function of genes that are considered to have agronomical importance. TILLING can also be applied in creation

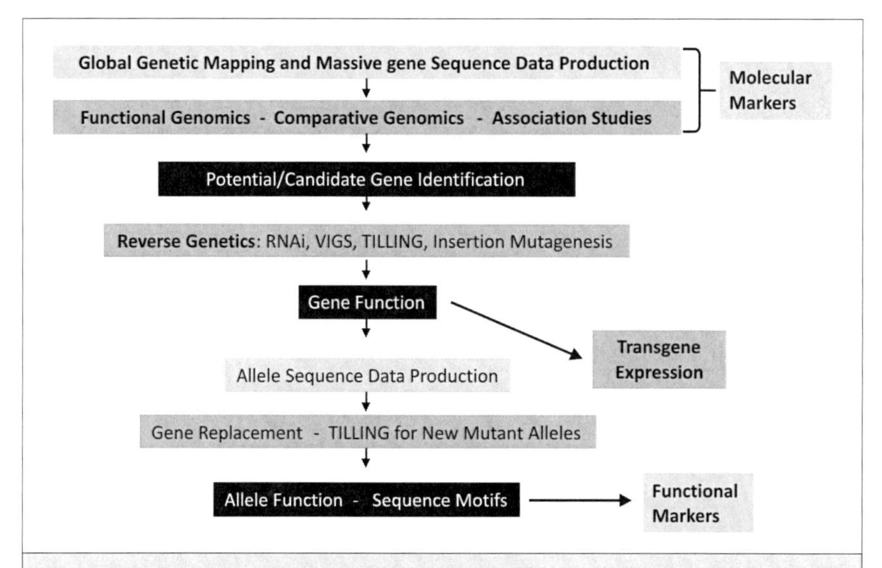

Fig. 2 Plant genomics approaches applicable to genetic improvement of crop plants. Three major outcomes of plant genomics, molecular markers, functional markers and transgene expression, are expected to be the key major players of plant breeding in 21st century.

or identification of new functional alleles or genetic variation with desirable phenotypes (e.g., Slade and Knauf 2005; Nieto et al. 2007). Plant genomics approaches applicable to genetic improvement of crop plants are outlined in Fig. 2. The National Research Initiative (NRI) of the Plant Genome program of the United States Department of Agriculture (USDA) and the Plant Genome program of the National Science Foundation (NSF) have been supporting the scientific community to create genomics resources. Under the Applied Plant Genomics Coordinated Agricultural Project (CAP) program of the USDA-NRI, genomics resources (genomics tool box) were specifically created to assist breeders in expediting breeding programs. Together with the on-going plant genomics research in the last decade, CAP programs generated an immense amount of resources. In Table 1, Uniform Resource Locators (URL) for websites containing genomics resources (tool boxes) for plant breeding are presented. These URLs identify experimental plant populations, markers and primers for marker amplification, protocols for conducting molecular analyses, literature relating to various traits, etc.

Table 1 Genomics resources and tools for plant breeding in the 21st century.

Crop Species	*Uniform Resource Locator*
Solanaceious	*https://www.msu.edu/~douchesd/SolResources.html*
Wheat	*https://www.msu.edu/~douchesd/SolResources.html*
Barley	*http://www.hordeumtoolbox.org*
Soybean	*http://soybase.org/community.php*
Maize	*http://www.maizegdb.org/*
Rice	*http://www.uark.edu/ua/ricecap/database.htm*

3 IMPLEMENTATION OF GENOMICS IN PLANT BREEDING

3.1 Management and Exploitation of Genetic Resources

Genetic diversity is prerequisite for the development of improved breeding materials. Although plant breeders prefer to use adapted and elite germplasm to develop new varieties, genetic variation has in several cases proven to be lacking in elite germplasm, particularly for disease resistance. Management and availability of genetic resources become increasingly important for the following reasons: (i) due to an increasing human population, doubling of seed yield is required in the next about 3-4 decades, (ii) novel applications of plant materials such as in the area of biorenewables require new types of germplasm, (iii) the projected climate change affects plant production systems globally, and (iv) new epidemic diseases are threatening crop production such as wheat stem rust.

The main activities of the about 150 gene banks worldwide can be divided into genebank management and crop breeding (Spooner et al. 2005). Genetic resources can be either maintained in situ in original habitats, or collected in ex situ gene banks. The major genomic tools to support these activities for managing and utilizing genebanks are molecular markers (de Vicente 2007). Therefore, the establishment of complementary DNA banks has been discussed recently (de Vicente and Anderson 2005).

Gene bank management can be divided into (i) taxonomic issues, (ii) germplasm characterization, (iii) acquisition and collection of materials, and (iv) maintenance of genetic integrity of genetic resources (Spooner et al. 2005). For several aspects related to description of genetic resources, molecular markers have proven to be superior to phenotypic descriptors, due to lack of environmental impact and the large number of 'neutral' markers as compared to the limited number of confounded phenotypic traits. Though results obtained with molecular markers partly differ from classical grouping, they have proven useful to resolve evolution and origin of several species, to establish relationships of populations within species, and to describe populations (Spooner et al. 2005). For efficient taxonomic characterization of new species, 'DNA barcoding' based on few universal genomic regions present in all species such as internal transcribed spacers have been used but found too variable within species (Spooner 2008). With the advent of low-cost and high-throughput sequencing technologies, analysis of whole plastomes, for example, seems to be both more powerful and feasible. For within species population characterization various genomic fingerprinting methodologies have been successfully applied (Spooner et al. 2005).

Knowledge on the diversity and the relationship of populations provides important information to decide, where to direct in situ conservation efforts, and which populations to collect to add useful genetic variation to existing ex situ germplasm collections. Both for in situ and ex situ collections, molecular markers are useful to maintain the integrity of original populations during seed (or plant) multiplication. Another important application is the identification and elimination of redundant populations. In order to maximize the value of germplasm collections given limited funding, the concept of 'core collections' has been developed (Frankel 1984). Essentially, the goal of core collections is to represent a maximum of genetic diversity within a species with a minimum of repetitiveness. Molecular markers combined with phenotypic data seem to be appropriate for establishing respective core collections (Spooner et al. 2005). Moreover, synchronization of collections in different gene banks might also be facilitated by the use of markers. DNA banks would collect and store genomic DNA for the available germplasm colletions in genebanks, and allow to readily access these resources with molecular tools (e.g., to identify particular donor populations).

The second main purpose of genebanks besides conservation is to make genetic resources available for crop breeding. Molecular markers have been used to identify exotic populations for maximizing genetic variation in elite x exotic combinations. Tanksley and Nelson (1996) proposed advanced backcrossing schemes for marker-assisted introgression of exotic chromosome segments into elite germplasm. This or similar methods have meanwhile proven to be successful to identify even yield or quality related genes and chromosome segments in a number of wild species including rice (Robin et al. 2003), barley (Pillen et al. 2003), and tomato (Frary et al. 2004). Alternatively, genes of interest can be identified within exotic populations, and transferred into elite germplasm either by transgenic or marker-aided approaches. Recent success stories include the *Phytophtora infestans* resistance gene isolated in wild potato, meanwhile successfully transferred into cultivated potato (Ballvora et al. 2002). In rice, a wide-incompatibility gene was isolated (Chen et al. 2008a), facilitating the exploitation of rice genetic resources. Different ploidy levels might inhibit use of genetic resources. Again, molecular markers are useful to determine ploidy levels, discriminate between allo- and autopolyploidy, and may be useful to identify ancestral parents to design species resynthesis programs.

Recent progress in developing low-cost and high-throughput methods for large-scale sequencing as well as high-throughput genotyping, will benefit both management and exploitation of genetic resources. With the availability of sequence and genotype data at low costs, the above mentioned topics related to systematics as well as gene bank management will increasingly involve these technologies. With rapidly increasing knowledge on the function of genes by use of functional genomics approaches particularly in model species such as *Arabidopsis thaliana* and rice, identification of respective genes in crop species will be facilitated either by use of sequence homology-based gene isolation (candidate gene approach), or by exploitation of the relationship of genomes (synteny) (Devos 2005). However, major progress in the area of genetic resources can likely be expected from the ability to functionally describe genetic variation based on association genetic approaches (Thornsberry et al. 2001). This will in the long run be helpful to value genetic resources and to allow more targeted identification of novel and useful alleles.

Recurrent Selection

Hallauer and Miranda (1988) discriminated plant breeding into long-, mid-, and short-term activities. In the short-term programs aiming at immediate variety development lead to the most rapid gains in selection experiments, but yield plateaus at suboptimal levels. In contrast, mid- and long-term selection experiments aiming at the improvement of source populations prior to variety development offer the potential to deliver source materials for higher yielding varieties. Reciprocal recurrent breeding programs are meanwhile established approaches securing long-term success of breeding programs.

Within the toolbox of genomics, in particular the usefulness of molecular markers has been discussed for improving the efficiency of selection programs (Lande and Thompson 1990). Marker-assisted selection (MAS) has been shown to be more efficient than conventional phenotypic selection for traits with low heritability (Lande and Thompson 1990). Meanwhile, marker-assisted recurrent selection (MARS) has been successfully implemented in breeding programs of Monsanto and other companies for different species (Eathington et al. 2007). Simulation studies for moderate population sizes (100 individuals) indicated that selection response was maximal for oligogenic inherited traits (10 QTLs), when QTL positions have been identified before applying MARS, and when markers were available for the QTL directly as compared to flanking markers (Bernardo and Charcosset 2006). In case of multigenic traits, selection for only those QTL with large effects seemed most efficient, even if all minor QTL positions were known.

More recently, genomic selection has been suggested as an even more efficient approach than MARS for marker-aided recurrent selection programs (Meuwissen et al. 2001; Bernardo and Yu 2007). Different from MARS, markers are not pre-selected based on associations with QTL for subsequent selection experiments. Instead, genomic selection exploits the availability of multiple low-cost markers, as developed in current genomics projects. During selection cycles, breeding values are assigned to markers and used for selection in the next cycle(s). In different scenarios for multigenic inherited traits, genomic selection was superior to MARS in a simulation study (Bernardo and Yu 2007), in particular with costs per data point below two cents. In addition, use of marker haplotypes rather than single markers in order to identify genome regions identical by descent increased genomic selection procedures (Calus et al. 2008).

A limitation for broad application of marker-aided recurrent selection procedures is the poor transferability of marker-QTL associations across different populations (e.g., Lübberstedt et al. 1998; Mihalevic 2005). This is caused by allelic, epistatic, and QTL x environment interactions. To overcome this limitation, the 'mapping as you go' (MAYG) approach has been proposed (Podlich et al. 2004). Essentially, this approach requires re-estimation of QTL effects in subsequent cycles of selection, in order to determine gene effects in altering genetic background. MAYG was more efficient than MARS in a simulation study (Podlich et al. 2004), however, at the expense of additional genotyping efforts.

In conclusion, especially with the advent of large numbers of markers at low cost by genomics programs, marker-aided selection procedures have become increasingly attractive. However, so far the amount of literature on MAS procedures is lacking far behind publications on detection of QTL-marker associations (Xu and Crouch 2008).

3.2 Generation of Genetic Variability

The first step in any breeding program is to create genetic variation. Genomics contributes by (i) broadening genetic variation available within a species by making genes available from other non-crossable species or organisms, (ii) by creating novel genetic variation within species, and (iii) by assisting in the identification of complementary parent genotypes to derive useful segregating populations for subsequent variety development. Worldwide, access to functionally characterized genes from any organism, likely had the largest impact on modern plant breeding in the past decade. Since the first commercialization of a transgenic plant in 1996, the acreage of transgenic crops has increased to 120 millon hectares, with mainly two groups of 'traits' from foreign organisms: herbicide tolerance and insect resistance (Bt) (James 2007). More recently, 'cisgenic' modifications based on transformation of intraspecific genes have been advocated (Schouten et al. 2006; Schubert 2006), both to overcome concerns regarding health and ecological effects of transgenes, and to exploit a n increasing body of characterized genes within major crop species, available from functional genomics studies.

As mentioned above, in several major crop species insertion mutations are induced at a large scale, in order to evaluate the function of genes in vivo. To create these knock-out mutations, intraspecific transposons are employed in species like maize. In species without any active endogenous transposons either *Agrobacterium tumefaciens* T-DNA insertions or transposons introduced from other species are used to create insertions. More recently, efficient procedures for down-regulation of genes based on RNAi derived from intraspecific gene sequences have been established. In all cases, knock-out alleles might represent novel genetic variation of interest for breeding programs. Similarly, TILLING populations have been established to induce point mutations for subsequent forward and reverse genetic approaches, which in addition might contain novel and useful variants for breeding programs. In tomato, for example, a 'natural' TILLING variant has been identified, which can substitute the transgenic Flavr savr tomato, and might thus be more acceptable by consumers world-wide (*http://abstracts.aspb.org/pb2003/public/P72/1452.html*). Finally, with an increased understanding of sequence-function relationships, valuable alleles might be identified in respective TILLING populations in a more targeted way by reverse genetic approaches.

When considering the whole genome of parental lines to be crossed for generating segregating populations, the quantitative genetic 'usefulness' concept applies (Schnell and Utz 1975). A recent study concluded that only the midparent values are the best phenotypic predictors of the usefulness of a line combination (Utz et al. 2001). Molecular markers were not able to predict the expected genetic variation based on the genetic distance of parental lines in a companion study (Bohn et al. 1999). Alternatively, other studies have suggested to make use of QTL information for parent selection (Zhong and Jannink 2007 and references therein). They also showed that a higher marker coverage of the genome might be needed in this context as employed by Bohn et al. (1999). However, whereas parent selection is limiting the success of all subsequent steps in variety development, systematic approaches to help finding optimal parent combinations are largely missing. Further decreasing sequencing costs enabling routine whole-genome sequencing combined with increasing knowledge on sequence – function relationships for multiple genes holds promise to overcome this problem in the long run.

3.3 Variety Development

After initial crosses to generate genetic variation, parent genotypes need to be developed for some, but not all breeding categories. In case of clone breeding, for example, all offspring of the initial crosses represent potential varieties. They 'only' need to be multiplied and extensively tested prior to variety registration trials. In contrast, both line and hybrid breeding require the development of pure breeding lines, which either represent potential varieties (line breeding) or parents of hybrid varieties. One of the major advances to accelerate this process for several major crop species was the development of doubled haploid production technologies (e.g., Röber et al. 2005), though based on conventional genetic or biotechnological methods rather than genomics approaches.

Marker-assisted selection (MAS) has been discussed for many years

(Lande and Thompson 1990) to be a useful tool to at least complement phenotypic selection. One major obstacle for traits with low heritability is requirement of extensive phenotypic evaluations in order to establish QTL-marker associations. This raises the question of the additional benefit of employing molecular markers. For simple inherited traits, often low-cost phenotypic assays are available, again raising the question on why to use marker assays. In addition, the transferability of QTL from one to other populations, or by using different testers in case of hybrid breeding has shown to be very limited (e.g., Lübberstedt et al. 1997, 1998). However, decreasing costs for high density marker assays or genome sequencing enable a better understanding of the pedigree and haplotype structure within elite germplasm (Jannink et al. 2001; Cavanagh et al. 2008), the effect of genome regions can be re-evaluated by strategies as MAYG (Podlich et al. 2004), mentioned above, and ultimately breeding values assigned to haplotypes (Bernardo 2007; Calus et al. 2008). Major breeding companies announced meanwhile that MAS has become routine and a key technology in their breeding programs (*http://todayinscience.wordpress.com/2007/11/08/accelerated-yield-technology-from-pioneer-hi-bred-2/*; Eathington et al. 2007).

In the long run, the number of characterized sequence polymorphisms (QTN or QTIndels) causal for trait variation will substantially increase by, e.g., association studies or TILLING (see for examples, Bagge et al. 2007). Functional markers (Andersen and Lübberstedt 2003) derived from respective polymorphisms will increase the power of MAS (Bernardo 2007) by avoiding the type III error of marker misclassification due to recombination between marker and QTL allele, and be accompanied by decreasing costs per marker data point in the recent years (Bagge and Lübberstedt 2008).

A variant of MAS is marker-assisted backcrossing (MAB) of single or multiple genes or genome regions from a donor into an elite recurrent parent line. MAB consists of two steps, foreground selection to ensure transmission of the genome region(s) or gene(s) of interest into the next generation, and background selection against donor chromosome segments (Frisch and Melchinger 2001). The goal is to establish near-isogenic lines carrying the gene(s) of interest in a short time. With advances in genomics, markers derived from the target gene(s) of interest will increasingly be available for foreground selection, as opposed to genetically linked markers. In addition, genome sequencing projects will reveal flanking sequences and enable the development of closely linked markers. Taken together, the precision of MAB by use of target-gene derived markers will be increased, whereas selection of rare recombinants adjacent to the target gene will be facilitated to eliminate undesirable side effects by linkage drag.

Advances in high-throughput DNA isolation methods starting from partial seed materials paired with low-cost marker assays dramatically increased the efficiency of MAB. The major limitation for rapid recovery of the recurrent parent genome in a few backcross (BC) generations and identification of rare recombinants in the vicinity of the target gene(s) is the evaluation of a large BC population for linked molecular markers (Frisch and Melchinger 2001). In addition, transfer of not only one or few genes from the same donor, but multiple genes from different donors into the same recurrent parent require more sophisticated crossing schemes (Fehr 1991). However, efficient marker procedures will facilitate these more laborious procedures to achieve introgressions within a short time. The same applies to introgression of single or multiple transgenes.

3.4 Examination of Experimental Varieties

Once parents of potential varieties have been developed, their value to create varieties has to be determined. In case of line and clone breeding, this relates to extensive field trials and phenotypic evaluation of multiple experimental line or clone varieties. The major traits such as seed yield require extensive field testing. However, knowledge of functional alleles, e.g., for particular resistances might obleviate the necessity of respective resistance tests, and thus reduces costs for the testing in specific environments or under controlled conditions. Accumulating knowledge on beneficial alleles for several agronomic traits will either reduce the number of experimental lines to be tested for seed yield or make breeding of varieties across all traits more efficient.

The value of genotypes or lines used to generate synthetic or hybrid varieties depends only partially on their *per se* performance. In contrast, their value depends primarily on their combining abilities within a population (synthetics) or with lines from another heterotic group (hybrids). The most promising experimental varieties consisting of heterozygous genotypes (hybrids) or populations (synthetics) show maximum hybrid vigor for yield characters.

Molecular markers have been successfully used to describe the genetic distance of inbred lines and to assign lines to heterotic groups (Melchinger 1999). However, so far the success of predicting hybrid performance, heterosis, or specific combining ability of interpool hybrids has been limited (Melchinger 1999). Nevertheless, Vuylsteke et al. (2000) and Schrag et al. (2006) demonstrated, that these predictions can be improved, if genetic effects of markers (due to marker-QTL associations) are taken into account rather than treating all markers equally. Thus, with increasing knowledge of gene (allele) and haplotype effects on the traits of interest, including allelic interactions, better predictions for hybrid performance might be obtained in future. However, epistatic interactions have so far been ignored with regard to prediction of hybrid performance, specific combining ability (SCA), etc., which was found to be important in a recent study on biomass yield in *Arabidopsis* (Kusterer et al. 2007).

Meanwhile, several studies have been published (reviewed by Hochholdinger and Hoecker 2007) on the use of genomic approaches to study heterosis. In particular, microarray or RT-PCR-based transcript profiling studies were performed on different organs and stages in *Arabidopsis* and maize. Within some of these studies, expression patterns across different inbred-hybrid triplets showed very limited overlap (Hoecker et al. 2006; Uzarowska et al. 2007). In conclusion, since the genes expressing 'heterotic patterns' varied substantially between triplets, transcript-based prediction of optimal inbred combinations to maximize expression of heterosis as well as hybrid performance for agronomic traits seems rather unlikely. It remains to be shown, whether metabolite signatures have high predictive power for biomass yield related characters (Meyer et al. 2007), and are better suited as 'bio-markers' for hybrid performance.

Neither the dominance nor the overdominance hypotheses were preferentially supported by the different transcript expression profiling studies (Hochholdinger and Hoecker 2007), whereas the epistasis hypothesis was not tested. Allele-specific transcript expression analyses of older versus more recent hybrids revealed an increase in dominant as compared to overdominant expression patterns. Another important finding in this context, resulting from sequencing of large genome segments in maize, is the frequent lack of colinearity of allelic genome regions. For example, Brunner et al. (2005) found several genes either lacking or present in large colinear regions of different inbreds. Thus, hybrids formed by crossing different inbreds with non-colinear genomes differ not only at the level of allele combinations, but also at the level of gene combinations. Hybrids would carry more genes than either of the non-colinear parental inbreds, albeit partly in hemi- rather than heterozygous state. Again, it remains to be shown, whether accumulating genomic sequence data from multiple inbreds in maize will be useful to predict, which combination of inbred line genomes, and thus gene combinations, will maximize hybrid performance. Furthermore, it is not yet clear, whether a similar high level of non-colinearity is also present in other plant species.

3.5 Variety Description, Registration, and Protection

The knowledge on the presence of particular alleles or allele combinations in varieties can be used for their description. For cereals, presence of specific resistance alleles in cereal varieties is included, e.g., in the Danish variety list (Willas 2006). Although based on pedigree information so far, it will likely be extended to information based on alleles functionally characterized by molecular approaches. An obvious advantage with accumulating functional allele information and availability of high-throughput genotyping of sequencing methods is the possibility for multiparallel validation of the presence of respective alleles by authorities involved in variety testing and registration. Moreover, respective information would be useful to determine DUS (distinctiveness, uniformity, stability) criteria, which are essential for admission of new varieties in several UPOV (Union for the Protection of New Varieties in Plants) member countries (Button 2008). Once varieties have been registered and are distributed to farmers, molecular markers, as provided by genomics projects, might be useful for protection of varieties. Especially the concept of 'essentially derived varieties' (EDVs) has been extensively discussed, in particular, how markers can be used to discriminate between sufficiently distinct and derived varieties (Heckenberger et al. 2005; Button 2008).

4 CONTRIBUTION TO TRAIT IMPROVEMENT

4.1 Plant Genomics for Input Trait Improvement

Plant genomics has been playing a major role in crop improvement. Two approaches, (i) transformation technology for expression of transgenes, and (ii) marker-assisted selection to integrate genes for genetic improvement of crop plants benefit from genomics research. Genetic improvement in crops can be accomplished in input or/and output traits (*http://www.ext.vt.edu/pubs/biotech/443-002/443-002.html*). Input traits lower the cost of cultivation and improve yield by reducing the losses from various biotic and abiotic stresses. Examples of input traits are: (i) resistance to insects and pathogens, (ii) tolerance to broad-spectrum herbicides, and (iii) tolerance to abiotic stresses such as heat, cold, drought, and salinity. On the other hand, output traits are those that enhance yield and quality of harvested products. Examples of output traits are: (i) oil with reduced trans-fatty acids, (ii) higher vitamin contents, (iii) better taste and flavor, (iv) increased self-life, (v) ornamental flowers with novel colors and fragrance, and (vi) low phytic acid contents in seeds, etc.

4.1.1 Improvement of Input Traits through Genetic Transformation Technology

With the discovery of genetic transformation, genetic improvement of crops for targeted traits became feasible. By 1996, genetically modified (GM) plants through transformation were commercially grown. In the following decade, by 2007, GM crops including soybeans, corn, canola and cotton were grown at over 282 million acres in 23 countries including 12 developing and 11 industrial countries (James 2007). It is estimated that during 1996-2007 the net benefits to the GM crop growers was US$34 billion (James 2007). One very important aspect of cultivation of GM crop is its role in reducing emission of greenhouse gas. GM crops, resistant to both insects and herbicides, require less use of farm equipments, no-till or reduced tillage and thereby contribute to reduced emission of greenhouse gas. Brookes and Barfoot (2008) estimated that in 2006 growing of GM crops reduced production of 14.7 billion kg CO_2, equivalent to 6.5 million fewer cars on the streets.

GM Crops with Tolerance to Broad-spectrum Herbicides

Resistance to the broad-spectrum herbicide glyphosate has been the most widely adopted input trait among GM crops developed to date (Castle et al. 2006). The trait was first introduced into soybean. Now, over 85% of the US and 56% of global soybeans are glyphosate resistant. The trait was also introduced into cotton, canola, and corn. Glyphosate kills most weed species and its application to glyphosate resistant or Roundup Ready® (Monsanto, St. Louis, USA) cultivars resulted in elimination of crop losses from competition of crops with weed species and reduction in tillage for controlling weeds. Thus, in addition to reduction in cost of cultivation, the possible soil erosion from tillage is reduced. Roundup Ready® crops were developed by use of a modified *Agrobacterium* gene, enoding the herbicide insensitive enzyme CP4 enolpyruvate-sikimate-3-phosphate synthase targeted to chloroplasts.

Additional herbicide resistance genes encoding either herbicide insensitivity or degrading enzymes for new herbicides have been discovered for creating GM crop species with new types of herbicide resistance. These new systems are, however, not as successful as Roundup Ready® brands (Castle et al. 2006). Research in this area is ongoing, and emphasis has now been given on creating non-target-site mechanisms, in which herbicides are modified to nontoxic compounds or compartmentalized in vacuoles using suitable transporters (Yuan et al. 2007).

GM Crops with Resistance to Pathogens

A major commercial success of GM crops with disease resistance is the viral resistance created using either viral coat proteins or replicase genes. This has been accomplished for commercial cultivation in several crop species including squash, potato, papaya (Castle et al. 2006). Success towards creating crop species through transformation with resistance to other pathogens such as bacterial, fungal, oomycetes or nematodes has been rather limited. Despite the fact that mechanisms of resistance against these pathogens are being known (Hammond-Kosack and Parker 2003; Chisholm et al. 2006; Bent and Mackey 2007; de Wit 2007; Genger et al. 2008) and several studies revealed candidates genes for creating possible disease resistant cultivars against these pathogens (e.g., Bhaskar et al. 2008; Cao et al. 2008; Jha et al. 2008; Lee et al. 2008; Yang et al. 2008), we are still far from developing a GM disease resistant crop species against any of these pathogens for commercial cultivation. The main challenge in creating GM disease resistant crops against non-viral pathogens is in the correct manipulation of disease resistance mechanisms (Gurr and Rushton 2005).

GM Crops with Resistance to Insects

One of the remarkable examples of GM crop is the transgenic crops carrying *Bacillus thuringiensis (Bt) cry* genes that produce a δ-endotoxin, commonly known as *Bt* toxin, with high toxicity against a broad range of insects pests, but not mammals or other organisms. This gram-positive bacterium was discovered over a century ago and the insecticidal property of its crystal proteins was realized only in the middle of the last century (Federici 2005). The first *cry* gene encoding crystal protein with insecticidal properties was isolated in 1981 (Schnepf and Whiteley 1981). Transgenic maize, cotton and potato carrying the *cry* gene were first introduced for commercial cultivation in 1996. Now second generation GM maize and cotton carrying stacks of *cry* genes and herbicide resistance genes have been released for commercial cultivation (Castle et al. 2006). Studies conducted in China for 15 years suggested that growing of *Bt* toxin containing GM cotton along with non-transgenic corn, peanuts, soybeans, and vegetables not only reduced the incidence of cotton bollworm (*Helicoverpa armigera*) in GM cotton, but also resulted in a marked decrease in outbreaks of this pest in multiple non-transgenic host crops (Wu et al. 2008).

The effort to discover new bacterial genes for insecticidal proteins led to discovery of additional novel genes, which could be useful in future GM insect resistant crops. The *tcdA* gene of the gram-negative bacterium *Photorhabdus luminescens* encodes toxin A that is highly toxic to a variety of agriculturally important pests. Transgenic studies in *Arabidopsis* for five selfing generations suggested the potentiality of this toxin in controlling southern corn rootworm (*Diabrotica undecimpunctata howardi*) (Liu et al. 2003).

GM Crops with Tolerance to Abiotic Stresses such as Heat, Cold, Drought, and Salt

GM crops created to adapt abiotic stresses will play a major role in meeting the food demands of the ever growing world population. Such crops will allow us to increase the area of cultivable land, and thereby, increase the total crop harvest. Currently no GM crops with resistance to abiotic stresses are available for commercial use. The progress towards creating GM crops with enhanced tolerance to abiotic stresses has been enormous (e.g., multiple abiotic stresses: Nakashima et al. 2007; Gutha and Reddy 2008; Hu et al. 2008; Tang et al. 2008; Wang et al. 2008; Xu et al. 2008; drought tolerance: Zhang et al. 2005; Valliyodan and Nguyen 2006; Xu et al. 2008; Chen et al. 2008b; salt tolerance: Singla-Pareek et al. 2003, 2008; Sanan-Mishra et al. 2005; Vashisht and Tuteja 2006; Uddin et al. 2008; cold tolerance: Jang et al. 2007; Peng et al. 2007; Chen et al. 2008b; Kim et al. 2008). It is expected that within the next few years GM crops with drought tolerance and then salt tolerance will be available for commercial production (James 2007; *http://www.agbioworld.org/biotech-info/articles/biotech-art/crushingcost.html*).

4.1.2 Improvement of Input Traits through Marker-assisted Selection

Marker-assisted selection (MAS) has recently been applied in improving rice and pearl millet for drought, and wheat for salinity tolerance through pyramiding QTLs for these input traits (reviewed in Witcombe et al. 2008). Jena and Mackill (2008) have recently reviewed the progress and prospects of MAS in rice for biotic and abiotic input traits. MAS has also been successfully applied for improvement of disease resistance in soybean (Concibido et al. 1996, 2004; da Silva et al. 2007; Saghai Maroof et al. 2008). Pyramiding disease resistance genes through MAS has been successful in rice for blast and bacterial blight resistance (Huang et al. 1997; Hittalmani et al. 2000; Singh et al. 2001; Narayanan et al. 2004), powdery mildew, leaf rust and stripe rust resistance in wheat (Liu et al. 2000; Kuchel et al. 2007) and soybean mosaic virus in soybean (Saghai Maroof et al. 2008).

4.2 Plant Genomics for Output Trait Improvement

Although several output traits could be addressed in this section (see above), the focus is on forage quality, where a number of genomic approaches have been applied and because of its close relationship to characters affecting biomass to biofuel conversion, which is of major current interest. Forage quality is determined by cell wall properties. Digestibility of cell walls is limited by content and composition of the lignin fraction.

The lignin biosynthetic pathway is well characterized (Boudet et al. 1995). Enzymes involved in the phenylpropanoid pathway are phenylalanine ammonia-lyase (PAL), cinnamate hydroxylase (C4H), coumarate hydroxylase (C3H), caffeic O-methyltransferases (COMT), ferulate hydroxylase (FA5H), and hydroxycinnamate CoA ligases (4CL). In total 34 genes have been identified in *Arabidopsis thaliana*, coding for enzymes in monlignol biosynthesis (Raes et al. 2003). End products of this common pathway, the hydroxycinnamoyl CoAs, are the precursors of flavonoids, stilbenes, phenolamides as well as lignins. Subsequently, cinnamoyl CoA reductase (CCR) and cinnamyl alcohol dehydrogenase (CAD) are involved in synthesis of the monomers p-Coumaryl, Coniferyl, and Sinapyl alcohol. In maize, genes for COMT (Collazo et al. 1992) and CAD (Halpin et al. 1998) have been isolated. Defective alleles of both genes have been shown to correspond to brown midrib mutations (COMT: *bm3*; CAD: *bm1*) (Barriere and Argillier 1993). Independent studies on *bm1* and *bm3* proved the concept of increasing silage quality by altering lignin biosynthesis (Barriere and Argillier 1993). Meanwhile, several additional genes affecting the biosynthesis of cell-walls and thus potentially digestibility, forage quality, etc., have been identified by various genomic approaches in maize (*http://*

www.polebio.scsv.ups-tlse.fr/MAIZEWALL/) and other crop species (*http://cellwall.genomics.purdue.edu/*).

Once qualified candidate genes affecting forage quality have been isolated, they can be used to (i) create new genetic variation in elite germplasm of forage crops, or (ii) exploit existing genetic variation in a more targeted way. Transgenic approaches have been successfully employed for improving cell wall digestibility in different forage crops, either by overexpression of novel genes, or by suppression of genes using antisense or RNAi technology (Spangenberg et al. 2001). A major research area is the down-regulation of genes from the lignin biosynthesis pathway (Ralph et al. 2004). Down-regulation of the maize *bm3* orthologue coding for COMT successfully altered lignin content and composition in grasses (Chen et al. 2003). However, implementation of transgenic approaches is not accepted in several countries, and requires costly risk evaluation (Wang et al. 2003). An alternative, at least for the knock-out approach, is the generation of new genetic variation by TILLING (McCallum et al. 2000, e.g. for maize: *http://genome.purdue.edu/maizetilling/*).

Breeding of quantitative traits such as forage quality can be improved by use of MAS. The major drawback of anonymous genetic markers (random DNA markers such as microsatellites, AFLP, RFLP, etc.) is that their predictive value depends on the known linkage phase between marker and target locus alleles (Lübberstedt et al. 1998). Thus, QTL mapping is necessary for each cross de novo, as different subsets of QTL are polymorphic in individual populations, and linkage phases between marker and QTL alleles disagree even in closely related genotypes. In contrast to anonymous genetic markers, functional markers (FM) are derived from polymorphic sites within genes causally involved in phenotypic trait variation (Andersen and Lübberstedt 2003). FM development requires functionally characterized genes and allele sequences from such genes. Secondly, mutations affecting plant phenotype within these genes must be identified.

Several candidate genes in relation to forage quality are meanwhile available from, e.g., maize (Guillaumie et al. 2007). Reports on association studies for genes involved in cell wall biosynthesis revealed promising targets for identification of polymorphic sites associated with cell wall quality, and thus for FM development. In a study of Guillet-Claude et al. (2004a), polymorphisms in the COMT/*bm3* and the CCoAOMT2 coding genes showed significant association with maize digestibility. A PAL gene was investigated in a set of 32 European elite inbred lines (Andersen et al. 2007). A one-bp deletion in the second exon of PAL, introducing a premature stop codon, was associated with higher plant digestibility. Similarly, a polymorphism in the maize peroxidase gene *ZmPox3*, based on an insertion of a MITE inducing a premature stop codon, was significantly associated with maize digestibility (Guillet-Claude et al. 2004b). Thus, availability of qualified candidate genes in the lignin pathway can be converted into informative molecular markers based on association studies.

5 CONCLUSIONS AND FUTURE PROSPECTS OF PLANT GENOMICS IN PLANT BREEDING

Isolation and functional characterization of multiple genes in several plant species and other organisms, as well as an increased understanding of the molecular basis of agronomic traits enable expression of new traits, which would have otherwise not been in the natural range of genetic variation within a species. Examples include herbicide resistance as mentioned above, new colors in ornamentals, and in the long run potentially nitrogen fixation in major non-legume crop species.

The main contributions of genomics to plant breeding are quantitative in nature. Several of the above mentioned applications of genomics increase the efficiency of plant breeding, e.g., accelerate the development of improved lines in MAS or backcross programs, and help to minimize masking environment or genotype x environment effects for identification of superior genotypes. Increased use of genomics approaches is largely driven by continuously decreasing costs for sequencing (per bp) and genotyping (per datapoint), while respective methods become increasingly powerful (bp per run; multiplexing of datapoints) (e.g., Bagge et al. 2008).

While genomics approaches become more efficient and cheaper over time, collection of phenotypic data becomes relatively more expensive compared to genotyping. For this reason, 'phenomics' was highlighted as one of the future needs, requiring a community approach to address all aspects of phenotyping (e.g., *www.maizegdb.org/AllertonReport.doc*). Just as the terms 'genome' and 'proteome' signify all genes and proteins of an organism, 'phenome' represents the sum total of its phenotypic traits. Different from traditional field trials, phenomics employs multiple controlled environmental conditions besides field trials, combined with advanced, preferentially non-destructive phenotyping methods. The first public large-scale phenomics activity in plants the 'Australian Plant Phenomics Facility' has recently been announced (*http://www.*

plantphenomics.org.au/). This facility received major public funding (AU $40 million) and will start operations in 2009 (*http://www.adelaide.edu.au/space/plant/*). A variety of assays for crops grown at laboratory to field scale are included. Moreover, private breeding companies established proprietary high-throughput phenotyping facilities (e.g., *http://www.pioneer.com/*).

Another limitation of exploiting the information on functionally characterized genes is the high costs for approval of new transgenic events. Although new transgenic events will continue to play an important role for major crop species (e.g., Monsanto Research Pipeline 2008), their number is very limited and far below the possibilities provided by functional and comparative genomics. Thus, unless the costs for approving transgenic events becomes lower in future, the major 'outlet' for use of genomics tools and information in plant breeding will be gene-derived or functional markers in the long run. An interesting development in this connection might be optimized 'cisgenic' procedures based on homologous recombination, as available in yeast and mammals. It would be a perfect tool to combine knowledge of functional alleles with transgene technology, and should stimulate discussion on simplified approval procedures.

References

Andersen JR, Lübberstedt T (2003) Functional markers in plants. Trends Plant Sci 8: 554-560

Andersen JR, Zein I, Krützfeldt B, Eder J, Ouzunova M, Wenzel G, Lübberstedt T (2007) Polymorphisms at the Phenylalanine Ammonia-Lyase locus are associated with forage quality in European maize (*Zea mays* L.). Theor Appl Genet 114: 307-319

Bagge M, Lübberstedt T (2008) Functional markers in wheat—Technical and economic aspects. Mol Breed 22: 319-328

Bagge M, Xia X, Lübberstedt T (2007) Functional markers in wheat. Curr Opin Plant Biol 10: 1-6

Ballvora A, Ercolano MR, Weiss J, Meksem K, Bormann CA, Oberhagemann P, Salamini F, Gebhardt C (2002) The R1 gene for potato resistance to late blight (*Phytophthora infestans*) belongs to the leucine zipper/NBS/LRR class of plant resistance genes. Plant J 30: 361-71

Barrière Y, Argillier O (1993) Brown-midrib genes of maize: a review. Agronomie 13: 865-876

Belo A, Zheng P, Luck S, Shen B, Meyer DJ, Li B, Tingey S, Rafalski A (2008) Whole genome scan detects an allelic variant of *fad2* associated with increased oleic acid levels in maize. Mol Genet Genom 279: 1-10

Bent AF, Mackey D (2007) Elicitors, effectors, and *R* genes: the new paradigm and a lifetime supply of questions. Annu Rev Phytopathol 45: 399-436

Bernardo R, Charcosset A (2006) Usefulness of gene information in marker-assisted recurrent selection: a simulation appraisal. Crop Sci 46: 614-621

Bernardo R, Yu J (2007) Prospects for genomewide selection for quantitative traits in maize. Crop Sci 47: 1082-1090

Bhaskar PB, Raasch JA, Kramer LC, Neumann P, Wielgus SM, Austin-Phillips S, Jiang J (2008) *Sgt1*, but not *Rar1*, is essential for the *RB*-mediated broad-spectrum resistance to potato late blight. BMC Plant Biol 8: 8

Bohn M, Utz HF, Melchinger AE (1999) Genetic similarities among winter wheat cultivars determined on the basis of RFLPs, AFLPs, and SSRs and their use for predicting progeny variance. Crop Sci 39: 228-237

Bonierbale MW, Plaisted RL, Tanksley SD (1988) RFLP maps based on a common set of clones reveal modes of chromosomal evolution in potato and tomato. Genetics 120: 1095-1103

Botstein D, White RL, Skolnick M, Davis RW (1980) Construction of a genetic linkage map in man using restriction fragment length polymorphisms. Am J Hum Genet 32: 314-331

Boudet AM, Lapierre C, Grima-Pettenati J (1995) Biochemistry and molecular biology of lignification. New Phytol 129: 203-236

Brookes G, Barfoot P (2008) Global impact of biotech crops: Socio-economic and environmental effects, 1996-2006. AgBioForum 11: 21-38

Brunner S, Fengler K, Morgante M, Tingey S, Rafalski A (2005) Evolution of DNA sequence nonhomologies among maize inbreds. Plant Cell 17: 343-360

Button P (2008) New developments in the international union for the protection of new varieties of plants (UPOV). In: Lübberstedt T, Studer B, Graugaard S (eds) Proc 27th EUCARPIA Symp on Improvement of Fodder Crops and Amenity Grasses, pp 32-45

Calus MPL, Meuwissen THE, de Roos APW, Veerkamp RF (2008) Accuracy of genomic selection using different methods to define haplotypes. Genetics 178: 553-561

Cannon SB (2008) Legume comparative genomics. In: Stacey G (ed) Genetics and Genomics of Soybean. Series: Plant Genetics and Genomics: Crops and Models 2. Springer, New York, NY, USA, pp 35-54

Cao Y, Yang Y, Zhang H, Li D, Zheng Z, Song F (2008) Overexpression of a rice defense-related F-box protein gene *OsDRF1* in tobacco improves disease resistance through potentiation of defense gene expression. Physiol Plant 134(3): 440-52

Castle LA, Wu G, McElroy D (2006) Agricultural input traits: past, present and future. Curr Opin Biotechnol 17: 105-112

Cavanagh C, Morell M, Mackay I, Powell W (2008) From mutations to MAGIC: Resources for gene discovery, validation and delivery in crop plants. Curr Opin Plant Biol 11: 215-221

Chen L, Auth CK, Dowling P, Bell J, Wang ZY (2003) Improving forage quality of tall fescue by genetic manipulation of lignin bioynthesis. In: Hopkins A, Wang Z-Y, Mian R, Sledge M, Barker RE (eds) Molecular Breeding of Forage and Turf—Developments in Plant Breeding. Kluwer Acad Publ, Dordrecht, Netherlands, pp 181-188

Chen J, Ding J, Ouyang Y, Du H, Yang J, Cheng K, Zhao J, Qiu S, Zhang X, Yao J, Liu K, Wang L, Xu C, Li X, Xue Y, Xia M, Ji Q, Lu J, Xu M, Zhang Q (2008a) A triallelic system of S5 is a major regulator of the reproductive barrier and compatibility of indica-japonica hybrids in rice. Proc Natl Acad Sci USA 105(32): 11436-11441

Chen QF, Xiao S, Chye ML (2008b) Overexpression of the Arabidopsis 10-kilodalton acyl-coenzyme A-binding protein ACBP6 enhances freezing tolerance. Plant Physiol 148: 304-315

Chisholm ST, Coaker G, Day B, Staskawicz BJ (2006) Host-microbe interactions: shaping the evolution of the plant immune response. Cell 124: 803-814

Choi HK, Kim D, Uhm T, Limpens E, Lim H, Mun JH, Kalo P, Penmetsa RV, Seres A, Kulikova O, et al. (2004) A sequence-based genetic map of *Medicago truncatula* and comparison of marker colinearity with *M. sativa*. Genetics 166: 1463-1502

Collazo P, Montoliu L, Puigdomenech P, Rigau J (1992) Structure and expression of the lignin O-methyltransferase gene from *Zea mays* L. Plant Mol Biol 20: 857-867

Concibido VC, Young ND, Lange DA, Denny RL, Orf JH (1996) RFLP mapping and marker-assisted selection of soybean cyst nematode resistance in PI 209332. Crop Sci 36: 1643-1650

Concibido VC, Diers BW, Arelli PR (2004) A decade of QTL mapping for cyst nematode resistance in soybean. Crop Sci 44: 1121-1131

Conley EJ, Nduati V, Gonzalez-Hernandez JL, Mesfin A, Trudeau-Spanjers M, Chao S, Lazo GR, Hummel DD, Anderson OD, Qi LL, Gill BS, Echalier B, Linkiewicz AM, Dubcovsky J, Akhunov ED, Dvorak J, Peng JH, Lapitan NL, Pathan MS, Nguyen HT, Ma XF, Miftahudin, Gustafson JP, Greene RA, Sorrells ME, Hossain KG, Kalavacharla V, Kianian SF, Sidhu D, Dilbirligi M, Gill KS, Choi DW, Fenton RD, Close TJ, McGuire PE, Qualset CO, Anderson JA (2004) A 2600-locus chromosome bin map of wheat homoeologous group 2 reveals interstitial gene-rich islands and colinearity with rice. Genetics 168: 625-637

Cooper M, Smith OS, Graham G, Arthur L, Feng L, Podlich DW (2004) Genomics, genetics, and plant breeding: A private sector perspective. Crop Sci 44: 1907-1913

Crane CF, Beversdorf WD, Bingham ET (1982) Chromosome pairing and associations at meiosis in haploid soybean (*Glycine max*). Can Genet Cytol 24: 293-300

da Silva MF, Schuster I, da Silva JFV, Ferreira A, de Barros EG, Moreira MA (2007) Validation of microsatellite markers for assisted selection of soybean resistance to cyst nematode races 3 and 14. Pesq Agropec Bras 42: 1143-1150

de Greef W (2005) GM crops: The crushing cost of regulation: *http://www.agbioworldorg/biotech-info/articles/biotech-art/crushingcosthtml*

De Vicente MC (2004) The evolving role of gene banks in the fast-developing field of molecular genetics. Issues in Genetic Resources No 11

De Vicente MC, Andersson MD (2005) DNA banks—providing novel options for gene banks? IPGRI—Topcial Reviews in Agricultural Biodivsersity

de Wit PJ (2007) How plants recognize pathogens and defend themselves. Cell Mol Life Sci 64: 2726-2732

Devos KM (2005) Updating the 'Crop circle'. Curr Opin Plant Biol 8: 155-162

Devos KM, Gale MD (1997) Comparative genetics in the grasses. Plant Mol Biol 35:3-15

Doganlar S, Frary A, Daunay MC, Lester RN, Tanksley SD (2002) A comparative genetic linkage map of eggplant (*Solanum melongena*) and its implications for genome evolution in the solanaceae. Genetics 161: 1697-1711

Druka A, Potokina E, Luo Z, Bonar N, Druka I, Zhang L, Marshall DF, Steffenson BJ, Close TJ, Wise RP, Kleinhofs A, Williams RW, Kearsey MJ, Waugh R (2008) Exploiting regulatory variation to identify genes underlying quantitative resistance to the wheat stem rust pathogen *Puccinia graminis* f. sp. *tritici* in barley. Theor Appl Genet 117: 261-272

Eathington SR, Crosbie TM, Edwards MD, Reiter RS, Bull JK (2007) Molecular markers in a commercial breeding program. Crop Sci 47: S154-S163

Federici BA (2005) Insecticidal bacteria: An overwhelming success for invertebrate pathology. J Invertebr Pathol 89: 30-38

Fehr WR (1991) Principles of Cultivar Development. Vol 1: Theory and Technique. Maximillan Publ, New York, USA, pp 172-198

Flint-Garcia SA, Thornsberry JM, Buckler ESt (2003) Structure of linkage disequilibrium in plants. Annu Rev Plant Biol 54: 357-374

Frankel OH (1984) Genetic perspectives of germplasm conservation. In: Arber WK et al. (eds) Genetic Manipulation: Impact on Man and Society. Cambridge Univ Press, Cambridge, UK, pp 161-217

Frary A, Fulton TM, Tanksley SD (2004) Advanced backcross QTL analysis of a *Lycopersicon esculentum* x *L. pennellii* cross and possible orthologs in the Solanaceae. Theor Appl Genet 108: 485-96

Frisch M, Melchinger AE (2001) Marker-assisted backcrossing for simultaneous introgression of two genes. Crop Sci 41: 1716-1724

Gaut BS, Doebley JF (1997) DNA sequence evidence for the segmental allotetraploid origin of maize. Proc Natl Acad Sci USA 94: 6809-6814

Genger RK, Jurkowski GI, McDowell JM, Lu H, Jung HW, Greenberg JT, Bent AF (2008) Signaling pathways that regulate the enhanced disease resistance of Arabidopsis "defense, no death" mutants. Mol Plant-Micr Interact 21: 1285-1296

Guillaumie S, San-Clemente H, Deswarte C, Martinez Y, Lapierre C, Murigneux

A, Barrière Y, Pichon M, Goffner D (2007) MAIZEWALL. Database and developmental gene expression profiling of cell wall biosynthesis and assembly in maize. Plant Physiol 143: 339-363

Guillet-Claude C, Birolleau-Touchard C, Manicacci D, Fourmann M, Barraud S, Carret V, Martinant JP, Barrière Y (2004a) Genetic diversity associated with variation in silage corn digestibility for three O-methyltransferase genes involved in lignin biosynthesis. Theor Appl Genet 110: 126-135

Guillet-Claude C, Birolleau-Touchard C, Manicacci D, Rogowsky PM, Rigau J, Murigneux A, Martinant JP, Barrière Y (2004b) Nucleotide diversity of the *ZmPox3* maize peroxidase gene: Relationships between a MITE insertion in exon 2 and variation in forage maize digestibility. BMC Genet 5: 19

Gurr SJ, Rushton PJ (2005) Engineering plants with increased disease resistance: How are we going to express it? Trends Biotechnol 23: 283-290

Gutha LR, Reddy AR (2008) Rice DREB1B promoter shows distinct stress-specific responses, and the overexpression of cDNA in tobacco confers improved abiotic and biotic stress tolerance. Plant Mol Biol 68(6): 533-555

Hallauer A, Miranda JB (1988) Quantitative Genetics in Maize Breeding, 2nd edn. Iowa State Univ Press, Ames, Iowa, USA

Halpin C, Holt K, Chojecki J, Oliver D, Chabbert B (1998) Brown midrib maize (*bm1*)—A mutation affecting the cinnamyl alcohol dehydrogenase gene. Plant J 14: 545-553

Hammond-Kosack KE, Parker JE (2003) Deciphering plant-pathogen communication: fresh perspectives for molecular resistance breeding. Curr Opin Biotechnol 14: 177-193

Heckenberger M, Bohn M, Melchinger AE (2005) Identification of essentially derived varieties obtained from biparental crosses of homozygous lines: I. Simple sequence repeat data from maize inbreds. Crop Sci 45: 1120-1131

Hittalmani S, Parco A, Mew TV, Zeigler RS, Huang N (2000) Fine mapping and DNA marker-assisted pyramiding of the three major genes for blast resistance in rice. Theor Appl Genet 100: 1121-1128

Hochholdinger F, Hoecker N (2007) Towards the molecular basis of heterosis. Trends Plant Sci 12: 427-432

Hoecker N, Keller B, Piepho H-P, Hochholdinger F (2006) Manifestation of heterosis during early maize (*Zea mays* L.) root development. Theor Appl Genet 112: 421-428

Hougaard BK, Madsen LH, Sandal N, de Carvalho Moretzsohn M, Fredslund J, Schauser L, Nielsen AM, Rohde T, Sato S, Tabata S, Bertioli DJ, Stougaard J (2008) Legume anchor markers link syntenic regions between *Phaseolus vulgaris, Lotus japonicus, Medicago truncatula* and *Arachis*. Genetics 179: 2299-2312

Hu H, You J, Fang Y, Zhu X, Qi Z, Xiong L (2008) Characterization of transcription factor gene *SNAC2* conferring cold and salt tolerance in rice. Plant Mol Biol 67: 169-181

Huang N, Angeles ER, Domingo J, Magpantay G, Singh S, Zhang G, Kumaravadivel N, Bennett J, Khush GS (1997) Pyramiding of bacterial blight resistance genes in rice: Marker-assisted selection using RFLP and PCR. Theor Appl Genet 95: 313-320

Hyten DL, Choi IY, Song Q, Shoemaker RC, Nelson RL, Costa JM, Specht JE, Cregan PB (2007) Highly variable patterns of linkage disequilibrium in multiple soybean populations. Genetics 175: 1937-1944

James C (2007) Brief 37: Global Status of Commercialized Biotech/GM Crops: 2007: *http://wwwisaaaorg/resources/publications/briefs/37/defaulthtml*

Jang JY, Lee SH, Rhee JY, Chung GC, Ahn SJ, Kang H (2007) Transgenic Arabidopsis and tobacco plants overexpressing an aquaporin respond differently to various abiotic stresses. Plant Mol Biol 64: 621-632

Jannink J-L, Bink MCAM, Jansen RC (2001) Using complex plant pedigrees to map valuable genes. Trends Plant Sci 6: 337-342

Jansen RC, Nap JP (2001) Genetical genomics: The added value from segregation. Trends Genet 17: 388-391

Jena KK, Mackill DJ (2008) Molecular markers and their use in marker-assisted selection in rice. Crop Sci 48: 1266-1276

Jha S, Tank HG, Prasad BD, Chattoo BB (2008) Expression of *Dm-AMP1* in rice confers resistance to *Magnaporthe oryzae* and *Rhizoctonia solani*. Transgen Res 18: 59-69

Kim JS, Jung HJ, Lee HJ, Kim KA, Goh CH, Woo Y, Oh SH, Han YS, Kang H (2008) Glycine-rich RNA-binding protein 7 affects abiotic stress responses by regulating stomata opening and closing in *Arabidopsis thaliana*. Plant J 55: 455-466

Kuchel H, Williams KJ, Langridge P, Eagles HA, Jefferies SP (2007) Genetic dissection of grain yield in bread wheat. I. QTL analysis. Theor Appl Genet 115: 1029-1041

Kusterer B, Muminovic J, Utz HF, Piepho H-P, Barth S, Heckenberger M, Meyer RC, Altmann T, Melchinger AE (2007) Analysis of a triple testcross design with recombinant inbred lines reveals a significant role of epistasis in heterosis for biomass-related traits in Arabidopsis. Genetics 175: 2009-2017

Lagercrantz U (1998) Comparative mapping between *Arabidopsis thaliana* and *Brassica nigra* indicates that *Brassica* genomes have evolved through extensive genome replication accompanied by chromosome fusions and frequent rearrangements. Genetics 150: 1217-1228

Lande R, Thompson R (1990) Efficiency of marker-assisted selection in the improvement of quantitative traits. Genetics 124: 743-756

Lander E (1996) The new genomics: A global views of biology. Nature 274: 536-539

Lederberg J, McCray AT (2001) 'Ome sweet 'Omics—A genealogical treasury of words. Scientist 16: 8

Lee SC, Hwang IS, Choi HW, Hwang BK (2008) Involvement of the pepper antimicrobial protein *CaAMP1* gene in broad spectrum disease resistance. Plant Physiol 148: 1004-1020

Liu D, Burton S, Glancy T, Li ZS, Hampton R, Meade T, Merlo DJ (2003) Insect resistance conferred by 283-kDa *Photorhabdus luminescens* protein TcdA in *Arabidopsis thaliana*. Nat Biotechnol 21: 1222-1228

Liu J, Liu D, Tao W, Li W, Wang S, Chen P, Cheng S, Gao D (2000) Molecular marker-facilitated pyramiding of different genes for powdery mildew resistance in wheat. Plant Breed 119: 21-24

Lübberstedt T, Melchinger AE, Schön CC, Utz HF, Klein D (1997) QTL mapping in testcrosses of European flint lines of maize: I. Comparison of different testers for forage yield traits. Crop Sci 37: 921-931

Lübberstedt T, Melchinger AE, Fähr S, Klein D, Dally A, Westhoff P (1998) QTL mapping in testcrosses of flint lines of maize: III. Comparison across populations for forage traits. Crop Sci 38: 1278-1289

McCallum CM, Comai L, Greene EA, Henikoff S (2000) Targeted screening for induced mutations. Nat Biotechnol 18: 455-457

Melchinger AE (1999). Genetic diversity and heterosis. In: Coors JG, Pandey S (eds) *The Genetics and Exploitation of Heterosis in Crops*. CCSA, ASA, and SSA: Madison, WI, USA, pp 99-118

Meuwissen THE, Hayes BJ, Goddard ME (2001) Prediction of total genetic value using genome-wide dense marker maps. Genetics 157: 1819-1829

Meyer RC, Steinfath M, Lisec J, Becher M, Witucka-Wall H, Törjek O, Fiehn O, Eckardt Ä, Willmitzer L, Selbig J, Altmann T (2007) Heterosis associated gene expression in maize embryos 6 days after fertilization exhibits additive, dominant and overdominant pattern. Plant Mol Biol 63: 381-391

Mihaljevic R, Schön CC, Utz HF, Melchinger AE (2005) Correlations and QTL correspondence between line *per se* and testcross performance for agronomic traits in four populations of European maize. Crop Sci 45: 114-122

Nakashima K, Tran LS, Van Nguyen D, Fujita M, Maruyama K, Todaka D, Ito Y, Hayashi N, Shinozaki K, Yamaguchi-Shinozaki K (2007) Functional analysis of a NAC-type transcription factor OsNAC6 involved in abiotic and biotic stress-responsive gene expression in rice. Plant J 51: 617-630

Narayanan N, Baisakh, N, Oliva, NP, VeraCruz, CM, Gnanamanickam SS, Datta K, Datta SK (2004) Molecular breeding: marker-assisted selection combined with biolistic transformation for blast and bacterial blight resistance in Indica rice (cv. CO39). Mol Breed 14: 61-71

Nieto C, Piron F, Dalmais M, Marco CF, Moriones E, Gomez-Guillamon ML, Truniger V, Gomez P, Garcia-Mas J, Aranda MA, Bendahmane A (2007) EcoTILLING for the identification of allelic variants of melon eIF4E, a factor that controls virus susceptibility. BMC Plant Biol 7: 34

Peng Y, Lin W, Cai W, Arora R (2007) Overexpression of a Panax ginseng tonoplast aquaporin alters salt tolerance, drought tolerance and cold acclimation ability in transgenic Arabidopsis plants. Planta 226: 729-740

Pillen K, Zaccharias A, Leon J (2003) Advanced backcross QTL analysis in barley (*Hordeum vulgare* L.). Theor Appl Genet 107: 340-352

Podlich DW, Winkler CR, Cooper M (2004) Mapping As You Go: An effective approach for marker-assisted selection of complex traits. Crop Sci 44: 1560-1571

Raes J, Rohde A, Christensen JH, van de Peer Y, Boerjan W (2003) Genome-wide characterization of the lignification toolbox in Arabidopsis. Plant Physiol 133: 1051-1071

Ralph J, Guillaumie S, Grabber JH, Lapierre C, Barrière Y (2004) Genetic and molecular basis of grass cell-wall biosynthesis and degradability. III. Towards a forage grass ideotype. CR Biol 327: 467-479

Röber FK, Gordillo GA, Geiger HH (2005) *In vivo* haploid induction in maize—performance of new inducers and significance of doubled haploid lines in hybrid breeding. Maydica 50: 275-283

Robin S, Pathan MS, Courtois B, Lafitte R, Carandang S, Lanceras S, Amante M, Nguyen HT, Li Z (2003) Mapping osmotic adjustment in an advanced back-cross inbred population of rice. Theor Appl Genet 107: 1288-1296

Saghai Maroof MA, Jeong SC, Gunduz I, Tucker DM, Buss GR, Tolin SA (2008) Pyramiding of soybean mosaic virus resistance genes by marker-assisted selection. Crop Sci 48: 517-526

Sanan-Mishra N, Pham XH, Sopory SK, Tuteja N (2005) Pea DNA helicase 45 overexpression in tobacco confers high salinity tolerance without affecting yield. Proc Natl Acad Sci USA 102: 509-514

Sandhu D, Gill KS (2002) Gene-containing regions of wheat and the other grass genomes. Plant Physiol 128: 803-811

Schlueter JA, Dixon P, Granger C, Grant D, Clark L, Doyle JJ, Shoemaker RC (2004) Mining EST databases to resolve evolutionary events in major crop species. Genome 47: 868-876

Schnell FW, Utz HF (1975) F1-leistung und Elternwahl für die Züchtung von Selbstbefruchtern. In: Bericht über die Arbeitstagung der Vereinigung Österreichischer Pflanzenzüchter, BAL Gumpenstein, Gumpenstein, Austria, pp 243-248

Schnepf HE, Whiteley HR (1981) Cloning and expression of the *Bacillus thuringiensis* crystal protein gene in *Escherichia coli*. Proc Natl Acad Sci USA 78: 2893-2897

Schouten HJ, Krens FA, Jacobsen E (2006) Cisgenic plants are similar to traditionally bred plants. EMBO Rep 7: 750-753

Schrag TA, Melchinger AE, Sørensen AP, Frisch M (2006) Prediction of single-cross hybrid performance for grain yield and grain dry matter content in maize using AFLP markers associated with QTL. Theor Appl Genet 114: 1037-1047

Schubert D, Williams D (2006) 'Cisgenic' as a product designation. Nat Biotechnol 24: 1327-1328

Sherman JD, Stack SM (1995) Two-dimensional spreads of synaptonemal complexes from solanaceous plants. VI. High-resolution recombination nodule map for tomato (*Lycopersicon esculentum*). Genetics 141: 683-708

Shoemaker RC, Polzin K, Labate J, Specht J, Brummer EC, Olson T, Young N, Concibido V, Wilcox J, Tamulonis JP, Kochert G, Boerma HR (1996) Genome duplication in soybean (*Glycine* subgenus *soja*). Genetics 144: 329-338

Singh S, Sidhu JS, Huang N, Vikal Y, Li Z, Brar DS, Dhaliwal HS, Khush GS (2001) Pyramiding three bacterial blight resistance genes (*xa5, xa13* and *Xa21*) using marker-assisted selection into indica rice cultivar PR106. Theor Appl Genet 102: 1011-1015

Singla-Pareek SL, Reddy MK, Sopory SK (2003) Genetic engineering of the glyoxalase pathway in tobacco leads to enhanced salinity tolerance. Proc Natl Acad Sci USA 100: 14672-14677

Singla-Pareek SL, Yadav SK, Pareek A, Reddy MK, Sopory SK (2008) Enhancing salt tolerance in a crop plant by overexpression of glyoxalase II. Transgen Res 17: 171-180

Slade AJ, Knauf VC (2005) TILLING moves beyond functional genomics into crop improvement. Transgen Res 14: 109-115

Spangenberg GC, Kalla R, Lidgett A, Sawbridge T, Ong EK, John U (2001) Breeding forage plants in the genome era. In: Spangenberg G (ed) Molecular Breeding of Forage Crops—Developments in Plant Breeding. Kluwer Academic Publ, Dordrecht, Netherlands, pp 1-41

Spooner D (2008) DNA barcoding: An oversimplified solution to a complex problem. Botany 2008 (abstract): *http://www.botanyconference.org/Workshops/2008WKS.php*

Spooner D, van Treuren R, de Vicente MC (2005) Molecular markers for genebank management. IPGRI Tech Bull No. 10

Tang L, Kim MD, Yang KS, Kwon SY, Kim SH, Kim JS, Yun DJ, Kwak SS, Lee HS (2008) Enhanced tolerance of transgenic potato plants overexpressing nucleoside diphosphate kinase 2 against multiple environmental stresses. Transgen Res 17: 705-715

Tanksley SD, Nelson JC (1996) Advanced backcross QTL analysis: a method for the simultaneous discovery and transfer of valuable QTLs from unadapted germplasm into elite breeding lines. Theor Appl Genet 92: 191-203

Tanksley SD, Bernatzky R, Lapitan NL, Prince JP (1988) Conservation of gene repertoire but not gene order in pepper and tomato. Proc Natl Acad Sci USA 85: 6419-6423

Tanksley SD, Ganal MW, Prince JP, de Vicente MC, Bonierbale MW, Broun P, Fulton TM, Giovannoni JJ, Grandillo S, Martin GB, et al. (1992) High density molecular linkage maps of the tomato and potato genomes. Genetics 132: 1141-1160

Thornsberry JM, Goodman MM, Doebley J, Kresovich S, Nielsen D, Buckler ES (2001) *Dwarf8* polymorphisms associate with variation in flowering time. Nat Genet 28: 286-289

Tuskan GA, Difazio S, Jansson S, Bohlmann J, Grigoriev I, Hellsten U, Putnam N, Ralph S, Rombauts S, Salamov A, Schein J, Sterck L, Aerts A, Bhalerao RR, Bhalerao RP, Blaudez D, Boerjan W, Brun A, Brunner A, Busov V, Campbell M, Carlson J, Chalot M, Chapman J, Chen GL, Cooper D, Coutinho PM, Couturier J, Covert S, Cronk Q, Cunningham R, Davis J, Degroeve S, Dejardin A, Depamphilis C, Detter J, Dirks B, Dubchak I, Duplessis S, Ehlting J, Ellis B, Gendler K, Goodstein D, Gribskov M, Grimwood J, Groover A, Gunter L, Hamberger B, Heinze B, Helariutta Y, Henrissat B, Holligan D, Holt R, Huang W, Islam-Faridi N, Jones S, Jones-Rhoades M, Jorgensen R, Joshi C, Kangasjarvi J, Karlsson J, Kelleher C, Kirkpatrick R, Kirst M, Kohler A, Kalluri U, Larimer F, Leebens-Mack J, Leple JC, Locascio P, Lou Y, Lucas S, Martin F, Montanini B, Napoli C, Nelson DR, Nelson C, Nieminen K, Nilsson O, Pereda V, Peter G, Philippe R, Pilate G, Poliakov A, Razumovskaya J, Richardson P, Rinaldi C, Ritland K, Rouze P, Ryaboy D, Schmutz J, Schrader J, Segerman B, Shin H, Siddiqui A, Sterky F, Terry A, Tsai CJ, Uberbacher E, Unneberg P, Vahala J, Wall K, Wessler S, Yang G, Yin T, Douglas C, Marra M, Sandberg G, Van de Peer Y, Rokhsar D (2006) The genome of black cottonwood, *Populus trichocarpa* (Torr. & Gray). Science 313: 1596-1604

Uddin MI, Qi Y, Yamada S, Shibuya I, Deng XP, Kwak SS, Kaminaka H, Tanaka K (2008) Overexpression of a new rice vacuolar antiporter regulating protein OsARP improves salt tolerance in tobacco. Plant Cell Physiol 49: 880-890

Utz HF, Bohn M, Melchinger AE (2001) Predicting progeny means and variances of winter wheat crosses from phenotypic values of their parents. Crop Sci 41: 1470-1478

Uzarowska A, Keller B, Piepho H-P, Schwarz G, Ingvardsen C, Wenzel G, Lübberstedt T (2007) Comparative expression profiling in meristems of inbred—hybrid triplets of maize based on morphological investigations of heterosis for plant height. Plant Mol Biol 63: 21-34

Valliyodan B, Nguyen HT (2006) Understanding regulatory networks and engineering for enhanced drought tolerance in plants. Curr Opin Plant Biol 9: 189-195

Vashisht AA, Tuteja N (2006) Stress responsive DEAD-box helicases: a new pathway to engineer plant stress tolerance. J Photochem Photobiol B Biol 84: 150-160

Vuylsteke M, Kuiper M, Stam P (2000) Chromosomal regions involved in hybrid performance and heterosis: Their AFLP®-based identification and practical use in prediction models. Heredity 85: 208-218

Wang Q, Guan Y, Wu Y, Chen H, Chen F, Chu C (2008) Overexpression of a rice OsDREB1F gene increases salt, drought, and low temperature tolerance in both Arabidopsis and rice. Plant Mol Biol 67: 589-602

Wang Z-Y, Hopkins A, Lawrence R, Bell J, Scott M (2003) Field evaluation and risk assessment of trangenic tall fescue plants. In: Hopkins A, Wang ZY, Mian R, Sledge M, Barker RE (eds) Molecular Breeding of Forage and

Turf—Developments in Plant Breeding. Kluwer Academic Publ, Dordrecht, Netherlands, pp 367-379

Willas J (2006) Sorter af korn, bælgsæd, og olieplanter 2006. Grøn Viden 320, DigiSource Danmark A/S, Denmark

Wilson LM, Whitt SR, Ibanez AM, Rocheford TR, Goodman MM, Buckler ESt (2004) Dissection of maize kernel composition and starch production by candidate gene association. Plant Cell 16: 2719-2733

Witcombe JR, Hollington PA, Howarth CJ, Reader S, Steele KA (2008) Breeding for abiotic stresses for sustainable agriculture. Philos Trans Roy Soc Lond B Biol Sci 363: 703-716

Wu K, Lu YH, Feng HQ, Jiang YY, Zhao JZ (2008) Suppression of cotton bollworm in multiple crops in China in areas with Bt toxin-containing cotton. Science 321: 1676-1678

Wu X, Shiroto Y, Kishitani S, Ito Y, Toriyama K (2009) Enhanced heat and drought tolerance in transgenic rice seedlings overexpressing *OsWRKY11* under the control of *HSP101* promoter. Plant Cell Rep 28(1): 21-30

Xu Y, Crouch JH (2008) Marker-assisted selection in plant breeding: From publications to practice. Crop Sci 48: 391-407

Xu DQ, Huang J, Guo SQ, Yang X, Bao YM, Tang HJ, Zhang HS (2008) Over-expression of a TFIIIA-type zinc finger protein gene *ZFP252* enhances drought and salt tolerance in rice (*Oryza sativa* L.). FEBS Lett 582: 1037-1043

Yang S, Gao M, Xu C, Gao J, Deshpande S, Lin S, Roe BA, Zhu H (2008) Alfalfa benefits from *Medicago truncatula*: the *RCT1* gene from *M. truncatula* confers broad-spectrum resistance to anthracnose in alfalfa. Proc Natl Acad Sci USA 105: 12164-12169

Yuan JS, Tranel PJ, Stewart CN Jr (2007) Non-target-site herbicide resistance: A family business. Trends Plant Sci 12: 6-13

Zhang JY, Broeckling CD, Blancaflor EB, Sledge MK, Sumner LW, Wang ZY (2005) Overexpression of *WXP1*, a putative *Medicago truncatula* AP2 domain-containing transcription factor gene, increases cuticular wax accumulation and enhances drought tolerance in transgenic alfalfa (*Medicago sativa*). Plant J 42: 689-707

Zhong Y, Jannink J-L (2007) Using quantitative trait loci results to discriminate among crosses on the basis of their progeny mean and variance. Genetics 177: 567-576

16 People, Policy and Plant Genomics

Emma Frow and Steven Yearley
ESRC Genomics Policy and Research Forum,
University of Edinburgh, St John's Land, Holyrood Road,
Edinburgh EH8 8AQ, UK
Corresponding author: emma.frow@ed.ac.uk

1 INTRODUCTION

In this chapter we will venture away from the confines of laboratories, research institutes and plant breeding stations that engage in plant genomics research, to consider some of the social, economic, political and regulatory issues associated with plant genomics. Why might this be important? Plants are central to human societies: they provide us with clean air, food to eat, medicines, building materials and sources of energy in both developing and industrialized societies. Scientific research and new technologies that affect the way we understand and use plants can have consequences that extend far beyond the laboratory. This might seem obvious, given that the purpose of most research is to discover more about the world around us, ideally to the benefit of society at large. But experience tells us that the products of plant science research are not all seamlessly integrated into society: sometimes they are welcomed, sometimes they are resisted, and sometimes they bring to light values or assumptions that we live by, forcing us to reconsider our place in the natural world. In this chapter we will begin to explore the intricate and dynamic relationship between science and society.

Starting with an early example of the relationship between society and knowledge of plants, the domestication of crops and development of tools for land cultivation allowed communities to settle and flourish. As a consequence, issues such as land ownership began to take on increased economic and political importance. The social stratification arising from these developments was marked—in feudal times, a child born into a family owning large tracts of land would lead a very different life to a child from a peasant family. Indeed, plants and political economy were once inseparable. According to 18th century economists, land was the key productive asset, enabling food to be grown and life to be sustained. In the United States (US), land-grant universities established under legislation in the 1860s and 1880s encapsulated the idea that agricultural improvement and advances in national wellbeing could be achieved through agricultural education and research.

But during the 20th century, these ideas began to seem outmoded. Industry, commerce and finance became the keys to a successful economy, whereas agriculture seemed fated to produce only slow rises in productivity. Many once plant-based products became displaced by petroleum-based alternatives. The 'tiger' economies of south-east Asia flourished in the absence of modernized agriculture, and by the 1980s many Japanese policy makers had come to view the US, with its vast agricultural lands, as some kind of throwback to an earlier form of economic life. Government agencies began to pull out of investments in agriculture and plant sciences, and many European states began supporting farmers with subsidies that were entirely unrelated to increasing productivity or economic performance. Institutions with a focus on agriculture and plant research were no longer in the limelight in the industrialised world. The global swing towards market liberalization in the 1980s (pursued in Anglo-American countries more enthusiastically than elsewhere) led to research institutions operating with an increasingly commercial ethos. This also affected trends in plant science, not necessarily because it was thought the best model for such research, but because plant science was swept along by broader trends in research policy.

We write this chapter in what might prove to be another turning point. In recent years, research policy across the globe has turned to promoting the development of a bio-based economy (Section 2.4). In the 21st century, politicians around the world talk with growing frequency of food scarcity and the need to develop alternatives to fossil fuels. Plants—in the broadest sense—are seen as a locus for new hopes: as a source of products such as biofuels, as smart biological systems for the innovative production of chemicals, and as a renewable resource in a world of declining options. In this chapter, we shall review the new opportunities with which plants are being linked, examine some of the institutional, policy and regulatory structures shaping plant science, and discuss the broader social context in which contemporary plant genomics research is situated, including some recent and well-known controversies. From this vantage point, we can begin to identify opportunities and challenges for plant genomics in the 21st century.

2 PLANT GENOMICS FOR WHAT PURPOSE?

Much of the recent interest and investment in plant genomics initiatives (e.g., genome sequencing projects) is linked to the potential for plant genomics research to improve our understanding of plant development and function—and ultimately, to develop improved plant varieties and plant-derived products. One of the obvious links between plant genomics and society is through new products that enter the marketplace, and in this first section we explore some of these many potential applications.

2.1 Crop Improvement

Crop improvement through plant breeding is perhaps the most obvious area of application for plant genomics. The long history of crop improvement has predominantly focused on traits with direct relevance to the farming process, such as yield, ripening time, ease of harvesting and processing, and drought and pest resistance. These objectives are still the focus of much plant breeding, but challenges relating to environmental and human health are adding new classes of desirable traits to the wish-list for improved crop varieties. Fortunately, access to genome sequences, genomic technologies and crop genetic diversity is rapidly increasing the number of complex traits that might be investigated in the development of new crop varieties.

2.1.1 Climate-resilient Crops

Climate change stands to affect different parts of the world in different ways, but the development of crop varieties that are 'climate-proof' is high on the global food security agenda. Crops should ideally be suited to the particular environmental and climactic conditions of a given region, and resilient to fluctuations in weather patterns. Drought resistance, flood tolerance and salt tolerance are all traits being pursued with these goals in mind. For example, in 2006, researchers at the International Rice Research Institute in Philippines and the University of

California-Davis identified the *Sub1A* gene for flood-tolerance in rice (Xu et al. 2006). Submergence-tolerant varieties are now being developed in India and other South-East Asian countries.

2.1.2 Low-input Crops

Alongside pressures to increase productivity from existing agricultural land in order to feed a growing world population, there is also a recognized need to reduce the environmental 'footprint' of farming. Agriculture, and commercial agriculture in particular, is an energy- and resource-intensive enterprise, requiring inputs in the form of water, fertilizers, pesticides, and so on. In fact, it has been estimated that 14% of total global greenhouse gas emissions in 2000 were attributable to agriculture (Stern 2007). Crop varieties that are less dependent on inputs stand to be more environmentally sustainable and more cost-effective for farmers. Nitrogen use efficiency is one example of a trait being developed for these reasons.

2.1.3 Improving Nutritional Value

Plant breeding has traditionally focused on traits relevant to farmers (especially yield), but growing understanding of the link between diet and health is leading to consumer demand for more nutritious foods (Morris and Sands 2006). In a world where over 800 million people go hungry (Food and Agriculture Organization 2006), and ironically, there are also record levels of obesity, there is potential for new crop varieties with enhanced nutritional value to help improve global health. Conventional plant breeding, marker-assisted selection and transgenic strategies are being pursued to develop varieties with, for example, higher vitamin content, lower allergenic potential, and different fatty acid or starch compositions. For example, in the US, soybean varieties with low linolenic acid content have been welcomed into the market (Powell 2007). The oil produced from these seeds has no trans-fatty acids, which have been linked to heart disease.

Probably the best-known example to date of crop improvement for nutritional purposes is Golden Rice, a transgenic rice variety engineered to express high levels of β-carotene in the grains. Golden Rice, so called because of the orange color of its grains, is designed to help combat vitamin A deficiency, a leading cause of blindness in developing countries. Golden Rice is not a commercial success story though—roadblocks in the form of intellectual property conflicts, stringent testing requirements for genetically modified (GM) crops, and public resistance mean that, ten years after development, Golden Rice is not available to farmers (Al-Babili and Beyer 2005). As has become abundantly clear over the past decade, developing a new and improved crop variety does not guarantee that consumers will embrace it (more on that later, see Section 4.2).

2.1.4 Energy Crops

Energy and environmental issues are gaining visibility on the global stage. The desire of many national governments to reduce both their dependence on foreign oil reserves and their greenhouse gas emissions has led to a surge of interest in using crops as a source of renewable energy—producing bioenergy from plant biomass. Biomass can be combusted to produce heat and electricity, or be converted into liquid biofuel. Of the wide range of possible biofuels, the most common at present are bioethanol (produced from starch- or sugar-rich crops such as sugarcane, wheat and corn) and biodiesel (produced from vegetable oils including soybean, palm, sunflower and rapeseed oil). A number of food crops and non-food crops are currently being targeted for bioenergy production, and it seems likely that perennial species (including trees and grasses) will have an increasingly important role.

Plant genomics stands to make a useful contribution to bioenergy development. Because common domesticated crops have not been selected for efficient carbon capture, a number of traits might be targeted for enhanced biomass production. These include photosynthetic light capture efficiency (which currently averages less than 2%), photoperiod response, nitrogen metabolism efficiency, lignin and cellulose content, abiotic stress tolerance, and pest/disease resistance (Ragauskas et al. 2006).

2.2 Industrial Applications

2.2.1 'Green Chemistry' and the Chemical Industry

Famously, the first motor engine designed by Rudolf Diesel in the 1890s ran on peanut oil. In the early 20th century, plants were an important source of industrial materials, including glues, soaps, paints and dyes. However, by the 1960s, petroleum derivatives had largely replaced the use of bio-based materials and chemical products. With the growing push to find alternatives to petrochemicals, the pendulum is now swinging back again. Similar to the situation with bioenergy, this is both an economic response (owing to the rising cost of oil), and one that acknowledges the need to improve the environmental footprint of the chemical industry.

The availability of complete genome sequences offers great possibilities for new plant compounds to be identified and exploited by the chemical industry. In fact, it is believed that higher plants synthesize over 200,000 different primary and secondary metabolites, but despite this, the chemical potential of plants is largely unexplored. Plant-derived products are commonly used in cosmetics, fragrances, detergents and solvents, and of course as textiles. Products such as biodegradable plastics, flavouring agents, paints, packaging and construction materials are increasingly being developed. For example, DuPont's plant-based biofibre Sorona® is made from corn starch that is fermented using genetically modified organisms and then polymerized.

The use of biocatalysts from plants and microorganisms is another important growth area for the chemical industry. Industrial processing relies heavily on catalytic and conversion processes in the production of high-value materials, and increasingly exploits comparative and functional genomics for the identification of suitable biological enzymes.

2.2.2 Drug Development and 'Pharming'

Plant products have long been important to the pharmaceutical industry: about 25% of all commonly prescribed medicines are directly or indirectly derived from plants. Well-known examples include aspirin (originally extracted from willow bark) and artemisinin (an anti-malarial drug obtained from the shrub *Artemisia annua* or sweet wormwood). As well as sampling natural genetic diversity to find new therapeutic compounds, several dozen companies worldwide are currently working with plants or plant cell cultures as vehicles for producing vaccines and other pharmaceutical products—a process sometimes called 'pharming' (Fox 2006).

2.3 Environmental Applications

Plant genomics and plant biotechnology are seen as holding great promise for improving the environmental footprint of the global agricultural and industrial sectors. But in addition to crop breeding for low-input and more environmentally sustainable traits, there are a number of other environmental issues to which plant genomics research can be applied.

2.3.1 Phytoremediation

Phytoremediation makes use of the natural ability of plants to contain, degrade or remove toxic chemicals and pollutants from contaminated soils and water (Pilon-Smits 2005). Phytoremediation can be used in combination with other techniques to clean up metals, pesticides, solvents, explosives, crude oil and other contaminants. It is generally seen as a cost-effective and environmentally sound approach to remediation. Plant species popular for their phytoremediation properties include sunflower, poplar and duckweed. Genomic information and tools are being used to identify important genes and pathways in phytoremediation processes, to better understand plant–microbe interactions important for remediation, and to develop varieties (including transgenic varieties) tailored for particular clean-up functions.

2.3.2 Environmental Surveillance and Monitoring

The ability to diagnose, monitor and track the spread of plant pests, pathogens, and potentially invasive plant species is crucial in a globalized world increasingly dependent on trade in plants and plant products. Such threats to plant health are by no means new: the Irish potato famine of 1845 and the virtual collapse of the European wine industry in the second half of the 19th century are both related to the spread of plant pests—potato late blight in the case of the potato famine, and the aphid *Phylloxera* in European vineyards. But the speed and scale of the trade in plant goods for horticultural, agricultural and industrial purposes continues to grow rapidly. The scale of potential loss is correspondingly large. For example, a recently identified threat to the American cotton industry is in the form of the Q-biotype whitefly, which has been imported into the US on poinsettia plants and is thought to be one of the world's most destructive crop pests (Dalton 2006). Add to such examples the prediction that climate change will influence the distribution of pests and increase the probability of epidemics, and the need for efficient monitoring strategies becomes even more important (Brasier 2008).

Genomic approaches can be instrumental in the development of specific and efficient monitoring devices such as biosensors and bioassays. Detection and genotyping of plant viral pathogens can be approached using a variety of methods, including microarray analysis (Wang et al. 2002; Webster et al. 2004). Plant health inspectors at national borders now routinely make use of molecular diagnostics for the identification of both imported plant species and known pests and plant pathogens. Genomic information also feeds into predictive models for risk analysis and surveillance of potential pests and diseases.

More generally, the field of environmental genomics is concerned with trying to develop markers of environmental and species health through genomic profiling (Environment Agency 2003; Dix et al. 2006). This is both with respect to containing the spread and minimizing harm from diseases, invasive species and environmental pollutants, and for informing conservation strategies.

2.3.3 Informing Conservation Strategies

The protection of biological diversity lies at the heart of conservation. Biological diversity (or 'biodiversity') is a term used to represent the variety of life on earth, and includes the genetic diversity within a species, as well as the diversity across species and ecosystems. On the whole, high levels of diversity (including genetic diversity) are thought to lead to increased productivity, increased nutrient retention, and greater ecosystem stability and adaptability (Tilman 2000; Hooper et al. 2005).

Genetic studies can help to create pictures of pattern and process at species, community and ecosystem levels (Whitham et al. 2006). Detailed information regarding species taxonomy, hybridization patterns, genetic diversity, gene flow and minimum viable population size can now be obtained using genetic techniques. Genomic technologies including high-throughput sequencing, microsatellite analysis and non-invasive DNA sampling are also leading to increased use of genetic information in informing plant conservation strategies. As outlined above, the use of biomarkers or biosensors may also enable more sophisticated measurements of 'harm' at species and ecosystem levels, promoting early and well-targeted interventions.

Another approach to conservation involves the preservation of plant genetic diversity through ex situ collections. This has become the focus of several recent initiatives, including the Svalbard Global Seed Vault, built in the Norwegian permafrost, and the Millennium Seed Bank Project at Kew Gardens in London, the world's largest ex situ conservation project. The Svalbard bank, also referred to as the 'Doomsday vault' is collecting seeds from all the recognized varieties of about 150 food crop species (Hopkin 2008). There are currently about 1,400 seed banks worldwide. They aim not only to conserve plant varieties, but increasingly to promote the characterization and exploitation of plant genetic diversity (e.g., for crop improvement). However, it should be noted that ex situ collections do not conserve the ecosystems necessary for plants to thrive.

2.4 The Bio-based Economy

From the discussion above, it is clear that modern biotechnology is opening up new options for exploiting plant genetic diversity in productive ways: from cellulosic bioethanol to pharmaceutical 'biofactories', phytoremediation devices and hand-held diagnostic tools, some of the impending applications for plants extend far beyond their traditional uses. The social and political agenda driving the development of such applications has several interrelated motivations: the desire to reduce our dependency on oil, a growing awareness of the possible consequences of climate change and the need for more environmentally sustainable living, the possibility for economic growth through the development of 'green' technologies, and so on.

The concept of a 'knowledge-based bioeconomy' is gaining momentum among policy makers and research funders (OECD 2006). At its heart, a knowledge-based bioeconomy is a global economy that can meet its material needs using biological resources: the raw materials and basic building blocks for energy, industry, growth and well-being are derived from biological, renewable resources (mainly plants and microorganisms). Arguably, humans have always had some version of a bioeconomy, being largely dependent on biological resources for food, clothing, shelter, and so on—even the fossil fuel economy obtains energy from 'ancient sunlight'. However, current thinking emphasizes the use of cutting-edge science and technology to support the transition away from a petroleum-based economy to one dependent on bio-renewables.

Central to the goal of developing a bio-based economy is the concept of a 'biorefinery'. A biorefinery is a closed-loop system for biomass conversion that is able to extract high-value products, generate energy, and recycle any waste back into the natural environment (Fig. 1). The closed-loop or 'cradle-to-cradle' operation of a biorefinery is thermodynamically different from the petrochemical refinery system, which follows a largely linear process of oil extraction, conversion, consumption and waste in the form of pollution. Biorefineries depend on holistic and integrated strategies for plant use (Kamm and Kamm 2004). Based on this, we might suggest that it will become increasingly important to develop improved crop varieties on the basis of not just one or two but a host of interrelated, complex traits. How might this goal affect the research strategies adopted by individual laboratories?

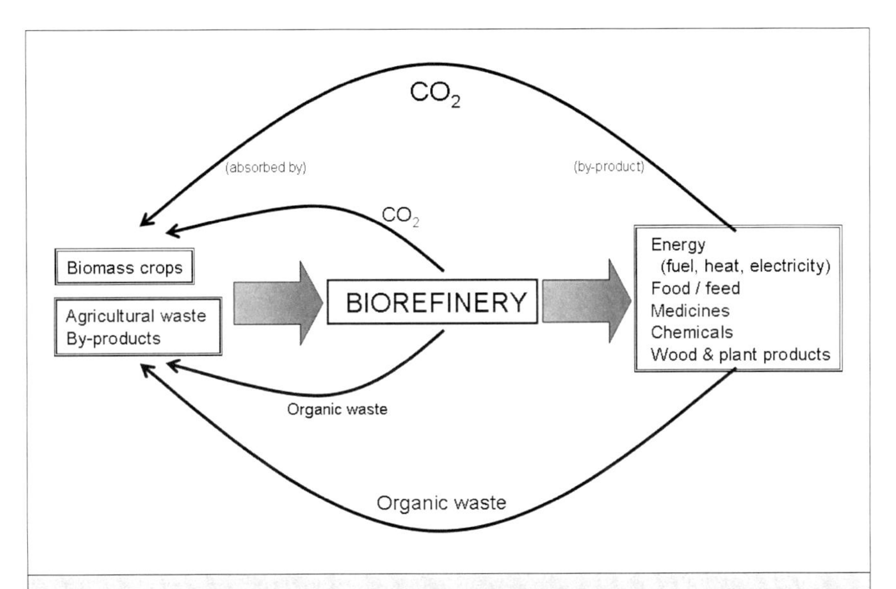

Fig. 1 A closed-loop biorefinery system. Biomass in the form of dedicated crops or agricultural waste and by-products are fed into biorefineries, which use a variety of processing mechanisms (combustion, gasification, enzymatic degradation, etc.) to convert the input biomass into outputs including energy, wood products, and high-value compounds for the chemical and pharmaceutical industries. Organic waste is recycled back into the agricultural or biorefinery systems. Such a closed-loop system is thought to result in minimal landfill waste and lower environmental impact than current petroleum-based refineries.

2.5 From Plant Genomics to People and Policy

Genomic approaches offer the potential to help understand the evolution and control of different plant traits, and to facilitate the use of genetic diversity across the wide range of applications described above. It is clearly an exciting time to be involved in plant genomics research. As with any field of scientific endeavour, some of the potential applications will prove more difficult to develop than others, some will be more economically attractive, some will have tangible environmental benefits, and others will almost certainly be associated with risks and negative consequences. Which applications can or should be developed? Who should develop them? Given the finite pot of money available for scientific research, which applications should be given priority and why?

Over the course of this chapter, we hope to show that these are not simply scientific questions—they have strong social and political dimensions too. Should research be funded on the basis of its potential to succeed? (This, it should be said, is notoriously difficult to predict.) Should it be funded on the promise of the intellectual property or economic revenue it stands to generate? Should a primary goal be to improve public health or environmental sustainability? How should we measure 'success'? And who should have a say in deciding what to spend public research money on? These types of questions see the debate move away quickly from a focus on research details into the realm of people and policy.

3 THE GLOBAL FUNDING AND RESEARCH LANDSCAPE FOR PLANT GENOMICS

The large number of possible applications for plant genomics research outlined in Section 2 makes it very difficult to provide a simple overview of the research and funding landscape for plant genomics. It is not even easy to specify exactly what counts as plant genomics research. Should it refer simply to genome sequencing projects? We take a broader definition of genomics, and would include for example structural, functional and comparative genomics, and more generally research that relies on genomic technologies and information. What about projects that fall under the heading 'plant biotechnology' as opposed to 'plant genomics'? Clearly biotechnology and genomics are not equivalent terms, but many biotechnology projects have considerable genomic components, and we would not want to ignore these. Also, should we consider research taking place in both publicly funded institutions and private organizations? Here we concentrate on providing an overview of broad trends in publicly funded plant genomics research.

3.1 Public Funding for Plant Genomics Research

Public funding for plant science has increased significantly since the mid-1990s, and a large proportion of this funding has gone into genomics research. In many ways, genomics has been a catalyst for renewed public investment into infrastructure and capacity-building in plant science, after decades of relative neglect in some parts of the world. A number of national and regional plant genomics funding programs have been established, and several international collaborations and consortia have also been set up, particularly in the context of genome sequencing projects (see Chapters 6-10 on sequencing initiatives). Here we provide a brief overview of trends in public funding for plant genomics in a few different regions of the world. A recurring theme in virtually all cases is the number of partners involved in setting up and funding large-scale plant genomics research programs.

3.1.1 United States

Until the late 1990s, crop plant genomics in the US was largely funded by the private sector, with little support from government sources. Partly in response to a 1995 proposal by the National Corn Growers Association for funding to sequence the corn genome, the National Plant Genome Initiative (NPGI) was established in 1998 (Pennisi 1998). Over the past ten years, the NPGI has handed out more than $800 million in research, infrastructure and training grants, and has come to be seen as the main program for basic plant science (Pennisi 2007). To date, the NPGI has had a strong focus on commercial crop plants, as well as some investment into research and sequencing of model organisms such as *Arabidopsis*. The US has been involved in many international crop genome sequencing projects, including maize, rice, sorghum, tomato, cotton and soybean.

Members of the NPGI's oversight body, the Interagency Working Group on Plant Genomics, represent a range of interests, and include among others the National Science Foundation, the US Department of Agriculture, the Department of Energy, the National Institutes of Health, the US Agency for International Development, and the US Forest Service (National Research Council 2008). Some of these members have also made significant independent investments into applied plant genomics research; for example, in June 2007 the Department of Energy announced the creation of three interdisciplinary Bioenergy Research Centres, receiving a combined $375 million over 5 years. The Environmental Protection Agency (EPA) is also investing in environmental genomics research initiatives, and established a Computational Toxicology Program in 2004.

3.1.2 Europe

Just as funding for plant sciences in the US began to increase dramatically in the late 1990s, European-level funding for plant science research dropped; this occurred in the context of public and political antipathy to plant biotechnology (see Section 4.2). The European Framework Programme (FP) grants provide about 5% of total public research funding across Europe, and are important in terms of setting strategic research priorities. With no dedicated pot of money for plant science in the fifth round of FP grants (FP5, 1999-2003), funding for plant genomics research fared poorly. In 2003, the European Commission initiated a stakeholder forum on plant genomics and biotechnology called 'Plants for the Future', with members from industry, academia and the agricultural sector. This forum has been largely successful at increasing the funding available for plant science in subsequent rounds of FP funding (FP6, 2003-2007; FP7, 2007-2013). The 'knowledge-based bio-economy' is seen as a central and coordinating theme in the FP7 funding strategy, and is being used to link research projects across a number of technology platforms. A European Research Area Network on plant genomics (ERA-PG) was also initiated in 2004 to promote networking and coordination activities, and as of 2008 has held two rounds of international collaborative research projects with a total budget of about $75 million (€ 50 million).

3.1.3 United Kingdom

The United Kingdom (UK) has seen significant changes to the structure and nature of plant science research funding over the past 50 years. In the early 20th century, a number of public plant breeding institutes were established, and until the 1960s plant breeding was largely a public sector activity. However, in the 1970s and 1980s, a number of government-funded plant science laboratories were closed or privatized. The most famous example of this was the selling of part of the Cambridge Plant Breeding Institute (PBI) to Unilever (now Monsanto). Currently, out of over 50 members of the British Society of Plant Breeders, only three are in the public sector (Agriculture and Environment Biotechnology Commission 2005).

At present, the principal source of public funding for plant science research in the UK is the Biotechnology and Biological Sciences Research Council (BBSRC), which subsumed the Agriculture and Food Research Council in 1994. In 2004, total BBSRC funding for plant science research was about $125 million (£65 million), accounting for over 50% of the non-industrial, national expenditure (BBSRC 2004). Research on individual plant species is dominated by *Arabidopisis*, with limited work on other species including wheat, *Brassica*, tomato, potato, maize and barley. The UK is involved in a number of international plant genome sequencing projects (see Chapters 6-10).

3.1.4 Australia

Similar to the US, the Australian government started making significant investments in genomics research in the late 1990s. The Australian Centre for Plant Functional Genomics (ACPFG), located in South Australia, has been one of the main beneficiaries—set up in 2003, it has to date received over $70 million (AUD $80 million) in research funding. Notably, this funding comes from public bodies and from a levy on grain growers by the Grains Research and Development Corporation. The research focus reflects the funding sources, and concentrates on functional genomics for developing wheat and barley varieties. The research facility houses both the ACPFG and commercial bioscience partners.

Other Australian initiatives include the Australian Genome Alliance, the New South Wales Centre for Agricultural Genomics (founded in 2004), and the DPI Plant Biotechnology Centre in Victoria (set up in 1995). All focus on plant genomics for agricultural purposes, particularly developing crop varieties to withstand environmental stressors such as drought and salinity, which are of growing national concern.

3.1.5 China

In contrast to many parts of the world, plant science research in China is almost entirely publicly funded. With issues such as food security, energy security and environmental health high on the national agenda, plant biotechnology has received strong public investment since the mid-1980s (amounting to over $100 million annually by 1999). China has an ambitious GM crop program, and with over 3.8 million hectares of GM crops (mainly cotton) grown in 2007, is the world's sixth largest grower of GM crops (Anon 2008). With public funding driving the research agenda, focus for the most part is on crops and pests relevant to the national agricultural system. China has strong research programs in functional genomics for rice and *Arabidopsis* (Huang et al. 2002), and also plays an active role in international plant genome sequencing projects. The Beijing Genomics Institute is an example of China's commitment to supporting genomic technologies: it was an empty building in 1998, but by 2002 had over 90 sequencing machines, a staff of 500, and had completed a shotgun sequence of the *indica* rice genome, published in *Science* (Normile 2002; Yu et al. 2002).

3.1.6 India

India has a long history of crop improvement and was the symbolic home to the Green Revolution in the 1960s, which saw large improvements in agricultural productivity through the introduction of new crop varieties, chemical products, machinery and irrigation practices (see Section 4.1). A lot of the funding for Indian plant science and agricultural research comes from government agencies, and there is also growing private investment by international companies, as well as a number of philanthropic initiatives (e.g., the M.S. Swaminathan Research Foundation). At the turn of the 21st century, India was spending approximately 0.3% of its agricultural gross domestic product (GDP) on agricultural research and plant biology. There have been calls to increase this funding, which is considered low even by developing country standards (which spend an average of 0.5% of agricultural GDP on R&D, Raghuram 2002).

This being said, a number of Centres for Plant Molecular Biology have been established since the early 1990s, together with a National Centre for Plant Genome Research, based in New Delhi. India has also been concentrating on developing National Gene Banks for germplasm conservation. Much of the plant science research in India is focused on tissue culture, biodiversity characterization, molecular taxonomy, and developing lower-input agricultural products. Transgenic crop production is also an area of growing research. Although India has yet to commercialize or import any GM food crop, it is the world's largest grower of GM cotton and has a number of field trials in place (Anon 2008). India is also contributing to genome sequencing efforts for species including rice, tomato and chickpea.

3.1.7 Multinational: CGIAR System

The Consultative Group on International Agricultural Research (CGIAR) was established in 1971, and aims to use science to promote sustainable agriculture in developing countries. This system currently supports 15 international research centres around the world (Table 1), and is supported by 64 members, including 21 developing countries, 26 developed countries, 4 co-sponsors, and 13 international organizations (including the World Bank and the United Nations Development Fund). Since 1971, the CGIAR system has received over $5.5 billion in funding.

Strategic goals of CGIAR efforts include genetic crop improvement, sustaining agricultural biodiversity (both in situ and through the mainte-

Table 1 CGIAR Centres.

Name	*Member since*	*Headquarters*	*Website*
Africa Rice Center (WARDA)	1975	Bouake, Cote d'Ivoire	*www.warda.org*
Biodversity International*	1974	Rome, Italy	*www.bioversityinternational.org*
International Center for Tropical Agriculture (CIAT)	1971	Cali, Colombia	*www.ciat.cgiar.org*
Center for International Forestry Research (CIFOR)	1993	Bogor, Indonesia	*www.cifor.cgiar.org*
International Maize and Wheat Improvement Center (CIMMYT)	1971	Mexico City, Mexico	*www.cimmyt.org*
International Potato Center (CIP)	1973	Lima, Peru	*www.cipotato.org*
International Center for Agricultural Research in the Dry Areas (ICARDA)	1975	Aleppo, Syria	*www.icarda.org*
International Crops Research Institute for the Semi-Arid Tropics (ICRISAT)	1972	Hyderabad, India	*www.icrisat.org*
International Food Policy Research Institute (IFPRI)	1979	Washington, DC	*www.ifpri.org*
International Institute of Tropical Agriculture (IITA)	1971	Ibadan, Nigeria	*www.iita.org*
International Livestock Research Institute (ILRI)	1994	Nairobi, Kenya	*www.ilri.org*
International Rice Research Institute (IRRI)	1971	Los Banos, Philippines	*www.irri.org*
International Water Management Institute (IWMI)	1991	Colombo, Sri Lanka	*www.iwmi.cgiar.org*
World Agroforestry Centre (ICRAF)	1991	Nairobi, Kenya	*www.worldagroforestrycentre.org*
WorldFish Center**	1977	Penang, Malaysia	*www.worldfishcenter.org*

*Formerly the International Plant Genetic Resources Institute (IPGRI).
**Formerly the International Center for Living Aquatic Resources Management (ICLARM).

nance of several ex situ genebanks), sustainable management of natural resources, promoting agricultural diversification and economic development, and supporting innovation. A CGIAR Genomics Task Force has recently been established to identify opportunities for enhancing genomic research and capabilities across the various research centres.

3.2 Private Sector Funding and Public-Private Partnerships

Great hope is being placed in the ability to use plant genome sequence information to develop many kinds of new products and processes. Whether the public sector has the capacity to deliver on the scale outlined in Section 2.4 is a different matter. If the transition towards an integrated bio-based economy is pursued in the coming decades, significant investment and commitment on the part of industry will be required. Indeed, many companies from a range of industrial sectors are starting to invest heavily in genomics-related research.

Arising from a wave of takeovers and industry consolidations in the 1990s, Joly and Lemarié (1998) identify the growth of multinational 'mega firms' with expertise in biotechnology, agrochemicals and seeds. Such companies (e.g., Monsanto, DuPont/Pioneer) have begun to exploit plant genomics for R&D, and are setting up networks with academic laboratories and start-up companies, into which they are investing hundreds of millions of dollars a year.

Notably, the industrial players investing in plant genomics are not restricted to seed and agbiotech companies. Energy and oil companies such as British Petroleum (BP) are also making significant investments. In 2006, BP (which now styles itself 'Beyond Petroleum') ran an international competition to fund a $500-million-dollar Energy Biosciences Institute in plant-based sources of energy. The grant went to the University of California at Berkeley, and in this new facility academic and industrial scientists will be working side-by-side. Chemical companies and car companies, among others, are also investing in plant and microbial biotechnology. New and unusual private partnerships are emerging: for example, in June 2006, British Sugar, BP and DuPont (representing the sugar, energy and chemical/seed industries, respectively) announced the joint construction of a biobutanol plant in the UK, using locally grown sugar beet as the feedstock.

Since the 1980s, an increasingly important model for private investment in R&D has been in the form of public–private partnerships, particularly between universities and industry. A recent report identifies over 1,000 university-industry R&D centres in the US as of 1990 (these were not limited to plant science) (Ervin et al. 2003), and this number is certain to be higher today. The BP initiative described above is just one such example relevant to plant genomics. In other parts of the world, large-scale public-private initiatives include the Génoplante program in France (created in 1999 to stimulate plant genomics research), and the Rural Industries Research and Development Corporation program in Australia.

The balance between public and privately funded research has implications for education and training capacities, and the types of products or applications developed. Tensions can arise in such partnerships owing to the sometimes conflicting goals and interests of public and private research, for example with regards to research strategies, commercialization, and intellectual property (Ervin et al. 2003). Who should be setting the agenda for research and education? Which groups benefit from the research findings? Does 'public good' research lose out in such partnerships?

3.3 Models of Innovation: From Genome Sequences to New Plant Products

If innovation in plant genomics to support a bio-based economy is seen as a good thing, how can it best be stimulated? Tackling this question requires a deeper understanding of what innovation involves, who the various actors and users are in innovation processes, and what kinds of impediments to innovation might exist.

As we have seen so far in this chapter, significant changes in funding patterns and research partnerships have occurred in plant sciences since the 1990s. The recent increases in public spending on plant biotechnology and genomics follow decades of privatization and declining state investment in plant and agricultural sciences in many parts of the world. Is increased spending sufficient to drive innovation? Thought is also being given to the way that plant science is funded and organized, with new research programs being established, a variety of industrial corporations investing, and new partnerships forming. Growing numbers of national or regional networks and 'technology platform' initiatives can also be seen.

Innovation in plant genomics for a bio-based economy clearly requires more than genome sequencing. What processes, institutional arrangements, policies and support systems might be needed to go from genome sequences to the integration of new bio-based products and processes in society? The dominant model in agricultural innovation, dating back to the 1960s and still a feature of many research policy documents, is 'technology transfer', which sees scientists as the core innovators and describes a 'pipeline' model of technology development, transfer and deployment (Hall 2007). Productive translation of research findings across various stages is often identified as a problem, accounting in part for the recent rise in funding for 'translational research', meant to help counteract this obstacle.

But the products resulting from technology-transfer models of innovation are not always well-suited to the needs or demands of the end user. This has been identified as a particular problem for many science-driven projects in the developing world (Ayele et al. 2006). In such instances, does the scientist always know best? Or might the user have complementary and equally valuable knowledge or insights that could feed into the innovation process? More participatory, iterative and networked models of innovation, characterized as part of broader 'innovation systems', are increasingly being identified as potentially productive approaches (Hall 2005; Leach and Scoones 2006; Cooke 2007; Editorial 2008). The rate of uptake by users/consumers is often higher, which is certainly one measure of success in innovation. What factors might contribute to a lack of user uptake—or indeed, even consumer resistance—to new biotechnologies? This question leads us to the next section.

4 THE SOCIAL CONTEXT FOR CONTEMPORARY PLANT GENOMICS

The inclusion of this chapter in a textbook on advanced plant genomics is evidence of the growing recognition by plant scientists that the social context for contemporary plant genomics can have significant effects on the progress of the discipline. In this section, we will consider a few examples that have shaped the current social, political and legal situations in which plant genomics research takes place.

4.1 The Green Revolution

The 'Green Revolution' was the 20th century's pivotal example of the way in which plant research could impact on agricultural productivity and thus on poverty, hunger and socio-economic development. Although India is the country most associated with the Green Revolution of the late 1960s and early 1970s, the effects were felt throughout south and south-east Asia, in Mexico, and to a lesser extent in Central and South America and in Africa.

The decisive technology in the case of the Green Revolution was the introduction of new seeds: high-yielding hybrid strains of wheat and of rice. The main centres of innovation were those that later became CGIAR establishments, notably CIMMYT in Mexico and IRRI in Philippines (Table 1). Their research attracted significant funding from the Rockefeller Foundation and, for both supporters and detractors, was seen as a demonstration of the way that market-based and 'capitalistic' approaches could mould agricultural development. Agricultural establishments and government agencies tended to back the technology as a way of producing a rapid and continuing increase in food production and supply, particularly as Indian governments had confronted famines in the early 1960s. But critics were quick to point to several features of the new technology that threatened to make it less than ideal.

First, the new varieties required inputs of fertilizer and irrigation if they were to be high-yielding. Moreover, they had to be purchased as seeds each year. The technology thus tended to favor the relatively well-off. These new varieties were no longer subsistence crops; farmers had to be part of the market system to even participate in this agriculture. Bigger farmers could get better access to credit and had larger holdings to benefit from economies of scale. The way the technology was introduced worked to deepen economic divides in the countryside, and in extreme cases led to poor farmers mortgaging their land to buy into the new technologies, with all the consequent risks for their future well-being. Research also focused on the seeds of crops that were grown in bulk and could be traded. Productivity increases in 'orphan crops' including pulses and coarse grains (see Section 6.2) did not match those in wheat and rice, and their availability tended to decrease on a per capita basis.

There were also ecological concerns. The new high-yielding varieties were impressively productive. But if all farmers used the same few strains, the result was unhealthy monocultures that were ideal for the spread of disease and pests. Rapid uptake of the early strains was followed in many cases only a few years later by drastically reduced harvests. Similarly, while traditional varieties had been bred to possess some drought or flood tolerance (or other adaptation to local conditions), the new homogenized varieties were typically less hardy under occasional extreme conditions. The sudden growth in fertilizer and pesticide use also had unchecked ecological implications.

A final socio-economic corollary of the new technology was that larger yields on larger farms encouraged a route towards farm mechanization. Land owners were motivated to invest in mechanical equipment not only for any advantages it might bring, but also because it insulated them from the threat of industrial action by farm labourers. Socio-economic relations in the countryside became more polarized into potential conflicts between farm managers and employees. At the same time, seed companies and the suppliers of agrochemicals and even farm machinery benefited from an expanded market; in the extreme, governments, industry and large farmers find that their interests align, in opposition to those of small farmers and landless labourers.

High-yielding varieties were nonetheless a policy success. In 1960, the total area under the high-yielding-varieties program in India was a negligible 1.9 million hectares. Since then growth has been spectacular, increasing to nearly 15.4 million hectares by 1970, 43.1 million hectares by 1980, and 63.9 million hectares by 1990 (Heitzman and Worden 1995). The rate of growth decreased significantly in the late 1980s, however, as additional suitable land was not available.

The Rockefeller Foundation (2006) now speaks of a 'new Green Revolution' for Africa, a continent that largely missed out on the Green Revolution in the 1960s and 1970s. What role is there for genomics in these efforts? An understanding of how to tailor technologies to local conditions and needs will be crucial. Developing improved varieties of staple crops (including orphan crops, see Section 6.2) is just one part of a complex picture involving infrastructure and capacity development, adoption of better farm management practices, and the establishment of effective partnerships and markets to support growth.

4.2 Genetically Modified Crops

The controversy surrounding genetically modified (GM) crops is probably the main example of disputes over plant biotechnology, and one that has sensitized the entire scientific community to the sometimes fractious relationship between biotechnology and society. Through the 1980s, companies and university scientists were working on developing genetically engineered products, and this area was routinely identified in science policy analyses as one of the potential hotspots for economic growth in the recession-affected economies of the late 1980s and early 1990s. What were some of the key issues and controversies to arise when GM products started to come to market in the 1990s? How did they play out in different parts of the world, and what can lessons can we learn?

4.2.1 Testing Novel GM Products

The principal focus for disputes became testing. Here was a new product, whether GM crop, animal or bacterium, that needed to be assessed. Of course, all major industrialized countries had procedures for testing new foodstuffs. But the leading question related to how novel were GM products taken to be. For some, the potential for the GM entity to reproduce itself or to cross with living relatives in unpredictable ways suggested that this was an unprecedented form of innovation that needed unprecedented forms of caution and regulatory care. On the other hand, industry representatives and many scientists and commentators claimed that it was far from unprecedented. People had been introducing agricultural innovations for millennia by crossing animals, allowing 'sports' to flourish, and so on. Modern (but conventional) plant breeding already used various chemical and physical procedures to stimulate mutations that might be beneficial. According to this view, regulatory agencies were already well prepared for handling innovations in living, reproductive entities. The ground for the regulatory battle was prepared to a large degree in the US, where the courts had endorsed the regulators' decision that it was products (particular foods, seeds, etc.) and not processes (the business of genetic modification) that should be at the heart of the testing (Kloppenburg 2004; Jasanoff 2005).

GM crops were first certified in the US, where they passed tests set by the Department of Agriculture, the Food and Drug Administration and the EPA. Although an early product, the Flavr Savr (sic) tomato, found little acceptance on the market, success came with GM corn, soybean, various beets and canola (rapeseed). Essentially, GM versions of these crops offered two sorts of putative benefits: genetic resistance to a pest, or tolerance to a particular proprietary weedkiller. The potential advantages of the former are evident (even if there is a worry about pests acquiring resistance); the supposed benefits of the latter are more roundabout (smaller quantities of weedkiller can be used at later stages in the growing season, as the crops are immune). As with the Green Revolution (Section 4.1), companies also benefit, as farmers are obliged to buy the weedkillers that match the seeds.

4.2.2 Europe and the Precautionary Principle

European companies were not far behind their American counterparts in bringing GM products to market, but European customers were far less accepting of the technology than those in North America. Social scientists have been interested in trying to explain the different responses that characterized European and North American polities. It is clear that European regulators have come to be more precautionary about this technology than American officials; the former claim to be acting in accordance with the precautionary principle, which in brief refers to the idea that one should act to mitigate likely risks even before one has overwhelming evidence for those risks (Dratwa 2002). But in practice, the precautionary principle itself does not tell the regulator how precautionary to be (Levidow 2001; Marris et al. 2005). Arguments about regulatory standards have simply switched to arguments about the meaning of precaution (Dratwa 2002). Discordant interpretations of precautionarity have taken a more precise form in disputes over the standard known as 'substantial equivalence'. As Millstone et al. (1999) pointed out, in order to decide how to test the safety of GM food, one has to make some starting assumptions. Precisely because GM crops are—by definition—different from existing crops at the molecular level, one needs to decide at what level one will begin to test for any differences that might give cause for concern or even rule out the new crop-technology. According to Millstone et al. (1999, p 525):

> "The biotechnology companies wanted government regulators to help persuade consumers that their products were safe, yet they also wanted the regulatory hurdles to be set as low as possible. Governments wanted an approach to the regulation of GM foods that could be agreed internationally, and that would not inhibit the development of their domestic biotechnology companies. The FAO/WHO [UN Food and Agriculture Organization/World Health Organization] committee recommended, therefore, that GM foods should be treated by analogy with their non-GM antecedents, and evaluated primarily by comparing their <u>compositional data</u> with those from their natural antecedents, so that they could be presumed to be similarly acceptable. Only if there were glaring and important compositional differences might it be appropriate to require further tests, to be decided on a case-by-case basis" (emphasis added).

Regulators and industry agreed on this criterion of substantial equivalence as the means for implementing such comparisons.

4.2.3 Substantial Equivalence

By the standard of substantial equivalence, if GM foods are compositionally equivalent to existing foodstuffs they are taken to be substantially equivalent in regard to consumer safety. Thus, GM soybeans were accepted for consumption after they passed tests focusing on a 'restricted set of compositional variables' (Millstone et al. 1999, p 526). However, and with just as much justification, regulators might have chosen to view GM foodstuffs as novel chemical compounds entering people's diets (see also Box 1). Before new food additives and other ingredients are approved, they are subject to extensive toxicological testing. These test results are then used very conservatively to set limits for 'acceptable daily intake' levels. With GM staple crops (grains and so on), the small amounts approved for acceptable daily intake would be commercially insufficient, but safety concerns would be strongly addressed.

The point here is not whether GM foods should be treated as food additives or pharmaceutical products, but that the decision to introduce the substantial equivalence criterion is not itself based on scientific research. Once this decision is taken, it forms the basis for subsequent research. For proponents of the technology, substantial equivalence is a straightforward and commonsensical standard. But the standard conceals possible debate about what the relevant criteria for 'sameness' are. As Millstone et al. (1999) point out, for some purposes the GM seed companies are keen to stress the distinctiveness of their products. For example, GM material can only be patented because it is demonstrably novel. How can one be sure that it is novel enough to merit patent protection, but not so novel that differences beyond the level of substantial equivalence may not turn out to matter a decade or two into the future?

4.2.4 Contamination and Labelling

Errors by manufacturers and suppliers have also stirred interest regarding GM foodstuffs. No matter how emphatic the assurances that the new plant technologies are well-tested and under control, there have been a series of problems with, for example, corn approved only for animal rations ending up in human foodstuffs, or with engineered traits arising in wild relatives (Yearley 2005). Such difficulties pose interesting challenges for one popular strategy for managing GM products, namely introducing labels and tracking systems. Of course, labelling and traceability rely on the adequacy of routine methods for identifying, tracing and containing the technology—and each of these issues has been disputed (Klintman 2002; Lezaun 2006).

Another major theme to have attracted the attention of social scientists has been the question of the reasons for public resistance and consumer anxiety. Actors within the controversy have clearly faced the same question, but they have tended to account for it in tendentious ways. Proponents of GM technology often blame public anxieties on scare tactics and protectionism, whereas opponents see corporate greed combated by the perspicacity of the public. Social scientists have taken a more even-handed approach, pointing principally to three factors. First, Europeans were being offered this new food technology in the wake of the BSE or 'mad cow' controversy. The changes in food-processing procedures that are now thought to have created the conditions for the release and spread of BSE prions had been pronounced safe by the same regulatory authorities. Particularly in Britain, the government initially insisted that the best scientific advice was that there was no danger to humans from the affected beef; subsequently in 1996 they announced a sudden change of mind. Thus, the idea of GM foodstuffs being considered safe by regulatory authorities and governmental advisers could easily be shrugged off and viewed with distrust. Events such as the Pusztai affair (Box 1) were drawn on to intensify this feeling that the scientific establishment was not to be trusted. In the absence of persuasive and comprehensive assurance, there was also a question about how ordinary citizens could make dietary decisions in their daily lives. In the UK, for example, the reputable supermarkets moved to institutionalize the reassurance that governmental agencies failed to provide by establishing their own private codes for GM-free produce (see Yearley 2005).

Another explanatory factor resulted from the fact that the European landscape has been shaped by centuries of agricultural practice. Natural heritage and farming have become inseparable. Thus there was concern from environmentalists, from countryside groups and even from official nature-protection bodies about the effect of this new technology on wildlife. Particularly in France, this was allied to the third explanatory issue: a desire to protect traditional rural lifestyles in the face of the perceived threats of globalization and economic liberalization. GM technologies were viewed as further evidence of American attempts to penetrate and reshape the European agricultural market.

4.2.5 International Trade and Regulation

The World Trade Organization (WTO) has had an important role in the conflict over trade in GM foods and seeds. American companies urged that European governments' resistance to GM imports should be combated by appeals to the WTO. A formal complaint was lodged in

Box 1 The 'Pusztai affair'.

Safety testing for GM foodstuffs was at the heart of the UK's widely publicized 'Pusztai affair'. Pusztai worked at a largely government-funded research establishment near Aberdeen in Scotland, and was part of a team examining ways of testing the food safety of GM crops. He and others were concerned that compositionally similar foodstuffs might not have the same nutritional or food-safety implications. The experiments for which he became notorious were conducted on rats. These laboratory rodents were fed three kinds of potatoes: non-GM potatoes, non-GM potatoes with a lectin from snowdrops added, and potatoes genetically modified to express the snowdrop lectin. It is well known that some lectins (e.g., those in red kidney beans) can cause problems when eaten. His results suggested that the rats fared worse on the GM lectin-producing potatoes than on either of the other diets, possibly implying that it was not the lectins that were causing the problem but some aspect of the business of genetic modification (see Ewen and Pustai 1999).

This controversy unravelled in a surprising way (Eriksson 2004). Pusztai announced his results in a reputable British television program, apparently intending not to argue against GM per se but to assert that more sophisticated forms of testing would be needed to address safety concerns fully—exactly the kinds of testing that he and colleagues might have been able to perform. But the headline message that came over was that GM foods might cause health problems when eaten. In a muddled and confusing way, Pusztai's conclusions came to be criticized by his own institute and he was ushered into retirement. The ensuing controversy and hasty exercise in news-management signally failed to concentrate on his findings and the details of his experimental design. Instead, people either championed Pusztai as a whistle-blowing researcher who was unjustly disciplined by his bosses for publicizing inconvenient findings, or dismissed him as a sloppy scientist who rushed into the public gaze with results that were unchecked and not refereed. On the face of it, it is a curious sociological phenomenon that such important studies have hardly been repeated, even if the 'Pusztai affair' lives on within the wider policy debate. The controversy, by focusing not on the experiment but on disputes over Pusztai's status as an expert in this field, took on a highly attenuated form.

2003. The American argument was that there is no scientific evidence of harm arising from GM food and crops, as these products have all passed proper regulatory hurdles in the American system and within the EU itself. Furthermore, in this view, labelling of GM produce in the European market (the procedure favored in many EU member states as a possible compromise way forward) would be discriminatory and an unfair trading practice, as it draws consumers' attention to an aspect of the product that has no relation to its safety (Klintman 2002). The label 'warns' the customer of GM content but, if that content is not dangerous, then all the label does is to penalize the US and other GM-using suppliers. According to this argument, the WTO should proscribe this labelling practice as an unjustified impediment to trade. In contrast, European consumer advocates argue that the American testing has not been precautionary enough, and that properly 'scientific' tests would require much more time and more diverse examinations than have been applied in routine American trials.

The difficulty here is that, by and large, the official expert scientific communities on opposing sides take diametrically opposing views. The American approach sees all products as having potential associated risks, and the art of the policy maker is to ensure that an adequate assessment of risk and benefit is made. European analysts are more inclined to argue that the very risk framework itself leaves something to be desired. For example, in the case of GM crops, they suggest that as yet there is no way of establishing the full range of possible risks, so no scientific risk assessment can be completed.

Within their separate jurisdictions, each of these opposing views can be sensibly and more-or-less consistently maintained. However, the differing views appear to be tantamount to incommensurable paradigms for assessing the safety and suitability of GM crops. There is no higher level of scientific rationality or expertise to which appeal can be made to say which approach is correct, and the WTO does not have its own body of 'super scientists' to resolve such issues. However, observers of the WTO fear that its dispute settlement procedures, although supposedly neutral and merely concerned with legal and administrative matters, tacitly favor the American paradigm. The resolution of this case in the USA's favor may thus not only affect policy towards GMOs but set a significant precedent for how disputed scientific views are handled before the WTO (see Murphy et al. 2006).

Although the GM case is mainly a bone of contention between wealthy northern countries, other nations have been caught up in the struggle. Facing food shortages caused by drought, in 2002 Zambia was offered food aid by the US that just happened to consist of genetically engineered cereals. Many felt that the US was using this case as a Trojan Horse to encourage the uptake of GM foods. However, Zambians realized there was a danger that their future exports to the EU would be threatened if they lost their GM-free status. The government thus equivocated over accepting the food aid, illustrating the extent to which the American–European struggle is associated with global food standards.

4.2.6 The Legacy of Controversies around GM Food and Crops

As important as the GM controversy may have seemed in Europe, at a global level fortune is currently favoring this technology. Genetically modified or genetically engineered crops are now—a mere decade on—ubiquitous at the planetary level. Corn, cotton, canola, sugar beet and soybean are now widely planted in North America, much of South America, China and Australia, and drought-resistant wheat is being tested in Australia.

There are signs, however, that the concerns of European consumers may be attracting attention elsewhere. For example, although China cultivates a large amount of GM cotton (largely Bt-cotton, which it regards as responsible for increasing yields and improving farm safety by decreasing pesticide use), the policy on GM rice has yet to be determined. Rice has enormous cultural and symbolic value in China, and opinion-leading sections of the population seem to be adopting precautionary attitudes (Keeley 2003). These are undoubtedly stoked by the realization that the regulatory system is far from robust, with non-approved GM rice strains widely available on the black market. Moreover, as a centre of rice biodiversity, arguments are also being raised about the risk of genetic contamination arising from widespread adoption of new GM varieties.

In Africa, civil-society actors also have an increasing voice, a fact that is being exploited both by GM enthusiasts and its detractors. 'Citizen groups' on both sides of the debate are being sponsored by national and international bodies, and the debate over GM is being overtaken by a related debate about which groups speak authentically with citizens' voices (Harsh 2005).

Within Europe there are signs of a fight-back by proponents of GM, taking advantage both of the perceived urgency of problems to do with food supply and security, and with the importance of non-food crops. 'Excessive' precaution is being opposed not by attacking precaution per se, but by suggesting that one also needs to take precautionary steps to secure the food supply. Insofar as anxieties about GM crops relate to dietary concerns, one might well anticipate less hostility towards GM trees for coppicing or GM grasses for biofuel production. Such crops may come to be offered as 'compromise' candidates for a European experiment with a form of GM that no-one will ever have to eat.

4.3 Bioprospecting and Intellectual Property

Bioprospecting—a term referring to the exploration of biodiversity for commercially valuable resources and products—has a long history. Quinine, coffee, tea, rubber, tobacco, and many other plants or plant products were brought back to Europe from early bioprospecting adventures (e.g., Drayton 2000; Merson 2000). Botanical gardens such as Kew Gardens in London often served as the repositories of specimens collected abroad, and became powerful institutions connected with the enterprise of Empire.

Today, the quest for commercially valuable plant products is increasingly focused at the molecular level. As discussed in Section 2, plant genetic diversity can be exploited for economic purposes by many industries, including the seed, horticulture, pharmaceutical, cosmetics, chemical and energy industries. However, as a general rule, those parts of the world richest in biodiversity are not wealthiest in terms of financial resources or scientific infrastructure. This raises difficult questions relating to the ownership and exploitation of genetic information from bioprospecting activities (ten Kate and Laird 1999). Where exactly is the 'value' to be found, who has the right to access genetic diversity, and who benefits from the development of new products? For example, the genetic diversity introduced and/or maintained through smallholder agriculture can be exploited by researchers in academia and industry for the development of new crop varieties. How should the benefits from products developed using such sources be distributed? The farmers arguably play a crucial role in ensuring the genetic diversity is there to exploit in the first place, but they do not typically patent their local varieties and might not share the advanced genetic understanding of crop breeding used by commercial breeders. In rainforests or others 'commons', the ownership of genetic resources is similarly problematic. This uncertainty has led to accusations of 'biopiracy' by groups or nations that feel their biodiversity is being exploited for commercial gain by unscrupulous entrepreneurs (Merson 2000).

A number of international mechanisms have recently been developed to regulate the use of and profit derived from plant genetic diversity. The 2004 International Treaty for Plant Genetic Resources for Food and Agriculture (ITPGRFA) was developed by an intergovernmental commission of the FAO, and provides a framework for the collection, characterization, evaluation, conservation and documentation of plant genetic resources for food and agriculture. A multilateral system of access and benefit-sharing is built into the treaty, whereby genetic resources are made available for research, breeding and training, but recipients are prevented from making intellectual property claims that would limit future access to the resources (Esquinas-Alcázar 2005). A proportion of the revenues derived from commercialization are returned to the treaty's strategy fund, to further support its implementation and development.

With respect to 'wild' biodiversity, the 1992 Convention on Biological Diversity (CBD) is an international agreement that asserts the sovereign right of nations over their own biological and genetic resources, and the authority of nations to determine external access to their genetic resources. The main objectives of the CBD are the conservation and sustainable use of biodiversity, and the fair and equitable sharing of benefits arising from the use of genetic resources. A number of countries have since developed national regulations and benefit-sharing mechanisms, which may involve granting access rights in exchange for financial, scientific, social and/or environmental support or compensation. A small number of North–South cooperative research agreements have been established, notably in Suriname and in Costa Rica (which set up the INBio institute for this purpose, and has had collaborations with the pharmaceutical company Merck). However, to date there remain few successful models of such partnerships.

4.4 Central Issues for Plant Genomics

Although the examples in Section 4 might not focus specifically on plant genomics, they certainly shape the context in which plant genomics is developing. Other developments in genomics and the life sciences more generally (e.g. stem cell technologies, the creation of genetic databases, etc.) also influence the broader context for plant genomics technologies. Drawing on a range of examples, we can identify a number of core issues that may feature in debates surrounding new plant genomics technologies:

- **Novelty.** Central to this issue are questions around what makes a product novel. Does novelty arise from the processes involved, or does it have more to do with characteristics of the final product? What are appropriate baselines for comparison and evaluation, and to what extent is this a technical or a legal matter?
- **Safety testing.** The precautionary principle demands that one should act to mitigate likely, significant risks even before one has overwhelming evidence for those risks. Is it ever possible to prove

that a new technology is safe and poses no risks? What are 'appropriate' levels of risk with regards to human health and the natural environment, and how might these be measured?

- **'Contamination'** of the human body and the natural environment through the spread or transfer of genes is sometimes raised as a concern. What are some of the different meanings and connotations of 'contamination', and how might technology and/or regulation help to minimize gene flow from new products?
- **Consumer choice and desires.** Are the new products entering the market seen as desirable or necessary in the eyes of consumers? Do they offer features that are otherwise unavailable? Is the option of consumer choice upheld? (For example, and linked to the issue of contamination, can consumers really choose between GM and non-GM produce?)
- **Social justice.** Who benefits from new technologies? Are some individuals or groups that should benefit being left out, and why? Should some actors (for example, European and Japanese farmers) be shielded from full-blooded market competition? These issues are closely linked to questions of access, ownership, and intellectual property around new biotechnologies.
- **Infrastructure.** Technology (or technology transfer) does not always provide solutions—issues of capacity, infrastructure and policy support affect the ways in which new technologies are taken up by different groups and nations.
- **International standards and regulations.** Global markets and trade mean that a degree of international harmonization regarding regulatory frameworks and safety standards is necessary. How can differences in national and sectoral approaches to regulation be reconciled, and who has the final say? How do different regulatory frameworks shape the research and funding landscape for new technologies?

As we have seen, particularly in the GM crop example (Section 4.2), improved scientific knowledge does not necessarily help us to resolve these issues. Sometimes the issue has very little to do with science, sometimes scientific information alone cannot determine what decision should be taken, and sometimes the science itself becomes a pawn in broader debates about risks, rewards or socioeconomic priorities (Jasanoff 1990, 2005).

5 GOVERNANCE AND REGULATORY FRAMEWORKS

What is the role for regulation when it comes to plant genomics? At a broad level, regulation is concerned with issues relating to safety (for example, regarding human health and the environment), access and distribution of benefits (as we saw for example in Section 4.3), and the coordination of activities across different levels and contexts (nationally, internationally, etc.). Ideally these goals would be promoted without unduly restricting or compromising basic science and innovation, but in practice this is a difficult balance to achieve. As discussed in relation to GM crops (Section 4.2), different countries may have different styles or approaches to regulation of biotechnology, influenced by factors including their political decision-making structures, cultural values, and different ways of interpreting scientific evidence (Jasanoff 2005).

As outlined in Section 2, improvements in our ability to exploit plant genomics for practical applications will almost certainly affect sectors as diverse as health, industry, environment, agriculture, energy and security. Rather than assuming traditional divisions between these sectors, a recent report by the Organization for Economic Cooperation and Development (2006) suggests that increasing convergence and interaction among research domains and technologies will take place in the development of a bio-based economy. This convergence is likely to challenge existing regulatory systems in many ways. To provide a simple example in the context of international trade, should biofuel crops count as energy or agricultural commodities? The ruling has implications in terms of trade rules, tariffs and so on, which are different for agricultural and energy commodities. In a different context, should plant-derived vaccines and pharmaceutical products be evaluated as traditional pharmaceuticals, as food-related products, or according to their method of production? New regulatory concepts, structures and processes are likely to be required as the bio-based economy develops.

As we saw for GM crops (Section 4.2), regulatory frameworks around genomics and biotechnology can encode moral or social values about what science should be allowed to progress. For example, the decision in many parts of the world to allow human therapeutic cloning but ban reproductive cloning is a reflection of what is deemed broadly acceptable at this point in time. What role is there then for public involvement in the regulation and governance of new technologies? Should the development of regulation fall strictly to policymakers (elected or unelected), and who should they consult during the decision-making process? In recent years, the idea that the general public is not sufficiently well-versed in scientific matters to inform regulation has been replaced with the idea that public 'engagement', consultation and two-way dialogue are necessary for allowing policymakers to develop regulatory policies that adequately reflect the relationships among science, culture, citizenship and responsibility. These relationships are not static, but continue to evolve as new knowledge and products are developed, and as new imaginations of the future take hold (e.g., Leach and Scoones 2003). A bio-based economy is one such transformative vision that arguably requires exploration and debate at a broad level (see also Section 6.6).

6 CHALLENGES FOR PLANT GENOMICS IN THE 21ST CENTURY

Returning to the applications outlined in Section 2, it is clear that plant genomics can be used for a variety of purposes. With growing concern around issues such as climate change, energy supply and security, food production, biodiversity loss, and dependence on fossil fuels, this is an exciting but also a challenging time for plant genomics. Are the challenges simply technical ones? No. In this final section, we will re-visit a few application areas for plant genomics. Rather than focusing on the technical hurdles, we will think about some of the challenges from social, economic, legal and political perspectives.

6.1 Energy Supply: Biofuels

Biofuels are not a new technology: during the oil price shocks of the 1970s, a number of countries started developing biofuel programs. Virtually all of these (with the notable exception of Brazil's bioethanol industry) were abandoned when oil prices dropped (Marris 2006). Several factors, including the rising cost of oil and climate-change related incentives for developing renewable energies, are once again making the economics of biofuel production more attractive. Will biofuels take off this time?

Decisions around which crop or plant species to focus on are not based on scientific merit alone—for example, corn is known not to be a particularly 'green' biofuel from the perspective of greenhouse gas emissions reduction or ethanol conversion efficiency (International Energy Agency 2004). The focus on producing bioethanol from corn in the US is almost certainly linked to the well-established and politically important corn-growing industry there.

The recent surge of interest in biofuel production is resulting in widespread changes in land use and export patterns, with implications for biodiversity and food security. Food crops such as wheat and corn are increasingly being used for biofuel production, and have been linked to rising food prices—and resulting civil unrest—around the world (Anon 2007a, b). Scientists and policy makers are placing great hope in technologies including genomics to help develop so-called 'next-generation' biofuels (produced from waste biomass or algae). The claim is that these biofuels will help to relieve pressures on land use and reduce competition between food and fuel crops. What new issues might be raised with the development of second- and third-generation biofuels?

6.2 Food Supply: Orphan Crops

The term 'orphan' crops describes a diverse range of crops that are regionally or locally important, but do not have large international markets and receive little support in terms of scientific research and investment. Examples of orphan crops include sorghum, pearl millet and tubers (such as cassava), which in Sub-Saharan Africa are more important than rice and wheat as staple crops. Although under-researched, orphan crops tend to be nutritious, adapted to harsh environments, and diverse in terms of their genetic profiles and economic niches (Nelson et al. 2004).

Genomic research is an expensive enterprise on the whole—how might research on orphan crops fare in a context of growing private investment in plant genomics initiatives, and intellectual property claims being made on biological material and information? (See the Golden Rice example in Section 2.1.) Should investment in research on orphan crops be evaluated according to economic return, or does it result in public goods of a different kind? There are frequent calls to promote public sector research on orphan crops (e.g., Naylor et al. 2004; Delmer 2005; Rockefeller Foundation 2006). To an extent, orphan crops might benefit from the transfer of relevant genomic information and tools from model species and major crops. More troublesome perhaps is how to support the innovation process as a whole, and to ensure that any insights derived through genomic approaches are connected to downstream breeding, farming and conservation efforts (see Section 3.3). Investment in infrastructure, capacity-building and participatory models of research will be crucial for plant genomics to make positive contributions to orphan crop development.

6.3 Plant Genomics and Conservation

Initiatives such as the Consortium for the Barcoding of Life (CBOL) aim to document life on earth through the identification of unique genetic 'barcodes' for each and every species. More broadly, genetics and genomics are seen as potentially useful tools for profiling structure and function at community and ecosystem levels, and for deriving measures or indicators of environmental harm that may help to inform conservation strategies (Witham et al. 2006, 2008). The growth of such projects indicates an increasingly scientific and molecular approach to understanding and protecting natural world, and has impacts on conservation policy and practice. Although undoubtedly valuable in some ways, what might some of the unintended consequences be?

To provide one example, the vast majority of people involved in conservation activities around the world are amateur naturalists of one form or another. Their direct, sensory experiences of and relationships with nature might seem at odds with more scientific and molecular representations of the natural world (Bäckstrand 2004; Ellis and Waterton 2004). The continued participation of amateur naturalists in conservation efforts is crucial. What data and whose knowledge should count when it comes to deciding conservation actions? These are highly political questions. Mechanisms to integrate different ways of 'knowing' the world—drawing on local knowledge, cultural memory, and measurements at genetic and other levels—will arguably be necessary for conservation efforts to be inclusive and productive.

6.4 Plant Genomics and Intellectual Property

The potential commercial applications of plant genomics for the bio-based economy are clear (see Section 2), and account for the growing private investment by many different industries in genomics-related networks and initiatives. The push to protect research findings by filing intellectual property claims is a growing feature in both public and private sector research. Patenting of DNA sequences and genes is now commonplace. Whole-genome patents have also been issued for a number of prokaryotes and viruses (O'Malley et al. 2005).

The precedent for patenting biological products and processes came with the landmark Diamond vs. Chakrabarty 1980 US Supreme Court case. Nonetheless, there is still debate over what aspects of biological material it is appropriate to patent. Patents are described by some as a necessary incentive for stimulating investment and innovation, and are hotly contested by others as being unethical and restricting access to what should be freely accessible technologies and products. For example, a recent document by civil society organization The ETC Group (2008) identifies over 500 patent applications by multinational seed and agrochemical corporations on so-called 'climate ready' genes. They express concern that these patents will "concentrate corporate power, drive up costs, inhibit independent research, and further undermine the rights of farmers to save and exchange seeds" (p 1) despite being marketed as solutions to the growing problem of climate change. How might intellectual property frameworks strike a balance between encouraging investment and innovation, and delivering public goods?

In the context of plant breeding, the principal mechanism for protecting new varieties is not by patenting but rather the Plant Variety Rights (PVR) system, established through the International Union for the Protection of New Varieties of Plants (UPOV) in 1961. This system for evaluating and registering new plant varieties is grounded in traditional approaches to plant breeding, and has not been readily able to accommodate new technologies such as genetic modification or more recent genomic approaches (e.g., marker-assisted breeding). Changes to the PVR framework are likely to be required in order to keep pace with advances in genetic technologies for plant breeding.

6.5 Synthetic Biology

Synthetic biology is an emerging discipline concerned primarily with the re-design of existing biological systems for useful purposes, and the design and construction of new biological parts and systems (Benner and Sismour 2005; Endy 2005). By applying an engineering-like approach to biology, synthetic biology researchers hope to manipulate biological systems in a more modular, predictable and controllable way than traditional genetic engineering approaches have to date. Genomics and large-scale DNA synthesis are core technologies supporting efforts in synthetic biology.

The commercial applications of synthetic biology are potentially significant—several current efforts focus on biosensor development, biofuel production and drug synthesis (e.g., for the anti-malarial compound artemisinin; Ro et al. 2006). Acknowledging that unrestricted access to smaller genetic component parts (e.g., 'BioBricks') is crucial for higher-level applications to be developed, key proponents of synthetic biology have established an open-access Registry for Biological Parts. However, alongside this open-access registry, numerous patents have already been filed for biological parts, minimal genomes and synthetic genomes, and for technologies such as genome synthesis, genome transplantation, and whole-genome engineering (ETC Group 2007a, c). Different models for intellectual property protection in synthetic biology are currently under active debate (Rai and Boyle 2007).

In addition to intellectual property considerations, a number of ethical and security issues have been raised with regards to synthetic biology. The potential for DNA synthesis to be abused for nefarious purposes has caused some concern, prompting discussion among scientists and policymakers about what appropriate regulatory frameworks might look like (see Bügl et al. 2007). Should scientists be left to regulate themselves, or is a degree of top-down control required? Ethical questions about what constitutes 'life' and whether synthetic biology researchers are 'playing God' have also featured in the public press (Editorial 2007, ETC Group 2007b). Researchers in synthetic biology have so far been willing to engage with social scientists, policymakers and civil society organizations to discuss these and other issues. Is this simply to try and ensure smooth sailing for synthetic biology—to avoid 'the next GM'? Will these discussions end up influencing the research questions pursued by scientists? Time will tell.

6.6 Emerging Issues: The Politics of Plants

As discussed in Section 2, this is an exciting time for plant genomics. The possible move to a more bio-based economy would represent a fairly fundamental transition from the thermodynamic flows linked to fossil fuel extraction, and may help to promote a more environmentally sustainable way of life. The opportunities for plant science to develop new uses for biomass are vast. However, the re-valuing of plants in terms of their technological potential is exposing tensions among the many different systems to which plants contribute—can they satisfy all of our demands in terms of food, energy, medicines, materials, biodiversity, and global environmental health?

Implicit in descriptions of the bio-based economy is that we are moving from a resource-limited economy (constrained by oil reserves) to one of potentially unlimited resource in the form of biomass. The million-dollar question is whether there is enough biomass to support the many environmental, social and economic objectives of the bio-economy. In the short term at least, this seems unlikely—plants are a renewable but finite resource, and certainly at present, land availability is a key limiting factor.

We suggest that a new 'politics of plants' is starting to emerge, as groups with different interests and operating at different scales (global, national, local, etc.) try to stake claims and compete for various aspects of biomass ownership, processing, and consumption (Frow et al. 2009). The implications are significant—the connections between energy, environment, food and health are becoming increasingly clear, and decisions taken in one sphere are likely to have consequences for the others. How can we achieve a reasonable balance among competing objectives? Who should have a say in what our future looks like? Although these questions might seem far-removed from concerns at the laboratory bench, they can have very tangible impacts in terms of influencing funding priorities and research strategies.

The 18th century economists who focused above all on land and agriculture may turn out to be more correct than has been acknowledged for two hundred years. But it is clear we cannot go back to old ways of valuing bio-resources—the genetic information that plants contain, and the genomic potential of previously under-valued botanical resources are increasingly seen as central to their worth. Exactly how this value can be realized, turned into products and appropriated is, of course, partly a question of science and technology; but it is just as much a matter of legal rules, international conventions and socio-economic practices. The new bioeconomy will be put together bit-by-bit, by plant scientists and also by lawyers, activists, entrepreneurs and ordinary citizens.

Acknowledgements

We would like to thank Nathan Gove for his help with background research for this chapter.

References

Anonymous (2007a) The end of cheap food. Economist 385 (6 Dec 2007): 11-12

Anonymous (2007b) Cheap no more. Economist 385 (6 Dec 2007): 83-85

Anonymous (2008) GM crops: A world view. Science 320: 466-467

Agriculture and Environment Biotechnology Commission (AEBC) (2005) What Shapes the Research Agenda in Biotechnology? Plant Breeding Case Study. Rep URN 05/1084: *http://www.aebc.gov.uk/aebc/subgroups/ra_plant_breeding.pdf* (cited 29 July 2008)

Al-Babili S, Beyer P (2005) Golden Rice—five years on the road—five years to go? Trends Plant Sci 10(12): 565-573

Ayele S, Chataway J, Wield D (2006) Partnerships in African crop biotech. Nat Biotechnol 24: 619-621

Bäckstrand K (2004) Scientisation vs. civic expertise in environmental governance: Eco-feminist, eco-modern and post-modern responses. Environ Polit 13(4): 695-714

Benner SA, Sismour AM (2005) Synthetic biology. Nat Rev Genet 6: 533-543

Biotechnology and Biological Sciences Research Council (BBSRC) (2004) A Review of BBSRC-Funded Research Relevant To Crop Science. A rep for BBSRC Council: *http://www.bbsrc.ac.uk/organisation/policies/reviews/scientific_areas/0404_crop_science.pdf* (cited 29 July 2008)

Brasier C (2008) The biosecurity threat to the UK and global environment from international trade in plants. Plant Pathol doi: 10.1111/j.1365-3059.2008.01886.x

Bügl H, Danner JP, Molinari RJ, Mulligan JT, Park H-O, Reichert B, Roth DA, Wagner R, Budowle B, Scripp RM, Smith JAL, Steele SJ, Church G, Endy D (2007) DNA synthesis and biological security. Nat Biotechnol 25: 627-629

Cooke P (2007) Growth Cultures: The Global Bioeconomy and its Bioregions. Routledge, London, UK

Dalton R (2006) The Christmas invasion. Nature 443: 898-900

Delmer DP (2005) Agriculture in the developing world: Connecting innovations in plant research to downstream applications. Proc Natl Acad Sci USA 102: 15739-15746

Dix DJ, Gallagher K, Benson WH, Groskinsky BL, McClintock JT, Dearfield KL, Farland WH (2006) A framework for the use of genomics data at the EPA. Nat Biotechnol 24: 1108-1111

Dratwa J (2002) Taking risks with the precautionary principle: Food (and the environment) for thought at the European Commission. J Environ Pol Plan 4: 197-213

Drayton R (2000) Nature's Government: Science, Imperial Britain and the 'Improvement' of the World. Yale Univ Press, New York, London

Editorial (2007) Meanings of 'life'. Nature 447: 1031-1032

Editorial (2008) A case for nurture. Nature 454: 918

Ellis R, Waterton C (2004) Environmental citizenship in the making: the participation of volunteer naturalists in biological recording and biodiversity policy. Sci Pub Pol 31: 95-105

Endy D (2005) Foundations for engineering biology. Nature 438: 449-453

Environment Agency (2003) Environmental genomics—An introduction. Environ Agency, Bristol: *http://publications.environment-agency.gov.uk/pdf/SGENOMICS-e-p.pdf* (cited 7 Aug 2008)

Eriksson L (2004) From Persona to Person: The Unfolding of an (Un)scientific Controversy. PhD Thesis. Cardiff Univ, Cardiff, UK

Ervin D, Lomax T, Buccola S, Kim K, Minor E, Yang H, Glenna L, Jaeger E, Biscotti D, Armbruster W, Clancy K, Lacy W, Welsh R, Xia Y (2003) University–industry relationships and the public good: Framing the issues in agricultural biotechnology. Pew Initiative on Food and Biotechnol Rep,

Washington DC: *http://www.agri-biotech.pdx.edu/Pew_UIRreport-FINAL%20use%20this.pdf* (cited 1 Aug 2008)

Esquinas-Alcázar J (2005) Protecting crop genetic diversity for food security: Political, ethical and technical challenges. Nat Rev Genet 6: 946-953

ETC Group (2007a) Extreme Genetic Engineering: An Introduction to Synthetic Biology. Rep, Jan 2007: *http://www.etcgroup.org/en/materials/publications.html?pub_id=602* (cited 7 Aug 2008)

ETC Group (2007b) Patenting Pandora's bug: Goodbye, Dolly...Hello, Synthia! J Craig Venter Institute seeks monopoly patents on the world's first-ever human-made life form. News release, 7 June 2007: *http://www.etcgroup.org/en/materials/publications.html?pub_id=631* (cited 7 Aug 2008)

ETC Group (2007c) Extreme monopoly: Venter's team makes vast patent grab on synthetic genomes. News release, 8 Dec 2007:*http://www.etcgroup.org/en/materials/publications.html?pub_id=665* (cited 7 Aug 2008)

ETC Group (2008) Patenting the "climate genes"...and capturing the climate agenda. Communiqué, Issue # 99: *http://www.etcgroup.org/en/materials/publications.html* (cited 7 Aug 2008)

Ewen SWB, Pusztai A (1999) Effect of diets containing genetically modified potatoes expressing *Galanthus nivalis* lectin on rat small intestine. Lancet 354: 1353-1354

Food and Agriculture Organization (2006) The State of Food Insecurity in the World 2006. Rep, Rome: *http://www.fao.org/docrep/009/a0750e/a0750e00.htm* (cited 7 Aug 2008)

Fox JL (2006) Turning plants into protein factories. Nat Biotechnol 24: 1191-1193

Frow, EK, Ingram D, Powell W, Steer D, Vogel J, Yearley S (2009) The politics of plants. Food Secur 1: 17-23

Hall A (2005) Capacity development for agricultural biotechnology in developing countries: An innovation systems view of what it is and how to develop it. J Int Dev 17: 611-630

Hall A (2007) Challenges to strengthening agricultural innovation systems: Where do we go from here? UNU-Merit Working Paper Series #2007-038. Maastricht, Netherlands: *http://www.merit.unu.edu/publications/wppdf/2007/wp2007-038.pdf* (cited 1 Aug 2008)

Harsh M (2005) Formal and informal governance of agricultural biotechnology in Kenya: Participation and accountability in controversy surrounding the draft Biosafety Bill. J Int Dev 17: 661-677

Heitzman J, Worden RL (eds) (1995) India: A Country Study. GPO for the Library of Congress, Washington: *http://countrystudies.us/india/* (cited 18 Aug 2008)

Hooper DU, Chapin FS, Ewel JJ, Hector A, Inchausti P, Lavorel S, Lawton JH, Lodge DM, Loreau M, Naeem S, Schmid B, Setälä H, Symstad AJ, Vandermeer J, Wardle DA (2005) Effects of biodiversity on ecosystem functioning: a consensus of current knowledge. Ecol Monogr 75: 3-35

Hopkin M (2008) Biodiversity: frozen futures. Nature 452: 404-405

Huang J, Rozelle S, Pray C, Wang Q (2002) Plant biotechnology in China. Science 295: 674-677

International Energy Agency (2004) Biofuels for transport: An international perspective. Paris, France: *http://www.iea.org/textbase/nppdf/free/2004/biofuels2004.pdf* (cited 1 Aug 2008)

Jasanoff S (1990) The Fifth Branch: Science Advisers as Policymakers. Harvard Univ Press, Cambridge, MA, USA

Jasanoff S (2005) Designs on Nature: Science and Democracy in Europe and the United States. Princeton Univ Press, Princeton, Oxford, UK

Joly P-B, Lemarié S (1998) Industry consolidation, public attitude and the future of plant biotechnology in Europe. AgBioForum 1(2): 85-90

Kamm B, Kamm M (2004) Principles of biorefineries. Appl Microbiol Biotechnol 64: 137-145

Keeley J (2003) Regulating biotechnology in China: The politics of biosafety. IDS Working Paper 208, Institute of Development Studies, Brighton: *http://www.ntd.co.uk/idsbookshop/details.asp?id=772* (cited 29 July 2009)

Klintman M (2002) The genetically modified food labeling controversy. Soc Stud Sci 32: 71-92

Kloppenburg JR (2004) First the Seed: The Political Economy of Plant Biotechnology. Univ Wisconsin Press, Madison, WI, USA

Leach M, Scoones I (2003) Science and citizenship in a global context. IDS Working Paper 205, Institute of Development Studies, Brighton: *http://www.ids.ac.uk/ids/bookshop/wp/wp205.pdf* (cited 1 Sept 2008)

Leach M, Scoones I (2006) The slow race: Making technology work for the poor. Democs, London, UK: *http://www.demos.co.uk/files/The%20Slow%20Race.pdf* (cited 18 Aug 2008)

Levidow L (2001) Precautionary uncertainty: Regulating GM crops in Europe. Soc Stud Sci 31: 845-878

Lezaun J (2006) Creating a new object of government: making genetically modified organisms traceable. Soc Stud Sci 36: 499-531

Marris C, Joly PB, Ronda S, Bonneuil C (2005) How the French GM controversy led to the reciprocal emancipation of scientific expertise and policy making. Sci Pub Pol 32: 301-308

Marris E (2006) Drink the best and drive the rest. Nature 444: 670-672

Merson J (2000) Bio-prospecting or bio-piracy: Intellectual property rights and biodiversity in a colonial and postcolonial context. Osiris 15: 282-296

Millstone E, Brunner E, Mayer S (1999) Beyond substantial equivalence. Nature 401: 525-526

Morris CE, Sands DC (2006) The breeder's dilemma—yield or nutrition? Nat Biotechnol 24: 1078-1080

Murphy J, Levidow L, Carr S (2006) Regulatory standards for environmental risks: understanding the US–European Union conflict over genetically modified crops. Soc Stud Sci 36: 133-160

National Research Council (2008) Achievements of the National Plant Genome Initiative and new horizons in plant biology. Natl Acad Press, Washington DC, USA

Naylor RL, Falcon WP, Goodman RM, Jahn MM, Sengooba T, Tefera H, Nelson RJ (2004) Biotechnology in the developing world: a case for increased investments in orphan crops. Food Pol 29: 15-44

Nelson RJ, Naylor RL, Jahn MM (2004) The role of genomics research in improvement of "orphan" crops. Crop Sci 44: 1901-1904

Normile D (2002) From standing start to sequencing superpower. Science 296: 36-39

O'Malley MA, Bostanci A, Calvert J (2005) Whole-genome patenting. Nat Rev Genet 6: 502-506

Organization for Economic Cooperation and Development (OECD) (2006) The Bioeconomy to 2030: Designing a Policy Agenda. OECD International Futures Programme Scoping Document, Paris, France: *http://www.oecd.org/dataoecd/48/1/36887128.pdf* (cited 8 Jul 2008)

Pennisi E (1998) A bonanza for plant genomics. Science 282: 652-654

Pennisi E (2007) The greening of plant genomics. Science 317: 317

Pilon-Smits E (2005) Phytoremediation. Annu Rev Plant Biol 56:5-39

Powell K (2007) Functional foods from biotech—an unappetizing prospect? Nat Biotechnol 25(5): 525-531

Ragauskas AJ, Williams CK, Davison BH, Britovsek G, Cairney J, Eckert CA, Frederick WJ Jr, Hallett JP, Leak DJ, Liotta CL, Mielenz JR, Murphy R, Templer R, Tschaplinkski T (2006) The path forward for biofuels and biomaterials. Science 311: 484-489

Raghuram N (2002) Indian plant biology enters the biotechnology era. Trends Plant Sci 7:2-94

Rai A, Boyle J (2007) Synthetic biology: Caught between property rights, the public domain, and the commons. PLoS Biol 5(3): e58 doi:10.1371/journal.pbio.0050058

Ro D-K, Paradise EM, Ouellet M, Fisher KJ, Newman KL, Ndungu JM, Ho KA, Eachus RA, Ham TS, Kirby J, Chang MCY, Withers ST, Shiba Y, Sarpong R, Keasling JD (2006) Production of the antimalarial drug precursor artemisinic acid in engineered yeast. Nature 440: 940-943

Rockefeller Foundation (2006) Africa's Turn: A New Green Revolution for the 21st Century. Rockefeller Foundation, New York, USA: *http://www.rockfound.org/library/africas_turn.pdf.* (cited 11 Aug 2008)

Stern N (2007) The Economics of Climate Change: The Stern Review. Cabinet Office—HM Treasury, London, UK

ten Kate K, Laird SA (1999) The Commercial Use of Biodiversity: Access to Genetic Resources and Benefit-sharing. Earthscan Publ, London, UK

Tilman D (2000) Causes, consequences and ethics of biodiversity. Nature 405: 208-211

Wang D, Coscoy L, Zylberberg M, Avila PC, Boushey HA, Ganem D, DeRisi JL (2002) Microarray-based detection and genotyping of viral pathogens. Proc Natl Acad Sci USA 99(24): 15687-15692

Webster CG, Wylie SJ, Jones MGK (2004) Diagnosis of plant viral pathogens. Curr Sci 86(12): 1604-1607

Whitham TG, Bailey JK, Schweitzer JA, Shuster SM, Bangert RK, LeRoy CJ, Lonsdorf EV, Allan GJ, DiFazio SP, Potts BM, Fischer DG, Gehring CA, Lindroth RL, Marks JC, Hart SC, Wimp GM, Wooley SC (2006) A framework for community and ecosystem genetics: From genes to ecosystems. Nat Rev Genet 7: 510-523

Whitham TG, DiFazio SP, Schweitzer JA, Shuster SM, Allan GJ, Bailey JK, Woolbright SA (2008) Extending genomics to natural communities and ecosystems. Science 320: 492-495

Xu K, Xu X, Fukao T, Canlas P, Maghirang-Rodriguez R, Heuer S, Ismail AM, Bailey-Serres J, Ronald PC, Mackill DJ (2006) *Sub1A* is an ethylene-response-factor-like gene that confers submergence tolerance to rice. Nature 442: 705-708

Yearley S (2005) Cultures of Environmentalism: Empirical Studies in Environmental Sociology. Palgrave Macmillan, Basingstoke, UK

Yu J, Hu S, Wang J, Wong GK, Li S, Liu B, Deng Y, Dai L, Zhou Y, Zhang X, Cao M, Liu J, Sun J, Tang J, Chen Y, Huang X, Lin W, Ye C, Tong W, Cong L, Geng J, Han Y, Li L, Li W, Hu G, Huang X, Li W, Li J, Liu Z, Li L, Liu J, Qi Q, Liu J, Li L, Li T, Wang X, Lu H, Wu T, Zhu M, Ni P, Han H, Dong W, Ren X, Feng X, Cui P, Li X, Wang H, Xu X, Zhai W, Xu Z, Zhang J, He S, Zhang J, Xu J, Zhang K, Zheng X, Dong J, Zeng W, Tao L, Ye J, Tan J, Ren X, Chen X, He J, Liu D, Tian W, Tian C, Xia H, Bao Q, Li G, Gao H, Cao T, Wang J, Zhao W, Li P, Chen W, Wang X, Zhang Y, Hu J, Wang J, Liu S, Yang J, Zhang G, Xiong Y, Li Z, Mao L, Zhou C, Zhu Z, Chen R, Hao B, Zheng W, Chen S, Guo W, Li G, Liu S, Tao M, Wang J, Zhu L, Yuan L, Yang H (2002) A draft sequence of the rice genome (*Oryza sativa* L. ssp. *indica*). Science 296: 79-92

17 Roadmap of Genomics Research in the 21st Century

Andrew H. Paterson
Plant Genome Mapping Laboratory,
University of Georgia, Athens GA, USA
E-mail: paterson@plantbio.uga.edu

1 INTRODUCTION

For one such as the author who was trained in classical plant breeding, the dramatic progress that has been made in revealing, understanding, and manipulating the hereditary blueprint of complex crop genomes continues to instill a sense of wonder. Only 20 years ago, the botanical research community was just gaining its first glimpse into plant genomes, often in the form of Southern blots segregating for restriction fragment length polymorphism (RFLP) loci. Today, at least ten of the genomes that we studied by what now seem rather crude techniques are fully sequenced, with many more slated for sequencing as described elsewhere in this volume (see Chapters 6-10 for details). Technological progress in many areas, such as dramatically accelerated sequencing technologies (Margulies et al. 2005; Shendure et al. 2005), targeted approaches for identification of mutants useful to determine functions of specific genes (Henikoff et al. 2004), and high-throughput methods to identify functional elements in genomic sequence (Birney et al. 2007), set the stage for choosing organisms for study based on their intrinsic ecological or evolutionary interest rather than because they are facile for genomics. This is not to downplay the importance of botanical models, in which the groundwork has been laid for many pan-taxon goals such as determining the functions of many thousands of genes (Alonso et al. 2003) and deducing the macro-evolutionary history of angiosperms (Bowers et al. 2003; Paterson et al. 2004). However, major gaps remain, for example, in relating genetic mechanisms to evolutionary outcomes, and in understanding how this relationship is mediated by ecological factors. Genomic models, selected for small genomes and short life cycles, present a biased picture of genome structure and evolution, and have intrinsic limitations as whole-organism-level study systems. Such gaps in knowledge will increasingly need to be filled by study of plants that are not traditionally viewed as botanical models. The ~200 or so crops that sustain humanity and our livestock will play a singularly important role in rounding out understanding of botanical diversity, since they combine economic importance with one or more attributes that distinguish them as a botanical model for some specific aspect of growth and development, such as single-celled seed-borne epidermal fiber of cotton, the subterranean pod containing oil-rich seeds of peanut, or the remarkable biomass productivity of *Miscanthus* (Heaton et al. 2004).

2 MAPPING AND SEQUENCING OF ANGIOSPERM GENOMES

There is good reason to expect that the genomes of most of our major crops will be fully sequenced early in the 21st century. Because plant genome sizes vary by nearly 2,000-fold, from 1C = 63 Mbp for the carnivorous plant *Genlisea margaretae* (Greilhuber et al. 2006) to 124,852 Mbp for the lily *Fritillaria assyriaca* (Bennett and Smith 1991), the decision to sequence one is presently a complex equation that integrates genome size with scientific/economic/social impact, phylogenetic distance from previously-sequenced plants (i.e. new information yield), relevant information from prior studies (such as genetic/physical maps or ESTs), sequencing/assembly/annotation costs, and the persuasiveness of individual (or groups of) investigators. In the aggregate, the genomes of 70 crops for which I found estimates of genome size total 1.48×10^{11} bp of DNA (Paterson 2006). Anticipating that these are representative of the remainder of the ~200 domesticates, to fully sequence only one genotype for each using present whole-genome shotgun or BAC-based technology (each assuming 8x redundancy of sequence coverage) would involve about 3.4×10^{12} bp of raw sequence, more than 72x the 4.9×10^{10} bp archived in GenBank as of this writing. In that new technologies (Shendure et al. 2004) together with ongoing efficiencies promise to sustain the sequence growth rates of about 60% per year that have been realized since the 1980s, the complete sequencing of these 200 domesticates would be predicted to take a remarkably short 14 years.

Many of these genomes will gain information from reduced representation approaches well in advance of their complete sequencing. *En masse* sequencing of 'expressed-sequence-tags' (ESTs or cDNAs), has been a natural and economical first step in plant gene discovery, with many angiosperms enjoying collections of 10^5 or more ESTs from diverse tissues, physiological states, and in some cases multiple genotypes, as detailed elsewhere in this volume. ESTs have provided new DNA markers, revealed hosts of candidate genes, permitted testing of evolutionary hypotheses, and are themselves a foothold toward sequencing a genome. Technically demanding approaches have been employed to generate large collections of full-length ESTs for *Arabidopsis* (Haas et al. 2003) and *Oryza* (Haas et al. 2003; Kikuchi et al. 2003) that are especially useful in genome annotation, clarifying exon-intron junctures and splice variants, identifying anti-sense RNA genes that may participate in gene regulation and imprinting, and in more definitive comparison of gene repertoires in diverse taxa.

With transcriptome coverage by ESTs in many angiosperms at or above the ~50% of genes beyond which the EST approach becomes inefficient (Soares et al. 1994), two new approaches show promise to further advance transcriptome coverage while also accessing introns and regulatory sequences from genomes for which complete sequencing is not yet justifiable. The generalization that expressed angiosperm DNA tends to be hypomethylated has been employed widely since the 1980s by using methylation-sensitive restriction enzymes to select for low-copy DNA clones suitable as locus-specific markers. 'Methylation filtration' (MF), host-cell-based selection against methylated DNA (Rabinowicz et al. 1999, 2003) likewise reduces the abundance of some repetitive DNA families in plant genomic DNA libraries. The second approach, 'Cot-based cloning and sequencing' (CBCS; Peterson et al. 2001, 2002a, b)), utilizes the principles and methods of DNA renaturation kinetics to fractionate a genome into subpopulations of DNA segments that differ in their iteration frequency, clone those respective subpopulations, and sequence each to a depth appropriate to cover its particular sequence complexity (readily estimated from Cot analysis; Britten and Kohne 1968; Goldberg 2001). Empirical comparisons of the two methods to one another are now possible in several species (Whitelaw et al. 2003; Springer and Barbazuk 2004; Lamoureux et al. 2005), generally showing a good degree of complementarity between the two.

3 DEDUCING THE EVOLUTIONARY HISTORY OF ANGIOSPERM GENOMES AND GENES: A FRAMEWORK FOR TRANSLATIONAL GENOMICS

The angiosperms, or flowering plants, sustain humanity by providing 'ecosystem services' including oxygen, food, feed, fiber, fuel, medicines, spirits, erosion and flooding control, soil regeneration, urban cooling and greenspace, wildlife habitat, and other benefits. Tracing to ancestors for which fossil evidence dates to 125-140 mya and 'molecular clocks' are converging on estimates of common ancestry 140-180 mya (Sanderson et al. 2004; Bell et al. 2005), radiation of the angiosperms during the mid-late Cretaceous produced most major lineages of flowering plants (Doyle and Donoghue 1993; Crane et al. 1995).

Shared ancestry provides a potential foundation for translational genomics, the application of hard-won gene functional information from botanical models to less well-studied crops, and indeed across much of the plant family tree. Genome sequences are revealing parallels in the arrangements of genes along the chromosomes of divergent plant lineages. Cereals have long been known to share a high level of conserved gene content and order (Paterson et al. 2005), and the sequences of several divergent dicots now show strong long-term conservation of gene order, albeit occasionally interrupted and altered by genome duplication (Bowers et al. 2003; Tang et al. 2008).

Already, with only a few genomes even in the draft sequence level, new multi-alignment approaches are beginning to reveal gene arrangements that are likely to trace to a common angiosperm ancestor (Ming et al. 2008; Tang et al. 2008). As we gain more and more information about the functions of specific genes in botanical models such as *Arabidopsis*, these alignments provide a mechanism by which to formulate well-supported hypotheses about the probable functions of genes in many additional taxa. One can envision a time when the collected functional genomics evidence from most (if not all) botanical models, might be tapped as a means to quickly learn much about the functions of many genes in our leading crops.

4 CATALOGING THE DIVERSITY IN THE GENE POOLS OF CROPS AND THEIR RELATIVES: A FOUNDATION FOR CROP IMPROVEMENT

The remarkable contributions of crop domestication and improvement to humanity have been made with virtually no knowledge of the exact genetic determination of the underlying traits. Genetic improvement of crops is an essential delivery system by which to meet many basic needs of the world's poor, and the roughly one million accessions in global genetic resources collections are the underpinning of future progress. The genome sequences will be a natural platform for formulation of hypotheses that translate burgeoning and hard-won functional data from botanical (or other) models into experiments that are testable in the field utilizing naturally occurring or induced genetic polymorphisms. However, such work will require the same level of information about the levels and patterns of genetic polymorphism in crops that is the focus of much effort to obtain in humans. Documenting the suite of naturally-occurring genetic polymorphisms in crop gene pools will yield many benefits—for example, revealing footprints of domestication (Wang et al. 1999) or signals of selection associated with specific genes (Thornsberry et al. 2001); and providing information about levels and patterns of linkage disequilibrium that will be essential to interpreting such clues (Remington et al. 2001; Hamblin et al. 2005). The predominantly self-pollinating nature of many crop species has the benefit that 'effective population size' is relatively small, and that most of the available allelic diversity can be sampled by analysis of 'core collections' that comprise modest subsets of global genetic resources collections (Brown 1989). More and better information about the distribution of variation over crop genes and genomes may also contribute to mechanistic understanding and eventual mitigation of the inherent 'instability' of ostensibly homozygous improved genotypes in many crops. Such instability adds substantially to the annual cost of producing and maintaining seed that accurately represent the desirable genotype that was selected, often requiring costly controlled pollination and other steps to ensure integrity such as 'rogueing' by skilled individuals to eliminate 'off-types' from seed production fields.

5 DETERMINISTIC CROP IMPROVEMENT?

With fully-sequenced genomes, improved understanding of gene functions, and thorough documentation of the levels and patterns of allelic diversity available, how will crop improvement change in the 21st century? Certainly the tools of the trade will be dramatically improved, and technological advances will permit tools such as DNA markers to be applied generally and routinely as a part of breeding programs. Indeed, the ability to identify and monitor functional polymorphisms in the genes directly, rather than in 'proxy' markers such as RFLPs and SSRs, may in some cases permit us to begin to use computational approaches to predict which alleles of a gene might have striking phenotypic effects and to test those effects empirically, in field populations. However, much of plant breeding is likely to remain empirical—this is particularly true in that many extensive mapping studies using improved experimental designs are revealing a greater importance of epistasis than we previously appreciated (Li et al. 1997; Yu et al. 1997; Yang 2004; Kroymann and Mitchell-Olds 2005; Malmberg and Mauricio 2005; Blanc et al. 2006). In other words, the effects of many alleles may be interdependent upon the functional status of other unlinked loci, forming complex networks that may confer a degree of 'buffering' to a phenotype. Such buffering may be highly desirable in the face of biotic or abiotic variations that threaten productivity or quality—but may also be a hindrance to making rapid gains from selection. Better understanding of the spectrum of gene expression patterns associated with allelic variations, and the deduction of gene regulatory networks, may provide a beginning toward identifying functional variants that are likely to have interdependent consequences (Gjuvsland et al. 2007).

The author finds attractive the possibility that better understanding of the evolutionary history of our major crops may guide further improvements. The genome sequences of extant organisms are, by definition, a 'success story'—Nature's record of what worked. In agriculture, 'synthetic' (man-made) polyploids are an enormously-important potential means to provide adaptive variation to the narrow gene pools of many crops (Zhang et al. 2005)—however these often fail, due to difficulties with producing novel polyploids or with incorporating their chromosomes (or segments thereof) into the genomes of crops. In nature, we now know that most polyploids lose one copy of most duplicated genes within a relatively short few million years after polyploid formation. Comparative analysis of independent polyploidizations in a wide range of taxa has suggested that gene retention/loss is far from random, perhaps guided by an underlying set of principles that favor rapid elimination of some duplicated genes and long-term retention of others (Paterson et al. 2006). Identification of genes that must be restored to 'singleton' status for a polyploid to be successful may open new doors into crop improvement. For example, genes containing protein functional domains enriched in *Arabidopsis* and *Oryza* singleton genes, respectively, are readily traced back to their specific host genes, which in turn can be used as probes to investigate additional taxa. Might one find orthologs of G-patch-containing proteins, for example, to be confined to singletons in recently-formed polyploids such as cotton, wheat, or canola, as they are in both *Arabidopsis* and *Oryza*? Could the lessons learnt from sequencing and analysis of paleopolyploid genomes permit us to devise strategies for genetic manipulation of newly-formed polyploids to improve our ability to use them in crop improvement?

The answer to this and many other questions, of course, remains uncertain. What is certain, however, is that we will have far more powerful resources to bring to bear on such questions.

References

Alonso JM, Stepanova AN, Leisse TJ, Kim CJ, Chen HM, Shinn P, Stevenson DK, Zimmerman J, Barajas P, Cheuk R, Gadrinab C, Heller C, Jeske A, Koesema E, Meyers CC, Parker H, Prednis L, Ansari Y, Choy N, Deen H, Geralt M, Hazari N, Hom E, Karnes M, Mulholland C, Ndubaku R, Schmidt I, Guzman P, Aguilar-Henonin L, Schmid M, Weigel D, Carter DE, Marchand T, Risseeuw E, Brogden D, Zeko A, Crosby WL, Berry CC, Ecker JR (2003) Genome-wide insertional mutagenesis of *Arabidopsis thaliana*. Science 301: 653-657

Bell CD, Soltis DE, Soltis PS (2005) The age of the angiosperms: A molecular timescale without a clock. Evolution 59: 1245-1258

Bennett M, Smith J (1991) Nuclear DNA amounts in angiosperms. Phil Trans Roy Soc Lond B 334: 309-345

Birney E, Stamatoyannopoulos JA, Dutta A, Guigo R, Gingeras TR, Margulies EH, Weng ZP, Snyder M, Dermitzakis ET, Stamatoyannopoulos JA, Thurman RE, Kuehn MS, Taylor CM, Neph S, Koch CM, Asthana S, Malhotra A, Adzhubei I, Greenbaum JA, Andrews RM, Flicek P, Boyle PJ, Cao H, Carter NP, Clelland GK, Davis S, Day N, Dhami P, Dillon SC, Dorschner MO, Fiegler H, Giresi PG, Goldy J, Hawrylycz M, Haydock A, Humbert R, James KD, Johnson BE, Johnson EM, Frum TT, Rosenzweig ER, Karnani N, Lee K, Lefebvre GC, Navas PA, Neri F, Parker SCJ, Sabo PJ, Sandstrom R, Shafer A, Vetrie D, Weaver M, Wilcox S, Yu M, Collins FS, Dekker J, Lieb JD, Tullius TD, Crawford GE, Sunyaev S, Noble WS, Dunham I, Dutta A, Guigo R, Denoeud F, Reymond A, Kapranov P, Rozowsky J, Zheng DY, Castelo R, Frankish A, Harrow J, Ghosh S, Sandelin A, Hofacker IL, Baertsch R, Keefe D, Flicek P, Dike S, Cheng J, Hirsch HA, Sekinger EA, Lagarde J, Abril JF, Shahab A, Flamm C, Fried C, Hackermuller J, Hertel J, Lindemeyer M, Missal K, Tanzer A, Washietl S, Korbel J, Emanuelsson O, Pedersen JS, Holroyd N, Taylor R, Swarbreck D, Matthews N, Dickson MC, Thomas DJ, Weirauch MT, Gilbert J, Drenkow J, Bell I, Zhao X, Srinivasan KG, Sung WK, Ooi HS, Chiu KP, Foissac S, Alioto T, Brent M, Pachter L, Tress ML, Valencia A, Choo SW, Choo CY, Ucla C, Manzano C, Wyss C, Cheung E, Clark TG, Brown JB, Ganesh M, Patel S, Tammana H, Chrast J, Henrichsen CN, Kai C, Kawai J, Nagalakshmi U, Wu JQ, Lian Z, Lian J, Newburger P, Zhang XQ, Bickel P, Mattick JS, Carninci

P, Hayashizaki Y, Weissman S, Dermitzakis ET, Margulies EH, Hubbard T, Myers RM, Rogers J, Stadler PF, Lowe TM, Wei CL, Ruan YJ, Snyder M, Birney E, Struhl K, Gerstein M, Antonarakis SE, Gingeras TR, Brown JB, Flicek P, Fu YT, Keefe D, Birney E, Denoeud F, Gerstein M, Green ED, Kapranov P, Karaoz U, Myers RM, Noble WS, Reymond A, Rozowsky J, Struhl K, Siepel A, Stamatoyannopoulos JA, Taylor CM, Taylor J, Thurman RE, Tullius TD, Washietl S, Zheng DY, Liefer LA, Wetterstrand KA, Good PJ, Feingold EA, Guyer MS, Collins FS, Margulies EH, Cooper GM, Asimenos G, Thomas DJ, Dewey CN, Siepel A, Birney E, Keefe D, Hou MM, Taylor J, Nikolaev S, Montoya-Burgos JI, Loytynoja A, Whelan S, Pardi F, Massingham T, Brown JB, Huang HY, Zhang NR, Bickel P, Holmes I, Mullikin JC, Ureta-Vidal A, Paten B, Seringhaus M, Church D, Rosenbloom K, Kent WJ, Stone EA, Gerstein M, Antonarakis SE, Batzoglou S, Goldman N, Hardison RC, Haussler D, Miller W, Pachter L, Green ED, Sidow A, Weng ZP, Trinklein ND, Fu YT, Zhang ZDD, Karaoz U, Barrera L, Stuart R, Zheng DY, Ghosh S, Flicek P, King DC, Taylor J, Ameur A, Enroth S, Bieda MC, Koch CM, Hirsch HA, Wei CL, Cheng J, Kim J, Bhinge AA, Giresi PG, Jiang N, Liu J, Yao F, Sung WK, Chiu KP, Vega VB, Lee CWH, Ng P, Shahab A, Sekinger EA, Yang A, Moqtaderi Z, Zhu Z, Xu XQ, Squazzo S, Oberley MJ, Inman D, Singer MA, Richmond TA, Munn KJ, Rada-Iglesias A, Wallerman O, Komorowski J, Clelland GK, Wilcox S, Dillon SC, Andrews RM, Fowler JC, Couttet P, James KD, Lefebvre GC, Bruce AW, Dovey OM, Ellis PD, Dhami P, Langford CF, Carter NP, Vetrie D, Kapranov P, Nix DA, Bell I, Patel S, Rozowsky J, Euskirchen G, Hartman S, Lian J, Wu JQ, Urban AE, Kraus P, Van Calcar S, Heintzman N, Kim TH, Wang K, Qu CX, Hon G, Luna R, Glass CK, Rosenfeld MG, Aldred SF, Cooper SJ, Halees A, Lin JM, Shulha HP, Zhang XL, Xu MS, Haidar JNS, Yu Y, Birney E, Weissman S, Ruan YJ, Lieb JD, Iyer VR, Green RD, Gingeras TR, Wadelius C, Dunham I, Struhl K, Hardison RC, Gerstein M, Farnham PJ, Myers RM, Ren B, Snyder M, Thomas DJ, Rosenbloom K, Harte RA, Hinrichs AS, Trumbower H, Clawson H, Hillman-Jackson J, Zweig AS, Smith K, Thakkapallayil A, Barber G, Kuhn RM, Karolchik D, Haussler D, Kent WJ, Dermitzakis ET, Armengol L, Bird CP, Clark TG, Cooper GM, de Bakker PIW, Kern AD, Lopez-Bigas N, Martin JD, Stranger BE, Thomas DJ, Woodroffe A, Batzoglou S, Davydov E, Dimas A, Eyras E, Hallgrimsdottir IB, Hardison RC, Huppert J, Sidow A, Taylor J, Trumbower H, Zody MC, Guigo R, Mullikin JC, Abecasis GR, Estivill X, Birney E, Bouffard GG, Guan XB, Hansen NF, Idol JR, Maduro VVB, Maskeri B, McDowell JC, Park M, Thomas PJ, Young AC, Blakesley RW, Muzny DM, Sodergren E, Wheeler DA, Worley KC, Jiang HY, Weinstock GM, Gibbs RA, Graves T, Fulton R, Mardis ER, Wilson RK, Clamp M, Cuff J, Gnerre S, Jaffe DB, Chang JL, Lindblad-Toh K, Lander ES, Koriabine M, Nefedov M, Osoegawa K, Yoshinaga Y, Zhu BL, de Jong PJ (2007) Identification and analysis of functional elements in 1% of the human genome by the ENCODE pilot project. Nature 447: 799-816

Blanc G, Charcosset A, Mangin B, Gallais A, Moreau L (2006) Connected populations for detecting quantitative trait loci and testing for epistasis: an application in maize. Theor Appl Genet 113: 206-224

Bowers JE, Chapman BA, Rong JK, Paterson AH (2003) Unravelling angiosperm genome evolution by phylogenetic analysis of chromosomal duplication events. Nature 422: 433-438

Britten RJ, Kohne DE (1968) Repeated sequences in DNA. Science 161:529

Brown AHD (1989) Core collections—A practical approach to genetic-resources management. Genome 31: 818-824

Crane PR, Friis EM, Pedersen KR (1995) The origin and early diversification of angiosperms. Nature 374: 27-33

Doyle JA, Donoghue MJ (1993) Phylogenies and angiosperm diversification. Paleobiology 19: 141-167

Gjuvsland AB, Hayes BJ, Omholt SW, Carlborg O (2007) Statistical epistasis is a generic feature of gene regulatory networks. Genetics 175: 411-420

Goldberg RB (2001) From cot curves to genomics. How gene cloning established new concepts in plant biology. Plant Physiol 125: 4-8

Greilhuber J, Borsch T, Muller K, Worberg A, Porembski S, Barthlott W (2006) Smallest angiosperm genomes found in Lentibulariaceae, with chromosomes of bacterial size. Plant Biol 8: 770-777

Haas BJ, Volfovsky N, Town CD, Troukhan M, Alexandrov N, Feldmann KA, Flavell RB, White O, Salzberg SL (2003) Full-length messenger RNA sequences greatly improve genome annotation. Genom Biol 3: Research 0029

Hamblin MT, Salas-Fernandez MG, Casa AM, Mitchell SE, Aquadro CF, Paterson AH, Kresovich S (2005) Patterns of short- and medium-range linkage disequilibrium in *Sorghum bicolor* show little correlation with local rates of recombination. Genetics (in press)

Heaton E, Voigt T, Long SP (2004) A quantitative review comparing the yields of two candidate C-4 perennial biomass crops in relation to nitrogen, temperature and water. Biomass Bioenerg 27: 21-30

Henikoff S, Till BJ, Comai L (2004) TILLING. Traditional mutagenesis meets functional genomics. Plant Physiol 135: 630-636

Kikuchi S, Satoh K, Nagata T, Kawagashira N, Doi K, Kishimoto N, Yazaki J, Ishikawa M, Yamada H, Ooka H, Hotta I, Kojima K, Namiki T, Ohneda E, Yahagi W, Suzuki K, Li CJ, Ohtsuki K, Shishiki T, Otomo Y, Murakami K, Iida Y, Sugano S, Fujimura T, Suzuki Y, Tsunoda Y, Kurosaki T, Kodama T, Masuda H, Kobayashi M, Xie QH, Lu M, Narikawa R, Sugiyama A, Mizuno K, Yokomizo S, Niikura J, Ikeda R, Ishibiki J, Kawamata M, Yoshimura A, Miura J, Kusumegi T, Oka M, Ryu R, Ueda M, Matsubara K, Kawai J, Carninci P, Adachi J, Aizawa K, Arakawa T, Fukuda S, Hara A, Hashizume W, Hayatsu N, Imotani K, Ishii Y, Itoh M, Kagawa I, Kondo S, Konno H, Miyazaki A, Osato N, Ota Y, Saito R, Sasaki D, Sato K, Shibata K, Shinagawa A, Shiraki T, Yoshino M, Hayashizaki Y, Yasunishi A (2003) Collection, mapping, and annotation of over 28,000 cDNA clones from japonica rice. Science 301: 376-379

Kroymann J, Mitchell-Olds T (2005) Epistasis and balanced polymorphism influencing complex trait variation. Nature 435: 95-98

Lamoureux D, Peterson DG, Li W, Fellers JP, Gill BS (2005) The efficacy of Cot-based gene enrichment in wheat. Genome 48: 1120-1126

Li ZK, Pinson SRM, Park WD, Paterson AH, Stansel JW (1997) Epistasis for three grain yield components in rice (*Oryza sativa* L). Genetics 145: 453-465

Malmberg RL, Mauricio R (2005) QTL-based evidence for the role of epistasis in evolution. Genet Res 86: 89-95

Margulies M, Egholm M, Altman WE, Attiva S, Bader JS, Bemben LA, Berka J, Braverman MS, Chen YJ, Chen Z, Dewell SB, Du L, Fierro JM, Gomes XV, Godwin BC, He W, Helgesen S, Ho CH, Irzyk GPI, Jando SC, Alenquer ML, Jarvie TP, Jirage KB, Kim JB, Knight JR, Lanza JR, Leamon JH, Lefkowitz SM, Lei M, Li J, Lohman KL, Lu H, Makhijani VB, Mcdade KE, Mckenna MP, Myers EW, Nickerson E, Nobile JR, Plant R, Puc BP, Ronan MT, Roth GT, Sarkis GJ, Simons JF, Simpson JW, Srinivasan M, Tartaro KR, Tomasz A, Vogt KA, Volkmer GA, Wang SH, Wang Y, Weiner MP, Yu P, Begley RF, Rothberg JM (2005) Genome sequencing in microfabricated high-density picolitre reactors. Nature 437: 376-380

Ming R, Hou S, Feng Y, Yu QY, Dionne-Laporte A, Saw J, Senin P, Wang W, Salzberg SL, Tang H, Lyons E, Rice D, Riley M, Skelton R, Murray J, Chen C, Eustice M, Tong E, Albert H, Paull RE, Wang ML, Zhu Y, Schatz M, Nagarajan N, Agbayani R, Guan P, Blas A, Wang J, Na JK, Michael T, Shakirov EV, Haas B, Thimmapuram J, Nelson D, Wang X, Bowers JE, Suzuki J, Tripathi S, Neupane K, Wei H, Singh R, Irikura B, Jiang N, Zhang W, Wall K, Presting G, Gschwend A, Li Y, Windsor AJ, Navajas-Perez R, Torres MJ, Feltus FA, Porter B, Paidi M, Luo MC, Liu L, Christopher D, Moore PH, Sugimura T, dePamphilis C, Jiang J, Schuler M, Mitchell-Olds T, Shippen D, Palmer JD, Freeling M, Paterson AH, Gonsalves D, Wang L, Alam M (2008) The draft genome of the transgenic tropical fruit tree papaya (*Carica papaya* Linnaeus). Nature 452: 991-997

Paterson AH (2006) Leafing through the genomes of our major crop plants: Strategies for capturing unique information. Nat Rev Genet 7: 174-184

Paterson AH, Bowers JE, Chapman BA (2004) Ancient polyploidization predating divergence of the cereals, and its consequences for comparative genomics. Proc Natl Acad Sci USA 101: 9903-9908

Paterson AH, Freeling M, Sasaki T (2005) Grains of knowledge: Genomics of model cereals. Genome Res 15: 1643-1650

Paterson AH, Chapman BA, Kissinger J, Bowers JE, Feltus FA, Estill J, Marler BS (2006) Convergent retention or loss of gene/domain families following independent whole-genome duplication events in *Arabidopsis*, *Oryza*, Saccharomyces, and Tetraodon. Trends Genet 22: 597-602

Peterson DG, Schulze SR, Sciara EB, Lee SA, Bowers JE, Nagel A, N. J, Tibbitts DC, Wessler SR, Paterson AH (2001) Integration of Cot analysis, DNA cloning, and high-throughput sequencing facilitates genome characterization and gene discovery: *http://www.ncbinlmnihgov/entrez*

Peterson DG, Schulze SR, Sciara EB, Lee SA, Bowers JE, Nagel A, Jiang N, Tibbitts DC, Wessler SR, Paterson AH (2002a) Integration of Cot analysis, DNA cloning, and high-throughput sequencing facilitates genome characterization and gene discovery. Genome Res 12: 795-807

Peterson DG, Wessler SR, Paterson AH (2002b) Efficient capture of unique sequences from eukaryotic genomes. Trends Genet 18: 547-550

Rabinowicz PD, Schutz K, Dedhia N, Yordan C, Parnell LD, Stein L, McCombie WR, Martienssen RA (1999) Differential methylation of genes and retrotransposons facilitates shotgun sequencing of the maize genome. Nat Genet 23: 305-308

Rabinowicz PD, McCombie WR, Martienssen RA (2003) Gene enrichment in plant genomic shotgun libraries. Curr Opin Plant Biol 6: 150-156

Remington DL, Thornsberry JM, Matsuoka Y, Wilson LM, Whitt SR, Doebley J, Kresovich S, Goodman MM, Buckler ES (2001) Structure of linkage disequilibrium and phenotypic associations in the maize genome. Proc Natl Acad Sci USA 98: 11479-11484

Sanderson MJ, Thorne JL, Wikstrom N, Bremer K (2004) Molecular evidence on plant divergence times. Am J Bot 91: 1656-1665

Shendure J, Mitra RD, Varma C, Church GM (2004) Advanced sequencing technologies: Methods and goals. Nat Rev Genet 5: 335-344

Shendure J, Porreca GJ, Reppas NB, Lin XX, McCutcheon JP, Rosenbaum AM, Wang MD, Zhang K, Mitra RD, Church GM (2005) Accurate multiplex polony sequencing of an evolved bacterial genome. Science 309: 1728-1732

Soares MB, Bonaldo MD, Jelene P, Su L, Lawton L, Efstratiadis A (1994) Construction and characterization of a normalized cDNA library. Proc Natl Acad Sci USA 91: 9228-9232

Springer NM, Barbazuk WB (2004) Utility of different gene enrichment approaches toward identifying and sequencing the maize gene space. Plant Physiol 136: 3023-3033

Tang H, Bowers JE, Wang X, Ming R, Alam M, Paterson AH (2008) Synteny and colinearity in plant genomes. Science 320: 486-488

Thornsberry JM, Goodman MM, Doebley J, Kresovich S, Nielsen D, Buckler ES (2001) Dwarf8 polymorphisms associate with variation in flowering time. Nat Genet 28: 286-289

Wang RL, Stec A, Hey J, Lukens L, Doebley J (1999) The limits of selection during maize domestication. Nature 398: 236-239

Whitelaw CA, Barbazuk WB, Pertea G, Chan AP, Cheung F, Lee Y, Zheng L, van Heeringen S, Karamycheva S, Bennetzen JL, SanMiguel P, Lakey N, Bedell J, Yuan Y, Budiman MA, Resnick A, Van Aken S, Utterback T, Riedmuller S, Williams M, Feldblyum T, Schubert K, Beachy R, Fraser CM, Quackenbush J (2003) Enrichment of gene-coding sequences in maize by genome filtration. Science 302: 2118-2120

Yang RC (2004) Epistasis of quantitative trait loci under different gene action models. Genetics 167: 1493-1505

Yu SB, Li JX, Tan YF, Gao YJ, Li XH, Zhang QF, Maroof MAS (1997) Importance of epistasis as the genetic basis of heterosis in an elite rice hybrid. Proc Natl Acad Sci USA 94: 9226-9231

Zhang PZ, Dreisigacker S, Melchinger AE, Reif JC, Kazi AM, Van Ginkel M, Hoisington D, Warburton ML (2005) Quantifying novel sequence variation and selective advantage in synthetic hexaploid wheats and their backcross-derived lines using SSR markers. Mol Breed 15: 1-10

Subject Index

454 sequencing 132, 311, 433
  platform 132
4DTV 259

ABI SOLiD 134, 317
Abiotic stress 313
ACPFG 541
Activation tagging 367
Acyclic graph 460
Agrobacterium 9, 34, 177, 377, 511
  A. tumefaciens 34
Albugo 193
Allelic diversity 575
Allohexaploid 313
Allozyme 74
Amplicon 134
Amplified fragment length polymorphism (AFLP) 24, 191, 279, 313
Aneuploid 27
Annotation 40, 214, 224
  functional annotation 224
Arabidopsis 5, 33, 70, 91, 110, 156, 157, 175, 176, 177, 226, 246, 355, 576
  A. lyrata 414
  A. thaliana 91, 157, 176, 226, 414
  Biological Resource Center 177
  database 176
  Genome Initiative (AGI) 176, 180
Artificial intelligence 448
Association mapping 320, 415
Average-linkage clustering 474

β-glucoronidase 377

Backcross 508
Bacterial artificial chromosome (BAC) 5, 26, 31, 33, 34-36, 39, 40, 113, 127, 179, 206, 211, 213, 232, 285, 306, 572
  anchoring 39
  based physical mapping 316
  by-BAC 206, 247, 322
  clone 36, 211
  contig 34, 40, 296
  end 211, 297
  end sequence 311
  end-sequencing 245
  filter 35
  fingerprinting 289
  insert 39
  library 31, 316
  mapping 253
  sequence 211, 213, 287, 297
  vector 33
Banana 337
Barley 385
Bayesian networks 451, 460
BBSRC 540
BGI 207
Bi-clustering 480
Binary BAC 33
BioBricks 563
Biodiversity 535
Bioeconomy 536
Biofactory 536
Biofuel 560
Bioinformatics 9, 448
Biological diversity 535

Biomass 564
Bioprospecting 556
Biorefinery 536
Biosynthesis 264
Biotic stress 313
BLAST 140, 191
BLASTN 253
BLASTP 222, 226, 257
Brachypodium 104, 311, 387
  B. distachyon 44, 312
Brassica 110, 162, 193
  B. oleracea 110
  B. rapa 162, 193
BrGSP 162
British Petroleum 544

Candidate gene 72, 294, 503
CAPS 179
Capsella 94
Carotenoid 291
CASAVA 153
CBOL 562
cDNA 105, 220, 256, 289, 314, 366, 430, 573
  alignment 220
  library 105
  microarray 108
CENSOR 224
CentO 232, 233
  arrays 233
  repeats 232
CGIAR 542
Chilling requirement 295
Chinese spring 316
ChiP 135
  based sequencing 135
Chloroplast DNA 78
Chromosome 37, 45
  landing 45
  walking 37
CIMMYT 321, 547
Citrus
  C. sinensis 336
Climate change 530
Climate-resilient 530
Clone-by-clone 126
Cloning 198
Clustering 469
Co-expression 91
  assay 461
Cochlearia 226
  C. officinalis 226
Colinearity 44, 69
Community sequencing program 330
Comparative genome hybridization 403
Complete-linkage clustering 475
Computability 448
Computational biology 448
Conditional probability 461
Consed 140
Consensus map 313
Conservation 5
Conserved orthologous set 72, 76
Contig 165
  map 165
Cot 312
Cross validation 464
Cycle sequencing 122, 123

DArT 313
dCAPS 330
Decision stumps 455
Decision trees 452
Deletion 27, 44, 112, 334
  bin 315
  mapping 314
  stock 27
Demography 87
Dideoxy 404
Diploidization 44
Directed edges 460
Distance metric 470
DNA 12
  banking 12
  barcoding 502
  microarray 108
  renaturation 573
  sequence 12
  sequencing 12
DOE 177, 245
DOE-JGI 159

Domestication 12
Drosophila 376, 406
  D. melanogaster 25, 406
  D. pseudoobscura 406
  D. simulans 406
  D. yacuba 406
Duplication 44, 69, 70, 228, 229
  segmental duplication 229
  tandem duplication 229
DuPont 544

Eco-genomics 90
Ecosystem 306
Ecotilling 327, 385
Eigencondition 484
Eigengene 484
Eigenvector 482
End-to-end sequencing 5
Endodormancy 291
Endoduplication 286
Enhancer 369, 371
  trap 371
Environmental monitoring 534
Environmental surveillance 534
ERA-PG 540
Escherichia 33, 105
  E. coli 105
Ethylmethane sulphonate 178
Eucalyptus 94
Euchromatin 324
Euclidean distance 470
EuGène 255
Evolution 188
Exon 40
  trap 370
  trapping 40
Expectation-maximization 464
Expressed sequence tag (EST) 3, 26, 72, 104, 105, 245, 279, 280, 306, 314, 573, 430
  chip 291
  EST-SSR 313
  library 430
  mapping 280
  sequences 105, 287
Expression QTL 318, 499

FAO 550
Fast-neutron 334
FGENESH 216, 220
FgenesH 255
FIBR 265
Filtration 124
FingerPrinted contig 127
Fingerprinting 210
FL-cDNA 217
Fluorescence in situ hybridization 30, 276, 323
FPCV4.7 49
Fragaria 337
Functional genomics 6, 289, 318, 499
Functionalization 229
Fuzzy clustering 480

Gain-of-function 368
Garbage in, garbage out 487
GBrowse 222
GDR 286
GenBank 110, 226
Gene 6, 67, 87, 107, 114, 188, 189, 229, 259, 412
  content 255
  conversion 412
  discovery 6, 114
  duplication 188
  expression 107, 452
  flow 87
  ontology 259, 451
  silencing 189, 328
  trees 67
GeneBuilder 221
GeneChip 436
GeneMarkHMM 220
Genetic 5, 12, 25, 26, 74, 75, 81
  diversity 501
  drift 81
  fine mapping 280
  hitchhiking 74
  mapping 26, 246, 280
  map 5, 25, 246
  resource 12
Genetically modified (GM) 548
  crop 511, 549

organism 355
plant 511
Genewise 255
Genome 2, 4, 228, 230, 246
discovery 2
genome-wide 230
mapping 4
polymorphism 2
reorganization 228
sequencer 20/FLX 49
sequencing 4, 246
-wide polymorphism 411
Genomics 1, 31
Genotyping 3, 314, 320
GENSCAN 219
Germin like protein 319
Glimmer 220
GLocate 219
Glycine
G. max 330
GMOD 222
GO-Slim 259
GrailExp6 255
Gramene 218, 440
Grape 160, 288, 335
Green chemistry 532
Green fluorescence protein (GFP) 319, 377
Green revolution 546

Haplotype 68, 408
diversity 408
trees 68
HapMap 408
HAPPY 30
mapping 30
Hardy-Weinberg equilibrium 73
HarvEST 440
Helianthus 70
Heterochromatin 324
HICF 286
Hierarchical clustering 472
High-resolution 42
mapping 42
Hudson-kreitmam-agaudé test 84
Hybridization 38
overgo 38
Hypergeometric test 485
Hyperplane 458

IGGP 335
IGROW 317
Illumina genome analyzer 132
Illumina/Solexa 317
iMap 49
InDels 324
INRA 317
Insertion 378
Intellectual property 356, 556
International HapMap project 407
Inversion 44
IRGSP 126, 157, 206
IRRI 547
ISSR 313
ITMI 316
ITPGRFA 556
IWGSC 164, 317

Jackknifing 466
JAZZ assembler 251
JGI 245, 286, 313

K-means clustering 478
Kernel trick 456

Leguminosae 329
Leucine-rich repeat 184
Ligation chain reaction 403
Lily 426
Linkage
disequilibrium (LD) 113, 320, 400, 498
Loss-of-function 368
Lotus 330
L. japonicus 330
Luciferase 377

Machine learning 451
Macrosynteny 297
MADS box 230, 286
Maize 311, 383
Map-based cloning 45

Mapping 25
  population 25
Mapping as you go 504
Marker 42
Marker-assisted backcrossing 507
Marker-assisted recurrent selection 504
Marker-assisted selection (MAS) 13, 287, 314, 321, 504, 507
MaskerAid 224
Material transfer agreements 356
Medicago 104, 288, 297, 329, 331
  M. truncatula 331
Metabolomics 8, 293
Methyl filtration 312
Methylation filtration 161
MIAME 438
Micro-synteny 297
Microarrays 3, 92, 108, 289, 318, 436
MicroRNA 320, 435
Minimum tiling paths 157
miRNA 435
Miropeats 224
Mis-expression 371
MITE 224, 364, 516
Mitochondrial DNA 77
Molecular pharming 9
Monsanto 206, 544
MPSS 229, 432
  library 229
mRNA 114, 230, 430
  metabolism 114
Multiple hypothesis correction 486
Musa
  M. acuminata 337
Mutagenesis 14, 326, 334, 363
  insertional mutagenesis 363
Mutation 2, 73, 121
  discovery 2
  rate 73

Natural selection 83
NCBI 195, 257, 286
nCounter 436
Nearly neutral theory 80
Neofunctionalization 71
Neutral theory 80
Next-generation sequencing 130
Nicotiana 322
NIH 177
Nipponbare 209
  assembly 209
  genome project 209
Northern blot 425
NPGI 539
NSF 177, 265, 500
Nucleotide 74
  binding site 184

OMAP 223
Open reading frame (ORF) 219, 225, 231, 322
ORNL 159
Orphan crop 564
OryGenes 363
Oryza 5, 157, 232, 233, 573
  O. brachyantha 233
  O. indica 207
  O. japonica 207
  O. javanica 207
  O. sativa 157, 205, 232
Overfitting 454
Overgo 38, 285
  probing 38

P1-derived artificial chromosome (PAC) 31, 127, 212, 232
Paleopolyploid 577
Papaya 334
PASA 220
Peach 166, 276
Pearson correlation 470
Pharming 533
Phaseolus 331
Phaseomics Global Initiative 331
Phenome 6
Phenomics 8
Phenotyping 9
Phrap 140
Phred 140
Phylogenetic tree 68
Physical map 5, 25, 234, 252

Physical mapping 289
Physical/genetic map 295
Phytoremediation 534
PILER 224
Pioneer 544
Plant variety rights 563
Plum pox virus 287
Polonator G.007 49
Polymerase chain reaction 107
PolyPhred 140
Polyploid 313
Polyploidization 44, 69
Polyploidy 4
Populus 159, 243, 248
  P. alba 248
  P. deltoides 248
  P. tremula 248
  P. tremuloides 248
  P. trichocarpa 159
Positional cloning 41, 45
Positive predictive value 466
Positive selection 83
Potato 322
Precipitation 124
Precision-recall 467
Principal component analysis 483
Promoter 370, 371
  trap 371
Proteome 292
Proteomics 8, 292
Pruning 454
Prunus 276
  P. persica 276
Pyrosequencing 130, 131

Quadratic programming 458
Quantitative trait loci (QTL) 7, 26, 192, 234, 280, 315, 321, 499

Radiation hybrid 29
  mapping 29, 315
Random amplified polymorphic DNA (RAPD) 191, 279
RAP 218
Rapidly amplified cDNA end (RACE) 434
RCS2 232
rDNA 276
Re-sequencing 7, 93, 404
Receptor kinase 185
Receptor-like kinase 410
Recombinant inbred 179
Recombinant inbred line 321
RECON 224
Recurrent selection 503
RepeatFinder 224
RepeatMasker 224
RepeatScout 224
Repetitive sequences 223
Reporter gene 371
REPuter/REPfind 224
Restriction fragment length polymorphism (RFLP) 24, 227, 279, 313
Retrotransposon 223, 364
Reverse genetics 333, 499
Reverse northern blot 426
Rhizobium 331
Ribonuclease 429
  protection assay (RPA) 429
Rice 157, 383
RISC 189
RITS 189
RNA 215, 426
  dot blot 426
  interference (RNAi) 319, 328, 333, 499, 505
  miRNA 215
  smRNA 215
  snRNA 215
  tRNA 215
Rosaceae 275, 335
Rubus 289

Saccharomyces 232
Sanger sequencing 123
Saturation 42
Scaffold 253
Self organizing map 480
Self-incompatibility 277
Sequence tagged site 25, 313
Sequencing 5, 74, 123, 130, 134, 251, 280, 404

by-synthesis 134
template 251
Serial analysis of gene expression 107, 431
SHARCGS 405
Short vegetative phase 286
Shotgun 5, 37, 129, 246
sequence 246
sequencing 5
subcloning 37
Silencing 319, 380
Simple sequence repeat (SSR) 3, 25, 72, 254, 279, 313
Single copy nuclear loci 72, 75
Single nucleotide polymorphism (SNP) 25, 77, 82, 113, 121, 191, 311, 313, 381, 400
discovery 3, 113
Single-feature polymorphism 403
Single-linkage clustering 474
Singular value decomposition 481
Sinorhizobium 331
SOL 162
Solanaceae 321
Solanum 162, 321
S. lycopersicum 162
Solexa 49, 132, 433
sequencing 132
SOLiD 132, 434
SOLiDTM system 49
Solid-phase reversible immobilization 126
Sorghum 161, 312
S. bicolor 161
Soybean 288, 330, 384
Spearman correlation 470
Species trees 67
Specific combining ability 509
Splice acceptor site 372
Splicing 230
alternative splicing 230
SRAP 313
SSAKE 405
SSLP 179
STM 313
Strawberry 336
STRUCTURE 90
Structure learning 464
Subcloning 39
Subfunctionalization 71, 230
Substitution 74, 78, 228
Support vector 458
Support vector machine 455
Suppression subtractive hybridization 106, 333
Symbiosis 333
Syngenta 207
Synteny 70, 503
Synthetic biology 563
Systematic acquired resistance 371

T-DNA 178, 363, 377
tagging 333
TAIR 157, 177, 439
TBLASTX 296
Telosomics 27
Terminator 125
Tetraploid 316
TIDR 311
TIGR 114, 187
TILLING/Tiling 127, 320, 327, 334, 355, 382, 499, 506
library 320
path 127
Tobacco 322
Tobacco Genome Initiative 322
Tobacco rattle virus 328
Tomato 162, 322
Training accuracy 453
Transcript 40, 104, 427
Transcription 71
factor 71
Transcriptional profiling 329
Transcriptome 2, 91, 104, 291
sequencing 2
Transcriptomics 7, 291
Transformation 378
competent artificial chromosome (TAC) 31, 33, 330
library 33
Transgene 9
Transgenics 9, 295, 503

Transition 86
Translocation 44
Transposable elements (TE) 215, 216, 221
  excision 216
  related gene 221
Transposon 224, 364, 380
  activation 380
  tagging 364
TRAP 313
Triticeae 313
Triticum 318

UGMS 315
Unigenes 289
Untranslated region 104
URL 500
USDA 177, 313, 355, 500
UTR 216

Vicia faba 35
Virus-induced gene silencing 328, 499
Vitis 160, 257
  Vitis vinifera 160

Wheat 164, 313, 385
Whole-genome 69, 403
  arrays 403
  duplication 262
  shotgun 312
WTO 554

YAC 33, 179
  library 33
Yeast artificial chromosome 26, 127

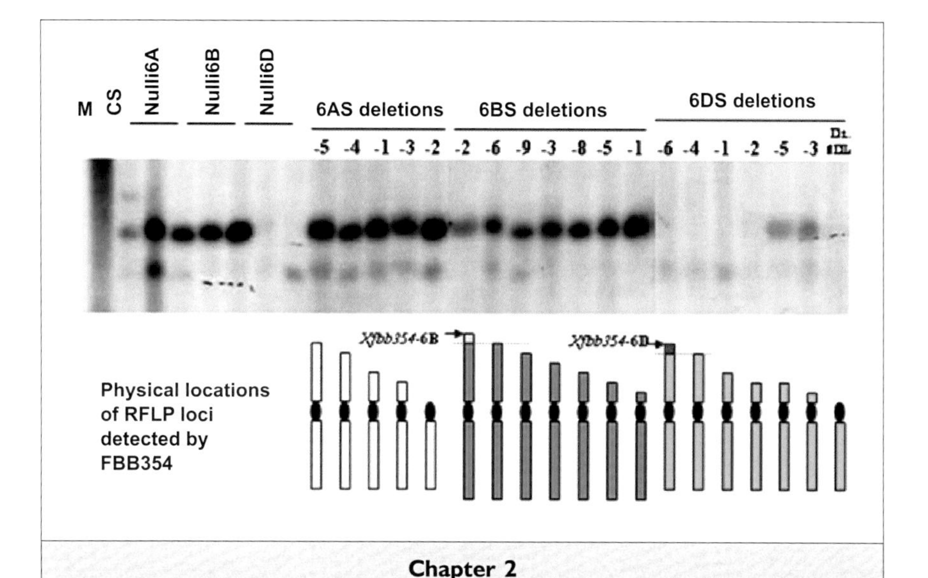

Chapter 2

Fig. 2 Physical mapping of RFLP loci with deletion lines in wheat (deletion mapping). RFLP marker FBB354 was used as the probe in Southern hybridization with a blot containing restriction enzyme digested genomic DNAs from wheat group-6 chromosome aneuploid stocks and deletion lines. Two loci of FBB354 can be physically located in the telomeric regions of 6BS and 6DS respectively based on the hybridization patterns. CS = Wheat cultivar "Chinese Spring", Nulli = Nullisomic, T = tetrasomic, d.t = ditelosomic.

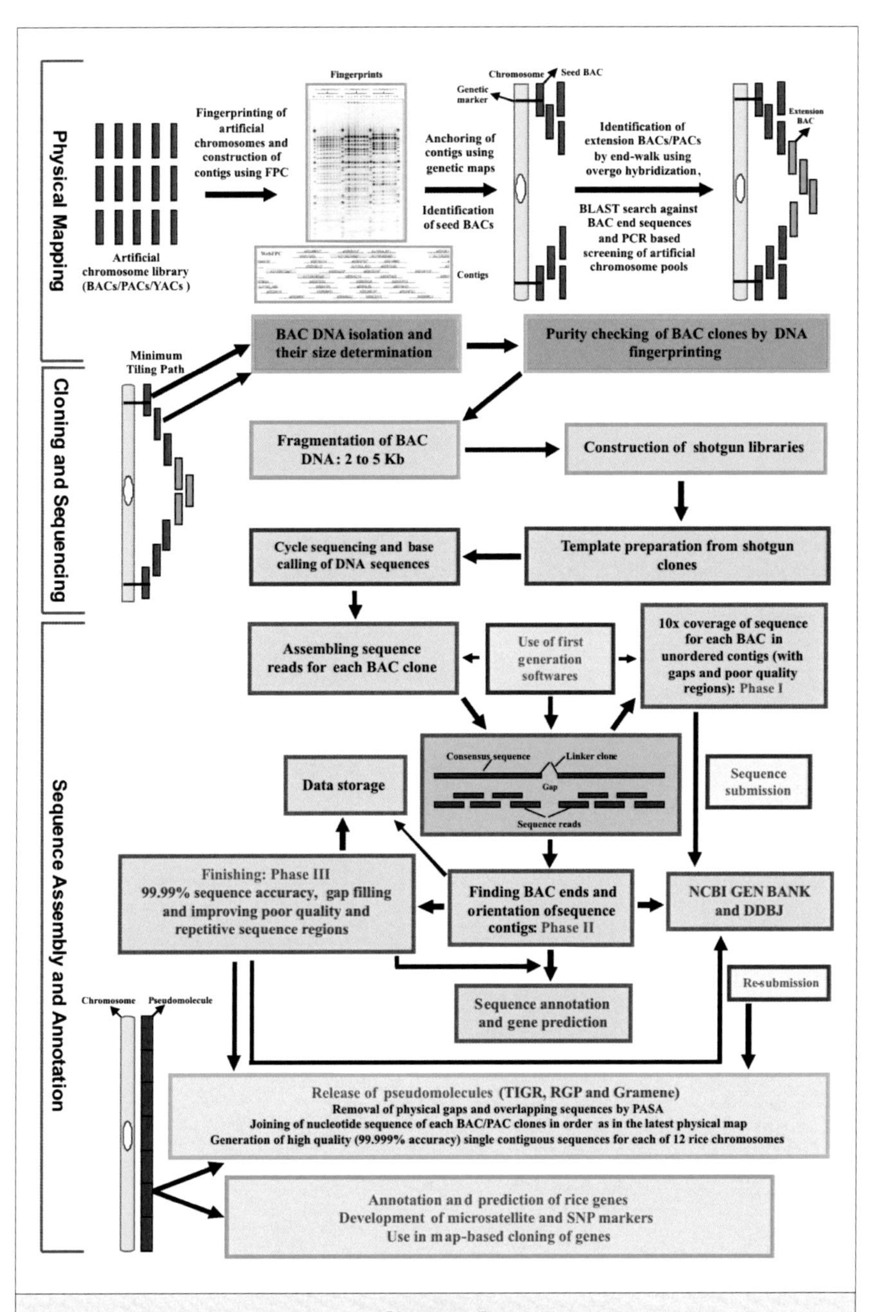

Chapter 5

Fig. 1 Clone-by-Clone approach for whole genome sequencing in rice.

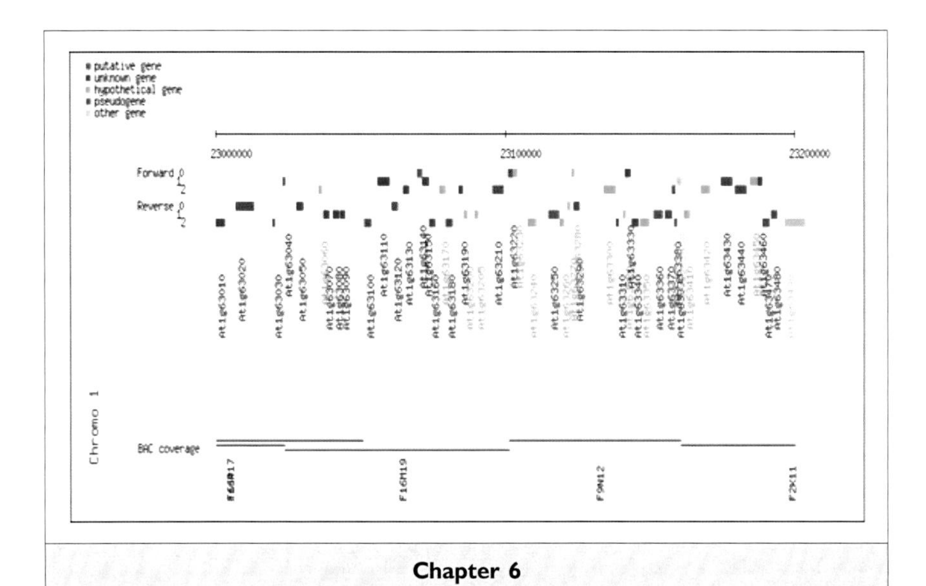

Chapter 6

Fig. 4 Gene density on Chromosome 1.

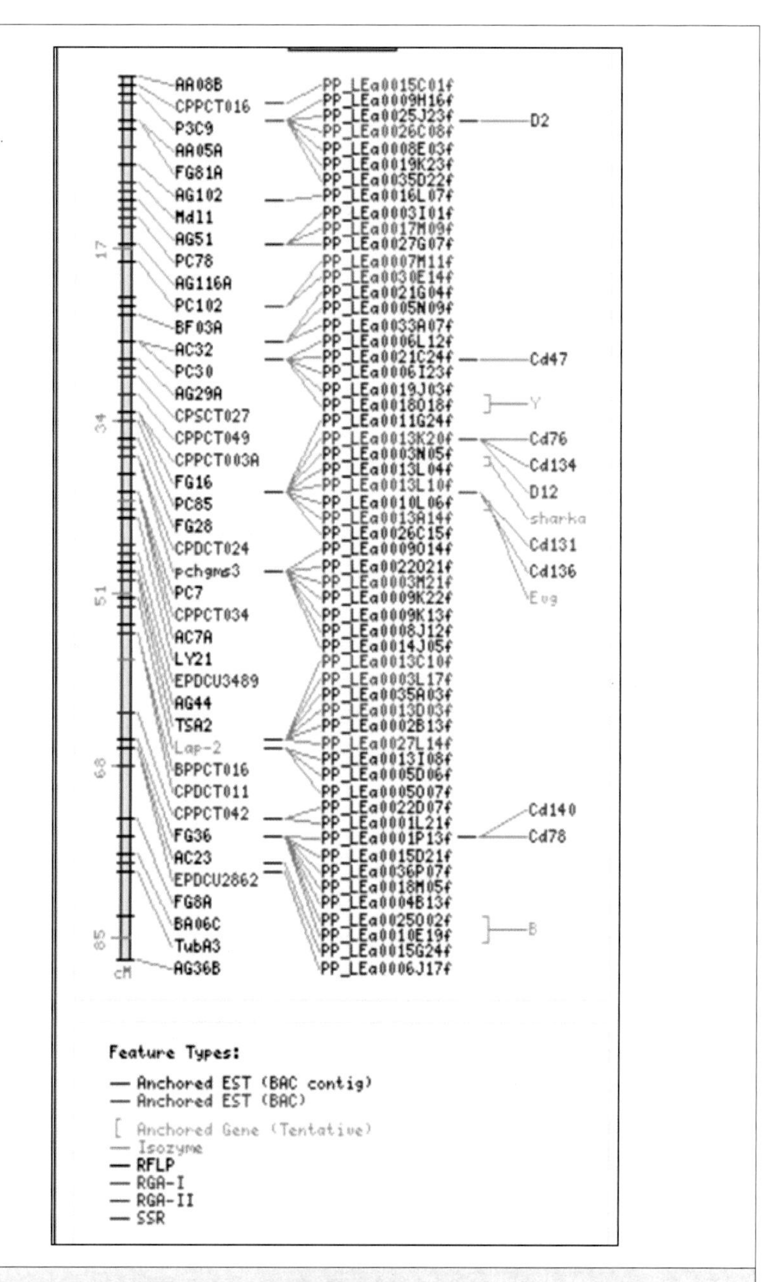

Chapter 9

Fig. 4 LG I of the Peach transcriptome map in GDR. The PPV resistance region has been defined in the mapping analysis of crosses shown in Table 1.

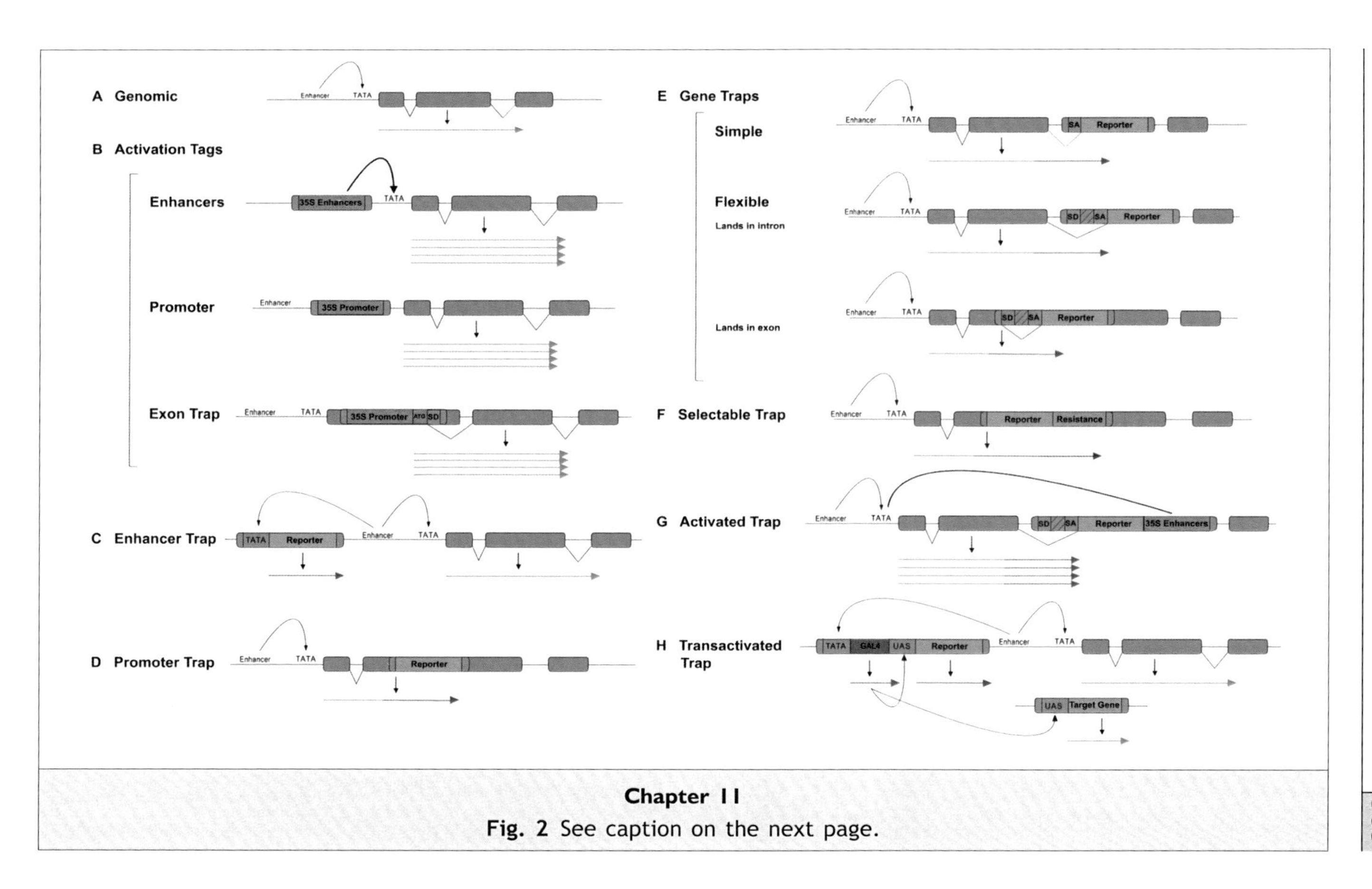

Chapter 11

Fig. 2 See caption on the next page.

Fig. 2 Diversity of T-DNA constructs used for insertional mutagenesis in plants (Adapted from Springer et al. 2000). Green boxes are exons; lines are introns; grey boxes delineateT-DNA borders; thatched grey boxes are introns within the T-DNA; "SA" is splice acceptor site; "SD" is splice donor site; "UAS" is upstream activation sequence; arrows represent transcripts that are produced.

(A) Typical gene structure.

(B) Activation tagging vectors. 35S enhancers are most commonly used to promote transcription, but the 35S promoter can also be used. An endogenous gene is overexpressed when the 35S enhancers interact with the endogenous promoter elements, or when the 35S promoter drives transcription. In an exon trap, an ATG and splice donor (SD) within the T-DNA may allow for an in-frame gene fusion.

(C) Enhancer trap. The reporter gene is expressed when an endogenous enhancer element (downstream or upstream) activates the minimal promoter (TATA).

(D) Promoter trap. The reporter gene is expressed when the T-DNA lands within the exon of an endogenous gene. The transcript and translated product are fusions between the endogenous gene and the reporter gene.

(E) Gene Traps. In a simple gene trap, the reporter gene is expressed when the T-DNA inserts within the intron of an endogenous gene, providing a splice acceptor site (SA) that functions with an endogenous SD site. In a flexible gene trap, the reporter gene is expressed when the T-DNA lands in either an exon or an intron, because the construct contains both SA and SD sites.

(F) Selectable Trap. In this version of a promoter trap, the resistance gene is included downstream of the reporter gene and creates a fusion product, such that antibiotic selection can be applied to identify successful promoter trap insertions.

(G) Activated Trap. In this version of a gene trap, the reporter gene is expressed to a greater extent due to the activity of the 35S enhancers.

(H) Transactivated Trap. In this version of an enhancer trap, the GAL4/UAS system is used to ectopically express target genes in the same pattern as any given trap line. GAL4 is overexpressed when an endogenous enhancer element (downstream or upstream) activates its minimal promoter (TATA). GAL4 binds to upstream activator sequences (UAS), which activate the reporter gene in the same transcriptional pattern as GAL4. A target gene of interest can be introduced under the control of a UAS element, which will also be expressed in the same pattern as GAL4 and the reporter gene.

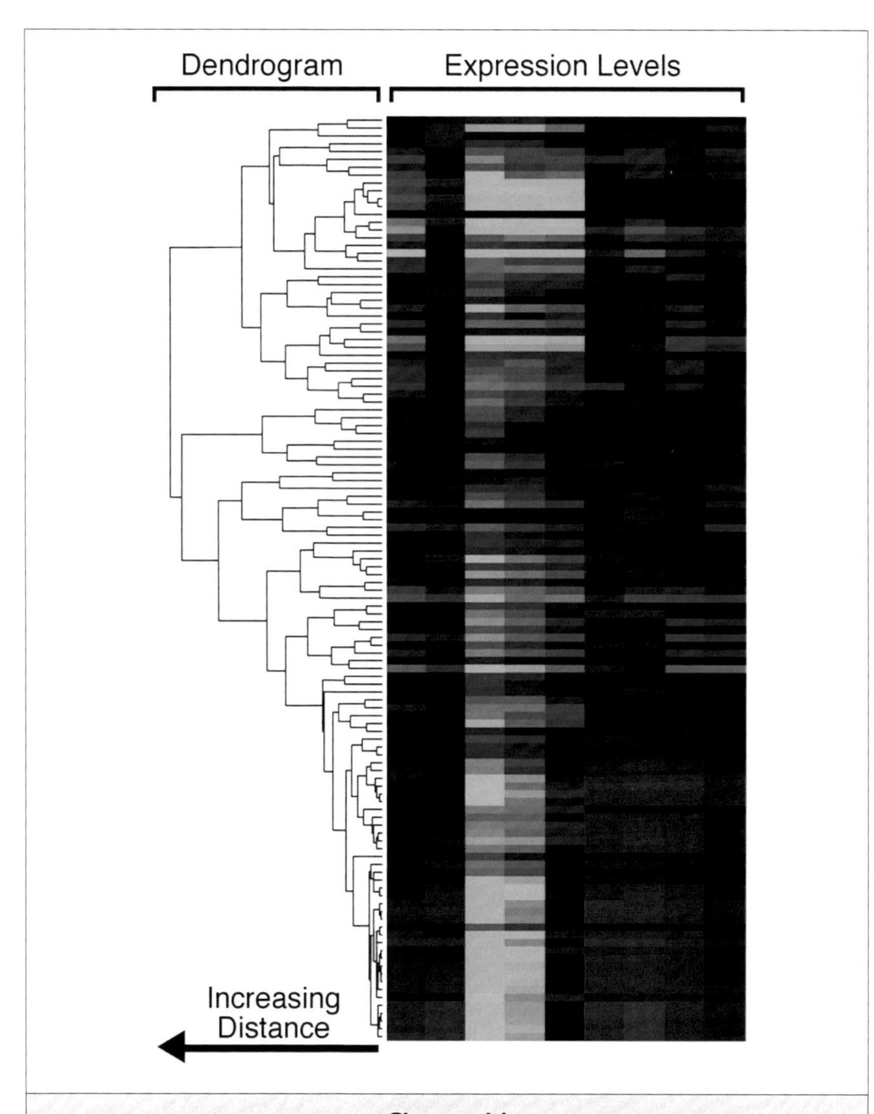

Chapter 14

Fig. 12 Example of a hierarchical dendrogram. Data vectors correspond to rows, and these vectors were hierarchically clustered. The tree structure shown on the left, called a dendrogram, encodes the distances between vectors clustered together. The greater the height of the branches in the tree structure that link vectors together, the greater the distance between them. For example, the short branches near the bottom show that these vectors are very similar, while the vectors at the top and bottom are more dissimilar since they are connected only at the top of the dendrogram.

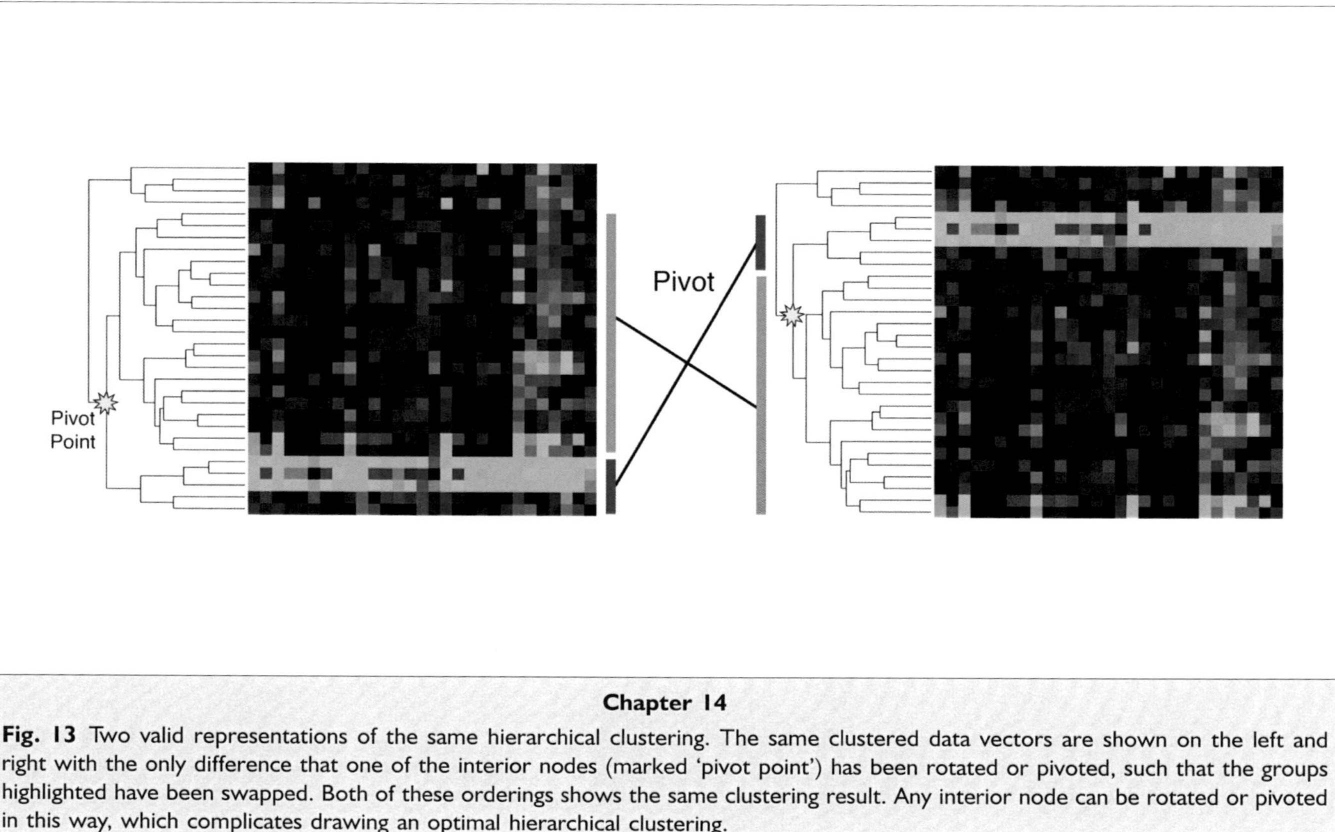

Chapter 14

Fig. 13 Two valid representations of the same hierarchical clustering. The same clustered data vectors are shown on the left and right with the only difference that one of the interior nodes (marked 'pivot point') has been rotated or pivoted, such that the groups highlighted have been swapped. Both of these orderings shows the same clustering result. Any interior node can be rotated or pivoted in this way, which complicates drawing an optimal hierarchical clustering.

Printed and bound by CPI Group (UK) Ltd, Croydon, CR0 4YY
21/10/2024
01777112-0013